Contents

Appendices

Introduction

Welcome to the challenging and exciting world of engineering! This book has been written to help get you through the mandatory units of the BTEC Level 3 Certificate and Diploma awards in Engineering. It provides the essential underpinning knowledge required of a student who wishes to pursue a career in any branch of engineering.

This book has been written by two highly experienced further and higher education lecturers. With over 50 years of practical teaching experience they have each contributed their own specialist knowledge which has been designed to cover the mandatory units of the revised and updated BTEC Engineering programme. Throughout the book has been adopted, a common format and approach has been adopted, with numerous student activities, examples, end of unit review questions and key points.

About the BTEC Level 3 Certificate and Diploma

The BTEC Level 3 Certificate and Level 3 Diploma qualifications have long been accepted by industry as appropriate qualifications for those who are about to enter industry or who are receiving training at the early stages of employment in industry. At the same time, these qualifications have become increasingly acceptable as a means of gaining entry into higher education.

BTEC Level 3 programmes in engineering attract a very large number of registrations per annum such that there is in excess of 35,000 students currently studying these qualifications in the UK by both part-time and full-time modes of study.

The BTEC Level 3 syllabus was recently reviewed and extensively updated and new programmes have been launched with effect from September 2010. The new scheme is likely to be adopted by *all* institutions that currently offer the programme as well as a number of others who will be offering BTEC qualifications for the first time.

Many organizations have contributed to the design of the new BTEC Level 3 Engineering programme including the Qualifications and Curriculum Authority (QCA), the Engineering Council and several Sector Skills Councils (SSC).

The Engineering Council continues to view the BTEC Level 3 Certificate/Diploma as a key qualification for the sector. They also recognize that BTEC Level 3 qualifications are frequently used as a means of entry to higher education courses, such as HNC/HND programmes and Foundation Degree courses.

To assist tutors and lecturers, this book is supported by a tutor support DVD, ISBN 978-0-12-382202-4, containing sample schemes of work and lesson plans, information on assessment, worked solutions and sample answers to selected activities, answers to all end of unit review questions, ready-made assignments and other teaching resources such as handouts, as well as electronic files for all illustrations in the book.

How to Use This Book

This book covers all of the mandatory units that are common to nearly all BTEC Level 3 Engineering programmes. Each unit contains *Text, Key points, Test your knowledge questions, Examples, Activities* and *End of unit review questions*.

The *Test Your Knowledge questions (TYK)* are interspersed with the text throughout the book. These questions allow you to check your understanding of the preceding text. They also provide you with an opportunity to reflect on what you have learned and consolidate this in manageable chunks.

Most *Test your knowledge questions* can be answered in only a few minutes and the necessary information, formulae, etc., can be gleaned from the surrounding text. *Activities*, on the other hand, make excellent vehicles for gathering the necessary evidence to demonstrate that you are competent in key skills. Consequently, they normally require a significantly greater amount of time to complete. They may also require additional library or resource area research time coupled with access to computing and other information technology resources.

Many tutors will use *Test your knowledge questions* as a means of reinforcing work done in class while *Activities* are more likely to be 'set work' for students to do outside the classroom. Whether or not this approach is taken, it is important to be aware that this student-centred work is designed to complement a programme of lectures and tutorials based on the BTEC syllabus. Independent learners (i.e. those not taking a formal course) will find complete syllabus coverage in the text.

The worked examples not only show you how to solve simple problems but also help put the subject matter into context with typical illustrative examples. In order to successfully tackle this work you will need to have a good scientific calculator (and get to know how to use it). You will also require access to an Internet-connected computer with word processing, spreadsheet, drawing and CAD software.

Finally, here are some general points to help you with your studies:

- Allow regular time for reading – get into the habit of setting aside an hour, or two, at the weekend. Use this time to take a second look at the topics that you have covered during the week or that you may have not completely understood.
- Make notes and file these away neatly for future reference – lists of facts, definitions and formulae are particularly useful for revision!
- Look out for the inter-relationship between subjects and units – you will find many ideas and a number of themes that crop up in different places and in different units. These can often help to reinforce your understanding.

- Don't expect to find all subjects and topics within the course equally interesting. There may be parts that, for a whole variety of reasons, don't immediately fire your enthusiasm. There is nothing unusual in this; however, do remember that something that may not appear particularly useful now may become crucial at some point in the future!
- However difficult things seem to get – don't be tempted to give up! Engineering is not, in itself, a difficult subject, rather it is a subject that *demands* logical thinking and an approach in which each new concept builds upon those that have gone before.
- Finally, don't be afraid to put your new ideas into practice. Engineering is about *doing* – get out there and *do* it!

Good luck with your BTEC Engineering studies!

Mike Tooley and Lloyd Dingle

A free student website is available at www.key2study.com/btecnat and features:

- Interactive quizzes with automatic marking and feedback
- A free comprehensive 2D CAD package for downloading
- A variety of spreadsheet tools for solving common engineering problems
- Useful engineering data summaries
- Extensive Visio symbol libraries for engineering drawing/CAD
- Drawing templates and sample drawings in industry-standard format
- Additional material to support learning activities and assignments
- Model answers to selected activities in Unit 1 can be found at: www.key2study/btecnat

Acknowledgements

Photo of circuit diagram (Unit 2)
©iStockphoto.com/kr7ysztof

Photo of magnetic levitation train (Unit 6)
©iStockphoto.com/FreezingRain

Photo of crash test dummy (Unit 7)
©iStockphoto.com/jacus

Health and safety in the engineering workplace affects **all** personnel, no matter what their involvement. Ensuring a healthy and safe working environment is not only the province of management - who are responsible for supplying the personal protective equipment and warning notices shown in the photograph; but also that of every individual. All staff and all visitors must use appropriate personal protective equipment and ensure that they are conversant with the safety warning notices, in order to keep themselves and others around them, out of danger, at all times.

Health and Safety in the Engineering Workplace

The health, safety and general welfare of people working or operating within any engineering context is of prime importance. Individuals should expect to be able to carry out their workplace engineering activities, in a safe environment, that do not compromise their general well-being or their ability to successfully complete their work. Organisations are responsible for ensuring that such an environment exists for their employees, visitors and the general public and, indeed, many pride themselves on making their own workplace superior to others and using this to their own competitive advantage, when recruiting staff. Thus, health and safety in the workplace is about the control and risk reduction measures that need to be put in place in order to best protect the health and safety of all people who carry out workplace activities and also those visitors and others who might be affected by these activities.

In this unit you will be given an introduction to some of the important legislation and regulations upon which health and safety measures in the workplace are based. Health and safety law is being continually updated with more legislation emanating directly from the European Union (EU) and it is the responsibility of organisations to keep abreast of and act on all new initiatives that affect workplace activities. You will be made aware of the latest health and safety legislation and regulations and of your individual roles and responsibilities in order to comply with them. You will learn about the hazards and significant risks associated with the layout, equipment and engineering activities that are undertaken in the workplace. You will learn about the control measures taken to minimise or eliminate risk and the process of risk assessment itself. Finally, the principles and methods we adopt to report and record workplace accidents and incidents in accordance with legal requirements will be covered.

Throughout this unit you will be asked to complete a number of activities to enhance your understanding, *selected answers to which will be found on the website that accompanies the book at* http://www.key2study/btecnat.

BTEC National Engineering. DOI: 10.1016/B978-0-12-382202-4.00001-5

Legislation and Regulations

Introduction to health and safety legislation

Health and safety *legislation* comes about, like all legislation (laws) through Acts of Parliament (statutes). These '*statutes*' generally give powers to ministers and/or secretaries of state to make '*statutory instruments*' which provide more detailed '*regulations*' for implementing the statute. One of the first modern Acts of Parliament (statute) that was directly related to health and safety was the Factories Act of 1937 and later its successor the current *Factories Act of 1961*. In this Act the general health and safety provision for factories were laid down, such as the requirements for cleanliness, sanitation, temperature, ventilation and lighting as well as the provision of a safe means of access and egress and the care and use of dangerous substances. In addition this Act also stipulated the requirements for welfare, such as the provision of drinking water, washing facilities and rest rooms. After the 1961 Factories Act an Act was passed by Parliament known as the *Offices, Shops and Railway Premises Act 1963*, where similar statutory requirements (laws) were passed on the health, safety and welfare provision for these premises.

The next health and safety legislation of note was the *Fire Precautions Act 1971*, in which law appertaining to fire certification, buildings that require certificates, premises involving excessive risk, building and other regulations about fire precautions, enforcement of these regulations and offences, penalties and legal proceedings were some of the important regulations introduced by this Act.

As a result of the Robins Report in 1972, that proposed substantial changes in the law on health and safety, the *Health and Safety at Work Act 1974 (HSWA)* came into being; this Act now forms the basis of British health and safety law. This Act sets out the general duties and responsibilities that employers have towards employees and members of the public, and employees have to themselves and to each other. We will look at this Act and its associated regulations in some detail, since this is the Act upon which much of health and safety in the engineering workplace is based.

One other piece of legislation that is of importance to the activities of engineers, other industries and the public at large is the introduction of integrated pollution control measures under the *Environmental Protection Act 1990* and its associated regulations.

KEY POINT

The Health and Safety at Work Act 1974 forms the basis of British health and safety law

Roles and responsibilities of the Health and Safety Executive

In order to ensure that the above legislation relating to health and safety in the workplace is implemented and enforced in a fair and consistent manner, the Department for Work and Pensions has two agencies responsible for health and safety in Great Britain directly related to the Health and Safety at Work Act (HSWA): *Health and Safety Executive (HSE)* and *the Health and Safety Commission (HSC)*. The HSC works to ensure that relevant legislation (including EU Directives) is appropriate and understood by conducting and sponsoring research; and submitting proposals for new or revised regulations and approved codes of practice. *The HSE is the operating arm of the Commission. It advises and assists the Commission in its functions and has specific responsibility, shared with local authorities, for enforcing health and safety law.*

Certain areas of risk or harm directly or indirectly related to work activities are covered by legislation other than the HSWA or its related regulations and are not dealt with by the HSE. These include consumer and food safety, marine, railway and aviation safety, all having their own legislative bodies. For example, aviation safety law is enforced/overseen in this country by the Civil Aviation Authority (CAA) and due to the international nature of the Aviation Industry, the CAA in turn work with and ensure the implementation of relevant and recent European Laws laid down and overseen by the European Aviation Safety Agency (EASA).

The combined HSC/HSE consult fully with people affected by their legislative proposals and adopt various approaches based on assessing and controlling risk, where they consider action is necessary to supplement existing arrangements; they offer three main options: Guidance, Approved Codes of Practice (ACOP) and Regulations.

HSE publishes *guidance* on a range of subjects that can be general or more specific to the heath and safety problems of an industry or to a particular process used in a number of industries. The main purposes of this guidance are to help people understand what the law says (including the interpretation of EC Directives), to help people comply with the law and to give technical advice. By virtue of its very name, guidance is not compulsory but if followed it will normally allow industries to comply with the law. Examples of HSE guidance include:

- basic advice on first aid at work (2002)
- five steps to risk assessment (1998)
- a guide to the Reporting of Injuries, Diseases and Dangerous Occurrences Regulations (RIDDOR) (1995)
- a short guide to Personal Protective Equipment at Work Regulations (1992).

The need for clarification of the regulations by offering practical examples of good practice has been recognised by HSC/HSE and as a result they have produced a series of related *Approved Codes of Practice* that provide the necessary guidance on what is reasonable and practical in order to comply with the law. Following Approved Codes of Practice (ACOP) is not mandatory by law but in the event of employers being prosecuted for a breach of health and safety law, the fact that they had not followed the advice given in the ACOP would find them at fault, unless they could show that they had equally good or superior practices in place that adequately demonstrated compliance. It is for this reason that ACOP are sometimes referred to as semi-legal or quasi-legal in nature. Examples of ACOP include:

- Provision and Use of Work Equipment Regulations (ACOP L22 1998)
- The Control of Substances Hazardous to Health (ACOP L5 2002)
- Gas Safety (Installation and Use) Regulations (ACOP L56 1998).

As mentioned earlier, **Regulations** are laws (statutory instruments) that are made up under Health and Safety Acts approved by Parliament that follow proposals and recommendations from HSC; this applies to both British Law and EC Directives. The Health and Safety at Work Act is the primary piece of legislation upon which health and safe regulations are based. Regulations may be issued in a form that identifies what must be achieved to minimize health and safety risk, or if the risks are considered very serious the regulations will dictate the specific actions that must be taken to alleviate these risks.

The HSE has an **inspection** role, where under the Health and Safety at Work Act 1974 and the Health and Safety (Enforcing Regulations) 1998, they share the enforcement of health and safety legislation at different premises between local authorities and/or themselves. Field operations inspectors work within different directorates and are responsible for different industries and areas of safety; these include nuclear, where inspectors are responsible for the safety of nuclear installations; construction; hazardous installations such as for petroleum, chemicals, biological agents; pipeline and road transport of hazardous substances, and mining operations and exploration activities. HSE inspectors have the power and ways to seek improvements if they are not satisfied with the level of the safety standards being achieved; these include:

- verbal or written information and advice
- improvement or prohibition notices
- prosecution in criminal courts
- implementing investigations for manslaughter (if thought appropriate) and/or for accidents/incidents in order to prepare legal action or learn lessons.

Local authorities have primary responsibility for the enforcement of health and safety legislation (in particular environmental health) in premises that include shops, offices, retail and wholesale distribution outlets, care homes and leisure facilities. Although these premises are not directly within the remit of the HSE, they may have some enforcement responsibilities over and above that of the local authority and a system of *flexible warrants* was introduced in 2006 to account for this fact.

The Health and Safety at Work Act

Introduction

The Health and Safety at Work Act (1974) is the major piece of legislation upon which all workplace health and safety regulations are based. It differs fundamentally in principle from any previous health and safety legislation. The underlying reasoning behind the Act was the need to foster a much greater awareness of the problems that surround health and safety matters and, in particular, a much greater involvement of those who are, or who should be, concerned with improvements in occupational health and safety. Consequently, the Act seeks to promote greater personal involvement coupled with the emphasis on individual responsibility and accountability.

You need to be aware that the Health and Safety at Work Act applies to *people*, not to premises. The Act covers all employees in all employment situations. The precise nature of the work is not relevant from the point of view of the Act, neither is its location. However, in this unit we will be concentrating on the hazards and risks associated with some of the more specific engineering activities that take place in the engineering workplace. The Act also requires employers to take account of the fact that other persons, not just those that are directly employed, may be affected by work activities. The Act also places certain obligations on those who manufacture, design, import or supply articles or materials for use at work to ensure that these can be used in safety and do not constitute a risk to health.

Key features of the HSWA 1974

After the introduction, the main body of the HSWA consists of four major parts followed by several schedules where, for example, schedule 3 contains the subject matter of the health and safety regulations. *Figure 1.1* shows the nature of these sections and the sort of subject matter they contain. Part I of the Act is concerned with the legislation relating to Health, Safety and Welfare in connection with Work as well as the Control of Dangerous Substances and Certain Atmospheric Emissions. The area connected with HSWA is of utmost importance to us and some of the topics contained within it are now covered in more detail.

Figure 1.1 summarizing the HSWA 1974 is based on information found in the 'statutory law database (SLD)', which may be viewed by going onto either www.statutelaw.gov.uk or onto the HSE website at www.hse.gov. uk, which guides you to the Government's statute law site. The SLD site is amended regularly as the statutory laws are updated; this occurs after the amended legislation has been fully approved, verified and written. Thus, there is sometimes a time lag between the latest current legislation and that shown on the database; this problem is overcome by consulting the *update status warning*, given at the beginning of the particular Act, which allows you to view the latest amendments.

In Part I of the Act, the preliminary subject matter explains the aims and objectives of the Act, which were summarised at the beginning of this section. In the 'general duties' section, you will find the duties of employers to their employees and to those affected by the activities of the employer's business, as well as the duties of employees at work. The duties of employers and employees are summarized in the Act, as given below.

Duties of employers

It is the *duty of the employer* to ensure, so far as is reasonably practicable, the health, safety and welfare at work of all the employees; also that all plant and systems are maintained in a manner so that they are safe and without risk to health. The employer is also responsible for:

- the absence of risks in the handling, storage and transport of articles and substances
- instruction, training and supervision to ensure health and safety at work

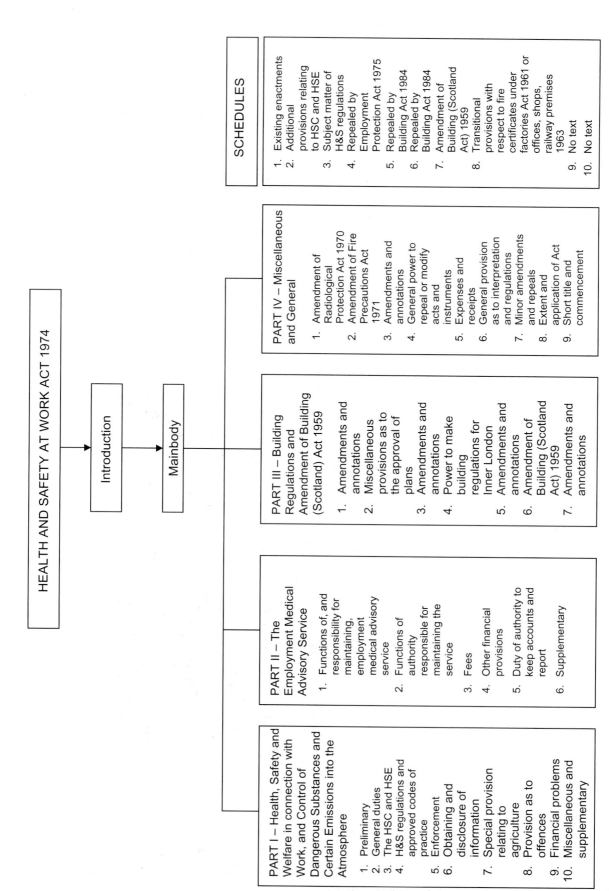

Figure 1.1 The Health and Safety at Work Act 1974

- the maintenance of the workplace and its environment to be safe and without risk to health
- where appropriate, to provide a statement of general policy with respect to health and safety and to make arrangements for safety representatives and safety committees.

In addition to having responsibilities to the employee, the employer has responsibilities to other persons such as the general public. It is therefore also the *duty of the employer* to:

- conduct his undertakings in such a way so as to ensure that members of the public (i.e. those not in his employ) are not affected or exposed to risks to their health or safety
- give information about such aspects of the way in which he conducts his undertakings to persons who are not his employees as might affect their health and safety.

Figure 1.2 shows typical health and safety notices (together with relevant safety procedure documents) prominently displayed in an engineering workshop.

Figure 1.2 Health and safety notices and procedures

Duties of employees

It is the *duty of every employee* whilst at work to take all reasonable care for the health and safety of himself and other persons who may be affected by his Acts and omissions.

Employees are required to:

- cooperate with the employer to enable the duties placed on him (the employer) to be performed
- have regard of any duty or requirement imposed upon his employer or any other person under any of the statutory provisions
- not interfere with or misuse anything provided in the interests of health, safety or welfare in the pursuance of any of the relevant statutory provisions.

It is the duty of *each person* who has control of premises, or access to or from any plant or substance in such premises (i.e. employer and employees), to take all reasonable measures to ensure they are safe and without risk.

The important aspects of the roles and responsibilities of the Health and Safety Commission (HSC) and the Health and Safety Executive (HSE) were briefly covered in the introduction, as were their powers of enforcement. To ensure that you are fully conversant with these and other important areas of Part I, there follow two activities that require you to investigate the Act further.

> **KEY POINT**
>
> The HSWA 1974 imposes duties/responsibilities on employees as well as employers

Activity 1.1

Visit the HSE website and find the relevant information on health and safety legislation and regulation and then answer the following:

(a) What are the primary functions of the Health and Safety Commission (HSC) and the Health and Safety Executive (HSE)?
(b) Explain the differences and relationship between 'Guidance', 'Approved Codes of Practice (ACOP)' and 'Regulations'.
(c) Describe four activities that are regulated and controlled under the Environmental Protection Act 1990.

Activity 1.2

Consult the Health and Safety at Work Act and answer the following questions:

(a) What are the general duties imposed on manufacturers regarding articles and substances for use at work?
(b) What general powers do HSE inspectors have under the Act?
(c) What is an improvement notice?
(d) What is a prohibition notice?
(e) Who issues such notices?
(f) Who can be prosecuted under the Act?

Some important health and safety statutory instruments

Statutory instruments or *regulations* are laws approved by Parliament as mentioned earlier. They are usually made under the HSWA 1974, following proposals from the HSC. In order to help us comply with these regulations and help us to recognize workplace hazards and so eliminate or reduce risk, the HSE have produced a range of Guidance leaflets and Approved Codes of Practice (ACOP) to assist us.

We will now spend some time considering the key features of those regulations that impact directly on the management and implementation of our activities in the engineering workplace.

The Management of Health and Safety at Work

The Management of Health and Safety at Work Regulations 1999 have been introduced to ensure that managers take all necessary actions to identify workplace hazards and implement procedures to eliminate or at least minimise the risk associated with these hazards. In order to achieve their goals and ensure that the workplace is a safer and healthy environment to be in, managers need to put in place detailed and appropriate arrangements for the management of health and safety as shown in Regulation 5. In essence this regulation breaks these arrangements into *five significant steps*, which if followed conscientiously will ensure success.

KEY POINT

The Management of Health and Safety at Work Regulations 1999 is the key piece of legislation that managers and other responsible people must abide by to ensure the health and safety of their employees and all people on their premises

Activity 1.3

Visit the HSE website and consult the Guidance and ACOP information on the Management of Health and Safety at Work Act 1999 and answer the following:

(a) Explain the necessity for the successful management of workplace health and safety and detail the possible consequences for the company, its employers and employees, if health and safety is poorly managed.

(b) Detail the five steps recommended for managing health and safety and describe the actions that need to be taken for each step.

Having completed activity 1.3 you will have noticed the importance placed on *risk assessment* (Regulation 3), which has become the major tool for managing and implementing good health and safety practice at work. Risk assessment is so important that later in this unit we dedicate a complete outcome to the subject so that you will be able to carry out a risk assessment related to your engineering work and know about the control measures needed to minimise risk. You will also have noticed that employers have responsibility for informing and training their employees in all relevant aspects of health and safety related to their place of work and surrounds, such as first-aid procedures, fire drills and precautions, and working with engineering machinery. All these areas and others will be covered in detail when we consider the individual regulations applied to hazards and risk in the engineering workplace.

Activity 1.4

By visiting the same HSE information sources as those used for activity 1.3, answer the following:

(a) Explain the employee's main duties and responsibilities to their employer to ensure that their employer's duties under the management of health and safety regulations are met.

(b) Explain the duty that employers have to *temporary workers* and *young persons* within their employ.

(c) Who may be issued with *exemption certificates* and from what regulations may they be exempt?

Workplace Health, Safety and Welfare

Employers have a general duty under section 2 of the HSWA 1974 to ensure, so far as is reasonably practicable, the health, safety and welfare of their employees and for people in control of non-domestic engineering facilities; they have a further responsibility (under section 4 of the HSWA) towards people who are not employees but use their premises. ***The Workplace (Health, Safety and Welfare) Regulations 1992*** cover a wide range of basic health, safety and welfare issues concerned with workplace facilities that need to be addressed by employers for the benefit of their workers and, where appropriate, other people who use their premises. Thus, issues to be considered include:

Health

- Ventilation
- Temperature
- Working in extremes of temperature
- Lighting
- Cleanliness and waste materials
- Room dimensions and space
- Workstation suitability
- Working with VDUs.

Safety

- Maintenance
- Floors and traffic routes
- Falls into dangerous substances
- Transparent or translucent door, gates or walls and windows
- Windows
- Doors and gates
- Escalators and moving walkways.

Figure 1.3 shows the safety sign for a drop hazard with a notice placed on a section of highway staging warning people to be aware of the different floor levels while working at height.

Welfare

- Sanitary conveniences and washing facilities
- Drinking water
- Accommodation for clothing and facilities for changing
- Facilities to rest and eat meals.

KEY POINT

The Workplace (Health, Safety and Welfare) Regulations 1992 are concerned with the facilities provided for the health, safety and welfare of the people in the workplace

Figure 1.3 Drop hazard

Activity 1.5

By consulting The Workplace (Health, Safety and Welfare) Regulations 1992 and any other relevant HSE guidance materials answer the following questions related to engineering workshops:

(a) What do the regulations say about what must be done or provided with respect to:
  (i) the safe movement of people and vehicles in and around the workshop
  (ii) workstation seating
  (iii) workshop cleanliness.
(b) What briefly does the law say about workshop temperature and lighting and what to avoid with respect to lighting?
(c) What briefly does the law say about the provision of toilets and washing facilities?

The health and safety issues concerned with those who regularly work with visual display units, e.g. computer screens, have their own set of regulations known as ***Health and Safety (Display Screen Equipment) Regulations 1992***. It is appreciated that if you are primarily engaged in engineering workshop production, manufacture or maintenance, you are unlikely to spend too much time in front of a computer screen. However, your duties may call upon you to carry out tasks, such as stores procurement and issue, where you are required to spend a considerable time using a computer. Under these circumstances you will need to understand the duties of your employer and your own responsibilities when using this equipment. Also, those engineers engaged in design work or diagnostic maintenance may well use computers for a considerable part of their working day and need to become familiar with the regulations. Activity 1.6 has been designed to familiarize you with key features of these regulations, so you are advised to complete it in the best way that you can.

Activity 1.6

Go onto the HSE website and look at the Display Screen Equipment Regulations 1992 and related guidance material and answer the following:

(a) Who are affected by the regulations?
(b) Briefly explain the nature of the typical health disorders that *may* affect those who spend most of their working day in front of display screen equipment.
(c) What are the main points that employers have to consider in order to comply with the regulations?
(d) What practical measures can you take at your workstation, to help yourself avoid potential health problems?

Noise

Engineering workshops and other engineering areas often contain many sources of noise such as metal work equipment, power presses, power tools, CNC machinery, hydraulic and pneumatic equipment, loading and unloading activities and vehicle movements. These noise distractions have obvious health and safety implications and therefore this important aspect of workplace health, safety and welfare is covered by its own separate legislation known as the ***Control of Noise at Work Regulations 2005***. *Figure 1.4* shows the warning signs for ear protection and eye protection when operating a grinding wheel. The noise given off by this machine when in operation is sufficient to warrant the mandatory use of ear defenders, while at the same time the spark and chip hazards resulting from the grinding of metal also require the mandatory use of eye protection.

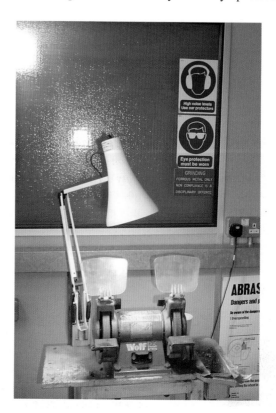

Figure 1.4 Ear and eye protection for the operation of a grinding wheel

Under the noise regulations employers are required to:

- assess the risk to their employees from noise at work
- take action to reduce the noise exposure that produces those risks
- provide their employees with hearing protection if noise exposure cannot be reduced by any other methods
- make sure the legal limits on noise exposure are not exceeded
- provide employees with relevant information, instruction and training
- carry out health surveillance where there is a risk to health.

The regulations *do not apply* to members of the public exposed to noise from their non-work activities, or making an informed choice to go to noisy places. Neither do they apply to low-level noise that is a nuisance but causes no risk of hearing damage.

Activity 1.7

Go onto the HSE website and consult the 'Noise at Work (INDG 362)' leaflet and then attempt the following tasks:

(a) **Explain what is meant by 'dB(A)' and 'dB(c)', with respect to noise measurement.**

(b) **Explain what is meant by 'noise action levels' and 'limit values'.**

(c) **Give two examples of machinery or equipment in your workplace, where ear protection or some other action is required to conform to the regulations.**

(d) **What should employers do to ensure that hearing protection is used effectively?**

We will be returning to the subject of noise when you are asked to consider risk assessment control measures later in the unit.

Personal Protective Equipment

Personal Protective Equipment (PPE) is defined in the PPE at Work Regulations 1992 as '*all equipment (including clothing affording protection against the weather) that is intended to be worn or held by a person at work which protects him against one or more risks to health and safety*'. This includes, for example, safety helmets, visors, goggles, gloves, high-visibility clothing, dust masks, safety footwear and safety harnesses. The regulations *do not cover* hearing protection which is already covered by the Control of Noise at Work Regulations, nor do they cover respiratory protective equipment which is covered in a range of other specialist regulations, e.g. COSHH regulations associated with diving safety and working with asbestos.

Figure 1.5 shows a pair of Perspex safety goggles which might typically be worn with the dust mask when carrying out grinding, sanding or cutting operations on a range of workshop machinery. *Figure 1.6* shows a man clad in a face shield with visor, long gauntlets, apron and safety shoes where all this protective clothing might be worn together when dealing with chemicals or substances at extremes of temperature, such as very hot or cold gases or liquids.

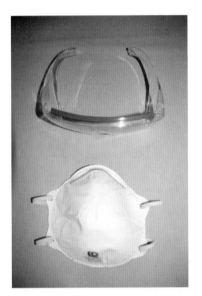

Figure 1.5 Safety goggles and dust mask

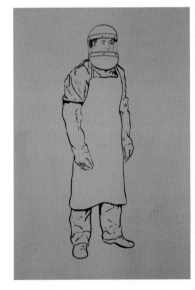

Figure 1.6 Protective clothing

It is the responsibility of employers to assess the suitability and select the most appropriate PPE for their employees where, for example, they will need to consider whether it is appropriate for the risks involved and whether it prevents or adequately control these risks. Also, the ease of adjustment for the wearer, the demands it places on the wearer when working and its compatibility with other essential PPE need to be considered.

KEY POINT

The Personal Protective Equipment Regulations 1992 do not cover noise; this is covered by its own legislation, The Control of Noise at Work Regulations 2005

Activity 1.8

Go onto the HSE website and look at the PPE at Work Regulations 1992 and other related guidance and answer the following:

(a) What do managers need to consider when assessing the suitability of PPE?
(b) Determine the hazards and types of PPE that may be used to reduce or eliminate their effects on the human body. You should consider these hazards with respect to human eyes, head, body, breathing, hands, arms, feet and legs.

Control of Substances Hazardous to Health

The **Control of Substances Hazardous to Health Regulations 2002 (COSHH)** require employers to assess the risks from hazardous substances that pose a threat to health and take appropriate precautions.

In engineering workshops there are particular substances that present a hazard to health when carrying out engineering operations. These include dusts from mechanical cutting, shaping and abrasive blasting; gas and fumes from soldering, brazing, welding and cutting; mists, bacteria and fungi in metalworking fluids; mists, sprays and fumes from lubricants, adhesives, paints, degreasing and stripping fluids; splash hazards and toxic fumes from plating, pickling fluids, hot salt baths and fluid treatment products.

In order to control these hazards, a variety of measures may be adopted including dust fume or vapour extraction, use of respirators, maintenance of fluids, use of barrier creams, skin checks (for example for dermatitis) and other regular employee surveillance checks such as for breathing difficulties and asthma by health professionals, if the hazards deem this necessary.

Figure 1.7 shows a belt-type abrading machine fitted with a dust extraction system to help prevent dust inhalation by the operator. There can also be seen further safety notices concerned with the safe operation of the machine and the need for ear and eye protection.

In order to identify the nature of the hazards produced by substances that pose a threat to health, manufacturers/suppliers are required under the COSHH regulations to clearly label their products and give full written details of the hazards. In addition, those products deemed 'dangerous for supply' must have a *safety data sheet* attached to them. To provide positive identification and to give a clear visual indication of the general nature of the threat to health associated with the product/substance, both the UK and the EU use a number of standard hazard symbols.

Figure 1.8 shows the four warning signs – poison/toxic, fire, explosive and corrosive hazards. The skull and crossbones is the general sign for

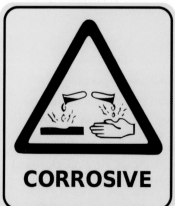

Figure 1.8 Hazard warning symbols

Figure 1.7 Abrading machine with dust extraction system

toxic substances, while *very toxic* substances have the abbreviation 'T+' positioned in the top left-hand corner of the symbol. The fire or general *flammability* symbol will have the abbreviation 'F+' in the top left-hand corner to indicate that the substance is *extremely flammable*. The *explosive* symbol may just have the abbreviation 'E' rather than the single word description; this may also be the case for the *corrosive* symbol, where the abbreviation 'C' may replace the single word description.

Figure 1.9 demonstrates the Europe-wide use of hazard warning symbols where the word descriptions are in French but the symbol and the abbreviation have the same meaning. You can see the abbreviation 'O' to indicate an *oxidising* hazard, that is, where chemicals react to give out heat. Next you will note the two symbols with a black cross where the abbreviation 'Xn' indicates that the substance may cause *harm* or damage to health and the abbreviation 'Xi' indicates that the substance is an *irritant* where, for example, it may cause inflammation to the skin or other tissue. Finally the warning symbol picturing a dead tree and fish is the *dangerous to the environment* symbol, which has the abbreviation 'N'.

From 2009, **International symbols** started to replace European symbols. Some of them are identical or very similar to the European symbols except they sit within a red diamond-shaped border on a white background. You can view some of these new international symbols by visiting the HSE website and looking at the web-friendly version of the leaflet, *Working with substances hazardous to health*. All hazard symbols, no matter in what national or international format, have specific meanings; if you are in any

O - Comburant

Xn - Nocif

Xi - Irritant

N - Dangereux pour l'environnement

Figure 1.9 Hazard warning symbols with descriptions in French

doubt about the dangers posed by chemicals or other substances, ***always read the label*** on the product and the accompanying safety certificate as appropriate.

Activity 1.9

Visit the COSHH section of the UK Government's Health and Safety Executive website (you will find this at www.hse.gov.uk/coshh). View or download a copy of 'COSHH: a brief guide to the regulations' and use it to identify the EIGHT steps that help to ensure that a company complies with the COSHH legislation.

Activity 1.10

While on the same site as that for activity 1.9, investigate the new European regulation known as REACH and answer the following:

(a) What is meant by REACH and what are the main functions of this regulation?
(b) Detail 10 pieces of information that need to appear on a REACH safety data sheet.

Before we leave the subject of substances hazardous to health, we need to know the hazards and dangers associated with handling ***asbestos*** and working in or around areas where asbestos is present. This substance has been related to some rather nasty health problems and you are asked in activity 1.11 to investigate the issues surrounding asbestos.

Activity 1.11

Asbestos is a material that has relatively only recently been identified as being dangerous to health. Use the HSE website and/or other information sources to answer the following questions:

(a) What is asbestos?
(b) Why is asbestos dangerous to health?
(c) What diseases are caused by exposure to asbestos?
(d) What occupations are particularly at risk from asbestos?
(e) What legislation applies to the use and handling of asbestos?

Electricity at work

Electrical installations, plant and equipment in the workplace should be treated with utmost respect because electricity can kill! You should always take note of warnings that are present concerning electricity (see *Figure 1.10*.)

The Electricity at Work Regulations 1989 made under the HSWA 1974 set out the duties for those people responsible for electrical systems, electrical equipment and conductors as well as for the work activities on or near electrical equipment. Thus employers need to:

- develop a suitable system of maintenance for both fixed installations and portable electrical equipment

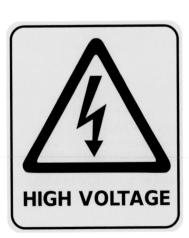

Figure 1.10 Electricity warning sign for high voltage

- ensure that electrical contractors are competent to carry out the work for which they have been contracted
- select equipment that is suitable for the job
- check that wiring and equipment is sound and properly installed
- ensure that those employees who are electricians, are competent.

The primary purpose of the regulations *is to prevent death or personal injury from electrical causes in connection with work activities*. Where the death or injuries result from:

- electric shock
- electric burn
- fires of electrical origin
- electric arcing
- explosions initiated or caused by electricity.

The hazards that are likely to cause these injuries or harm can come from human contact with live parts (normal mains voltage 230 V AC can kill); faults, such as short circuits or poorly maintained insulation, which can cause fires and explosions; or fires where electricity could be the source of ignition, e.g. a piece of portable electrical equipment that sparks and arcs in a potentially flammable or explosive atmosphere such as around petroleum fuel storage tanks and pumps.

You will be required later in this unit to consider in more detail the *hazards* (anything that can cause harm) and *risks* (the chances that someone will be harmed) associated with electricity in the engineering workplace.

Activity 1.12

Go onto the HSE website and consult the Electricity at Work Regulations 1989 and answer the following questions:

(a) What is meant by an adverse or hazardous environment?
(b) With respect to regulation 2, briefly define electrical systems, electrical equipment and conductors.
(c) Detail four ways in which electrical fires may be started.

Fire regulations and process precautions

Communities and Local Government (CLG) have policy responsibility for general fire safety in all non-domestic premises in England. The Scottish Government and Welsh Assembly have similar responsibilities for Scotland and Wales.

General fire safety in England and Wales is delivered through compliance with the Regulatory Reform (Fire Safety) Order 2005. In Scotland, fire safety duties are contained in Part 3 of the Fire (Scotland) Act 2005 and the Fire Safety (Scotland) Regulations 2006. Fire and Rescue Authorities (FRA) are the principal enforcers and have a statutory duty to enforce the requirements of the legislation.

Figure 1.11 Typical fire point

HSE no longer have responsibility for premises under the Fire Certificate (Special Premises) Regulations 1976; this has been revoked with effect from October 2006. In the majority of cases, responsibility for 'special premises' has transferred from the HSE to the relevant local FRA.

The role of the HSE in fire safety is now primarily concerned with *process fire precautions*; these are the special fire precautions required in any workplace in connection with the particular process being carried out there. Thus in an engineering workshop, there will be specialist process fire precautions associated with welding, brazing and soldering; the handling and storage of fuels, oils and other flammable liquids and for working with flammable gases, vapours and dust such as in a woodworking shop.

Figure 1.11 shows a typical fire point with a mandatory sign showing that eye protection must be worn when using the extinguishing equipment. The 'break glass' emergency fire-warning device can also be seen, together with a warning notice explaining that these water-based extinguishers may be used for wood, paper, etc., but *must not be used for electrical fires*.

Activity 1.13

1. Go onto the HSE website and look up the guidance leaflet 'INDG 227' *Safe working with flammable substances* and answer the following:
   (a) Detail the five safety principles associated with flammable substances that may be remembered using the acronym VICES.
   (b) Detail (in bullet point form) the additional specific precautions that are necessary (apart from those identified in part (a)) to ensure that the risks from flammable liquids, flammable solids and flammable gases are controlled.
2. By looking at the fire extinguishers in your local fire point, researching in your local library or otherwise, answer the following questions:
   (a) What is the colour code of a water-type extinguisher and what sort of materials may be extinguished using this type of extinguisher?
   (b) You are required to extinguish an electrical fire, what type of extinguisher would you use and how is it colour coded?

First aid at work

The essential aspects of first aid that employers have to address are set out in the **Health and Safety (First-Aid) Regulations 1981**.

Employers need to ensure that they provide equipment and facilities that are adequate and appropriate to enable first aid to be rendered to their employees if they are injured or become ill at work. In addition, depending on the nature and size of their company, employers also need to ensure that an adequate and suitable number of qualified persons are present and are able to render first aid to their employees.

Figure 1.12 shows a typical first-aid station equipped with a first-aid kit, an eye irrigation kit and certain notices detailing information that may be needed in an emergency. Among these notices should be a first-aid

Figure 1.12 A typical first-aid station

kit contents list that can be amended as the contents are used and the kit replenished accordingly.

The aim of first aid is to reduce the effects of injury or illness suffered at work whether caused by the work itself or not. As mentioned above, the first-aid provision itself needs to be adequate and appropriate for the circumstances. What this means in practice is that sufficient first-aid equipment, facilities and personnel should be available to:

- give immediate assistance to casualties with both common injuries or illness, as well as those likely to arise from specific hazards at work
- summon an ambulance, paramedics or other professional help.

A first-aid incident/accident *record book* should be kept to record the circumstances and actions taken every time a person suffers an injury and first aid needs to be administered. This book may be the same as that used to detail recordable incidents, diseases and occurrences under RIDDOR (as detailed in the final section of this unit).

The hazards and risks associated with a number of engineering workplace processes and equipment will be considered in detail in the next two sections and the identification of these hazards and subsequent risks will enable you to put yourself in the position of the employer and decide, for example, on the nature and amount of first-aid provision that is necessary in your workplace.

Figure 1.13 shows an eyewash facility needed in an area where there is a possibility of foreign debris such as dust, swarf or other contaminants entering the eye as a result of workplace activities. Also in the figure the edge of a water fountain providing clean filtered drinking water can be seen.

Figure 1.13 Eyewash facility

Activity 1.14

Go onto the HSE website and look up the Health and Safety (First-Aid) Regulations 1981, Approved Codes of Practice and guidance L74 and answer the following:

(a) Produce a checklist of eight, important points that employers should consider when assessing their need to justify the level of first-aid provision in their workplace.
(b) Under what circumstances will a 'first-aid room' be considered necessary?
(c) What are the essential differences in the responsibilities of *an appointed person* and a *first aider*?

One other important duty employers (or other people responsible for premises) have is to report certain accidents and incidents at work under the ***Reporting of Injuries, Diseases and Dangerous Occurrences Regulations 1995*** (**RIDDOR**). You are asked to complete activity 1.15 next that introduces you to the RIDDOR reporting system, which we will return to later in this unit.

Activity 1.15

Look at the HSE guidance leaflet L73 'A guide to the Reporting of Injuries, Diseases and Dangerous Occurrences Regulations 1995' and the leaflet entitled 'What is RIDDOR?' (available online from the Health and Safety Executive website at www. hse.gov.uk/coshh) and use them to answer the following questions:

(a) What is RIDDOR?
(b) Who should report under RIDDOR?
(c) What must be reported under RIDDOR?
(d) When should incidents be reported under RIDDOR?
(e) Give FOUR examples of major injuries that are reportable under RIDDOR.
(f) Give FOUR examples of dangerous occurrences that are reportable under RIDDOR.

We now leave the subject of legislation and regulation to concentrate on the hazards and possible risks that directly affect us in the engineering workplace. Much of the legislation and many of the resulting regulations will impact on the methods we use to assess risk and the control measures we need to adopt to reduce or eliminate these risks when we consider them later in this unit.

Hazards and Risk in the Engineering Workplace

Before we consider the identification and control of the types of hazard and the risks they pose, we first need to formally define these terms, both of which were introduced informally earlier in this unit.

A *hazard is anything that might cause harm*, e.g. working on ladders, spilt liquids, flammable substances or machine cutting tools.

Risk is the chance of harm being done, as well as how serious that harm may be.

Introduction to workplace activities and regulation

Before looking at workplace activities in detail, you should be aware of the important regulations that cover the provision and use of workplace equipment. They are known as the *Provision and Use of Work Equipment Regulations (PUWER) 1998*. The primary aim of these regulations is to ensure that work equipment (including mobile and lifting equipment) should not result in health and safety risks, regardless of its age, condition or origin. The types of equipment covered by the regulations include those that are specific to engineering and engineering workshops. Regulations on the inspection, maintenance and protection measures for equipment as well as regulations for the control and reduction of risk are covered. One important subset of the equipment covered under PUWER is that of lifting equipment and the lifting operations that make use of this equipment. To cater for this important category, separate regulations known as the **Lifting Operations and Lifting Equipment Regulations (LOLER) 1998** have been produced. There is also a lot of HSE guidance material and Approved Codes of Practice that is targeted to specific specialist areas of equipment that you will be directed to during your study of this section. Two pieces of guidance material produced by the HSE are:

1. **Health and Safety in Engineering Workshops**
2. **Using Work Equipment Safely (INDG 229).**

Both of these will prove to be useful sources of information for identifying workplace hazards and their associated risks.

Engineering workshops and other workplace areas like hangars, production lines, motor transport, manufacturing and maintenance facilities all have a vast and diverse amount of engineering activities for which there are numerous hazards that have the potential to cause harm and pose varying degrees of risk.

Thus in an engineering workshop, you may see taking place any one or more of the following types of activity:

- Manual lifting and handling
- Lifting and lowering operations using winches, hoists, jacks, trestles, lifting trucks and powered lifting platforms
- Working at height using, for example, platforms, ladders, highway staging, scaffolding, gantries and cherry pickers
- Vehicle movements
- Work on electrical installations and equipment
- Use of portable electrical equipment such as power drills and hand lamps
- Use of electrically, hydraulically or pneumatically powered tools and equipment
- Machining operations such as milling, turning, grinding, boring, punching and pressing
- Welding, brazing and soldering operations
- Fabrication and repair operations such as bending, sawing, guillotining, filing, drilling, riveting and adhesive bonding
- Use of flammable substances such as lubricants, oils, degreasing agents, petroleum spirit, paints, doping agents and compressible gases (oxygen, hydrogen, etc.)
- Painting, spraying and powder coating operations.

We will consider the hazards and risk potential of many of the above, using examples, case studies and activities that you will be asked to complete. In considering these hazards and associated risk, we are in effect carrying out a *risk assessment* and an invaluable piece of guidance to help us is contained in the HSE document, **Five steps to risk assessment**, which may be downloaded from the HSE website at: www.hse.gov.uk/risk/fivesteps.htm

However, before we consider these hazards and risks in detail, you need to familiarize yourselves with the more common safety signs and symbols you will see around and in your work area. The health and safety regulations

(a)

(b)

(c)

(d)

FIRST-AID BOX

Figure 1.14 Types of safety sign ((a) prohibition, (b) warning, (c) mandatory, (d) emergency/first aid)

require employers and other responsible people to convey health and safety information in a variety of ways, one essential component of which is the use of safety signs.

Safety signs

A **safety and/or health sign** provides information or instructions about safety or health at work by means of a signboard, a colour, an illuminated sign or acoustic signal, a verbal signal or hand signal. A **signboard** is a sign that provides information or instructions by a combination of shape, colour and a symbol or pictogram.

Signboards can be of the following types:

(a) Prohibition sign
(b) Warning sign
(c) Mandatory sign
(d) Emergency escape or first-aid sign
(e) Firefighting sign.

An example of the shape, colour and form of each type of sign is shown in *Figure 1.14*.

(a) **Prohibition signs** are those signs that *prohibit behaviour likely to increase or cause danger*. For example, the *no access for pedestrians* sign shown in *Figure 1.14a*. The intrinsic features of this sign include a round shape and black symbol or pictogram on a white background, with red edging and a red diagonal line. The red part of the sign must take up at least 35% of the area of the sign.

(b) **Warning signs** are those signs that *give warning of a hazard or danger*. For example, the *low-temperature* sign shown in *Figure 1.14b*. The intrinsic features of this sign are its triangular shape, black symbol or pictogram on a yellow background and black edging. The yellow part of the sign must take up at least 50% of the total area of the sign.

(c) **Mandatory signs** are those signs that *prescribe specific behaviour*. For example, the *safety helmets must be worn* sign shown in *Figure 1.14c*. The intrinsic features of this sign are its round shape, white symbol or pictogram on a blue background and white border. The blue part of the sign must take up at least 50% of the total area of the sign.

(d) **Emergency escape or first-aid sign** are signs *that give information on emergency exits, first aid or rescue facilities*. For example, the *first-aid box* sign shown in *Figure 1.14d*. The intrinsic features of this sign are its square or rectangular shape, white symbol or pictogram on a green background with a white border. The green part of the sign must take up at least 50% of the total area of the sign.

(e) **Firefighting signs** are those signs that give information on the location of firefighting equipment and first-aid firefighting points, with supplementary 'this way' signs to direct people to the firefighting equipment (for example, the fire extinguisher sign shown in *Figure 1.11*). The intrinsic features of this sign are its square or rectangular shape, white symbol or pictogram on a red background and its white border. The red part of the sign must take up at least 50% of the total area of the sign.

> **KEY POINT**
>
> There are five types of health and safety signs: prohibition, warning, mandatory, emergency escape/first-aid and firefighting signs

You should now gain some practice in being able to read the signboards in and around your workplace. If you are unsure of their meaning, you should consult the HSE website and look at the web-friendly download version of *The Health and Safety (Safety Signs and Signals) Guidance on the Regulations leaflet L64.* You will have an opportunity for further recognition practice when you attempt the multiple-choice quiz at the end of this unit.

Lifting and working at height

There are essentially two major types of lifting operations, those that involve manual handling and those that employ lifting equipment. The regulations and guidance for the former may be obtained by consulting the *Manual Handling Regulations 1992*, and guidance on these regulations may be found in the HSE leaflet L23. Lifting operations that involve the use of equipment and/or vehicles is covered by the *LOLER 1998* regulations mentioned earlier, with both the Approved Codes of Practice (ACOP) and Guidance material being given in the HSE leaflet L113. *The Working at Height Regulations 2005* cover all the general legislation and risks associated with working at height. There is also a lot of specialist ACOP and Guidance material provided by the HSE, e.g. the many leaflets too numerous to mention that cover all aspects of working at height on building sites.

You will be asked to investigate the hazards associated with manual handling as an activity, but first we will consider the first of three case studies on a lifting operation that involves the use of equipment and the further complication of working at height!

Case studies on the identification and control of workplace hazards and risk

Case study 1 – Lifting operations and working from height

Scenario

Figure 1.15 shows a photograph of a lifting operation that involves the raising of an auxiliary engine into the tail section of an aircraft using a cradle and hand winches. The winch cable attachment points are located inside the tail section of the aircraft itself.

In order to get the engine and cradle assembly up to the fuselage tail section on top of the highway staging (scaffolding), a gantry (overhead crane) or mobile crane is used.

Figure 1.16 gives you some idea of the height of the working platform, showing the view from the tail section platform itself. The engineers have to climb up the ladder sections of the highway staging until they reach their working section, as shown in *Figure 1.15*.

As one of the health and safety representatives for this hangar, you are responsible for assessing workplace hazards and making recommendations for controlling them. In this capacity, you should now carry out the following tasks.

Figure 1.15 Engine lifting operation at height

Figure 1.16 View from the tail section of the aircraft

Tasks

1. Define *hazard* and *risk* in terms of health and safety.
2. Describe the methods/procedures you will adopt to identify the hazards associated with all the processes involved in lifting this engine into the tail section of the aircraft.

3. Identify the hazards associated with the whole engine lift process and for the hazards that offer most risk explain how they may be controlled. *Note*: the organizational and management aspects of controlling the risk posed by these hazards *need not* be considered at this stage.

Answers

1. A ***hazard*** is anything that might cause harm, e.g. working on ladders, spilt liquids, flammable substances or machine cutting tools. ***Risk*** is the chance of harm being done, as well as how serious that harm may be.

2. Prior to considering the hazards associated with these lifting operations, I would consult the relevant ACOP and Guidance material provided by the HSE, including there information on:

   - The five steps to risk assessment
   - Lifting Operation and Lifting Equipment Regulations
   - Safe use of Lifting Equipment ACOP and Guidance L 113
   - Working at Height Regulations 2005

   Then following HSE guidance I would use the following *procedure*:

   - Walk around and under the aircraft on the hangar floor, where people work, noting things that might pose risk, particularly from falls from above, as well as other possible hazards in the area. Being advised on the types of hazard by consulting the HSE regulations and guidance given above.
   - Talk with the supervisors and engineers who carry out this type of work-based task, while at the same time working at height.
   - Look at the workplace accident book to determine the types of problem we have had in the past.
   - Write down on a risk assessment form or similar the people who could be harmed by the hazards, how they may be harmed and the adequacy of the control measures that are needed to meet health and safety requirements.
   - Discuss my findings with the workplace supervisors and senior health and safety representative and after considering their advice, put my findings into practice.

3. The hazards associated with the engine lift operation fall into two major categories, *those associated with the lift and lifting equipment itself* and *those associated with working at height*. The first general point to make is that there is no alternative but to have to work at height; the aircraft engineers need to be close to the tail section in order to offer up the auxiliary engine that needs to be fitted.

   Lift and lifting equipment

   - People may be struck by engine that is slung under crane jib or gantry
   - Crush injury or death if engine slips in sling or possible cable break when hoisting
   - Serious injury or death if people are run over or knocked by moving vehicle, or jib or gantry cable and hoist assembly
   - Head injury/other injury from falling components from engine or from overhead gantry cab or mechanism
   - Slip or trip hazard from people involved with lifting operation
   - Collision injury with other parts of aircraft and/or ground equipment (cuts, knocks, etc.) from those involved in lifting operation
   - Crush injury or death if engine slips in cradle, when being fitted into tail section
   - Crush injury or death from any possible crane or overhead gantry malfunction or careless operation.

Working at height

- Death or injury from fall
- Injury from slips and trips
- Injury to others from dropped tools and equipment
- Head injuries from scaffolding and aircraft structure above
- Eye injuries from protruding objects on aircraft/engine
- Crush injury or death if engine slips in cradle or swings as a result of hoist cable failure
- Crush injury or death from any possible lifting cradle failure.

The above provides a relatively comprehensive list of the hazards that exist throughout the whole engine lift process. These hazards all have the potential to do serious harm or even cause death, so all these hazards must be controlled in an effort to minimize or if possible eliminate the risks involved. The table below shows the hazards I have identified set against those that might be harmed, and the control measures I think necessary to control the risk they pose. Normally you would complete a *risk assessment form* indicating your findings and necessary actions. We will be looking at this form later, when you are asked to carry out a complete risk assessment.

Hazards	Who might be harmed and how	Control measures
Operating overhead gantry or mobile crane to move and lift heavy loads, e.g. aircraft engine	Those technicians and engineers involved in the lift operation and those people walking or working in and around the lift area. These people may suffer crush injury, other serious injury or death if struck by a moving vehicle if there is a malfunction of the crane or overhead gantry, or if struck by an unstable load, detached load, jib or gantry hoist.	1) Cordon-off area and restrict access to mobile crane vehicle and/or overhead gantry area while lift operations are taking place. 2) Post warning notices and safety signs and use audible warnings as necessary. 3) Ensure that there is a purpose-built fully approved ladder way up to the gantry cab, with a suitable railed platform to enable operators to enter and exit gantry cab safely. 4) Ensure that only suitably qualified and trained technicians operate the vehicle crane or gantry and connect/disconnect the engine to cradle and lifting slings. 5) Ensure that crane lifting vehicle, gantry, cables, pulleys and associated lifting slings and other equipment are serviceable, are of the correct type and are capable of safely supporting the load from the engine and its cradle. 6) Ensure that safety helmets and other appropriate safety clothing are worn by those involved or around during the lift. 7) Ensure that there is two-way radio communication or that lifting signals can be clearly seen between the gantry and/or crane operator and those on the ground.

(Continued)

Hazards	Who might be harmed and how	Control measures
Lifting engine into aircraft tail section	The technicians and engineers involved in the lift and those working on or walking under the highway staging may suffer head, eye or other serious injury by an engine cradle cable failure, malfunction or fracture or by protruding objects from the engine, aircraft or by falling objects from the highway staging platform.	1) Only qualified and fully trained aircraft technicians should be responsible for the lift of the engine into the aircraft. 2) Suitably designed rigid and stable highway staging should be used; with a platform of sufficient strength and size to support the weight of the engine and cradle prior to lifting into aircraft. 3) 'No entry' notices should be posted around the highway staging and access limited only to those involved with the task. 4) Safety helmets and other suitable and appropriate personal protection equipment should be used by all involved with the lift and/or on the highway staging.
Working at height	Death or serious injury from fall, slips or trips on highway staging or fall from gantry operating cab. Head and eye injuries from protruding objects and foreign debris on the sections of the highway staging.	1) Inertia reels and harnesses or other personal safety equipment should be used as necessary. 2) Highway staging platforms should be fitted with guardrails and kick boards around perimeter to prevent falls and prevent tools, etc. being kicked or dropped from the platform. 3) Highway staging should be clean, oil and grease free and with all foreign debris removed. Safety warning notices should be posted as to the hazards when climbing and working on the staging. 4) Gantry door should have safety locking system fitted, to prevent inadvertent opening. 5) Safety helmets and other appropriate safety equipment should be worn by those on the highway staging.

In *Figure 1.15* the guardrails and perimeter kick boards can be clearly seen and it is quite apparent that the highway staging is fit for purpose and of the right design. If you look back to *Figure 1.3* you can see the drop hazard sign attached to the bottom ladder section of the highway staging.

The above table gives a reasonably comprehensive picture of the types of hazard and control measures associated with this particular task. However, this list cannot be considered complete because for a full risk assessment the management and organisation control measures and all necessary follow-up actions need to be included.

In the above case study, only lifting operations that required the use of specialist lifting equipment were considered. In the activity that follows you will be asked to consider the hazards and necessary control measures for manual lifting operations.

Activity 1.16

You work in a medium size workshop where metal fitting and machining activities take place. This requires the manual handling of sheet metal from delivery vehicles to racks and the lifting and stacking of component parts into storage compartments that are up to 3 m in height.

(a) Identify and list all the hazards associated with both of these manual lifting operations, detailing who may be harmed and how.
(b) Describe in bullet point form the control measures you would want to see in place to minimize or eliminate the risks posed by these lifting operations.

Case study 2 – Engineering workshop machinery

In this case study we are going to consider the hazards and control measures associated with a variety of engineering workshop machinery. Machines used for cutting, sawing, grinding, bending, milling and turning metal and other materials pose their own hazards, which although similar in nature require a variety of control measures to ensure that the risks from them are minimised.

Scenario

You work as a general fitter and machinist in a medium-sized engineering workshop. Your workshop has a range of lathes, milling machines, pillar drills, guillotines and grinding machines similar to those shown in *Figures 1.4, 1.17* and *1.18*. If you look back to *Figure 1.4* you will see a picture of a typical metal grinding wheel, while *Figure 1.17* shows a section of a typical machine shop with a number of lathes and vertical milling machines and *Figure 1.18* shows a modern manual milling machine complete with a digital read-out (DRO) facility.

Figure 1.17 Typical traditional machine shop

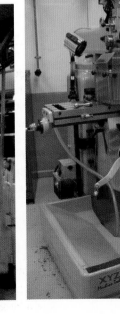

Figure 1.18 A modern milling machine with DRO

You have been asked by your supervisor to walk around and identify all the hazards associated with the machines that are in your workshop and report back on your findings. You know that to do the job properly and to fully comply with the regulations you will need to consult with the workshop health and safety representative and carry out the following tasks.

Tasks

1. Detail the regulations, ACOP and Guidance material you would consult to assist you in identifying the hazards and assessing the risks that the machinery poses.
2. Produce a list of the hazards posed by the types of machinery you have in your workshop.
3. Produce a table with the same headings as that for case study 1 on lifting operations and working from height that details the hazards, who might be harmed and how and the control measures you would adopt to minimize or eliminate risk.

Answers

1. Mandatory and useful sources of reference that I would consult include:
   Provision and Use of Work Equipment Regulations (PUWER) 1998, ACOP and Guidance on Regulations L22
   Management of Health and Safety at Work Regulations 1999
   HSE document *Five steps to risk assessment*
   HSE publication, *Health and Safety in Engineering Workshops*
   HSE guidance on *Safety in the use of abrasive wheels leaflet HSG17*
   HSE publication, *Essentials of Health and Safety at Work.*
2. General hazards for all this machinery with moving parts and cutting or abrading surfaces include:
   - possible entanglement of loose clothing (such as neck ties and loose sleeves), hair and dangling jewellery
   - crush injuries from work piece or platform (lathes, CNC machines, etc.)
   - strike injuries from moving parts, ejected material or work piece movement or detachment
   - cuts and severing injuries due to cutting tools, blades and ejected swarf
   - shock hazard from electrical supply
   - health hazard from machining fluids
   - noise hazard particularly from abrading and powered sawing machines
   - vibration hazard particularly from grinding machines.
3. When considering the control measures that need to be adopted, I have not only taken into account the immediate actions necessary as a result of the hazards posed by the machines. I have also taken into account those secondary measures remote from the machines that need to be put in place to ensure that risk is minimized or eliminated not only for the fitters and operators, but also for all people employed in the workshop or visiting who might be put at risk.

Hazards	Who might be harmed and how	Control measures
From machines for: Turning Milling Drilling Sawing Grinding Guillotining	Employee fitters and operators. People passing machines and internal or external visitors. 1) Crush injuries and suffocation from entanglement of loose clothing and other articles 2) Crush injuries from work piece or moving platform 3) Strike injuries from moving parts, ejected material or work piece detachment or movement 4) Cuts and severing injuries from cutting tools, cutting blades or ejected swarf 5) Friction burns and lacerations from grinding surfaces and sanding belts 6) Shock hazard or other injury from electrical supply 7) Breathing difficulties, asthma or dermatitis from machining fluids 8) Ear problems from noise of abrading, power saw and other machines 9) Repetitive strain or other problems from vibration hazard from grinding machines and others	*General* Fitters and operators trained and competent in the use of the machines and able to carry out pre-use checks First-aid station and/or first-aid room, near to hand. Qualified first aider or appointed person is on call, as appropriate. All machines regularly maintained and kept in good order. Maintenance contractors and other visiting tradesmen are made fully aware of all local health and safety hazards and are qualified and competent to undertake the contracted maintenance. All machines installed and sited in accordance with manufacturers' instructions. All machines provided with easily accessible on/off buttons and emergency stop provision, as per design. Machines are so positioned that there is sufficient space for marked pedestrian walkways that keep people away from the immediate hazards posed by the machines. *Specific* (1), (2), (3), (4) and (5) Ensure operators and fitters are trained in machine safety, emergency shutdown procedures, pre-use checks and daily or periodic checks. Post warning signs about dangers of loose clothing and entanglement dangers. Ensure fitters and operators are aware of the need to tie back hair or wear a cap, ensure overalls are securely fastened at cuff and neck. Ensure all hazardous parts of the working sections of the machine are suitably guarded. Use interlock or moveable guards as appropriate. Post safety signs for eye protection, safety boots and other PPE. Use jigs, fixtures and clamps to secure the work piece. Provide extraction equipment for abrasive wheel and sanding machines. Ensure only trained and competent people are allowed to change abrasive wheel and carry out maintenance on abrading machines. Guard opening over wheel or sanding belt, restricted to that needed to enable the work to be done. (6) All electrical power supplies to machines is of approved type, is of the lowest-voltage design and is regularly inspected and maintained. Emergency electrical shutdown provision and on/off buttons are easily accessible and clearly marked. Electrical hazard sign is appropriately positioned, with written warning of danger as well as symbol. Isolation switches, emergency stops switches and circuit breakers are in good working order and isolation procedures are clearly understood by operators, fitters and those qualified to undertake maintenance. (7) Use masks and gloves as appropriate when handling machining fluid. Do not use unrefined mineral oils or mildly refined distilled oils. Avoid or minimize processes that cause mist, fumes, vapour and splashing from the machining fluids. Regularly change machining oils to prevent contamination. Remove and clean contaminated clothing at the earliest opportunity. (8), (9) Ensure that abrading machines are balanced and bearings are sound. Wear ear defenders that reduce noise but enable operator to maintain awareness. Post warning signs for noise and vibration hazard. Consider a quieter machine or change the process for a quieter one. Use anti-vibration mountings where possible.

The table not only shows the specific control measures that are directly related to the care and operation of the machine but also the general control measures that ensure the health and safety of all people that work in or frequently visit the workshop.

We have so far looked at two, fairly diverse engineering workshops, where in the first the emphasis was on maintenance and the second was on machine production. To give you a little more practice in hazard recognition and control measures, you should attempt the next activity, where you are asked to investigate the hazards that exist and the control measures needed to make welding and brazing activities safe.

Activity 1.17

By consulting the appropriate Regulations and HSE ACOP and Guidance material or otherwise, attempt the following tasks:

(a) Identify the hazards that exist in a typical engineering workshop where welding/brazing activities take place and for the hazards associated with these activities, detail those people that might be harmed and how.
(b) Suggest 10 control measures that should be in place or put into place to help reduce or eliminate the risk from welding/brazing activities.

Before we leave the study of hazards, risk and control measures, we move on to full *risk assessment* and the methods and requirements for *accident/incident recording*. We will consider one more case study associated with electricity that provides a further illustration of the ways in which we identify hazards, determine the associated risks and recommend control measures.

Case study 3 – Electricity

Scenario

A small electrical and electronics repair company has one primary workshop where they undertake a variety of corrective repairs and maintenance to electrical machines, such as portable motors, generators and other electrical appliances as well as to audio-visual electronic equipment that has been returned for assessment and repair from retailers.

A part of this workshop has been set aside for *electrical testing*, where the test section has three major functions:

1. Diagnostic and fault-finding tests to ascertain the initial condition and nature of the unserviceability of the returned equipment.
2. Routine safety checks during repair.
3. Quality assurance and pre-issue checks and tests on equipment after repair.

The equipment that is repaired ranges from very low-voltage electronic components up to single-phase and three-phase mains voltage (230 V AC and 400 V AC) electrical machines. *Figure 1.19* shows the type of electronic work that is carried out by the company. One photograph shows a resistance check being carried out on a very low-voltage printed circuit board, and the other an automatically controlled conveyor system demonstrator is awaiting repair.

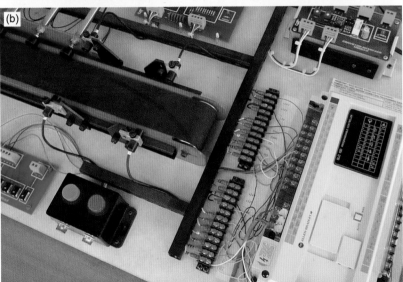

Figure 1.19 (a) Circuit board resistance check; (b) an automatically controlled conveyor system

There are at any one time up to six technicians and one supervisor working in the electrical test section of the workshop and up to forty-five technicians, six supervisors and one duty manager working in the main repair and assembly section of the workshop.

Tasks

By interrogating the appropriate Health and Safety Regulations and HSE ACOP and Guidance material or by using an alternative investigative approach, attempt the following tasks.

1. Detail the *general* hazards associated with the workshop building electrical installation, plant and equipment and describe the types of injury that may occur as a result of these hazards.
2. Detail 10 control measures that should be implemented by management and acted upon by employees to help mitigate or eliminate the risk associated with the building installation and equipment.

3. Summarize the most important issues that need to be taken into account according to the *Electricity at Work Regulations 1989* when carrying out *electrical testing using test equipment*.

Answers

Mandatory and useful sources of reference consulted in the process of completing the tasks included:

- *The Electricity at Work Regulations 1989*
- *HSE memorandum of guidance on the Electricity at Work Regulations 1989*
- *Electricity at work: Safe working practices HSE leaflet HSG85*
- *Safety in electrical testing at work: HSE leaflet INDG 354*
- *Safety in electrical testing: Products on production lines HSE leaflet EIS38*
- *Electrical test equipment for use by electricians: HSE guidance note GS 38*
- *Electrical safety and you: HSE guidance leaflet INDG 231*
- *Essentials of health and safety at work: HSE guidance book ISBN 0717661792.*

1. The primary hazards and injuries associated with electricity and electrical installations are:
   - *Contact with live parts*, particularly live conductors made live by faulty wiring and connections causing burns and electric shock that *can kill* at normal mains voltage of 230 V AC or below! Also injuries from falling and other secondary injuries that result from original electric shock hazard.
   - *Electrical fires*, started by poor installation and faulty equipment, again that can cause burns, smoke inhalation and suffocation, as well as the release of dangerous toxic/carcinogenic substances (causing health problems such as cancer) from burning insulation, encased electric circuit boards or other components and equipment.
   - *Explosion and/or fire*, where electrical arcing/sparking could be the source of ignition in a potentially flammable atmosphere, e.g. in and around open fuel tanks or fuel/oil storage facilities. This may result in death from explosion hazard or serious burn injuries from fire.
2. Managers/employers should ensure that:
   - All employees are suitably trained to carry out mandatory checks prior to the operation of equipment that is powered from a fixed electrical installation.
   - All electrician employees are suitably qualified, trained and competent to work on the electrical installations and associated electrical equipment that is to be used in the workshop.
   - All employees are aware of the hazards associated with the electrical supplies to equipment and have been trained in electric shock first aid, including the methods used to isolate the supply and make safe in normal and emergency circumstances.
   - A suitable system of maintenance for fixed installations has been developed and is being followed.
   - All electrical installation contractors are suitably qualified and competent to undertake their contracted work on electrical installations throughout the workshop.
   - Only select installed electrical equipment that is suitable for the jobs for which it was designed.
   - The law on '*live*' electrical work is fully understood and is being applied to all workshop electrical installation work by qualified and competent employee electricians and third-party electrical contractors.
   - The lowest possible voltage equipment is used from fixed electrical supplies wherever possible.

- All electrical installation safety devices, such as circuit breakers, fuses and residual current devices, are serviceable and correctly rated for the circuit they protect.
- Any new electrical installation has been properly designed, installed and safety tested against laid-down standards.
- All unserviceable electrical installations and equipment are electrically isolated and taken out of service until remedial repair, replacement and/or maintenance can be carried out.
- All appropriate safety and warning signs have been appropriately positioned and all specialist and other particular electrical maintenance requirements are written into local workshop procedures and orders.

3. The important issues to be considered are similar to those identified above for electrical installations and equipment in the answer to task 2. More specifically, the issues relating to electrical testing *using test equipment* that need to be addressed are summarized below.

- Test equipment so far as is reasonably practical should be constructed, maintained and used in a way to prevent danger. For example, use of the low ohms meter, with an equivalent series resistor (ESR) capability in *Figure 1.19*, where a clear warning to discharge the capacitor before use will help to prevent electric shock to the operator and also safeguard the circuit being tested.
- Obey the law on 'live' work. So no live work, unless it is unreasonable to work 'dead' *and* it is reasonable to work 'live' *and* suitable precautions are taken to prevent injury.
- Work must be carried out in a safe manner. So, the control of risks while working and control of test areas should be considered. Also suitable tools, clothing and insulation barriers such as rubber mats should be used as appropriate. Adequate information on the test equipment and the types of use for it should be provided. The work station/area should be provided with suitable lighting and access as well as having sufficient space to enable the testing to be completed safely and efficiently.
- Those involved with the testing work must do all in their power to prevent danger and injury to themselves and others around them.
- Employees involved with the testing work must have adequate training, skill and experience to perform the tasks asked of them, with adequate supervision where appropriate.

Activity 1.18

The maintenance electricians employed by the small electrical and electronics company identified in Case study 3 are not only responsible for maintaining electrical installations but also for the maintenance of *portable equipment* such as electric kettles, electrically powered cleaning equipment and portable electric hand tools used by other employees in the workshop.

By consulting the HSE booklet HSG107 Maintaining portable and transportable electrical equipment or otherwise:

(a) Detail the important points that should be covered in a visual user check.
(b) What should be covered on a formal visual inspection?
(c) What should be covered on a combined inspection and test of portable electrical equipment and who should carry it out?

Risk Assessment, Accidents and Incidents

In this final part of the unit we will be learning how to carry out all phases of a risk assessment using a case study and by considering the work we have already done in previous case studies where we identified particular engineering work hazards and their associated control measures. In the second and final area of study we will consider the methods we use to report and record accidents and incidents that may occur in the engineering workplace.

Risk assessment in the engineering workplace

Introduction

You have already had practice in identifying particular engineering workplace hazards and suggesting control measures to help eliminate or reduce the risk from these hazards for a variety of activities. Also, as an exercise you were asked to look at the **HSE leaflet INDG 163** *Five steps to risk assessment*. It is now time to consider these five steps in detail, since by following the guidance in this leaflet you will cover all the stages that are necessary for a successful risk assessment.

Why do we need to carry out a risk assessment? What is a risk assessment? You may already know the answer to these questions by virtue of the topics we have already covered and the activities you have been asked to complete, but here is a reminder.

A *risk assessment* needs to be carried out by those in authority (managers/employers) in order to offer the best protection for their employees, their company business and in order to comply with the law. Risk assessment is simply a step-by-step methodical examination process used to identify sources of harm that, if followed, helps those in authority to concentrate on the risks that really matter in a particular engineering work environment and put the measures in place to eliminate or at least minimize these risks.

The five recommended steps to help us assess risk in the engineering workplace are:

- Step 1 – Identify the hazards
- Step 2 – Decide who might be harmed and how
- Step 3 – Evaluate the risks and decide on precautions
- Step 4 – Record your findings and implement them
- Step 5 – Review your assessment and update if necessary.

If you look back at Case study 1, you will see the procedure that was used in the answer to question 2 in order to *identify the hazards*. This same procedure was used to identify the hazards in case studies 2 and 3 and hopefully you adopted it when completing the activities that followed each of these case studies, where you were not only asked to identify the hazards but also to decide *who might be harmed and how*. In addition you were asked to identify and describe the control measures that needed to be in place or put in place in order to reduce risk. In other words, *evaluate the risks and decide on the precautions.* You have, therefore, already had some practice in carrying out the *first three steps* to a full risk assessment.

As a manager, employer or person responsible for the health and safety of the workforce, visitors and indeed all people who may be put at risk by company activities and facilities, you need to consider not only the immediate risks posed by the hazards at the point of work but also all the secondary issues that may create risk. We have looked at some of these secondary issues in the previous case studies but not in any great depth. Neither have we been asked to look at the last two very important steps in the process (*recording and implementing findings* and *reviewing the assessment and updating as necessary*). In the next case study we will follow the whole process and attempt to produce a complete and satisfactory risk assessment that not only complies with the law but also offers the best possible workplace environment for all concerned. We will use the engineering facilities and activities of a *motor vehicle repair, bodywork and maintenance workshop*, as our example throughout.

Before we embark on this case study, I have listed below some of the important legislation that is relevant to our motor vehicle workshop example risk assessment. The ACOP and Guidance directly associated with specific facilities and activities will be given as these motor vehicle (MV) workshop activities are considered.

Some relevant legislation

- *Workplace (Health, Safety and Welfare) Regulations 1992*
- *Management of Health and Safety at Work Regulations 1999*
- *Manual Handling Operations Regulations 1992 (amended 2002)*
- *Provision and Use of Work Equipment Regulations (PUWER) 1998*
- *Lifting Operations and Lifting Equipment Regulations (LOLER) 1998*
- *Control of Substances Hazardous to Health Regulations (COSHH) 2002*
- *Control of Lead at Work Regulations 1998*
- *Road Vehicles (Brake Linings Safety) Regulations 1999*
- *Pressure Systems Regulations 2000*
- *The Dangerous Substances and Explosive Atmospheres Regulations 2002*
- *The Personal Protective Equipment Regulations (PPE) 2002*
- *The Control of Vibration at Work Regulations 2005*
- *The Control of Noise at Work Regulations 2005*
- *Working at Height Regulations 2005.*

The legislation, responsibilities, recording and reporting of accidents, incidents and emergencies at work are so important that it will be covered separately in its own section, after we have covered risk assessment. For completeness the important legislation covering accidents, incidents and other emergencies that impact on all work is given below; you have in fact considered all this legislation earlier in this unit.

- *The Reporting of Injuries, Diseases and Dangerous Occurrences Regulations (RIDDOR) 1995*
- *Health and Safety (First-Aid) Regulations 1981.*

Other important aspects of the law on accidents, emergencies and health hazards are covered in the *Management of Health and Safety at Work Regulations 1999* and the *COSHH Regulations 2002.*

Case study 4 – Risk assessment of an MV servicing, repair and bodywork engineering workshop

Scenario

You work as a senior supervisor in a large garage that has workshop facilities for motorcar servicing and general maintenance, general repairs and reconditioning and bodywork repairs and paint spraying. As someone in a senior position and the company health and safety representative, with current knowledge and understanding of health and safety legislation and procedures, you have been tasked with carrying out *a full risk assessment* on all three areas of activity within the garage.

Figure 1.20 shows a view of the workshop used for general servicing and scheduled maintenance, with the hydraulic lifting ramps to the left and the pedestrian access corridor running down the middle, vehicles awaiting servicing and undergoing other maintenance activities are positioned on the right of shot. The rolling road, diagnostic test set and other miscellaneous equipment are out of shot.

Figure 1.20 MV workshop arrangement for general servicing and maintenance

Figure 1.21 shows a view of the workshop used for general repairs with drilling machinery in the background and the cars and engines subject to repair and reconditioning in the centre and foreground. Other fitting, repair, test and lifting equipment are out of shot. The bodywork repair section (with welding equipment and composite repair materials) and the spray booths are not shown in the figures.

The garage employs twenty fitters in the general servicing and routine maintenance workshop as well as two supervisors; it has two under-18 apprentices, employed on work experience. In the general repairs and engine-reconditioning workshop, there are fifteen fitters working on early and late shifts, two supervisors and two under-18 apprentices on work experience, who only work on the early shift. The bodywork and spraying section have twelve fitters and one supervisor

Figure 1.21 MV workshop arrangement for general repairs and engine conditioning

employed on bodywork repairs and four full-time paint sprayers (operating within two spray booths), one who Acts as the local supervisor for spraying operations when on shift. The garage has a separate stores area with two storemen, a staff rest room, toilets and washroom facilities, a first-aid station and the requisite mandatory fire points. There is also a front of house reception and waiting area for customers. In addition to all these staff and facilities, there is a general manager with overall responsibility for the day-to-day operations of the garage, assisted by the senior supervisor (you) and two joint owners of the business, who concern themselves with strategic issues and the business finance.

Note

The format of this case study is slightly different from the previous three, in that each of the allocated tasks is written separately complete with a suggested answer and interspersed with activities you will be asked to complete in order for us to achieve the overall risk assessment for the garage premises and activities.

Case study 4 – Task 1

By using the procedure for the identification of hazards suggested in the HSE guidance leaflet INDG 163 (Five steps to risk assessment) and other appropriate information sources, determine and detail the hazards that you think should be taken into account for the motorcar servicing/general maintenance and general repairs/reconditioning workshops within the garage.

Task 1 – Answer

Apart from consulting the legislation/regulations given earlier in this section and leaflet INDG 163, I would also read the relevant parts of the following ACOP and Guidance material prepared and presented by HSE.

- *Health and safety in motor vehicle repair: HSG67*
- *Essentials of health and safety at work: ISBN 0717661792*
- *Health and safety in engineering workshops.*

The views, opinions and expert guidance of the workshop supervisors would also be sought as well as those from the garage first aider and professional bodies directly associated with the motor trade. All of their guidance and suggestions would be taken into account as well as those I formulated personally from my comprehensive walkaround.

From the information gathered and from personal knowledge of the premises and work activities, I have compiled the following account of the hazards applicable to one or other or both of the workshops.

1. Lifting operations and equipment (hydraulic ramps, slings, hoists, jacking and manual handling)
2. Trip, slip and fall hazards (workshop equipment, floors, spilt liquids, inspection pits, servicing platforms, replenishing stores)
3. Electrical equipment (power supplies, fixed electrical equipment, diagnostic test equipment, portable power tools and lead lamps, soldering irons, heaters, steam and water pressure cleaners)
4. Fire/burns (petrol, oils, flammable vapours, running engines, explosion, hot materials and equipment, cold burns air-conditioning)
5. Health (oils, cleaning fluids, brake and clutch linings, batteries, battery charging, sealants, exhaust emissions)
6. Vehicle movement and workplace transport (in/out of workshops, road testing, rolling road and brake tests)
7. Compressors (pressurized containers, air under pressure, pneumatic power tools, tyre inflation)
8. Hand and powered fitting tools and machinery.

Activity 1.19

By using the reference sources given in the answer to task 1 of this case study and/or other information sources, determine and detail the particular hazards that you think should be taken into account for the motorcar bodywork repair shop and for the painting and spraying activities that take place within the garage.

In completing task 1 and activity 1.19 we have completed *step* 1 of the *five steps to risk assessment* for our garage, where you may or may not have included the *first-aid provision* for the garage. In the brief we are told that there is a first-aid room and a registered first aider available in the event of accident or emergency. We may assume that the first-aid room Acts as the first-aid post for the garage and that within the room there are facilities for dealing with wounds and burns, eyewash equipment, toilet facilities and some form of couch/bed that can be used by casualties for rest, treatment or rehabilitation.

We will now attempt to complete steps 2 and 3 of the risk assessment process for our garage in accordance with HSE guidance.

In evaluating the risks and deciding on the necessary precautions (step 3), you may feel that this exercise is rather artificial when presented here as a case study, because in a real situation there will be *precautions/procedures already in place* and therefore your risk assessment would generally only look at what *further action* might need to be taken to further reduce the risk of harm from the identified hazards. As part of the requirements for this unit I am sure you will be asked to carry out a risk assessment for your own place of work or for the facilities you are using, as part of your engineering training and this will add a sense of reality to the process. That said, the process of assessing the risks and deciding what should be done to minimize the chances of harm to personnel and equipment is *a very useful exercise in its own right* that will help you when you are in a position that requires you to carry out your own risk assessment.

Case study 4 – Task 2

By consulting with fellow supervisors, fitters, the appropriate legislation and HSE ACOP/Guidance material and using your engineering experience, answer the following questions:

2.1)   Decide for the hazards identified in task 1 who might be harmed and how.

2.2)   Evaluate the risks and describe the precautions that need to be put in place for the hazards identified in task 1.

Task 2 – Answers

The answers to questions 2.1 and 2.2 are best given collectively in tabular form as shown here.

Hazards	Who might be harmed and how?	What precautions need to be taken (i.e. what actions need to be put in place)
Lifting operations involving lifting equipment or mobile lifting vehicles. Falling loads	Those fitters and other employees involved with the lifting operation or those people walking or working in and around the lift area. Crush injuries, other serious injury or death if struck by moving ramps or moving lift vehicles, detached suspended loads, moving suspended load or vehicle falling-off jacks or trestles.	• All lifting equipment, slings and hoist to be pre-use checked and routinely checked and serviced by qualified person. • Post warning notices, use cordons and restrict access to lifting areas as necessary. • Provide training for fitters on use and safety associated with lifting equipment in regular use (e.g. hydraulic ramps, hoists, lifting vehicles, slings and jacks). • Jacks and axle stands to be regularly checked for serviceability and maximum load never exceeded. • Axle stands used to support vehicle once jacked to height.
Manual handling and lifting	Fitters and stores personnel, lifting components in/out of vehicles or to and from overhead storage racks. Back injuries, muscle and ligament injuries, toe and foot injuries.	• Training given to personnel likely to be involved with manual lifting operations. • Heaviest stores/components to be stored as close to ground level as possible. • Take note of stores and component packaging weight labels. • Lifting vehicle to be used to move heavy and bulky stores between racks and work areas. • Instruction given on manual handling of awkward components and assemblies, e.g. wheel assemblies, gearbox and engine block.

(Continued)

Hazards	Who might be harmed and how?	What precautions need to be taken (i.e. what actions need to be put in place)
Trips, slips and falls	All employees and visitors, walking through and around workshop areas of the garage and those employees using platforms and ladders. Limb, head and facial injuries, broken bones, internal injuries.	• Wear personal protective equipment. • Clearly marked walkways through workshop areas. • Post 'do not walk' and other prohibition signs near work areas that pose a trip, slip or fall hazard. • Clean up fuel, oil and grease spillages immediately. • Use drip trays and other vessels to contain spillages. • Ensure inspection pits are covered when not in use and written warnings can be clearly seen. • Ensure ladders, steps and platforms are correctly maintained, serviceable before use and stowed away when not in use. • Platforms and storage staging fitted with kick boards, hand rails and ladder ways as appropriate.
Electrical equipment (power supplies, fixed electrical equipment, diagnostic test equipment, portable power tools and lead lamps, soldering irons, heaters, steam and water pressure cleaners)	All fitters and other employees that use electrical equipment. Electric shock injury or possible fatalities. Burn or scald injuries from poorly maintained or faulty equipment, especially portable equipment like lead lamps and power tools.	• All installed equipment and power supplies are regularly maintained by competent trained electricians. • All portable electrical equipment is tested and certified annually by a qualified electrician. • All portable electrical equipment is removed from service when defective and serviced, repaired or replaced by competent electrician. • All personnel are trained in the safety aspects/usage procedures of the electrical equipment they need to operate. • Low-voltage equipment such as power tools and lead lamps is used. • Pneumatic-powered tools such as drills and power drivers have been introduced. • PPE is issued and used when operating fixed and portable electrical equipment and machines as appropriate.
Fire/burns	Fitters, other staff and visitors could suffer severe or fatal burns or other fatal injuries, resulting from the burning building, from hot or burning vehicles or explosion of volatile fuels (including liquefied petroleum gas (LPG)), equipment and components or suffer cold burns from vehicle air-conditioning units.	• All personnel read and understand fire orders and emergency evacuation procedures; all exits from garage clearly marked. • Personnel trained by external experts (e.g. local fire service or suppliers) on identification and use of fire extinguishers. • Regular fire drills carried out, with roll calls by supervisors at fire assembly points. • Sufficient exits/entries and alternative exit/entry doors are provided for easy evacuation in the event of any fire emergency. • Dangers and hazards of working with LPG fully explained and understood and LPG vehicles dealt with in a separate part of workshop. • No smoking in all areas of the garage, strictly enforced, with no smoking notices posted. • Orders forbidding potential sources of ignition being carried in personal clothing (e.g. non-safety matches and tubes of lighter fuel).

- Fire points clearly labelled and fitted with appropriate first-aid firefighting appliances and equipment.
- Fire alarms fitted and maintained by contractor.
- Firefighting equipment regularly inspected and maintained by supplier.
- All spilt fuel, oils and greases cleaned up immediately spillage occurs.
- Only use portable electrical equipment designed for use in volatile or potentially explosive atmospheres.
- PPE used when work has to be carried out on vehicles with hot components and/or cold air-conditioning components.
- Strict procedures followed when fuelling/de-fuelling vehicle fuel tanks (e.g. always carried out in safe marked area outside building).
- Petrol never used as a component cleaning fluid. Paraffin bath used, masks and other PPE used by those operating paraffin bath.

Health (irritants, toxic liquids, cryogenic liquids and air pollutants)	Fitters and other employees may suffer from dermatitis, skin rashes, acid burns, breathing difficulties, asthma, asphyxiation or poisoning from hydraulic fluids, cleaning fluids, refrigerants, LPG, brake/clutch linings, spilt battery acid, battery charging or vehicle exhaust fumes.	• All oils, lubricants, LPG and other volatile substances stored within separate metal lockers in accordance with COSHH regulations. • All personnel working with fuels, oils, lubricants and refrigerants: where appropriate, use gloves, goggles, and other protective equipment and barrier creams to protect skin and other exposed parts. • An appropriate separate vented facility exists for battery charging and maintenance. • Dust masks and other appropriate PPE is used when servicing and handling batteries, clutch/brake linings and other possible harmful materials. • Vehicle exhaust is attached to extractor system when engine is running and extractor system is regularly maintained to ensure full serviceability at all times.
Vehicle movements and workplace transport	All personnel and visitors may suffer serious crush injuries or even fatalities from cars or other workplace vehicle movements in/out of the workshops, or when road testing vehicles or operating the rolling road and brake test machinery.	• Visitor parking area clearly marked out in such a position as to avoid or easily see vehicles being driven in/out of workshop. • Warning given to fitters in vicinity of vehicle and vehicle driven slowly and with care in/out of workshop. • Warning given and space provided for the slow and careful transfer of stores, using forklift truck or other loading vehicle. • Rolling road and brake testing rig only operated by qualified personnel and used within safe limits as recommended by manufacturers. • Suitable doors provided as part of the workshop for easy access and egress of vehicles and equipment. • Vehicles to be cleared by fitter and/or supervisor as being safe and road worthy prior to any road test. • Pedestrian walkways provided as necessary to safeguard personnel from vehicle movement activities. • Safety warning signs posted as appropriate.

(Continued)

Hazards	Who might be harmed and how?	What precautions need to be taken (i.e. what actions need to be put in place)
Hand and powered fitting tools and machinery	Fitters and other personnel using hand/powered tools and machinery may suffer cuts, abrasions, piercing, crush or severing injuries using tools such as chisels, hacksaws, spanners, hammers, pillar drills, guillotines, grinding wheels and other machinery.	• All fitters competent and able and willing to carry out checks on tools and machinery prior to use and report defects to those in authority. • All fitters familiar with local instructions and code of conduct for use and care of tools and operating each machine effectively and safely. • All unserviceable tools and machine malfunctions/faults reported to supervisor and offending item removed from service. • All machinery that is powered mechanically, electrically or pneumatically has suitable guards and failsafe devices fitted and operational. • Relevant safety notices and signs posted at each machine. • Abrasive wheel regulations followed and only trained personnel allowed to change grinding wheels. • Appropriate PPE provided at each separate machine and used by all fitters.
Compressors and compressed air	All fitters and other workers could suffer blast injuries if the compressed air receiver was to shatter sending out high-velocity debris, or if a wheel assembled with an inflated tyre was to shatter or the tyre was to blow, causing possible blast injury to internal organs or ear drums, or possible ingestion injury resulting from the body absorbing high-pressure jets of air through pierced skin.	• All workers made aware of the dangers of containers under pressure and the requirement for care when handling and the need to avoid impact damage. • Need for regular maintenance and pre-use checks of compressor, valves and hose assemblies. • All workers to be made aware of the dangers of pressurized air leaks from compressor reservoirs, inflated tyres and other pressurized containers. • Need to use inflation cages to inflate tyres, seat the bead and avoid the consequences of tyre blow out or wheel assembly fracture. • All workers trained in the care and use of pneumatic power tools and the need to exercise extreme caution when using them with other people in the vicinity. • Operating levers on airlines have a failsafe device fitted (e.g. dead man's handle).

Note:

As part of your risk assessment, you need to ensure that garage premises comply with *The Workplace (Health, Safety and Welfare) Regulations 1992*. You may assume that the garage has adequate facilities for personal hygiene and welfare, in that they have provided a separate staffroom/restroom, male and female toilets and wash facilities, fresh drinking water, workshop and restroom heating and personal lockers/storage bins.

In addition, you may also assume that the garage provides adequate trained staff (first aiders and appointed persons) and facilities for rendering emergency first aid and is able to comply in all respects with the *Health and Safety (First-Aid) Regulations 1981*. These facilities include a fully equipped first-aid room, sufficient trained personnel and an accident and incident recording book.

Activity 1.20

By consulting the appropriate legislation and HSE ACOP/Guidance material and any other information sources you think useful, answer the following questions:

(a) Decide for the hazards you identified in activity 1.19 (for the bodywork repair shop and paint spraying activities), who might be harmed and how.
(b) Evaluate the risks and describe the precautions that need to be put in place (actions that need to be taken) for the hazards identified in activity 1.19.

You may assume for this activity that the garage facilities, arrangements and procedures for the welfare of staff and for first-aid emergencies have been met and have already been risk assessed and found to be more than adequate.

We are now in a position to complete steps 4 and 5 of our risk assessment. *Step 4 requires us to record our findings and implement them*, while *step 5 requires you or another competent person to review your risk assessment and update it if necessary*.

Step 4 is best achieved by prioritizing what needs to be done based on the severity of the risk posed and the cost and time needed to find a solution. For example, the recommended actions I have given to reduce the risk from *trips*, *slips* and *falls* (Case study 4 task 2 answers) can be achieved mainly from 'good housekeeping', where employees are required to clean up spillages immediately they occur, use drip trays wherever practical, replace covers over inspection pits after use, check ladders, steps and platforms for serviceability before use and stow them away after use. These actions can be achieved by providing initial instruction for new and current employees and by adequate supervision to ensure that best practice is maintained. Management (you) would need to ensure that appropriate safety signs and other information was available and appropriately positioned, as well as ensuring that walkways were clearly marked and positioned for safe pedestrian movement. Good housekeeping and the maintenance of standards cost very little and can be achieved with nothing more than some initial instruction and then maintained by good supervision. The marking of walkways and the provision of safety signs and covers are also easily achieved with little expense and effort; whereas any improvements to my fire/burn precaution concerning the use of properly designed portable electrical tools and equipment in potentially explosive atmospheres, say by using pneumatically operated alternatives, would incur significant initial costs. Also, offering another alternative solution, such as reducing the concentration of potentially explosive vapours in areas in which the work is to take place or by limiting the occurrence of this type of work would require capital outlay and long-term planning to achieve these safety improvements. The above examples give you an idea of the sort of things you need consider when setting your priorities for putting the findings of your risk assessment into practice.

The date that you set for any review/follow-up (*step 5*) will depend on the actions taken as a result of the previous or initial risk assessment and it will also be dependent on any significant changes of equipment or workplace activity. In the case of our motor vehicle garage example, different review dates may need to be set for different and diverse areas of activity, for example, when considering the front of house reception and the paint spraying booths. To assist you with all five of the steps you are advised to follow, the HSE have provided a down-loadable template at the end of their leaflet *INDG 163 Five steps to risk assessment*. The tables and headings I have used for the first three steps of our risk assessment have been based on the first three columns of this very useful template, which if you are unfamiliar with you should download now by going to www.hse.gov.uk/publns/indg163.pdf. The headings and information for recording actions for steps 4 and 5 shown on this template are reproduced below for your convenience.

Step 4 How will you put the assessment into action?

Remember to prioritize. Deal with those hazards that are high risk and have serious consequences first.

Action by whom	Action by when	Done
_____	_____	_____
_____	_____	_____
_____	_____	_____
_____	_____	_____
_____	_____	_____
_____	_____	_____
_____	_____	_____
_____	_____	_____
_____	_____	_____
_____	_____	_____
_____	_____	_____
_____	_____	_____

Step 5 Review date   Review your assessment to make sure you are still improving or at least not sliding back
If there is a significant change in your workplace, remember to check your assessment and, where necessary, amend it.

Note that under all the headings in the HSE template guidance is given similar to that illustrated in steps 4 and 5 above.

By completing the action columns under step 4 and by providing a review date (step 5) based on our earlier discussion, we have effectively completed all the steps for our risk assessment of our garage workshops. We now go on to look at the last topic in this unit, concerned with the methods used to report on and record accidents and incidents.

Reporting and recording accidents and incidents

You were introduced to the ***Reporting of Injuries, Diseases and Dangerous Occurrences Regulations 1995*** earlier in this unit and if you attempted all the questions in activity 1.15 then you will, by now, have a good idea of what RIDDOR is all about! These regulations are made up under the HSWA 1974 and they apply a single set of reporting requirements to all work activities in Great Britain and to the offshore oil and gas industry. We will not concern ourselves too much with this latter part of the regulations, as those involved with the offshore industry will cover this section in detail as part of their health and safety training. The main purpose of the regulations is to generate reports to the HSE and local authority that are collectively known and referred to as ***the enforcing authorities***. The regulations apply to Great Britain (i.e. England, Scotland and Wales) but not to Northern Ireland, the Channel Islands or the Isle of Man, who have their own set of regulations.

A very useful publication that may be downloaded in PDF form from the HSE website and which provides a wealth of information on the duties of employers and employees under RIDDOR is the third edition published in 2008 of ***A guide to the Reporting of Injuries, Diseases and Dangerous Occurrences Regulation 1995, L73***. We will use this leaflet as our primary source of information and reference on accident and incident reporting and recording throughout this final section and try to answer the questions posed in activity 1.15 and other important questions raised by employers and others in a position of responsibility.

Let us start by trying to answer the question what is RIDDOR?

What is RIDDOR?

In order to answer this question, we need to ask a series of other questions that enable us to determine the who, what, why, when and where of incident reporting and recording.

Who should report the specified workplace incidents? The answer is simple, *employers, the self-employed and other people in control of premises*. They may report incidents using a variety of methods that include:

- Reporting online
- Reporting by telephone
- Reporting by e-mail
- Reporting by post.

There has also been set up an *Incident Contact Centre* (ICC) that Acts as a one-stop reporting service. So in the event of really serious incidents, that involve death, serious illness or disease, employers/owners, etc., can report the incident by telephone without undue delay. Telephone calls are recorded at the time of contact with the centre and a transcript is sent to the employers/management of the business concerned, who are then able to check it for errors and omissions.

What must I report? The short answer is, *deaths, major injuries, dangerous occurrences, over-3-day injuries, certain diseases and gas incidents.*
Note again that although the regulations apply to the reporting of offshore incidents, the specific criteria that apply to certain offshore activities will not be covered here. Now, the short answer to our question is all very well but it throws up yet another set of questions. For example, what types of injury are considered serious? You should have an idea of the answer to this question, since you were asked a similar question in activity 1.15. Similar questions arise when we need to decide what occurrences are considered dangerous and what diseases must be reported.

The regulations answer these questions in great detail, and in the HSE guide to these regulations a useful summary is given of reportable injuries, dangerous occurrences and diseases. To further familiarize you with deciding what incidents are reportable, you should now attempt activity 1.21.

Activity 1.21

Download and study the introduction section of the HSE RIDDOR guidance leaflet L73 and use it to answer the following questions:

(a) A fitter at work stubs his big toe against the base of his lathe that results in a visit to the hospital, where a hairline fracture to the big toe is detected. What reporting action under RIDDOR needs to be taken and why?

(b) An engine being hoisted to height (as in Case study 1) suffers a failure of the lifting sling that results in the engine overbalancing, dropping suddenly and swinging violently, but nobody is injured. What reporting action under RIDDOR needs to be taken and why?

(c) 5 kg of propane gas is released into the building where it is stored, but no injuries to personnel occur as a result of the incident. What reporting action needs to be taken under RIDDOR and why?

Why should I report? One compelling reason for reporting certain accidents, incidents and ill health is that in certain well-defined circumstances it is a *legal requirement*. Another very valid reason is that it assists the HSE and local authorities in identifying where and how risks arise, and as a result of further investigation it enables them to offer advice/guidance to help prevent them from occurring in the future.

When do I need to make a report? Although the regulations specify varying times for reporting different types of incident, it is advisable to report them as soon as possible and the quickest way to do this is via the ICC using the telephone. As mentioned earlier in the case of death, major injury or dangerous occurrences, the enforcing authority must be notified without delay. Cases of individuals being either absent from work or unable to complete their work satisfactorily for more than 3 days, as a result of an incident, must be notified to the enforcing authority within 10 days of the incident occurring. Cases of disease should be reported as soon as confirmation is received from a doctor that the employee is suffering from a reportable work-related disease.

Records and record keeping

Under RIDDOR legislation it is the duty of the employer, those in control of premises and the self-employed to keep records of any reportable injury, disease or dangerous occurrence. These records may be kept in any form, such as:

- keeping copies of the report form on file (F2508, 2508A)
- by use of completed accident book entries
- recording all essential details on computer
- maintaining a written log.

The record of each reportable incident needs to contain a minimum amount of details. Records must include the date and method of reporting; the date, time and place of the event; personal details of those involved; and a brief description of the nature of the event or disease.

Certain employers under the *Social Security (Claims and Payments) Regulations 1979* must keep the HSE version of the Accident Book (B1510). The Accident Book in these cases may be required to provide evidence in favour of or against any possible claimant. Employers under these circumstances need to identify which of the accidents recorded in the book are reportable under RIDDOR. For example, an accident involving a cut to a finger that required first aid would be recorded in the book but would not need to be reported to the enforcing authorities under RIDDOR. The recording of all accidents, with details of any subsequent first aid given, no matter how trivial, or what method of recording is used, is good practice and if injury occurs record keeping becomes a legal requirement. It thus provides the employer with readily available statistics that may be used to assess and reduce accidents rates, as well as providing readily available evidence of actions for any subsequent personal claims that may be made. If the Accident Book is chosen as the preferred method of recording accidents, then some other separate method would need to be chosen to record reported cases of disease, such as use of the HSE report form (F2508A).

> **KEY POINT**
>
> Always check with the enforcing authorities if you are unsure when an incident needs to be reported under RIDDOR

Activity 1.22

Go onto the HSE website and obtain a copy of the forms F2508 and F2508A, then answer the following questions:

(a) State the purpose of each form, namely F2508 and F2508A.
(b) Detail, in bullet point form, the information required by each part of the F2508 and describe how this information differs from that required on the F2508A.
(c) Where should the form be sent once completed?

One of the benefits of keeping accurate records of accidents, injuries, diseases and other occurrences to staff, no matter whether they are reportable or not, is so that we have readily available facts and information that may be used to help us assess the true costs of these incidents. In the final part of this section we are going to take a brief look at how we can assess these costs and determine their impact on the business.

The costs and causes of accidents and ill health

We have spent quite a lot of time in this unit emphasizing the importance of risk assessment in the engineering workplace, and the need for employers and those in a position of responsibility to ensure that they have a plan for continuous monitoring and improvement in order to help reduce risks. Not having a policy and implementation plan for controlling risk could prove to be extremely costly should the lack of such a plan result in an increased frequency of accidents and/or ill health in the workplace. To give you an idea of just how costly, let us look at some of the costs to a business that may result from an accident, occupational disease or dangerous occurrence.

Essentially there are two types of cost associated with work-based accidents or ill health, *direct or insured costs* and *indirect or uninsured costs*. Insurance policies do not cover everything and may only pay out for serious injury or damage to buildings, property and major equipment. So what are the uninsured or indirect costs? According to the HSC leaflet '*reducing risks–cut costs*', which may be downloaded from the HSE website, some or all of the following are uninsured costs that may affect a business:

- Lost time
- Extra wages, overtime payments
- Sick pay
- Production delays
- Fines
- Loss of contracts
- Legal costs
- Damage to products
- Clearing site
- Investigation time by management/employers
- Loss of business reputation.

In the following incident cost assessment table, which is my simplified version of that based on the incident cost calculator shown in the HSC leaflet '*reducing risks–cut costs*', I have put in some cost figures based on the following imaginary scenario.

Scenario

An employee was working at height on highway staging in a small engineering company that had 12 employees. He turned to leave his platform and fell through a gap that existed between his and an adjacent platform on the staging. He suffered a broken arm, broken leg and internal injuries to his spleen. He spent 14 days in hospital undergoing major surgery and was off work for 4 months. On his return to work he was placed on light duties in the stores for a further 4 months before he was able to resume his full maintenance role. The owner of the business (his employer) was prosecuted following the incident.

> **KEY POINT**
>
> The indirect costs of a work-related accident, illness or serious occurrence always far exceed the direct costs of the incident

ASSESSMENT OF INCIDENT COSTS

Incident details:

Date: 02/09/2010   Time: 11.20 am   Place: Company maintenance workshop

Mr Joseph Smith was carrying out routine maintenance work on the turbine engine in our main workshop. This required him to work at height from one of the upper platforms (3.5 m high) of our pre-positioned and pre-erected highway staging. As he turned from his platform on the staging, he fell through the gap that was present between the two top platforms, catching his arm during the fall and landing at the base of the platform where he impacted with the concrete floor and sustained his injuries. An ambulance was called immediately by a fellow fitter from his mobile phone and our company first aider was quickly in attendance.

Injured/ill person:

Name:	Joseph Smith
Address:	
Position:	Senior maintenance fitter
Injury/illness details:	Broken right arm, broken left leg, spleen, muscle and tissue trauma

Incident/accident costs: £

Direct/insurance costs

Increases in employers liability insurance premiums	11200
Excess on claims	800
Fines and damages awarded by court	22000
Court and legal representatives fees	5500

Indirect/non-insured costs:

Incident/treatment	First-aid treatment	60
	Transport to/from hospital	50
	Replacement of items (injured person/others)	245
	Other	
Lost time	Injured person	430
	Owner/manager	2700
	Supervisors	1200
	First aider/s	450
	Other	
Production/business	Lost production/overtime payments	6000
	Plant/vehicle damage/replacement	
	Sickness/injured person wages	12000
	Other	
Investigation	Owner/management	2400
	Safety advisor/representatives	360
	Dealing with enforcing authorities	1200
	Other miscellaneous expenses	660
	Total	**67255**

It can be seen from this table of costs that this accident, had it taken place, would have cost this small privately owned maintenance engineering company a great deal of money, worry and inconvenience, simply because of the incorrect assembly of a section of highway staging and lack of safety awareness on the part of management and others responsible for implementing and monitoring health and safety policy and procedures!

Now the above case is rather extreme and was deliberately made so for illustration purposes. However, it must be accepted that no matter how robust our plans are to reduce risk, there is always the possibility that employees may have an accident or suffer from ill health due to unforeseen circumstances and/or gaps in our risk management system. These gaps can only be plugged if robust plans are put in place to thoroughly investigate both the indirect and direct cause of the accident/incident and put measures in place to ensure that the cause of a particular accident/incident cannot occur again.

Indirect causes of accidents to employees may be personal or result from failures on the part of management; such causes include:

- lack of knowledge and training
- lack of skill
- lack of awareness of hazards
- emotional problems that impact on work
- resentment and lack of motivation
- physical and mental incapacities
- employees taken on by management (for design, production, maintenance, purchase). *Note*, these employees may initially be unqualified or inexperienced and *require* additional training and/or strict supervision in order to prevent them making mistakes (the lack of such training/supervision can lead to mistakes being made and subsequent accidents or serious incidents occurring)
- equipment wear through use
- abnormal wear and tear
- lack of regular servicing of plant/equipment.

Some of the more obvious *direct causes* of accidents/incidents include:

- individuals participating in horseplay
- disobeying operating instructions
- tampering with or deactivating safety devices
- operating equipment outside limits
- using faulty equipment
- inappropriate use of equipment
- not using PPE as required
- inadequate, poorly fitted or inoperative safety locks, guards and devices
- untidy work environment (poor housekeeping)
- having to work in hazardous atmospheres or with hazardous substances
- inadequate lighting
- excessively noisy atmosphere
- inadequate separation of hazardous substances.

When an engineering company is investigating accidents and dangerous occurrences that occur on their premises, they must ensure that they identify the cause of these incidents and put in measures to prevent a re-occurrence.

To complete your study of this unit and to enable you to test your knowledge there follows an End of Unit Review in the form of a multiple-choice test.

End of Unit Review

1. Which of the following Acts forms the basis of British health and safety law?

    (a) Factories Act of 1961

    (b) Offices, Shops and Railway Premises Act 1963

    (c) Health and Safety at Work Act 1974

    (d) Environmental Protection Act 1990.

2. Which of the following statements is true?

    (a) The HSE ensures that relevant legislation is appropriate

    (b) The HSE is the operating arm of the HSC

    (c) The HSE conducts and sponsors research

    (d) The HSE submits proposals for new or revised regulations.

3. The Health and Safety at Work Act places duties on:

    (a) employers and self-employed people only

    (b) employees and self-employed people only

    (c) self-employed people only

    (d) all people at work.

4. Why do you need to know the health and safety regulations that deal with your type of engineering work?

    (a) They place legal duties on you

    (b) They tell you how health and safety is managed in your workplace

    (c) They tell you how to write your own risk assessment

    (d) They explain how you should carry out your job.

5. You work in an area where the heating and lighting is poor. Which of the following regulations lays down the standards for heating and lighting in the workplace?

    (a) The Workplace Health, Safety and Welfare Regulation 1992

    (b) The Noise, Heating and Lighting Regulations 1995

    (c) The Environmental Protection Regulations 1990

    (d) The Personal Protective Equipment Regulations 1992.

6. Which one of the following abbreviations is used to show *average* noise levels?

    (a) dB(A)

    (b) dB(B)

    (c) dB(C)

    (d) dB(D).

7. Noise protection is required for employees that need to shout or raise their voices to converse, when:

    (a) 20 m apart

    (b) 10 m apart

    (c) 5 m apart

    (d) 2 m apart.

8. Which of the following statements is *true* when using a grinding wheel to grind steel?

    (a) Eye protection must be worn at all times during operation

    (b) Eye protection need not be worn if wheel is guarded

    (c) Ear protection is never necessary: only the eyes are at risk

    (d) It is essential that an air-fed helmet be worn during operation.

9. Which of the following statements is *false* with respect to the regulations covering Personal Protective Equipment (PPE)?

    (a) PPE should be properly assessed before use to ensure it is suitable

    (b) Employers can charge employees for PPE, provided it is returnable

    (c) PPE must be maintained and stored properly

    (d) PPE must be supplied with instructions on how to use it safely.

10. Exposure to asbestos fibres is dangerous because it can cause:

    (a) mesothelioma

    (b) lung cancer

    (c) asbestosis

    (d) all of the above.

11. Why should barrier creams and protective gloves be used when working with mineral oils?

    (a) They keep your hands clean

    (b) They provide protection against dermatitis

    (c) Barrier creams allow dirt to be removed easily

    (d) Protective gloves are environmentally friendly.

12. Which of the following statements are true with respect to the duties of employers under the Electricity at Work Regulations 1989?

    (a) Be able to administer first aid for electric shock

    (b) Select electrical equipment that is suitable for the job

    (c) Ensure that wiring and equipment is sound and properly installed

    (d) All of the above.

13. Who in England and Wales are responsible for premises under the Fire Certificate (Special Premises) Regulations 1976?

    (a) The Health and Safety Executive

    (b) The Health and Safety Commission

    (c) The Local Fire and Rescue Authority

    (d) The local police force.

14. Which of the following statements concerning first aid at work is *false*?

(a) An appointed workplace first aider should have a valid certificate of competence

(b) An appointed person must have a valid certificate of competence

(c) An appointed person is responsible for looking after the first-aid equipment and facilities

(d) An appointed person is responsible for calling the emergency services when required.

15. When must you record an accident in the accident book?

(a) Only if you have suffered internal injuries

(b) Only if you have to be off work for more than 2 days

(c) If you are injured in anyway at work

(d) Only if you have to go to hospital.

16. Which of the following is *not* a reportable injury under RIDDOR?

(a) Temporary loss of sight

(b) Ingestion of a biological agent

(c) Injury that rendered the victim unconscious

(d) A fracture to the forefinger.

17. Occupational asthma is a disease that affects your:

(a) muscles

(b) hearing

(c) breathing

(d) Joints.

18. Under RIDDOR how long must employers keep a record of a reportable death, injury, disease or dangerous occurrence?

(a) 6 months from date of incident

(b) 2 years from date of incident

(c) 3 years from date of incident

(d) 5 years from date of incident.

19. An HSE 2508 form is used to:

(a) report a notifiable disease

(b) report absence from work

(c) report an injury or dangerous occurrence

(d) report all workplace accidents.

20. Which of the following statements describes emergency escape signs:

(a) triangular in shape with a yellow background

(b) round in shape with a white symbol on a blue background

(c) square or rectangular in shape with white symbol on a green background

(d) square or rectangular in shape with a white symbol on a red background.

21. This sign tells us that a substance is:

Figure 1.22 Figure for E of U review question 21

(a) corrosive

(b) requires cleaning before it is safe

(c) harmful

(d) Irritant.

22. This sign tells us that a substance is:

Figure 1.23 Figure for E of U review question 22

(a) corrosive

(b) toxic

(c) harmful

(d) Irritant.

23. This sign tells us:

Figure 1.24 Figure for E of U review question 23

(a) a toxic substance is present

(b) an irritant is present

(c) a laser hazard exists

(d) a radiation hazard is present.

24.   This sign means

Figure 1.25 Figure for E of U review question 24

(a) Ear phones must *not* be worn

(b) Ear phones must be worn

(c) Safety clothing must be worn

(d) Ear protection must be worn.

25.   Signs that are round in shape and have black symbols on a white background with a red boundary are:

(a) warning signs

(b) prohibition signs

(c) mandatory signs

(d) firefighting signs.

26.   You need to manually lift a load, what is the best way to find out if the load is too heavy to lift?

(a) Try to raise one end of the load in a crouched position

(b) Assess its weight by its size

(c) Determine the weight of the load

(d) Grasp the load in both hands and quickly jerk it upwards.

27.   Which part of your body is likely to be injured if you lift a heavy load carelessly or incorrectly?

(a) Back

(b) Knees

(c) Arms

(d) Shoulders.

28.   You are guiding a suspended load onto a stand, that is being winched into position. What should you *never* do?

(a) give instructions to the winch man concerning its position

(b) stand underneath it

(c) use signals

(d) shout orders.

29.   You need to use a lifting sling to lift a heavy load. Who is responsible for carrying out pre-use checks on the sling before use?

(a) Your supervisor

(b) The workshop safety representative

(c) Your employer

(d) You.

30.   You are to work at height on an aircraft wing. What safety equipment should you use?

(a) Harness with inertia reel attachment

(b) Reinforced footware

(c) Joint protectors

(d) An inflatable safety jacket.

31.   What is the most important reason for keeping a clean and tidy workplace free of clutter and obstacles?

(a) So that you have a more hygienic place in which to work

(b) To help prevent slips, trips or falls

(c) To comply with your supervisor's instructions

(d) So that you are always prepared for inspection by your health and safety representative.

32.   You are required to use a lead lamp in and around a fuel tank that has been emptied and vented. Which of the following precautions should you take?

(a) Use a lead lamp that has been designed to be used in volatile atmospheres

(b) Move tank away from other workshop equipment

(c) Ensure a suitable fire extinguisher is near to hand

(d) All of the above.

33.   Which of the following statements associated with the operation of rotating machinery is *false*?

(a) Ensure that overalls are buttoned up and no loose clothing or hair is present prior to operating the machine

(b) All machines should be provided with easily assessible emergency stop provision

(c) Swarf clearance may take place while machine is operating

(d) Pre-use checks must be carried out by the operator prior to using the machine.

34.   Who is *not* responsible for carrying out a risk assessment in the workplace?

(a) Employees

(b) Managers

(c) Self-employed

(d) Owners.

35.   A hazard may be defined as:

(a) the probability that someone may be harmed

(b) an obstacle to movement

(c) the potential to cause harm

(d) the exposure to injury or loss.

36.   When using the *five-step* approach to risk assessment, step 2 requires us to:

(a) record findings and implement them

(b) decide who might be harmed and how

(c) identify the hazards

(d) review the assessment and update accordingly.

37.   Which of the following is considered to be an *indirect cause* of a workplace accident or incident?

(a) Lack of knowledge or training

(b) Disobeying operating instructions

(c) Participating in horseplay

(d) Operating equipment outside limits.

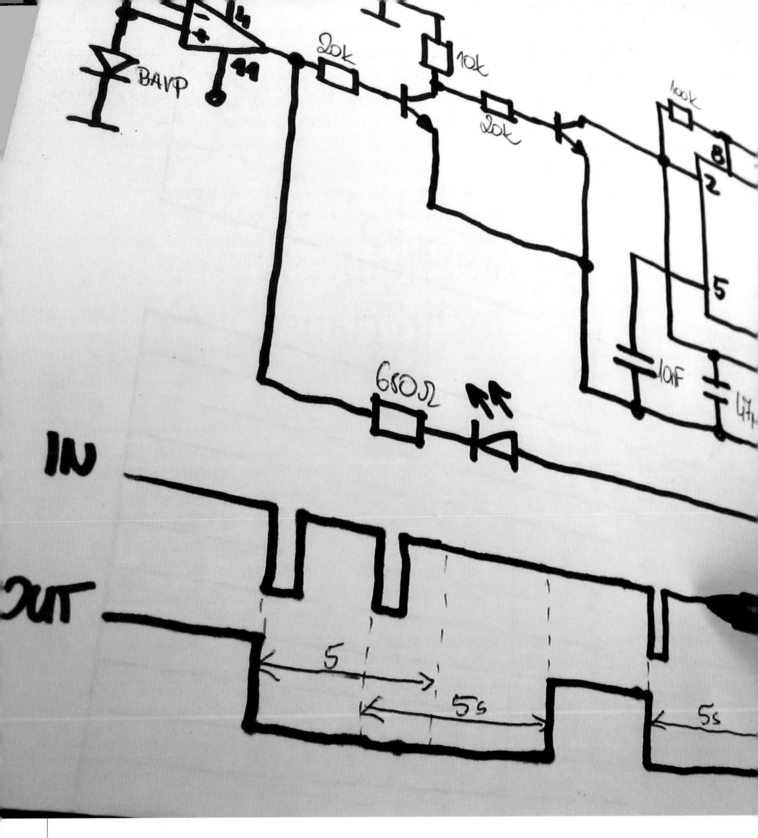

Engineers need to communicate their ideas using a variety of techniques including report writing and formal engineering drawing. They also need to be able to quickly sketch their ideas using standard symbols and conventions as shown in this circuit diagram for an electronic prototype. Being able to create and deliver formal presentations that are appropriate to both technical and non-technical audiences is yet another skill that engineers need to have.

Communications for Engineering Technicians

Effective communication skills are essential for anyone working in engineering. This unit will help you develop skills in using and interpreting information in a wide variety of forms. It aims to provide you with experience of speaking, reading and writing as well as graphical means of communication including drawing and sketching. These skills are essential not only for employment in engineering, but also as a basis for further study. The unit also gives you a variety of techniques used for obtaining, evaluating, processing and presenting information.

This unit is often assessed by means of a portfolio of evidence that you build as you progress through the course. You must ensure that you include a variety of sufficient and appropriate evidence within your portfolio. Assessment may also involve assignments and case study work carried out in conjunction with other units (such as Project and Business Systems) as well as sketches and drawings, presentations, data sheets, technical reports and letters that you have produced. You should begin to work on your portfolio as soon as you start the unit and then continue to collect evidence systematically as you progress through the unit. You should also cross-refer the evidence in the portfolio to the individual learning outcomes and the list of evidence requirements. Your tutor will be able to help you with this.

This unit has strong links with the Health and Safety in the Engineering Workplace and Project mandatory units. Wherever possible, you should apply the techniques that you have developed in this unit to work undertaken in the Project and Business Operations in Engineering units. There are also links to the optional unit on Engineering Drawing.

Being able to present your work and demonstrate proficiency in your own communications skills is an important aspect of this unit. To do this, you will need to have access to appropriate presentation and graphics software (such as Microsoft PowerPoint, Visio, Corel Draw, etc.) and spreadsheet software (such as Microsoft Excel) as well as hardware including scanners, laser and ink-jet printers, optical character recognition and speech recognition software, digital cameras and bar code readers. Your school or college will normally be able to supply you with everything that you need as well as guidance and help with its use.

BTEC National Engineering. DOI: 10.1016/B978-0-12-382202-4.00002-7

Information and Information Sources

The dictionary definition of 'communication' usually mentions something like 'the means by which we convey information'. There are, of course, many different ways of conveying information. You might consider, for example, the different ways that you could let your tutor know that you will not be attending college next week. You could speak to him in person, telephone him, leave a message on his answering machine, put a note on his desk, send him an e-mail message, or even write a letter and post it to him.

The method that you chose depends on a number of factors (not the least of which might be how friendly he is!). To be certain that he gets the message you might decide to speak to him in person (this might not be so easy if you do not have a good reason for not attending!). Alternatively, your college may require that you explain your absence in writing and they may have a form that you must complete. In deciding how to convey a simple message like this you probably need to think about several factors, including:

- How important is the information?
- How will I know that the information has been conveyed and understood?
- Do I need to keep a record of the information?
- How well do I know the person or persons with whom I am communicating?
- How urgent is the information?
- Is this a 'one-off' message or is it part of something much bigger?

Forms of communication

The forms of communication that we use in everyday life can be broken down into four main types, namely:

- written
- graphical
- verbal
- other (non-verbal).

Each of these main types of communication can be further subdivided. For example, graphical communication can take the form of drawings, sketches, block diagrams, exploded views, graphs, charts, etc. Some of these can be further divided. For example, there are many different types of graph and chart. We have shown some of these in *Figure 2.1*.

In everyday life, we usually convey information by combining different forms of communication. For example, when we speak to other people

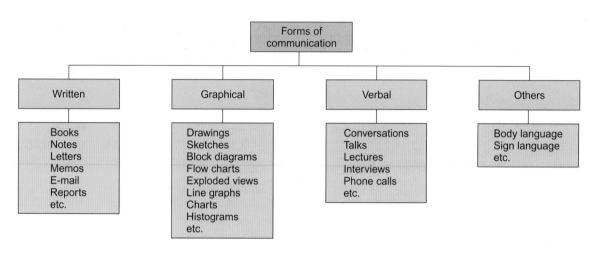

Figure 2.1 Forms of communication

UNIT 2

KEY POINT

Methods of communication like those shown in *Figure 2.1* are not mutually exclusive. Communication can be made more effective by combining different methods so that they reinforce the message or the information that you wish to convey

KEY POINT

The use of graphics to communicate ideas is essential for designers, engineers and draftspersons. Technical drawing are graphical representations of an idea or product that is to be processed, manufactured or constructed. Engineers use drawings to specify and transfer technical information

we often combine verbal with non-verbal (body language) forms of communication. Body language can help add emphasis to our words or can be used to convey additional meaning. Presentations to groups of people usually involve verbal communication supported by visual aids such as overhead projector transparencies, handouts or flipcharts. Technical reports invariably combine written text with diagrams and photographs.

Throughout this unit there are numerous opportunities to develop your own skills in obtaining, processing, evaluating and presenting information. In order to do this, you should be prepared to make use of appropriate technology (i.e. hardware and software). For this reason, we will not study hardware and software as a separate topic. Instead, we will assume that you will acquire familiarity with its use as you progress through the topics in this book.

Activity 2.1

From the types of communication shown in *Figure 2.1*, identify (with reasons) methods of communication that you consider most appropriate in each of the following situations:

(a) Making an appointment to see a doctor.
(b) Apologizing for forgetting your sister's birthday.
(c) Directing a friend to a restaurant in the next town.
(d) Selling your car.

Present your work in the form of a brief set of hand-written notes.

Activity 2.2

Complete the table shown below by placing a tick against the applicability of different forms of communication in relation to fitting a plug to the electric cable on a portable appliance:

Situation	Highly applicable	Possibly applicable	Not applicable
A written instruction sheet			
A verbal commentary supplied on a cassette tape			
A sequence of diagrams with brief text			
A flowchart that lists each of the steps required			
A videotape			

Copy the table into a word-processing package and then print out your work.

Activity 2.3

Describe TWO forms of non-verbal communication and give examples of a situation in which each might be used. Present your work as a single A4 word-processed page.

Information sources

Engineers use a wide variety of information in their everyday lives. This information is derived from a variety of different sources including:

- books
- application notes
- technical reports
- data sheets and data books
- catalogues
- engineering drawings
- CD-ROM
- databases
- websites.

Books

Books, whether they are paper-based publications or one of the new generation of electronic books (eBooks), provide information on an almost infinite number of subjects and a good technical library can be invaluable in any engineering context. All books contain summary information. This typically includes:

- the date of first publication
- the date(s) of any reprint(s) (with or without corrections)
- the date(s) of any subsequent edition(s)
- the date(s) of any reprint(s) of subsequent editions (with or without corrections)

- information concerning copyright
- British Library Cataloguing data
- an ISBN number
- other library (e.g. Library of Congress) cataloguing data
- information relating to the printing and binding of the book.

TYK 2.1

Take a look at the information that appears in the first few pages of this book and then answer the following questions:

1. What was the year of first publication?

2. Is this book a reprint?

3. Who owns the copyright?

4. What is the book's ISBN number?

5. Who has published the book?

When using a book as a source of information it is important to ensure that it is up to date. It is also necessary to ensure that the content is reliable and that there are no omissions or errors. Book reviews (often published in the technical press) can be useful here!

Electronic books are presented in electronic rather than paper format. The book actually comprises of a number of files which can either be downloaded into a PC, laptop, or pocket reader device or which can be supplied on a CD-ROM. In order to convert the files that make up the book into something that can be read on the screen, the PC, laptop or pocket reader requires appropriate software such as Adobe's Acrobat Reader or Microsoft's Reader. Popular file formats for eBooks include HTML (HyperText Markup Language – see page 79) and Adobe's PDF (Portable Document File) format (see *Figure 2.2*).

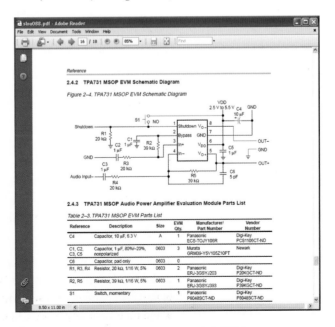

Figure 2.2 An extract from an electronic document in Adobe's PDF format

Application notes and technical reports

Application notes are usually brief notes (often equivalent in extent to a chapter of a book) supplied by manufacturers in order to assist engineers and designers by providing typical examples of the use of engineering components and devices. An application note can be very useful in providing practical information that can help designers to avoid pitfalls that might occur when using a component or device for the first time.

Technical reports are somewhat similar to application notes but they focus more on the performance specification of engineering components and devices (and the tests that have been carried out on them) than the practical aspects of their use. Technical reports usually include detailed specifications, graphs, charts and tabulated data.

Typical section headings used in application notes and technical reports include:

Summary	A brief overview for busy readers who need to quickly find out what the application note or technical report is about.
Introduction	This sets the context and background and provides a brief description of the process or technology – why it is needed and what is does. It may also include a brief review of alternative methods and solutions.
Main body	A comprehensive description of the process or technology.
Evaluation	A detailed evaluation of the process or technology together with details of tests applied and measured performance specifications. In appropriate cases comparative performance specifications will be provided.
Recommendations	This section provides information on how the process or technology should be implemented or deployed. It may include recommendations for storage or handling together with information relating to health and safety.
Conclusions	This section consists of a few concluding remarks.
References	This section provides readers with a list of sources of further information relating to the process or technology, including (where appropriate) relevant standards and legislation.

KEY POINT

Application notes explain how something is used in a particular application or how it can be used to solve a particular problem. Application notes are intended as a guide for designers and others who may be considering using a particular process or technology for the first time

Technical reports, on the other hand, provide information that is more to do with whether a component or device meets a particular specification or how it compares with other solutions. Technical reports are thus more useful when it comes to analysing how a process or technology performs than how it is applied.

Activity 2.4

Write an application note that explains the use of AA-size NiMh batteries as replacements for the conventional alkaline batteries used in a digital camera. You should carry out some initial research before starting to write your application note but the following are possible headings (these can be combined or expanded if you think it necessary):

- Executive summary
- Introduction
- Basic requirements for batteries used in digital cameras
- Comparison of three battery types (NiCd, NiMH and conventional alkaline types)
- Battery life and charging arrangements
- Analysis of costs
- Suitability
- Recommendations
- Reference data (including a list of manufacturers and suppliers).

Use a word processor to present your work in printed form.

Data sheets and data books

Data sheets usually consist of abridged information on a particular engineering component or device. They usually provide maximum and minimum ratings, typical specifications, as well as information on dimensions, packaging and finish. Data sheets are usually supplied free on request from manufacturers and suppliers. Collections of data sheets for similar types of engineering components and devices are often supplied in book form. Often supplementary information is included relating to a complete family of products. An example of a data sheet is shown in *Figure 2.3*.

Test your knowledge

TYK

TYK 2.2

Refer to the extract from the Howard Associates data sheet shown in *Figure 2.3* and use it to answer the following questions:
(a) What is the date of issue of the data sheet?
(b) Who owns the copyright of the data sheet?
(c) How many independent logic gates are contained in the device?
(d) What do the letters 'NC' mean?
(e) How many pins are there on an 'FK' package?
(f) Which two devices are suitable for operation over the 'military temperature range'?
(g) What is the absolute maximum supply voltage for these devices?
(h) What is the storage temperature range specified for these devices?
(i) What does the manufacturer specify as typical values of supply voltage and operating free-air temperature for these devices?

Howard Associates

DATA SHEET	54ALS08, 54AS08, 74ALS08, 74AS08 Quadruple 2-input AND gates

MAIN FEATURES

- Four independent 2-input positive logic AND gates
- Available in military (54) and commercial (74) versions
- Standard 14-pin DIL and 20-pin leadless packages
- Military version operates over a wide temperature range
- Complementary to 54/74ALS00 quad 2-input NAND gate

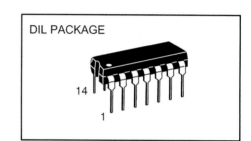

DIL PACKAGE

14

1

TRUTH TABLE
(each gate)

INPUTS		OUTPUT
A	**B**	**Y**
H	H	H
L	X	L
X	L	L

H = logic 1 (high)
L = logic 0 (low)
X = don't care (either high or low)

SYMBOL

1A 1   3 1Y
1B 2

2A 4   6 2Y
2B 5

3A 9   8 3Y
3B 10

4A 12   11 4Y
4B 13

ELECTRICAL CHARACTERISTICS (T_A = +25°C)

Parameter	Symbol	54ALS08			74ALS08			Unit
		Min.	Nom.	Max.	Min.	Nom.	Max.	
Supply voltage	V_{CC}	4.5	5.0	5.5	4.5	5.0	5.5	V
High-level input voltage	V_{IH}	2.0			2.0			V
Low-level input voltage	V_{IL}			0.8			0.8	V
High-level output current	I_{OH}			−0.4			−0.4	mA
Low-level output current	I_{OL}			4.0			8.0	mA
Operating free-air temperature	T_A	−55		125	0		70	°C

Data sheet reference: 95-072
page 1

Figure 2.3 An extract from a semiconductor data sheet

ABSOLUTE MAXIMUM RATINGS (T_A = +25°C)

Supply voltage, V_{CC}		7 V
Input voltage, V_I		7 V
Operating free-air temperature range, T_A:	54ALS08	−55°C to +125°C
	74ALS08	0°C to +70°C
Storage temperature, T_{stg}		−65°C to +150°C

PIN CONNECTIONS

(top view)

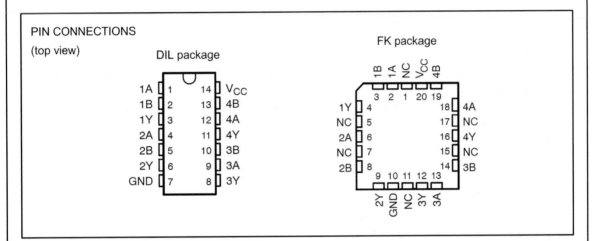

ELECTRICAL CHARACTERISTICS

Parameter	Test conditions		54ALS08			74ALS08			Unit
			Min.	†Typ.	Max.	Min.	†Typ.	Max.	
V_{IK}	V_{CC} = 4.5 V	I_I = −18 mA			−1.5			−1.5	V
V_{IH}	V_{CC} = 5 V	I_{OH} = −0.4 mA	3.0			3.0			V
V_{OL}	V_{CC} = 4.5 V	I_{OL} = 4 mA		0.25	0.4		0.25	0.4	V
		I_{OL} = 8 mA					0.35	0.5	V
I_I	V_{CC} = 5.5 V	V_I = 7 V			0.1			0.1	mA
I_{IH}	V_{CC} = 5.5 V	V_I = 2.7 V			20			20	μA
I_{IL}	V_{CC} = 5.5 V	V_I = 0.4 V			−0.1			−0.1	mA
I_O	V_{CC} = 5.5 V	V_O = 2.25 V	−20		−112	−30		−112	mA
I_{CCH}	V_{CC} = 5.5 V	V_I = 4.5 V		1.3	2.4		1.3	2.4	mA
I_{CCL}	V_{CC} = 5.5 V	V_I = 0 V		2.2	4.0		2.2	4.0	mA

†Typical values are at V_{CC} = 5 V, T_A = +25°C

Data sheet reference: 95-072
page 2

© Howard Associates 2007

Figure 2.3 (Continued)

Catalogues

Most of the manufacturers and suppliers provide catalogues that list their full product range. These often include part numbers, illustrations, brief specifications and prices. While catalogues are often extensive documents with many hundreds or thousands of pages, short-form catalogues are usually also available. These usually just list part numbers, brief descriptions and prices but rarely include any illustrations. A brief extract from a short-form catalogue is shown in *Figure 2.4*.

Diecast Boxes   IP65 Sealed/Painted

A range of high-quality diecast alumnium boxes with an optional grey epoxy paint finish to RAL7001. The lid features an integral synthetic rubber sealing gasket and captive stainless steel fixing screws. Mounting holes and lid fixing screws are outside the seal, giving the enclosure protection to IP 65.

Standard supply multiple = 1   Delivery normally ex-stock

| Size | | | | | | | Price each | | |
L	W	H	T	Finish	Manufacturer's ref:	Stock code	1-9	10-24	25+
90	45	30	3.0	None	1770-1541-21	DB65-01	£4.52	£3.95	£3.50
90	45	30	3.0	Grey	1770-1542-21	DB65-01P	£5.40	£4.90	£4.45
110	50	30	4.5	None	1770-1543-22	DB65-02	£5.25	£4.50	£4.15
110	50	30	4.5	Grey	1770-1544-22	DB65-02P	£6.42	£5.37	£4.95
125	85	35	5.0	None	1770-1545-23	DB65-03	£6.15	£5.17	£4.71
125	85	35	5.0	Grey	1770-1546-23	DB65-03P	£7.10	£6.05	£5.65

Figure 2.4 An extract from a short-form catalogue

Engineering drawings

Engineering drawings can be produced either manually or on a computer using suitable CAD software. Drawings produced manually can range from freehand sketches to formally prepared drawings produced with the aid of a drawing board and conventional drawing instruments. Later in this unit we will introduce you to the most common types of engineering drawing.

CD-ROM

Compact disks can provide storage for around 650 Mbytes of computer data. This is roughly equivalent to 250,000 pages of A4 text. It is, therefore, hardly surprising that the compact disk has now become firmly established as a storage medium for a variety of different types of information, including text, drawings, sound and video (i.e. multimedia). The latest PCs offer CD-ROM drives that can be used to store data on recordable CDs (CD-R) or rewritable CDs (CD-RW). High-end PCs now also have DVD (Digital Versatile Disk).

Activity 2.5

Complete the table shown below by placing a tick in the column against the most appropriate method for communicating the listed information (tick only once in each row):

	Application note	Data sheet	Shortform catalogue	Technical report
Summary of the precautions to be observed when handling a chemical etching fluid				
Cost of die-cast boxes supplied in various quantities, from 1 to 100				
Maximum working temperature for a power transistor				
Recommended printed circuit board layout for an audio amplifier				
Comparison of different types of surface finish for the interior of a domestic microwave oven				
Physical dimensions of a marine radar for fitting to a small boat				
Description of tests applied to an off-road vehicle				
Performance specification for a satellite TV aerial				

Copy the table into a word-processing package and then print out your work.

TYK 2.3

An engineering company manufactures a variety of different types of fastener including nuts and bolts, screws and washers. What typical information would appear in a short-form catalogue for this company?

TYK 2.4

Your company requires 15 diecast boxes suitable for enclosing a printed circuit board of thickness 3 mm measuring 80 mm × 35 mm. The tallest component stands 15 mm above the board and a minimum clearance of 5 mm is to be allowed all round the board. The enclosure is to be supplied ready for mounting the printed circuit board and should not need any further finishing other than drilling. Prepare a fax message to Dragon Components (the hardware supplier whose short-form catalogue extract appears in *Figure 2.4*) giving all the information required to fulfil your order.

Databases

A database is simply an organized collection of data. This data is usually organized into a number of records each of which contains a number of fields. Because of their size and complexity and the need to be able to quickly and easily search for information, a database is usually stored within a computer and a special program – a *database manager* or *database management system* (DBMS) – provides an interface between users and the data itself. The DBMS keeps track of where the information is stored and provides an index so that users can quickly and easily locate the information they require (see *Figure 2.5*).

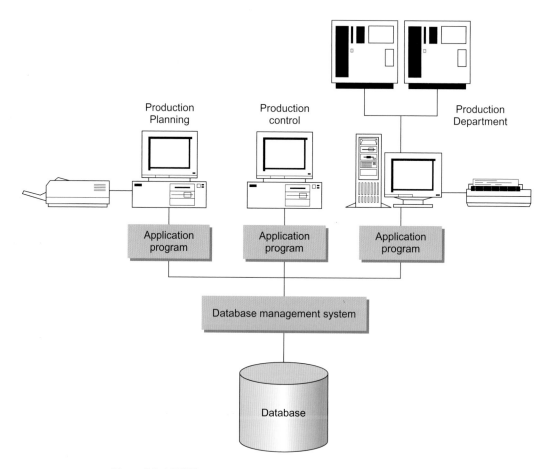

Figure 2.5 A DBMS

TYK 2.5

A manufacturer wishes to distribute a set of data sheets on a conventional CD-ROM. Each data sheet consists of an Adobe Acrobat file having a size of 2.8 Mbytes. In addition, the manufacturer has included an index and other files amounting to a total of 80 Mbytes. Approximately how many data sheets can be included on a single CD-ROM?

KEY POINT

A database consists of *records* arranged into one or more *tables*. Records are uniquely identified by their *key*. Each record is divided into a number of *fields*, one of which is the key

The database manager will also allow users to search for related items. For example, a particular component may be used in a number of different products. The database will allow you to quickly identify each product that uses the component as well as the materials and processes that are used to produce it.

The structure of a simple database is shown in *Figure 2.6*. The database consists of a number of *records* arranged in the form of one or more *tables*. Each record is divided into a number of *fields*. The fields contain different information but they all relate to a particular component. The fields are organized as follows:

Field 1 Key (or index number)
Field 2 Part number
Field 3 Type of part
Field 4 Description or finish of the part

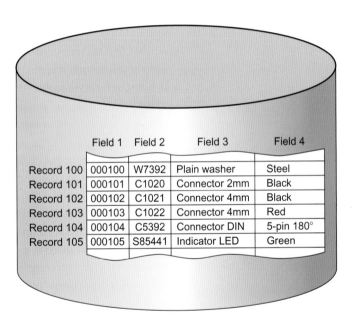

	Field 1	Field 2	Field 3	Field 4
Record 100	000100	W7392	Plain washer	Steel
Record 101	000101	C1020	Connector 2mm	Black
Record 102	000102	C1021	Connector 4mm	Black
Record 103	000103	C1022	Connector 4mm	Red
Record 104	000104	C5392	Connector DIN	5-pin 180°
Record 105	000105	S85441	Indicator LED	Green

Figure 2.6 The structure of a simple parts database

The *key* (in this case Field 1) allows us to uniquely identify the data stored in the record. Keys may be meaningful data (such as a product name) or they may be an index number that we have invented for the purpose. The important feature of a key is that it must be *unique*.

TYK

TYK *2.6*

A college uses a database to hold information on its students. In this database application, explain why a student's last name is not suitable for use as the key to a record.

The database management system is used to build and maintain the database. It also provides the interface between the user and the information stored in the database. Tasks performed by the DBMS include:

- adding new records
- deleting unwanted records
- amending records
- linking or cross-referencing records
- searching and sorting the database records
- printing reports of selected records.

Most engineering companies use several specialized databases in order to manage different functions. To help you understand this, consider the case of an engineering company that manufactures fork-lift trucks. The company might use the following databases to help it to organize the different aspects of its operation:

- a *product database* containing records of each vehicle manufactured and any modifications fitted
- a *manufacturing database* containing records of all components and materials used during manufacture
- a *customer database* containing records of all customers and vehicles supplied to them
- a *spare parts database* containing records of all spare parts held showing where they are stored and the quantity held.

Figure 2.7 shows how these four databases relate to three different departments within the company. Since these three functions cannot operate in isolation there is a need to exchange data between the three databases. To give you some idea of how this might work, let us assume that the sales department has been given the task of marketing a new fork-lift truck that has just become available. The manager of the sales department has been given a target that requires him to sell 10 fork-lift trucks by the end of the next quarterly period. He has been allocated a fixed budget specifically to meet the costs associated with this sales campaign.

The sales manager decides to produce a brochure giving details of the new fork-lift truck and offering a substantial discount to any previous customers who may wish to 'trade in' their existing fleet of fork-lift trucks. The brochure will be mailed to all UK customers who have purchased fork-lift trucks in the last 10 years. Sales staff will then follow this up with a telephone call to each named customer contact.

The information required to produce the new fork-lift truck brochure will be drawn from the company's product database. The mailing list, contact names and telephone numbers will be taken from a report generated from the customer database.

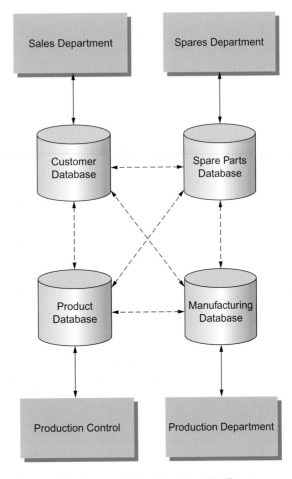

Figure 2.7 Four specialized databases and their relationships with different company functions

The structure and content of a typical record in the product database will include the following information:

Product reference number:	FLT1022
Product name:	Challenger
Vehicle chassis type:	BCX077
Quantity in stock:	7
Scheduled production (current period):	10
Scheduled production (next period):	12

The structure and content of a typical record in the customer database will include the following information:

Customer reference number:	13871
Company name:	Enterprise Air Freight
Address (line 1):	Unit 8
Address (line 2):	Bath Road Industrial Estate
Town/city:	Feltham
County or state:	Middlesex
Post code or zip code:	UB10 3BY
Country:	UK
Contact name:	David Evans

Contact title:	Purchasing manager
Contact salutation:	Dear Dave
Telephone number:	020-8979-7756
Fax number:	020-8979-7757
e-mail:	devans@enterprise.co.uk

TYK *2.7*

Explain the following terms when applied to a database:
(a) field
(b) record
(c) key
(d) relational.

TYK *2.8*

Give two examples of information that would be stored in a manufacturing database.

Nowadays, there is a trend towards integrating many of the databases within an engineering company into one large database. This database becomes central to all of the functions within the company. In effect, it becomes the 'glue' that holds all of the departments together. The concept of a centralized manufacturing database is a very sound one because it ensures that every function within the company has access to the same data. By using a single database, all departments become aware of changes and modifications at the same time and there is less danger of data becoming out of date.

TYK *2.9*

In setting up the mailshot, what fields within the customer database will be used to:
(a) determine which customers are included in the mailing list
(b) generate the label that will be attached to the information pack.

In a *relational database*, you are able to create views that display only selected fields in a data file or that combine fields from multiple data files. There are several good reasons for breaking a large amount of information into a number of smaller data files. In some cases, small data files will process more quickly than using one very large data file. Better security can be maintained by breaking a large data file into several smaller data files. By controlling access to each data file you can also control what users can see! In any event, when designing a database it is essential to take time planning the individual data files that will be needed and, in particular, considering the relative association that the records contained in the data files may have to each other.

TYK

TYK *2.10*

There is a need to ensure that information contained in a database is kept up to date. In the case of the customer database (see text), which records and field may need updating as a result of the mailshot and telephone sales campaign?

Figure 2.8 shows how information can be taken from separate data tables and combined into a single report. In this case, the matching field is the employee number and the report lists those employees earning a salary of more than £30,000 together with their National Insurance (NI) numbers.

The World Wide Web

The Internet is the name given to a huge network of computers all over the world that communicate with each other. The Internet makes it possible for you to access the World Wide Web, a vast collection of on-line information kept on numerous web servers throughout the world. The World Wide Web began in March 1989, when Tim Berners-Lee

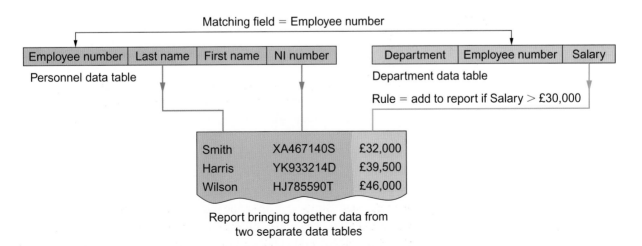

Figure 2.8 Matching data tables to produce sophisticated reports

of the European Particle Physics Laboratory at CERN (the European Centre for Nuclear Research) proposed the project as a means to better communicate research ideas among members of a widespread organization.

If you have an Internet connection to your PC, the Web is accessed via a program called a *browser* that helps you to navigate between the *pages* of information held on the Web. The browser available within the college is Microsoft's *Internet Explorer*.

Information on the Web is stored as pages of electronic text with *hyperlinks*. Clicking on a hyperlink will allow you to jump from one page to another, or from one place in a page to another place on the same page. Hyperlinks often appear in a different colour text and are usually underlined. They can also take the form of pictures and graphics that you can click on. A hyperlink can give you access to anything that can be stored electronically: text, graphics, video, sound, or a file that you can *download* to your PC. A typical web page is shown in *Figure 2.9.*

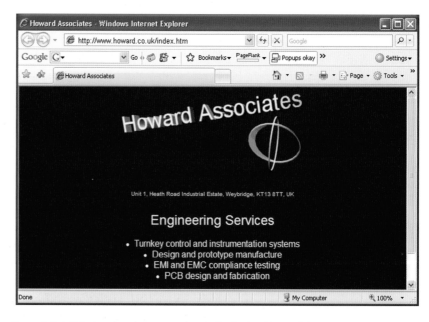

Figure 2.9 A typical engineering company's website displayed in a web browser

Web pages are written using the language that contains embedded commands. The important distinction between a web page and a document produced by a word processor is that, in the case of the web page, the instructions for displaying the document are imbedded in the document itself (they are called *tags*). The web browser reads the tags and then uses them to construct the page. The tags are defined by a language called *HyperText Markup Language* (HTML). *Figure 2.10* shows the source HTML code for the page shown in *Figure 2.9*. See if you can make sense of it!

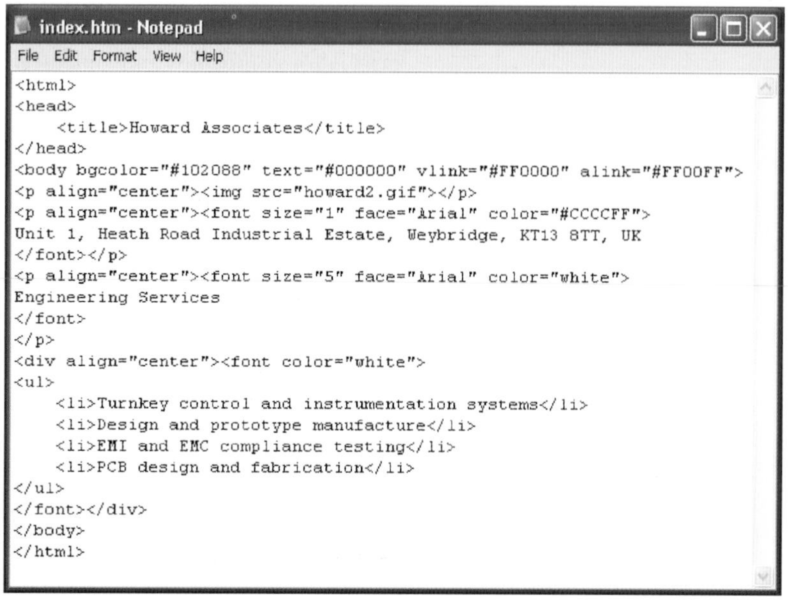

```
index.htm - Notepad
File   Edit   Format   View   Help
<html>
<head>
    <title>Howard Associates</title>
</head>
<body bgcolor="#102088" text="#000000" vlink="#FF0000" alink="#FF00FF">
<p align="center"><img src="howard2.gif"></p>
<p align="center"><font size="1" face="Arial" color="#CCCCFF">
Unit 1, Heath Road Industrial Estate, Weybridge, KT13 8TT, UK
</font></p>
<p align="center"><font size="5" face="Arial" color="white">
Engineering Services
</font>
</p>
<div align="center"><font color="white">
<ul>
    <li>Turnkey control and instrumentation systems</li>
    <li>Design and prototype manufacture</li>
    <li>EMI and EMC compliance testing</li>
    <li>PCB design and fabrication</li>
</ul>
</font></div>
</body>
</html>
```

Figure 2.10 The HTML code responsible for generating the page shown in Figure 2.9

Every page on the Web has its own unique address called a *Uniform Resource Locator* (URL). The URL specifies the protocol used to retrieve the data – usually either *HyperText Transfer Protocol* (HTTP) or *File Transfer Protocol* (FTP), the address of the machine, the path and filename of the page, e.g. `http://www.brooklands.ac.uk` where:

`http://`	specifies the hypertext transfer protocol
`www`	specifies a World Wide Web (WWW) site accessed via a program called a *web browser*
`brooklands.ac.uk`	is Brooklands College's *domain name* (the `.ac.uk` tells you that the domain is that of a UK-based academic institution).

The last part of a URL gives you some information about the type of site you are visiting as well as where the owner's business is located. For example, a .com site is a commercial site (often, but not exclusively, located in the USA). A .co.uk site is a commercial site located in the UK. Similarly, a site with a URL ending in .co.fr is likely to be a commercial site located in France. Different rules apply to different countries and registration authorities; however, the following is worth noting.

UK registrations

`co.uk` is for companies and general use

`org.uk` is for organizations and non-profit-making companies. However, anyone can register a `org.uk` domain

`net.uk` is for Internet Service Providers (ISP). The rules for `.net.uk` are very strict; the governing body for UK domain names will not allow a non-ISP to register these domains

`ltd.uk` is for UK limited companies. You must provide your company registration number

`plc.uk` is for UK public limited companies.

International registrations

`com` is for companies and general use

`net` is for ISP. The international rules are more relaxed than the UK in that anyone can register a `.net` domain

`org` is for organizations and non-profit-making companies. Once again anyone can register a `.org` domain.

Search engines

One of the most useful features of the Web is the availability of search engines that will help you to locate the information that you require. Search engines are just large computers that contain cross-referenced lists of URLs. To use a search engine, you simply enter the name of the search engine that you wish to use in your browser. Once the search engine's home page has loaded you can enter the text that you wish to search for (see *Figure 2.11*). To get the best out of a search engine you need to give some thought as to what it is you are searching for and

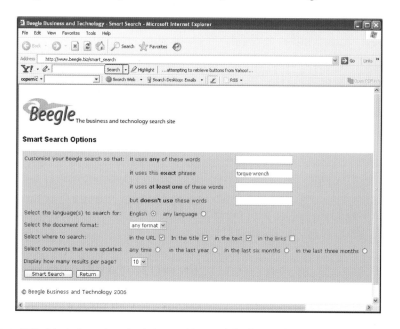

Figure 2.11 A typical search engine being used to search for the term 'torque wrench'. Note that the advanced search options provide you with a way of narrowing down your search to an exact phrase

how best to describe it unambiguously. If you give the search engine too little to go on, say just one word, you may be rewarded with a huge number of references to sites that contain references to the word that you have used. On the other hand, using words that are very specific may restrict the search to too few sites. In any event, it is always worth experimenting with a search engine. With a little trial and error you will soon get to know how to get the best out of it!

On-line services, directories and other reference sources

On-line services are services that add values to the World Wide Web. Originally, these services built and maintained trunk networks that could be used by their customers. They also added their own *content* (such as extensive databases, news and weather reports, software libraries, etc.) to the Web. Users who are prepared to pay for the service can access this material. Customers still pay for some on the on-line services but the trend, in recent years, has been to make added value services free.

Web directories, such as Yahoo, offer an editorially selected, topically organized list of Web sites. To accomplish that goal, these sites employ editors to find new Web sites and work with programmers to categorize them and build their links into the site's searchable index. To make things even easier, all the major search engine sites now have built-in topical search indexes, and most Web directories had added keyword search facilities.

Web-based business directories provide access to companies within a wide range of sectors. Using a business directory makes it easily possible to locate companies in a particular sector and in a specific location (e.g. a chemical supplier in the West Midlands).

In addition, on-line reference sources, such as Wikipedia, provide an excellent source of information which can be freely accessed and even edited and updated by users. A typical reference source Web page is shown in *Figure 2.12*. A word of caution is necessary here: when searching for information on the Web it is advisable to consult several sources and then compare your results before placing reliance on the information that you find. The reason for this is that almost all can place information on the Web, but not all of them are accurate and some of them can be deliberately misleading!

Figure 2.12 A web-based reference source being used to locate and display information on fire extinguishers

Intranets

Intranets work like the Web (with browsers, Web servers and Web sites) but companies and other organizations use them internally. Companies use them because they let employees share corporate data, but they are cheaper and easier to manage than most private networks because nobody needs any software more complicated or more expensive than a web browser, for instance. They also have the added benefit of giving employees access to the web. Intranets are closed off from the rest of the Net by *firewall* software, which lets employees surf the Web but keeps all the data on internal Web servers hidden from those outside the company.

TYK *2.11*

List TWO advantages and TWO disadvantages of a catalogue supplied on CD-ROM compared with one that is published on the World Wide Web.

Extranets

One of the most recent developments has been that of the *extranet*. Extranets are several intranets linked together so that businesses can share information with their customers and suppliers. Consider, for example, the production of a European aircraft by four major aerospace companies located in different European countries. They might connect their individual company intranets (or parts of their intranets) to each other, using private leased lines or even using the public Internet. The companies may also decide to set up a set of private newsgroups so that employees from different companies can exchange ideas and share information.

TYK *2.12*

Briefly explain what is meant by a URL. Give an example of a URL for a UK-based engineering company. What is the URL of the Web page shown in *Figure 2.9*?

E-mail

Like ordinary mail, e-mail consists of a message, an address and a carrier that has the task of conveying the message from one place to another. The big difference is that e-mail messages (together with any attached files) are broken down into small chunks of data (called *packets*) that travel independently to their destination along with innumerable other packets travelling to different destinations. The packets that correspond to a particular e-mail message may travel by several different routes and may arrive out of order and at different times. Once all the packets have arrived, they are recombined into their original form. This may all sound rather complicated but it is nevertheless efficient because it prevents large messages hogging all of the available bandwidth. To put this into context, a simple page of A4 text can be transferred half-way round the world in less than a minute! *Figure 2.13* shows a typical e-mail message written using Microsoft Outlook Express.

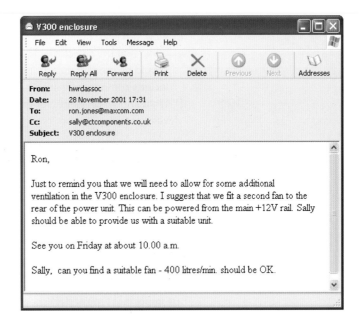

Figure 2.13 A typical e-mail message

TYK *2.13*

The URLs of three web sites are:

1. www.gloscat.ac.uk

2. www.sony.co.uk

3. www.btec.org.uk

What does this information tell you about each of the three sites?

Activity 2.6

Investigate the use of databases in your school or college. Working as part of a group, begin by finding out the name of the person who has overall responsibility for collecting and processing student data. Interview him or her and find out what information is held in the database and how it is organized. Also find out about the reports that are generated by the database and who has access to the information. Prepare a brief presentation to the rest of your class using appropriate handouts and visual aids.

Activity 2.7

Visit the Texas Instruments web site at `www.ti.com`

(a) Use the search facility to search for information on '74ALS08'.
(b) When the device has been located, open the 'Product Folder' to view more information.
(c) Click to download the full data sheet for the device (this is provided in the form of an Adobe Acrobat (.PDF) file).
(d) View and print the full data sheet.
(e) Given that the device employs 'totem-pole' outputs, sketch the circuit that should be used for measuring the switching characteristics of the device and include component values.
(f) What are typical values for the 'low-state' and 'high-state' output voltages?

Present your work in the form of a word-processed 'fact sheet'. Insert your circuit diagram into this document (either by scanning it or cutting and pasting it from a drawing package). Attach your 'fact sheet' to the printed data sheet.

Activity 2.8

One important type of data sheet is known as a Material Safety Data Sheet (MSDS). Use the World Wide Web to locate information on ferric chloride solution (printed circuit board etching fluid). To locate this information you will probably need to make use of one of several on-line MSDS databases. These include MSDSonline, MSDSSearch, MSDS Solutions and Oxford University. You should locate these (and other) sources using a popular search engine. Once you have identified several information sources use them to answer the following questions:

1. Is this material flammable?
2. Is this material corrosive?
3. What is the specific gravity of this material?
4. Is this material hazardous when in contact with the skin?
5. What does the material look like?
6. What precautions should be observed when handling and storing this material?

Present your work in the form of a word-processed 'fact sheet'.

Activity 2.9

Visit the Web site of Britten-Norman, a manufacturer of aircraft based on the Isle of Wight. The URL for Britten-Norman is:

`http://www.britten-norman.com`

Search the site for information on the Defender aircraft and obtain the e-brochure for the aircraft. Use the information to answer the following questions:

1. At what UK airport is the company based?
2. What is the full address of Britten-Norman?
3. What approval does the company hold for its maintenance, repair and overhaul facility?
4. When did the first Defender enter service?
5. Name two customers for the Defender aircraft.
6. What makes the Defender ideal for use in maritime counter-terrorism operations?
7. What is the maximum take-off weight for the Defender?
8. What is the maximum payload for the Defender?
9. What is the take-off distance for the Defender?
10. At what speed does the Defender's stall with flaps down and power off?

Write a brief word-processed report describing three innovative features and three specialized applications of the Defender aircraft. Include a picture of the aircraft in your report.

Activity 2.10

Use a search engine (such as Google) to locate information on electric drills. Visit the first four sites displayed as a result of your search and note down the URL of each of these sites. Summarize the contents of each of the sites by writing a paragraph describing each site. Then rate each site on a scale of 1–10 on the basis of content, presentation and ease of use. Summarize your results in a table.

Now repeat the activity using a web directory (such as Yahoo). Use the directory to navigate to four different sites giving details of electric drills. Once again, note down the URL of each site, summarize its contents and rate it on a scale of 1–10 (again presenting your results in the form of a table). Compare these two search methods.

Present your findings in the form of a word-processed 'fact-sheet'.

Activity 2.11

If you have not already done so, set up a web-based e-mail account in your own name. Note down all of the steps that you took to open the account including details of any electronic forms that you had to complete.

Present your findings in the form of a brief word processed article for your local paper showing how easy it is to open and make use of an e-mail account. You should assume that the reader is non-technical.

Presenting Engineering Information

Engineers rely heavily upon graphical methods of communication. Drawings and charts produced to international standards and using international symbols and conventions suffer no language barriers. They are not liable to be misinterpreted by translation errors. Graphical communication does not replace spoken and written communication but instead is used to simplify, reinforce and complement other means of communication.

Having now established the need for communicating engineering information, let us look at the various methods of graphical communication available. We can broadly divide engineering information into two categories. That which is mathematically based and that which is technically based. We will start by looking at ways of representing mathematical data.

Graphs

Just as engineering drawings are used as a clear and convenient way of describing complex components and assemblies, so can graphs be used to give a clear and convenient picture of the mathematical relationships between engineering and scientific quantities. *Figure 2.14a* shows a graph of the relationship between distance s and time t for the mathematical expression $s = \frac{1}{2} at^2$, where the acceleration $a = 10\,\mathrm{m\,s^{-1}}$.

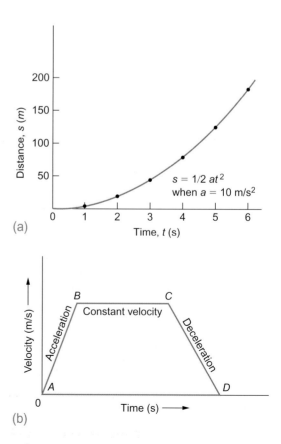

Figure 2.14 Line graphs

In this instance it is correct to use a continuous flowing curve to connect the points plotted. Not only do these points lie on the curve, but every corresponding value of s and t between the points plotted also lie on the curve.

However, this is not true for every type of line graph. *Figure 2.14b* shows a graph relating speed and distance for a journey. From A to B the vehicle is accelerating. From B to C the vehicle is travelling at a constant speed. From C to D the vehicle is decelerating (slowing down). In this example, it is correct to join the points by straight lines. This is because each stage of the journey is represented by a linear mathematical expression which is unrelated both to the previous and following stages of the journey.

Activity 2.12

Draw a graph showing how the following voltage varies with time:

Time, t (s)	0	0.1	0.2	0.3	0.4	0.5	0.6	0.7
Voltage, v (V)	0	1.95	2.95	3.46	3.72	3.86	3.93	3.96

What can you infer from the shape of the graph?

Present your findings in the form of a hand-drawn graph on graph paper. Label your drawing clearly.

Activity 2.13

Given the formula $N = \dfrac{1000S}{\pi d}$

where N = spindle speed in rev/min
S = cutting speed = 33 m/min
d = drill diameter in mm
π = 3.142

Use the formula to complete the table below, then enter the data into a spreadsheet package and use it to produce a graph relating drill diameter, d, and spindle speed, N.

From the graph determine the spindle speed of a drill having a diameter of 5 mm.

Drill diameter, d (mm)	2	4	6	8	10
Spindle speed, N (rev/min)					

Present your work in the form of a printed spreadsheet and graph.

Histograms

Histograms are used for plotting information where the change is discrete rather than continuous. To put this into context consider the number of National Diplomas awarded by a college over a 6-year period. The award of a National Diploma happens at the end of each academic year when a particular number of students achieve the qualification. This varies from year to year but, since it does not change on a continuous basis, it is inappropriate to illustrate the relationship using a line. Instead, we use a series of bars representing the number of National Diplomas awarded. Despite the fact that there is no line joining the bars together we can still clearly see a trend.

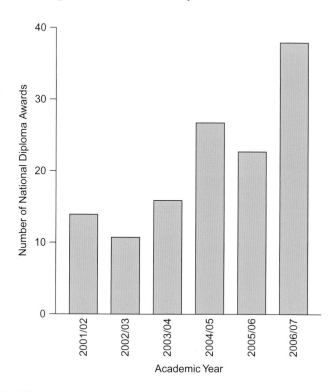

Figure 2.15 A histogram

Test your knowledge

TYK 2.14

Use *Figure 2.15* to answer the following questions:

1. In which academic year were the least number of National Diplomas awarded?

2. Assuming that the class size remained the same over all six years, in which year were the students most successful?

3. How many National Diplomas were awarded in 2004/05?

4. Expressed as a percentage, by how much did student achievement improve over the 2-year period, 2003/04 to 2004/05?

5. What is the total number of National Diplomas awarded over the 6-year period?

6. What is the average number of National Diplomas awarded each year for the 6-year period?

TYK

UNIT 2

Activity 2.14

The total number of machines manufactured by two competing engineering companies are shown in the table below:

Year	Machines produced by company A	Machines produced by company B
1995	135	250
1996	191	350
1997	188	325
1998	222	375
1999	253	290
2000	295	240
2001	320	216

Draw histograms to compare the performance of these two companies. What might you infer from these results?

Present your work in the form of hand-drawn histograms with a brief hand-written comment.

Bar charts

Like histograms, bar charts are also used for displaying statistical data, but are usually plotted horizontally. They are also often made to look more attractive by using 3-D drawing, as shown in *Figure 2.16*.

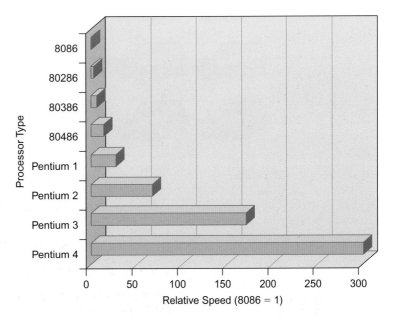

Figure 2.16 A 3-D bar chart

UNIT 2

Test your knowledge

TYK

TYK *2.15*

Figure 2.16 shows how the speed of a PC has increased with successive generations of processor. Use this bar chart to answer the following questions (answers need only be approximate):

1. How fast is a Pentium 3 when compared with the original 8086 processor?

2. What is the relative speed of a Pentium 4 processor when compared with a Pentium 1 processor?

3. What is the relative speed of a Pentium 1 processor when compared with an 80486 processor?

4. What is the percentage increase of speed between a Pentium 1 and Pentium 2 processor?

Ideographs (pictograms)

Ideographs (or pictograms) are frequently used to simplify statistical data so that it can be made meaningful to the general public. A typical example is the number of cars produced by a car manufacturer, over a 4-year period, as shown in *Figure 2.17*. In this example, each symbol represents 1000 cars. Therefore in 2003, 4000 cars were produced (four symbols each representing 1000 cars makes a total of 4000 cars).

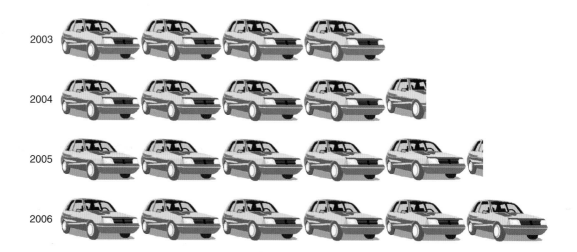

Car production = 1000

Figure 2.17 An ideograph

Some ideographs use the size of a symbol (rather than a number of symbols of identical size) to represent quantities. An example of this is shown in *Figure 2.18*.

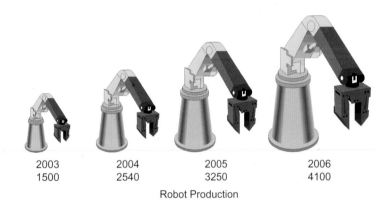

2003   2004   2005   2006
1500   2540   3250   4100

Robot Production

Figure 2.18 An alternative form of ideograph

Activity 2.15

Use an ideograph to compare the production of lager and bitter for a local brewery over a 6-week period (each symbol is to represent 1000 pints):

Week	Lager production (pints)	Bitter production (pints)
1	1750	2500
2	2500	4000
3	3500	6000
4	4000	6500
5	4250	3500
6	3500	5000

Use a simple drawing or art package to produce the ideograph and present the printed result.

Pie charts

Pie charts are used to show how a total quantity is divided up into its individual parts. Take a look at *Figure 2.19a*. Since a complete circle is 360°,

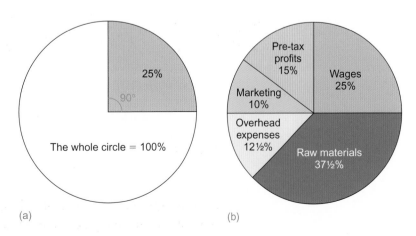

(a)   (b)

Figure 2.19 Pie charts

we can represent 25% of a complete circle as $\left(360° \times \dfrac{25}{100}\right) = 90°$

Figure 2.19b shows how the annual expenditure of an engineering company can be represented by a pie chart.

As with bar charts, pie charts can be presented in 3-D in order to make them more attractive. *Figure 2.20* shows a 3-D pie chart showing how the costs of a PC system unit are divided between the main components.

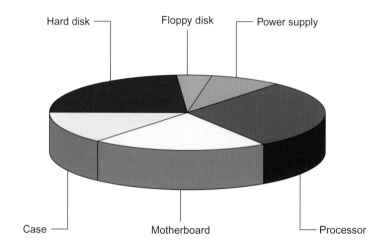

Figure 2.20 A 3-D pie chart

Activity 2.16

Draw pie charts to illustrate the breakdown of costs for four products:

Costs	Proportion of costs			
	Product A	Product B	Product C	Product D
Labour	45%	60%	25%	30%
Material	15%	15%	35%	40%
Overhead	25%	15%	25%	25%
Profit	15%	10%	15%	5%

Use the pie charts to answer the following questions:

(a) What can you infer from these results?
(b) Which product is the most profitable?
(c) Which product is the least profitable?
(d) If the total cost of all four products is the same, what recommendations would you wish to make?

Use a spreadsheet package to produce the pie-charts and insert these into a word-processed answer sheet.

Test your knowledge

?

TYK

TYK 2.16

Use *Figures 2.19b and 2.20* to answer the following questions:

1. What is the most expensive component in a PC system unit?

2. Which is the least expensive component in a PC system unit?

3. Roughly what proportion of the cost is attributable to the case?

4. What is the proportion of cost of the motherboard and processor taken together?

5. If the total cost of a system unit is £250, determine the approximate cost of the power supply.

Scatter diagrams

Scatter diagrams are used when we need to show that a change in one value is likely to result in a change in another. Scatter diagrams are usually produced by plotting corresponding pairs of values on an X–Y chart. This is usually done for a large number of observations (much larger than the number of values taken to plot a conventional line graph). The resulting grouping of dots on the diagram helps us to understand the *correlation* between the two values and whether there are any significant trends.

Figure 2.21 shows how the bond strength for an adhesive is related to the curing time. Here, the performance of 19 individual samples has been plotted and a clear trend can be seen.

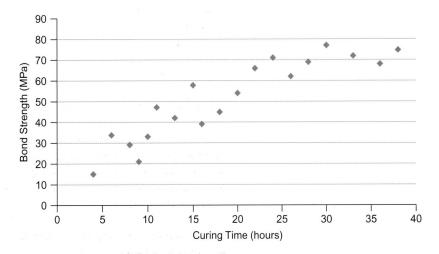

Figure 2.21 A scatter diagram

TYK 2.17

Use *Figure 2.21* to answer the following questions:

1. What was the maximum bond strength observed and at what curing time was this achieved?

2. What was the minimum bond strength observed and at what curing time was this achieved?

3. What can you infer about the relationship between bond strength and curing time for values of curing time up to 20 hours?

4. What appears to happen after 20 hours' curing time?

5. If the adhesive has a specified bond strength of 60 MPa, what minimum curing time should be recommended?

TYK 2.18

Figure 2.22 shows the results of an experiment on a semiconductor diode plotted as a graph. Use the graph to answer the following questions:

1. What current will flow through the diode when a potential difference of 0.15 V is applied to it?

2. What potential difference will appear across the diode when a current of 3.5 mA flows through it?

3. Resistance, R, can be determined by dividing potential difference, v, by current, i. What is the resistance of the diode at (a) $v = 0.2$ V and (b) $i = 3$ mA?

4. Describe the shape of the graph – what happens when the potential difference exceeds 0.2 V?

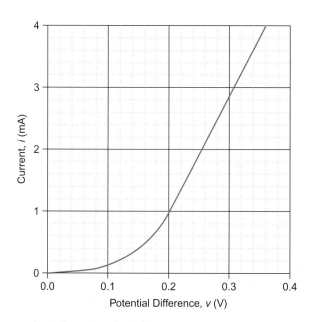

Figure 2.22 See TYK 2.18

Test your knowledge

TYK

TYK 2.19

Figure 2.23 shows production of TTL logic devices by a semiconductor manufacturer over a 5-year period. Use this to answer the following questions:

1. What was the total production of TTL logic chips in (a) 2002 and (b) 2003?

2. In which year was the production of chips a maximum?

3. In what years did production exceed 400,000 chips?

4. In what years did production fall below 500,000 chips?

5. What was the total production over the 5-year period?

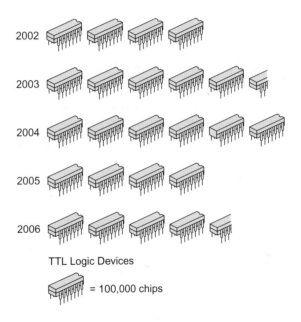

TTL Logic Devices

= 100,000 chips

Figure 2.23 See TYK 2.19

Activity 2.17

Complete the table below placing a tick in the column showing the most appropriate method of communicating the information in each situation listed:

Situation	Line graph	Histogram	Pictogram	Pie chart
Number of passengers carried by an airline each year in a 6-year period				
Variation of temperature in a room over a 24-hour period				
Monthly production of TV sets by a consumer electronics manufacturer				

KEY POINT

When selecting a method of graphical communication in order to convey information it is important to consider what sort of audience it is intended for. This means that you will often need to simplify the representation of the data so that it can be more easily grasped by the user. In every case, it is essential to avoid any possibility of misunderstanding the information that you are trying to represent

Situation	Line graph	Histogram	Pictogram	Pie chart
Average starting salary for a newly employed graduate over a 10-year period				
Market share of five companies that produce cars				

Copy the table into a word-processing package and then print out your work.

Activity 2.18

The following data relates to a 12-month operating period for a small engineering company:

Total income:	£7,200,000
Material costs:	£1,450,000
Labour costs:	£2,900,000
Overhead costs:	£1,150,000

Construct a 3-D pie chart (using an appropriate spreadsheet package) to show the relationship between the costs and the profit.

Activity 2.19

The following data was obtained during a number of observations of the working life of an optical projection unit when used in different ambient temperatures. Plot this data using a scatter diagram. What can you infer from these results?

Ambient temperature (°C)	Time to failure (hours)
15	2505
25	1950
20	2807
32	850
27	1365
19	2219
35	606
37	722
18	2605
12	3503
22	3083
21	2430
29	1971
22	2004
24	2696
20	3431
18	3122
16	3909

Use a spreadsheet to process the data and to create an X–Y scatter diagram. Then insert the scatter diagram into a word-processed document with your comments.

Design sketches

Engineers and designers frequently use quick hand-drawn sketches to illustrate design concepts as well as features associated with a particular product or process. In fact, being able to use sketches to illustrate your ideas is an essential part of becoming an engineer! Sketches are also an excellent way of communicating your ideas to other engineers and designers without having to resort to words or formal engineering drawings. When producing a design sketch there are no 'hard-and-fast' rules other than ensuring that the sketch is clear and unambiguous (i.e. not liable to be misinterpreted) and that it is adequately labelled.

Activity 2.20

Produce a design sketch for an entry-level digital camera. The camera is to incorporate an in-built flash unit and an LCD screen. The camera is to be powered by two AA batteries and is to have a USB connection for downloading images to a computer. Make sure that all of these features are identified in your sketch. Present your work in the form of hand-drawn sketches and hand-written notes.

Engineering Drawings

Like the graphs that we have just considered, there are many different ways of representing and communicating technical information. To avoid confusion, such information should make use of nationally and internationally recognized symbols, conventions and abbreviations. These are listed and their use explained in the appropriate British Standards. Such standards are lengthy and costly, but a summary is available for educational use. This document is entitled 'Engineering drawing practice for schools and colleges' and it has the British Standards reference number PP 8888. The document is abridged from the earlier British Standard BS 308.

Block diagrams

These show the relationship between the various elements of a system. *Figure 2.24* shows the block diagram for a simple radio receiver. This sort of diagram is used in the initial stages of conceptualizing a design or to provide an overview of the way in which an engineering system operates.

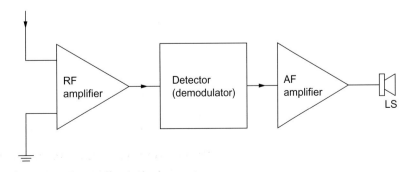

Figure 2.24 A block diagram

Activity 2.21

Construct a block diagram showing the arrangement of the braking system of a car. Label your drawing clearly. Present your results in the form of a printout from a simple drawing or technical illustration package.

Flow diagrams

Flow diagrams are used to illustrate a sequence of events. They are used in a wide variety of applications including the planning of engineering processes and the design of computer software. *Figure 2.25* shows a flowchart for the process of drilling a hole. The shape of the symbols used in this flowchart have particular meanings as shown in *Figure 2.26*. For the complete set of symbols and their meanings you should refer to the appropriate British Standard BS 4058.

Activity 2.22

Your bicycle tyre is flat and may have a puncture or may simply need reinflating. Draw a flowchart for checking the tyre and, if necessary, replacing the wheel. *Figure 2.27* will provide you with a starting point. Present your results in the form of a printout from a simple drawing or technical illustration package.

Circuit and related diagrams

Circuit diagrams are used to show the functional relationships between the components in an electric or electronic circuit. The components are represented by symbols and the electrical connections between the components drawn using straight lines. It is important to note that the position of a component in a circuit diagram does not represent its actual physical position in the final assembly. Circuit diagrams are sometimes also referred to as schematic diagrams or schematic circuits.

Figure 2.28a shows the circuit for an electronic filter unit using standard component symbols. *Figure 2.28b* shows the corresponding physical layout diagram with the components positioned on the upper (component side) of a PCB. Finally, *Figure 2.28c* shows the copper track layout for the PCB. This layout is developed photographically as an etch-resistant pattern on the copper surface of a copper-clad board.

The term 'wiring diagram' is usually taken to refer to a diagram that shows the physical interconnections between electrical and electronic components. Typical applications for wiring diagrams include the wiring layout of control desks, control cubicles and power supplies. Wiring

UNIT 2

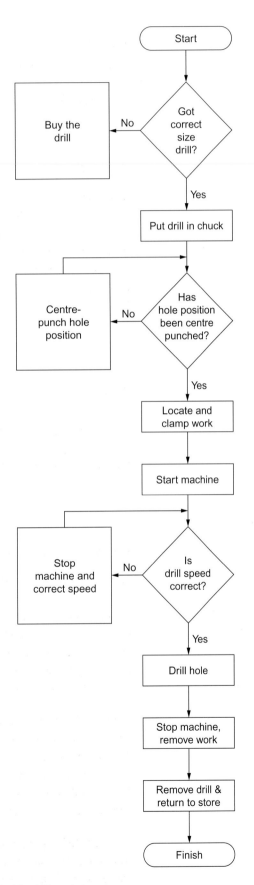

Figure 2.25 Flowchart for drilling a hole

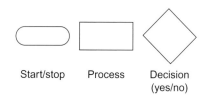

Figure 2.26 Some common flow chart symbols

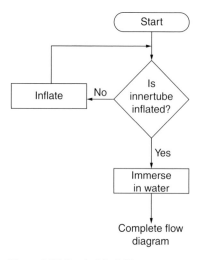

Figure 2.27 See Activity 2.22

diagrams are directly related to circuit schematics (circuit diagrams). As an example, architects use circuit schematics to show the electrical wiring and components inside a building or plant. They will also provide installation drawings to show where the components are to be sited. In addition, they may also provide a wiring diagram to show how the wires and cables are to be routed to and between the components. The symbols used in architectural installation drawings and wiring diagrams are not the same as those used in circuit diagrams.

Schematic circuit diagrams are also used to represent pneumatic (compressed air) circuits and hydraulic circuits. Pneumatic circuits and hydraulic circuits share the same symbols. You can tell which circuit is which because pneumatic circuits should have open arrow heads, while hydraulic circuits should have solid arrowheads. Also, pneumatic circuits exhaust to the atmosphere, while hydraulic circuits have to have a return path to the oil reservoir. *Figure 2.29* shows a typical hydraulic circuit.

Just as electrical circuit diagrams may have corresponding installation and wiring diagrams, so do hydraulic, pneumatic and plumbing circuits. Only this time the wiring diagram becomes a pipework diagram. A plumbing example is shown in *Figure 2.30*. As you may not be familiar with the symbols, we have named them for you. Normally, this is not necessary and the symbols are recognized by their shapes.

General arrangement drawings

Figure 2.31 shows the layout of a typical drawing sheet. To save time these are printed to a standardized layout for a particular company, ready for the draftsperson to add the drawing and complete the boxes and tables.

The basic information found on most drawing sheets consists of:

- drawing number and name of the company
- title and issue details
- scale
- method of projection (first or third angle)
- initials of persons responsible for drawing, checking, approving, tracing, etc. together with the appropriate dates
- unit(s) of measurement (inches or millimetres) and general tolerances
- material and finish
- copyright and standards reference
- guidance notes such as 'do not scale'
- reference grids so that 'zones' on the drawing sheet can be quickly found
- modifications table for alterations which are related to the issue number on the drawing and identified by the means of the reference grid.

The following additional information may also be included:

- fold marks
- centre marks for camera alignment when microfilming

KEY POINT

Layout drawings are made to develop the initial design of a unit or machine. A layout drawing must show all the information necessary to enable an assembly or detail drawing to be produced

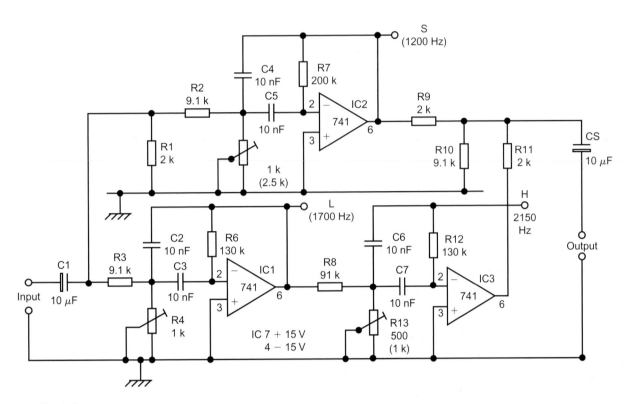

(a) Circuit diagram

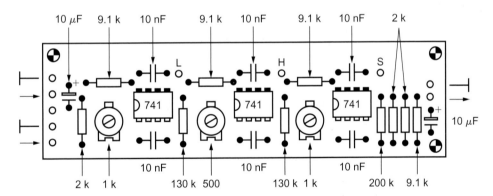

(b) Component layout diagram

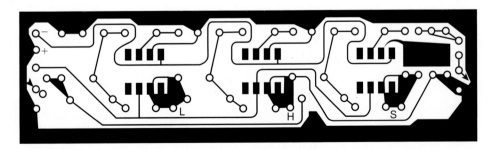

(c) PCB copper track layout

Figure 2.28 A typical electronic circuit diagram with corresponding layout diagram and printed circuit board (PCB) copper track layout

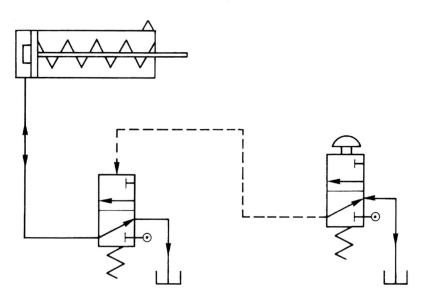

Figure 2.29 A typical hydraulic circuit

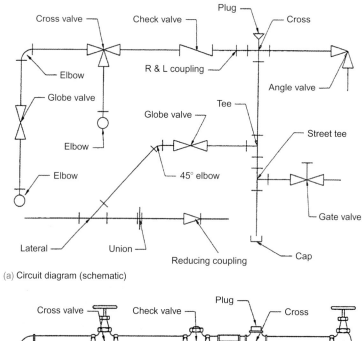

(a) Circuit diagram (schematic)

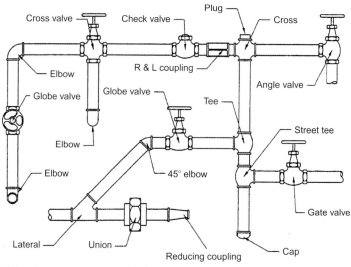

(b) Piping diagram

Figure 2.30 Typical plumbing circuit with corresponding piping diagram

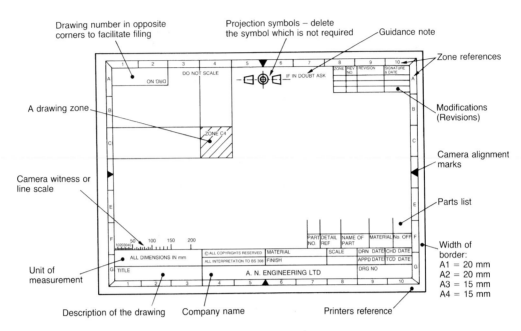

Figure 2.31 Layout of a typical drawing sheet

- line scale, so that the true size is not lost when enlarging or reducing copies
- trim marks
- orientation marks.

Figure 2.32 shows a typical general arrangement (GA) drawing. This shows as many of the features listed above as are appropriate for this drawing. It shows all the components correctly assembled together. Dimensions are not usually given on GA drawings although, sometimes, overall dimensions will be given for reference when the GA drawing is of a large assembly drawn to a reduced scale.

The GA drawing shows all the parts used in an assembly. These are listed in a table together with the quantities required. Manufacturers' catalogue references are also given for any components that are not being manufactured. The parts are usually 'bought-in' as 'off-the-shelf' parts from other suppliers. The detail drawing numbers are also included for components that have to be manufactured as special items.

Detail drawings

As the name implies, detail drawings provide all the details required to make the component shown on the drawing. Referring back to *Figure 2.32* we can see from the table that the detail drawing for the punch has the reference number 174/6. *Figure 2.33* shows this detail drawing. In this instance, the drawing provides the following information:

- shape of the punch
- dimensions of the punch and the manufacturing tolerances
- material from which the punch is to be made and its subsequent heat treatment

UNIT 2

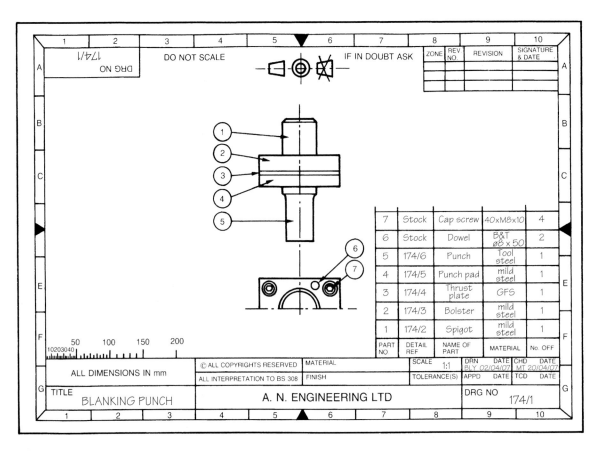

Figure 2.32 A typical GA drawing

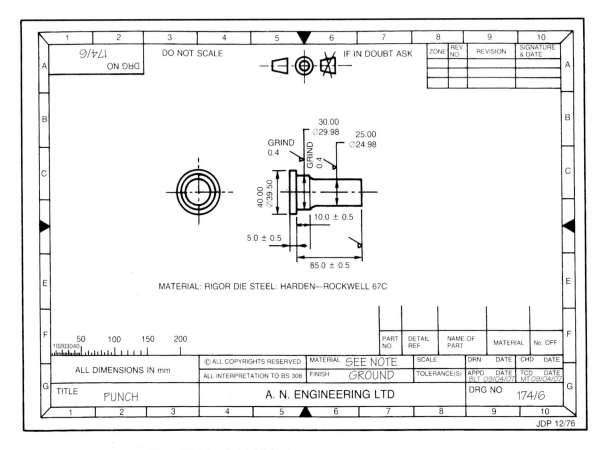

Figure 2.33 A typical detail drawing

- unit of measurement (millimetre)
- projection (first angle)
- finish
- guidance note 'do not scale drawing'
- name of the company
- name of the draftsperson
- name of the person checking the drawing.

It should go without saying that the amount of information given will depend upon the nature of the job. Drawings for a critical aircraft component, for example, will be much more fully detailed than those for a wheelbarrow component!

Production and storage of engineering drawings

Computer aided design and manufacture (CAD/CAM) has now largely replaced manual methods used for engineering drawing and data storage. CAD software (see *Figures 2.34 to 2.37*) is used in conjunction with a computer and the drawing produced on the computer screen is saved in a computer file on disk. Networked CAD/CAM and computer aided engineering (CAE) systems have made it possible to share data and drawings over a network. This allows many people to have access to the same data (usually stored on a powerful network server). This is now the most efficient and economical method of data storage. Very many complex drawings can be saved onto a hard disk or optical drive on a single file server. When required, the file can be recalled for immediate viewing on a computer screen and hard copy can be printed out to any desired scale at the touch of a key.

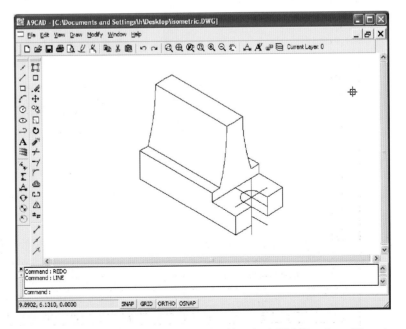

Figure 2.34 A conventional isometric drawing produced by a basic 2D CAD package (this package can be downloaded from our companion website, http://www.key2study.com/btecnat/)

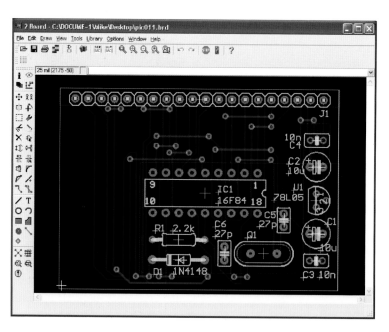

Figure 2.35 An auto-routed PCB layout for a microcontroller produced by a CAD PCB design package

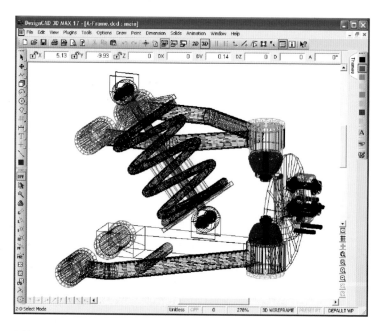

Figure 2.36 A wire frame drawing of a mechanical assembly drawn using a powerful 3-D CAD package

Despite the fact that computerized methods of data storage are now commonplace, it is important to know something about the older, traditional methods of storage. These are described briefly below:

- *Tracing linen* This was the traditional material making technical drawings. It was strong and durable and stood up well to the effects of the ultraviolet arc lamps used for making 'blue-print' copies. These got their name from the fact that the print appeared as white lines on a blue background.
- *Tracing paper* This is widely used in conjunction with manual drawing techniques. It is cheap and readily available. It is also easy to draw on. Unfortunately, the paper becomes brittle with age and requires careful

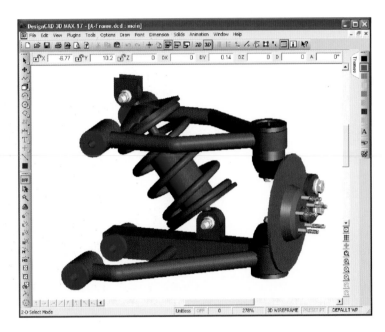

Figure 2.37 The fully rendered session of the mechanical assembly (note how shading is used to produce a realistic 3-D view)

handling. Therefore, it is not suitable where print copies have to be made frequently.

- *Tracing film* This is tough plastic film that is shiny on one side and matt on the other. You draw on the matt surface. No special techniques are required in its use and it stands up to repeated handling without deterioration. It is more expensive than tracing paper.
- *Microfilm* The storage of full size 'negatives', as the tracings are called, when produced on linen, paper or film takes up a lot of room. These large drawings can be reduced photographically onto 16 mm or 35 mm film stock for storage. This saves considerable space. The microfilm copies can be projection printed (enlarged) full size when required for issue.
- *Microfiche* Libraries, offices and stores use microfiche systems. Data is stored photographically in a grid of frames on a large rectangle of film. A desktop viewer is used to select and enlarge a single frame. The frame is then projected onto a rear projection screen for easy reading. This system is more likely to be used for storing literal and numerical data than for drawings.

Activity 2.23

(a) Compare the printed drawing sheets provided by your tutor with the model sheet shown in *Figure 2.31*. List any differences.

(b) Examine the general arrangement (GA) drawings provided by your tutor and compare them with the model GA shown in *Figure 2.32*. List any differences.

(c) Examine the detail drawings provided by your tutor and compare them with the model detail drawing shown in *Figure 2.33*. List any differences.

Present your results in the form of a set of brief handwritten notes.

Engineering drawing techniques

Engineering drawings can be produced using a variety of different techniques. The choice of technique is dependent upon a number of drawing factors such as:

Speed	How much time can be allowed for producing the drawing. How soon the drawing can be commenced.
Media	The choice will depend upon the equipment available (e.g. CAD or conventional drawing board and instruments) and the skill of the person producing the drawing.
Complexity	The amount of detail required and the anticipated amount and frequency of development modifications.
Cost	Engineering drawings are not works of art and have no intrinsic value. They are only a means to an end and should be produced as cheaply as possible. Both initial and ongoing costs must be considered.
Presentation	This will depend upon who will see/use the drawings. Non-technical people can visualize pictorial representations better than orthographic drawings.

Nowadays technical drawings are increasingly produced using computer aided drawing (CAD) techniques. Developments in software and personal computers have reduced the cost of CAD and made it more powerful. At the same time, it has become more 'user friendly'. Computer aided drawing does not require the high physical skill required for manual drawing, which takes years of practise to achieve. It also has a number of other advantages over manual drawing. Let us consider some of these advantages:

> **KEY POINT**
>
> Modelling is use to describe the design stage of a 3-D model or part. A model can be created physically or by using a 3-D solid modelling package on a computer

Accuracy	Dimensional control does not depend upon the draftsperson's eyesight.
Radii	Radii can be made to blend with straight lines automatically.
Repetitive features	For example, holes round a pitch circle do not have to be individually drawn, but can be easily produced automatically by 'mirror imaging'. Again, some repeated, complex features need only be drawn once and saved as a matrix. They can then be called up from the computer memory at each point in the drawing where they appear at the touch of a key.
Editing	Every time you erase and alter a manually produced drawing on tracing paper or plastic film the surface of the drawing is increasingly damaged. On a computer you can delete and redraw as often as you like with no ill effects.
Storage	No matter how large and complex the drawing, it can be stored digitally on floppy disk. Copies can be taken and transmitted between factories without errors or deterioration.

Prints   Hard copy can be produced accurately and easily on laser printers, flat bed or drum plotters and to any scale. Colour prints can also be made.

Pictorial techniques

Engineering drawings such as general arrangement drawings and detail drawings are produced by a technique called orthographic drawing using the conventions set out in BS 308. Since we will be asking you to make orthographic drawings from more easily recognized pictorial drawings, we will start by introducing you to the two pictorial techniques widely used by draftspersons (*Figures 2.38 and 2.39*).

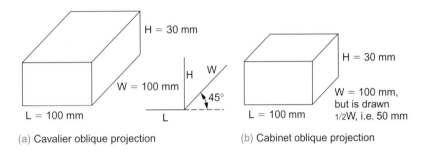

Figure 2.38   A simple oblique drawing

Oblique drawing

Figure 2.38 shows a simple oblique drawing. The front view (elevation) is drawn true shape and size. Therefore, this view should be chosen so as to include any circles or arcs so that these can be drawn with compasses. The lines forming the side views appear to travel away from you, so these are called 'receders'. They are drawn at 45° to the horizontal using a 45° set-square. They may be drawn full length as in cavalier oblique drawing or they may be drawn half-length as in cabinet oblique drawing. This latter method gives a more realistic representation, and is the one we will be using.

Activity 2.24

(a) Obtain a sheet of quadrille ruled paper (maths paper with 5 mm squares) and draw the box shown in *Figure 2.38* full size. Use cabinet oblique projection.

(b) Now use your compasses to draw a 50 mm diameter hole in the centre of the front (elevation) of the box.

(c) Can you think of a way to draw the same circle on the side (receding) face of the box? It will not be a true circle so you cannot use your compasses.

Present your results in the form of a hand-constructed drawing with hand-written notes.

Isometric drawing

Figure 2.39a shows an isometric drawing of our previous box. To be strictly accurate, the vertical lines should be drawn true length and the receders

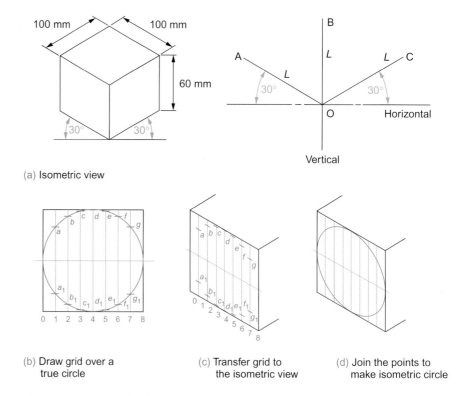

(a) Isometric view

(b) Draw grid over a
    true circle

(c) Transfer grid to
    the isometric view

(d) Join the points to
    make isometric circle

Figure 2.39 Isometric drawing

should be drawn to a special isometric scale. However, this sort of accuracy is rarely required and, for all practical purposes, we draw all the lines full size. As you can see, the receders are drawn at 30° to the horizontal for both the elevation and the end view.

Although an isometric drawing is more pleasing to the eye, it has the disadvantage that all circles and arcs have to be constructed. They cannot be drawn with compasses. *Figure 2.39b–d* shows you how to construct an isometric curve. You could have used this technique in Activity 2.24 to draw the circle on the side of the box drawn in oblique projection.

First, we draw the required circle. Then we draw a grid over it as shown in *Figure 2.39b*. Next number or letter the points where the circle cuts the grid as shown. Now, draw the grid on the side elevation of the box and step off the points where the circle cuts the grid with your compasses as shown in *Figure 2.39c*. All that remains is to join up the dots and you have an isometric circle as shown in *Figure 2.39d*.

Activity 2.25

(a) Draw, full size, an isometric view of the box shown in *Figure 2.39*. Isometric ruled paper will be of great assistance if you can obtain some.

(b) Draw a 50 mm diameter isometric circle on the TOP face of the box (remember that *Figure 2.39* shows it on the side of the box).

Present your results in the form of a hand-constructed drawing with hand-written notes.

Another way of drawing isometric circles and curves is the 'four-arcs' method. This does not produce true curves but they are near enough for all practical purposes and quicker and easier than the previous method for constructing true curves. The steps are shown in *Figure 2.40*.

1. Join points B and E as shown in *Figure 2.40b*. The line BE cuts the line GC at the point J. The point J is the centre of the first arc. With radius BJ set your compass to strike the first arc as shown.
2. Join points A and F as shown in *Figure 2.40c*. The line AF cuts the line GC at the point K. The point K is the centre of the second arc. With radius KF set your compasses to strike the second arc as shown. If your drawing is accurate both arcs should have the same radius.
3. With centre A and radius AF or AD strike the third arc as shown in *Figure 2.40d*.
4. With centre E and radius EH or EB strike the fourth and final arc as shown in *Figure 2.40e*.
5. If your drawing is accurate, arcs 3 and 4 should have the same radius.

AC = CE = EG = GA
AB = 1/2 AC
CD = 1/2 CE
EF = 1/2 EG
GH = 1/2 GA

(a) Draw an isometric grid of appropriate size

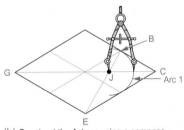

(b) Construct the 1st arc using a compass located as shown

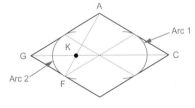

(c) Draw the 2nd arc using the construction process as shown

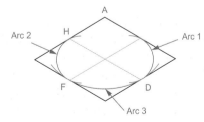

(d) Draw the 3rd arc through the point's shown

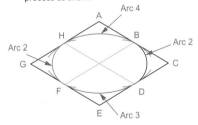

(e) Complete the process drawing the 4th arc from the opposite corner

Figure 2.40 The 'four-arcs' method

Activity 2.26

Use the technique just described to draw a 40 mm diameter circle on the side face of our box. Start off by drawing a 40 mm isometric square in the middle of the side face. Present your results in the form of a hand-constructed drawing with hand-written notes.

Activity 2.27

(a) *Figure 2.41a* shows some further examples of isometric drawings. Redraw them as cabinet oblique drawings.

(b) *Figure 2.41b* shows some further examples of cabinet oblique drawings. Redraw them as isometric drawings. Any circles and arcs on the vertical surfaces should be drawn using the grid construction method. Any arcs and circles on the horizontal (plan) surfaces should be drawn using the 'four-arcs' method.

Present your results in the form of a hand-constructed drawing with hand-written notes.

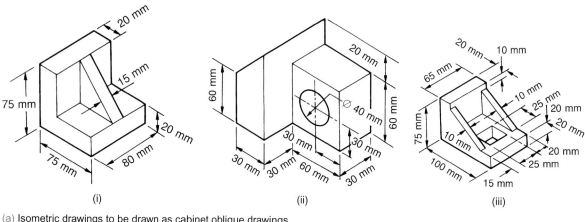

(a) Isometric drawings to be drawn as cabinet oblique drawings

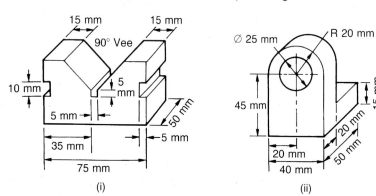

(b) Cabinet oblique drawings to be redrawn as isometric drawings

Figure 2.41 See Activity 2.27

Orthographic drawing

GA and detail drawings are produced by the use of a drawing technique called orthographic projection. This is used to represent 3-D solids on the 2-D surface of a sheet of drawing paper so that all the dimensions are true length and all the surfaces are true shape. To achieve this when surfaces are inclined to the vertical or the horizontal we have to use auxiliary views, but more about these later. Let us keep things simple for the moment.

First angle projection

Figure 2.42a shows a simple component drawn in isometric projection. *Figure 2.42b* shows the same component as an orthographic drawing. This time we make no attempt to represent the component pictorially. Each view of each face is drawn separately either full size or to the same scale. What is important is how we position the various views as this determines how we 'read' the drawing.

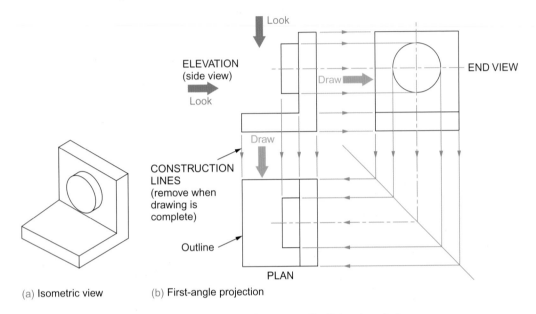

(a) Isometric view   (b) First-angle projection

Figure 2.42 An isometric view and its corresponding first angle projection

Engineers use two orthographic drawing techniques, either first angle or third angle projection. The former is called 'English projection' and the latter is called 'American projection'. The drawing in *Figure 2.42* is in first angle projection. The views are arranged as follows:

Elevation	This is the main view from which all the other views are positioned. You look directly at the side of the component and draw what you see.
Plan	To draw this, you look directly down on the top of the component and draw what you see below the elevation.
End view	This is sometimes called an 'end elevation'. To draw this you look directly at the end of the component and draw what you see at the opposite end of the elevation. There may be two end views, one at each end of the elevation, or there may be only one end view if this is all that is required to completely depict the component. *Figure 2.42* requires only one end view. When there is only one end view this can be placed at either end of the elevation depending upon which position gives the greater clarity and ease of interpretation. Whichever end is chosen the rules for drawing this view must be obeyed.

Use feint construction lines to produce the drawing as shown in *Figure 2.42b*. When these are complete, 'line-in' the outline more heavily. Carefully remove the construction lines to leave the drawing uncluttered, thus improving the clarity. *Figure 2.43* shows the finished drawing.

Activity 2.28

Figure 2.43 showed some components using pictorial projections. We now want you to redraw these components in first-angle orthographic projection. To start you off we have drawn the first one for you. This is shown in *Figure 2.44*. Note how we have positioned the end view this time so that you can see the web. Present your results in the form of a hand-constructed drawing with hand-written notes.

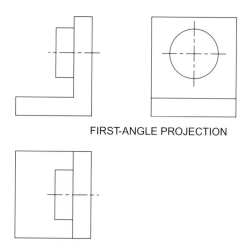

FIRST-ANGLE PROJECTION

Figure 2.43 Completed first angle projection

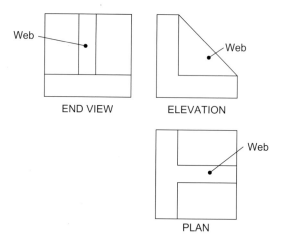

END VIEW   ELEVATION

PLAN

Figure 2.44 See Activity 2.28

Third angle projection

Figure 2.45 shows the same component, but this time we have drawn it in third angle projection for you. The views are arranged as follows:

Elevation   Again we have started with the elevation or side view of the component and, as you can see, there is no difference.

Plan   Again we look down on top of the component to see what the plan view looks like. However, this time we draw the plan view above the elevation. That is, in third angle projection we draw all the views from where we look.

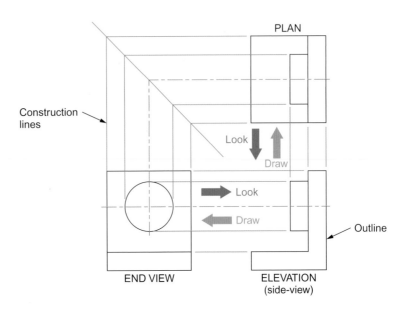

Figure 2.45 Third angle projection

End view	Note how the position of the end view is reversed compared with first angle projection. This is because, like the plan view, we draw the end views at the same end from which we look at the component.

Again use feint construction lines to produce the drawing as shown in *Figure 2.45*. Then 'line-in' the outline more heavily and carefully remove the construction lines for clarity, unless you have been instructed otherwise. *Figure 2.46* shows the finished drawing in third angle projection.

Activity 2.29

Figure 2.41 showed some components using pictorial projections. We now want you to redraw these components in third angle orthographic projection. To start you off, we have again drawn the first one for you. This is shown in *Figure 2.47*. Once again, note how we have positioned the end view so that you can see the web. Present your results in the form of a hand-constructed drawing with hand-written notes.

UNIT 2

Test your knowledge

?

TYK

TYK 2.20

(a) *Figure 2.48* shows some components drawn in first angle projection and some in third angle projection. We have not necessarily drawn all the views each time. Instead, we have only drawn as many of the views as are needed. State which is first angle and which is third angle.

(b) Two of the drawings are standard symbols for indicating whether drawing is in first angle or whether it is in third angle. Which drawings do you think are these symbols?

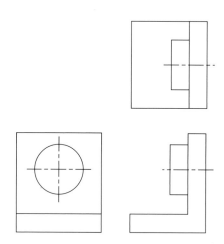

Figure 2.46 Completed third angle projection

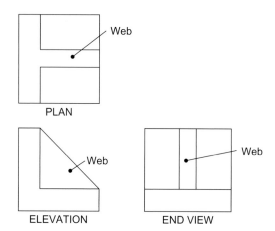

Figure 2.47 See Activity 2.29

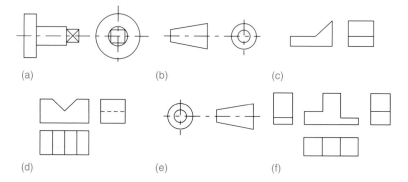

Figure 2.48 See TYK 2.20

Activity 2.30

Figure 2.49 shows pictorial views of some more solid objects.

Use a 2-D CAD package to draw these objects

(a) in first angle orthographic projection and label the views.
(b) in third angle orthographic projection and label the views.

Present your results in the form of printed drawings using appropriate drawing sheets (these will usually be supplied in the form of a template that is approved for use in your school or college).

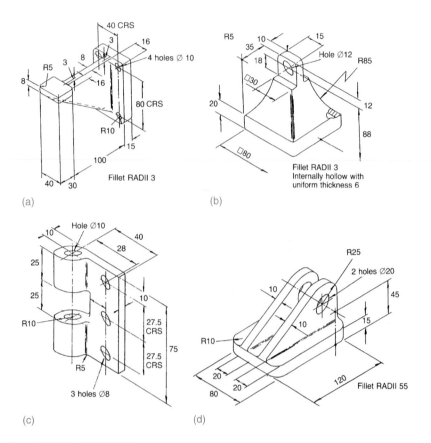

Figure 2.49 See Activity 2.30

Auxiliary views

In addition to the main views on which we have just been working, we sometimes have to use auxiliary views. We use auxiliary views when we cannot show the true outline of the component or a feature of the component in one of the main views. For example, when a surface of the component is inclined as shown in *Figure 2.50*.

Activity 2.31

Use a CAD package to draw the component shown in *Figure 2.51* in isometric projection given that each square has a side length of 10 mm.

Also draw the component in:

(a) first angle orthographic projection (only two views are required)
(b) third angle orthographic projection (only two views are required)
(c) cabinet oblique projection.

Present your work in the form of a portfolio of printed drawings. Clearly mark each drawing with the projection used.

Production of engineering drawings

Standard conventions are used in order to avoid having to draw, in detail, common features in frequent use. *Figure 2.52* shows a typical dimensioned engineering drawing. Some conventions can help us save

UNIT 2

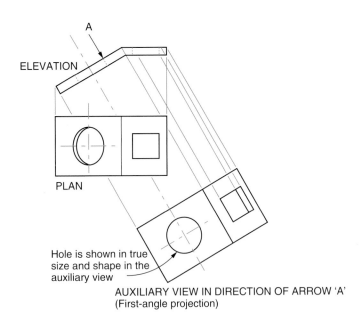

Figure 2.50 An auxiliary view

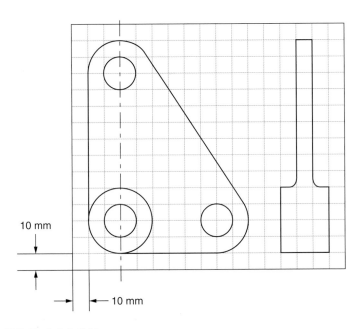

Figure 2.51 See Activity 2.31

a great deal of time and effort. For example, *Figure 2.53a* shows a pictorial representation of a screw thread while *Figure 2.53b* shows the standard convention for a screw thread. Clearly the latter is much quicker and easier to draw!

All engineering drawings should be produced using appropriate drawing standards and conventions for the following reasons:

Time   It speeds up the drawing process by making life easier for the draftsman as indicated above. This reduces costs and also reduces the 'lead time' required to get a new product into production.

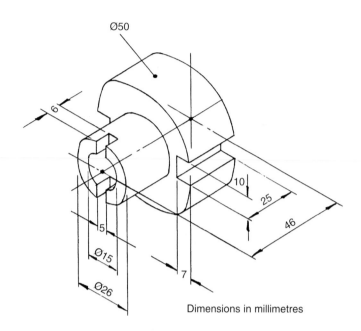

Dimensions in millimetres

Figure 2.52 A dimensioned drawing

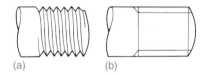

(a)   (b)

Figure 2.53 Screw threads

Appearance   It makes your drawings look more professional and improves the 'image' of yourself and your company. Badly presented drawings can send out the wrong messages and can call your competence into question.

Portability   Drawings produced to international standards and conventions can be read and correctly interpreted by any trained engineer, anywhere in this country or abroad. This avoids misunderstandings that could lead to expensive and complex components and assemblies being scrapped and dangerous situations arising. The only difficulties will arise from written notes that are language dependent.

Drawing conventions used by engineers in the UK are specified in BS 308. This is produced in three parts:

- Part 1 General principles.
- Part 2 Dimensioning and tolerancing of size.
- Part 3 Geometrical tolerancing.

These are all 'harmonized' with their appropriate ISO (International Standards Organization) counterparts.

As has been stated earlier, you will need to locate a copy of the British Standards Institution's publication PP 8888: Engineering Drawing Practice for Schools and Colleges. Also useful is PP 7307: Graphical Symbols for use in Schools and Colleges. This latter standard contains symbols for use in electrical, electronic, pneumatic and hydraulic schematic circuit diagrams. It also contains many other useful symbols.

UNIT 2

Other British Standards of importance to engineering draftspersons are:

BS 4500	ISO Limits and Fits (these are used by mechanical and production engineers).
BS 3939	Graphical symbols for electrical power, telecommunications and electronic diagrams.
BS 2197	Specifications for graphical symbols used in diagrams for fluid power systems and components.

Planning the drawing

Before we start the drawing and lay pencil to paper we should plan what we are going to do. This saves having to alter the drawing later on. We have to decide whether the drawing is to be pictorial, orthographic or schematic. If orthographic we have to decide on the projection we are going to use. We also have to decide whether we need a formal drawing or whether a freehand sketch is all that is required. If a formal drawing is needed then we have to decide whether to use manual techniques or CAD.

Paper size

When you start to plan your drawing you have to decide on the paper size. Engineering drawings are usually produced on 'A' size paper. Paper size A0 is approximately 1 square metre in area and is the basis of the system. Size A1 is half the area of size A0, size A2 is half the area of size A1 and so on down to size A4. Smaller sizes are available but they are not used for drawing. All the 'A' size sheets have their sides in the ratio of 1:12. This gives the following paper sizes:

A0	841 mm × 1189 mm
A1	594 mm × 841 mm
A2	420 mm × 549 mm
A3	297 mm × 420 mm
A4	210 mm × 297 mm

These relationships are shown diagrammatically in *Figure 2.54*.

The paper size you choose will depend upon the size of the drawing and the number of views required. Be generous, nothing looks worse than a cramped drawing and overcrowded dimensions. It is also false economy since overcrowding invariably leads to reading errors. As you will already have seen from some of the previous examples, the drawing should always have a border and a title block. This restricts the blank area available to draw on. *Figures 2.55 and 2.56* show how the views should be positioned. These layouts are only a guide but they offer a good starting point until you gain more experience. If only one view is required then it is centred in the drawing space available.

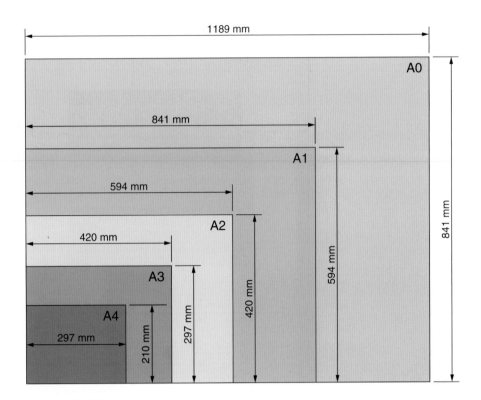

Figure 2.54 Paper sizes

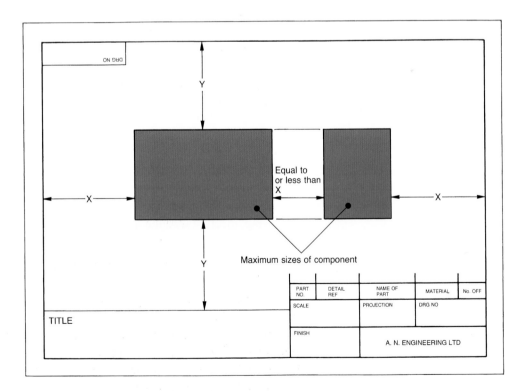

Figure 2.55 Positioning the drawing

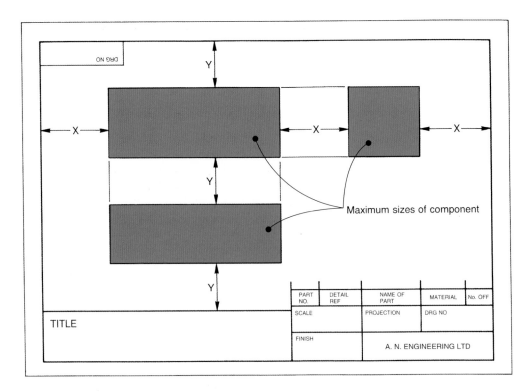

Figure 2.56 Positioning the drawing

Title block

A typical title block was shown in *Figure 2.32*. If you refer back to this figure you will see that it is expandable vertically and horizontally to accommodate any written information that is required. The title block should contain:

- The drawing number (which should be repeated in the top left-hand corner of the drawing).
- The drawing name (title).
- The drawing scale.
- The projection used (standard symbol).
- The name and signature of the draftsperson together with the date on which the drawing was signed.
- The name and signature of the person who checks and/or approves the drawing, together with the date of signing.
- The issue number and its release date.
- Any other information as dictated by company policy.

Scale

The scale should be stated on the drawing as a ratio. The recommended scales are as follows:

- Full size = 1:1
- Reduced scales (smaller than full size) are:

1:2	1:5	1:10
1:20	1:50	1:100
1:200	1:500	1:1000

(NEVER use the words full size, half size, quarter size, etc.)

- Enlarged scales (larger than full size) are:

2:1	5:1	10:1
20:1	50:1	100:1

Lines and linework

The lines of a drawing should be uniformly black, dense and bold. On any one drawing they should all be in pencil or in black ink. Pencil is quicker to use but ink prints more clearly. Lines should be thick or thin as recommended below. Thick lines should be twice as thick as thin lines. *Figure 2.57* shows the type of lines recommended in BS 308 for use in engineering drawing and how the lines should be used. This is reinforced in *Table 2.1*.

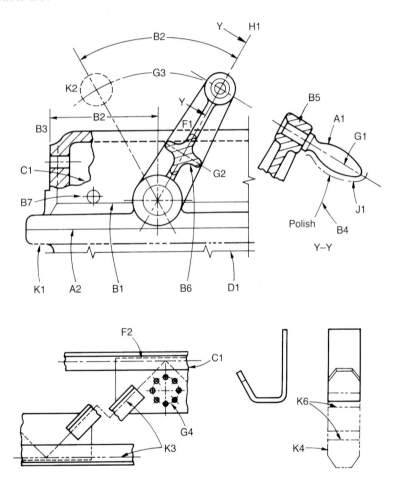

Figure 2.57 Use of various line types

Sometimes the lines overlap in different views. When this happens, as shown in *Figure 2.58*, the following order of priority should be observed.

- Visible outlines and edges (type A) take priority over all other lines.
- Next in importance are hidden outlines and edges (type E).
- Then cutting planes (type G).
- Next come centre lines (types F and B).
- Outlines and edges of adjacent parts, etc. (type H).
- Finally, projection lines and shading lines (type B).

UNIT 2

Table 2.1 Types of line (see Figure 2.57)

A	**Continuous thick**
A1	Visible outline
A2	Visible edge
B	**Continuous thin**
B1	Imaginary line of intersection
B2	Dimension line
B3	Projection line
B4	Leader line
B5	Hatching
B6	Outline of a revolved section
B7	Short centre line
C	**Continuous thin irregular**
C1	Limit of partial view
D	**Continuous thin zig-zag**
D1	Limit of partial view
E	**Dashed thick**
E1	Hidden outline
E2	Hidden edge
F	**Dashed thin**
F1	Hidden outline
F2	Hidden edge
G	**Chain thin**
G1	Centre line
G2	Line of symmetry
G3	Trajectory and locus line
G4	Pitch line and pitch circle
H	**Chain thin, thick at ends**
H1	Cutting plane
J	**Chain thick**
J1	Indicates special requirement
K	**Chain thin double dashed**
K1	Outline of adjacent part
K2	Outline of extreme position
K3	Centroid line
K4	Initial outline
K5	Part in front of cutting plane
K6	Bend line

Leader lines

Leader lines, as their name implies, lead written information or dimensions to the points where they apply. Leader lines are thin lines (type B) and they end in an arrowhead or in a dot as shown in *Figure 2.60a*. Arrowheads touch and stop on a line, while dots should always be used within an outline.

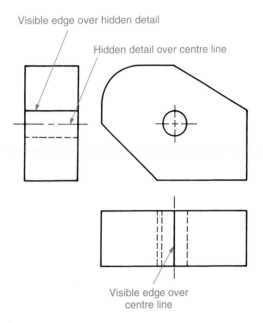

Figure 2.58 Line priorities

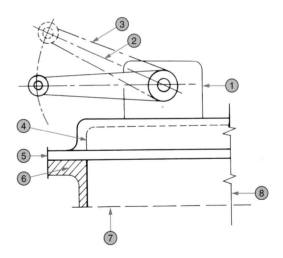

Figure 2.59 See TYK 2.21

- When an arrowed leader line is applied to an arc it should be in line with the centre of the arc as shown in *Figure 2.60b*.
- When an arrowed leader line is applied to a flat surface, it should be nearly normal to the lines representing that surface as shown in *Figure 2.60c*.
- Long and intersecting leader lines should not be used, even if this means repeating dimensions and/or notes as shown in *Figure 2.60d*.
- Leader lines must not pass through the points where other lines intersect.
- Arrowheads should be triangular with their length some three times larger than the maximum width. They should be formed from straight lines and the arrowheads should be filled in. The arrowhead should be symmetrical about the leader line, dimension line or stem. It is recommended that arrowheads on dimension and leader lines should be some 3–5 mm long.

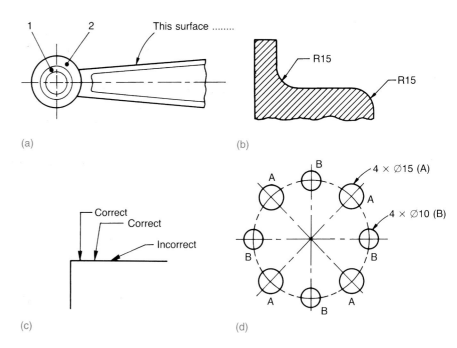

Figure 2.60 Examples of the use of leader lines

TYK 2.21

Figure 2.59 shows a simple drawing using a variety of lines. We have numbered some of the lines. List the number and, against each number, write down whether the type of line chosen is correct or incorrect. If incorrect state what type of line should have been used.

TYK 2.22

Figure 2.61 shows some applications of leader lines with arrowheads and leader lines with dots. List the numbers and state whether the application is correct or incorrect. If incorrect explain (with sketches if required) how the application should be corrected.

- Arrowheads showing direction of movement or direction of viewing should be some 7–10 mm long. The stem should be the same length as the arrowhead or slightly greater. It must never be shorter.

Letters and numerals

Style The style should be clear and free from embellishments. In general, capital letters should be used. A suitable style could be:

A B C D E F G H I J K L M N O P Q R S T U V W X Y Z 1 2 3 4 5 6 7 8 9 0

Size	The characters used for dimensions and notes on drawings should not be less than 3 mm tall. Title and drawing numbers should be at least twice as big.
Direction	Notes and captions should be positioned so *of lettering* that they can be read in the same direction as the information in the title block. Dimensions have special rules and will be dealt with later.
Location	General notes should all be grouped together and *of notes* not scattered about the drawing. Notes relating to a specific feature should be placed adjacent to that feature.
Emphasis	Characters, words and/or notes should not be emphasized by underlining. Where emphasis is required the characters should be enlarged.

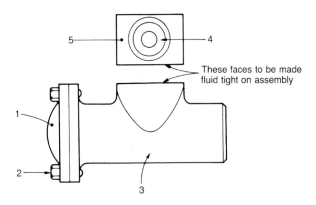

These faces to be made fluid tight on assembly

Figure 2.61 See TYK 2.22

Table 2.2 Types of line (see TYK 2.23)

Abbreviation or symbol	Term
AF	
ASSY	
CHAM	
	Countersink
CYL	
	Diameter (in a note)
	Figure
	Hexagon
MATL	
	Maximum
	Minimum
	Radius (in a note)
	Radius (preceding a dimension)
REQD	
RH	
SPEC	
Ø	

TYK 2.23

With reference to appropriate standards, complete Table 2.2.

TYK 2.24

With reference to appropriate standards, complete *Figure 2.62*. You must take care to use the same types of line as shown in the standard or the conventions become meaningless. This applies particularly to line thickness.

KEY POINT

A *dimension* is a numeric value expressed in appropriate units of measure and indicated on a drawing and in other documents along with lines, symbols and notes to define the size or geometric characteristic, or both, of a part or part feature.

KEY POINT

A *reference dimension* (usually without tolerance) is used for information only. It is considered auxiliary information and does not govern production or inspection operations. A reference dimension repeats a dimension or size already given or derived from other values shown on the drawing or related drawing. Reference dimensions are enclosed in brackets, for example (23.50).

KEY POINT

Tolerance is the total amount by which a specific dimension is permitted to vary. The tolerance is the difference between the maximum and minimum limits.

Symbols and abbreviations

If all the information on a drawing were written out in full, the drawing would become very cluttered. Therefore, symbols and abbreviations are used to shorten written notes. Those recommended for use on engineering drawings are listed in BS 308, and in the corresponding student version PP 7308.

Conventions

These are a form of 'shorthand' used to speed up the drawing of common features in regular use. The full range of conventions and examples of their use can be found in appropriate standards so we will not waste space by listing them here. However, by completing the next exercise you will use some of the more common conventions and this will help you to become familiar with them.

Dimensioning

When a component is being dimensioned, the dimension lines and the projection lines should be thin full lines (type B). Where possible dimensions should be placed outside the outline of the object as shown in *Figure 2.63a*. The rules are:

- Outline of object to be dimensioned in thick lines (type A).
- Dimension and projection lines should be half the thickness of the outline (type B).
- There should be a small gap between the projection line and the outline.
- The projection line should extend to just beyond the dimension line.
- Dimension lines end in an arrowhead that should touch the projection line to which it refers.
- All dimensions should be placed in such a way that they can be read from the bottom right-hand corner of the drawing.

UNIT 2

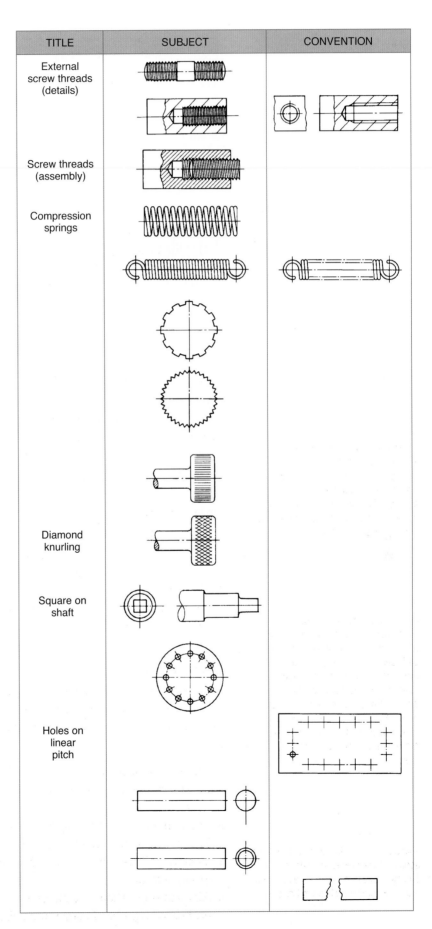

TITLE	SUBJECT	CONVENTION
External screw threads (details)		
Screw threads (assembly)		
Compression springs		
Diamond knurling		
Square on shaft		
Holes on linear pitch		

Figure 2.62 See TYK 2.24

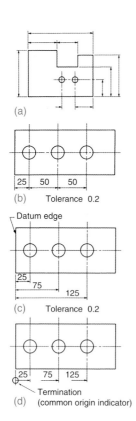

Figure 2.63 Dimensioning

KEY POINT

The *origin* or *datum point* is the name given to the point from where the location or geometric characteristics of a part are established. The correct identification of datums on a component and the related dimensioning can be vitally important in the manufacturing process (e.g. CNC machining).

KEY POINT

A *feature* is a general term applied to a physical portion of a part, e.g. a surface, hole or slot.

A *datum feature* is a geometric feature of a part that is used to establish a datum. For example, a point, line, surface, hole, etc.

The purpose of these rules is to allow the outline of the object to stand out prominently from all the other lines and to prevent confusion.

There are three ways in which a component can be dimensioned. These are:

- Chain dimensioning as shown in *Figure 2.63b*.
- Absolute dimensioning (dimensioning from a datum) using parallel dimension lines as shown in *Figure 2.63c*.
- Absolute dimensioning (dimensioning from a datum using superimposed running dimensions as shown in *Figure 2.63d*. Note the common origin (termination) symbol.

It is neither possible to manufacture an object to an exact size nor to measure an exact size. Therefore, important dimensions have to be *toleranced*. That is, the dimension is given two sizes: an upper limit of size and a lower limit of size. Providing the component is made so that it lies between these limits it will function correctly. Information on limits and fits can be found in BS 4500.

The method of dimensioning can also affect the accuracy of a component and produce some unexpected effects. *Figure 2.63b* shows the effect of chain dimensioning on a series of holes or other features. The designer specifies a common tolerance of ±0.2 mm. However, since this tolerance is applied to each and every dimension, the cumulative tolerance becomes ±0.6 mm by the time you reach the final, right-hand hole. Not what was intended. Therefore, absolute dimensioning as shown in *Figure 2.63c* and *d* is to be preferred in this example. With absolute dimensioning, the position of each hole lies within a tolerance of ±0.2 mm and there is no cumulative error. Further examples of dimensioning techniques are shown in *Figure 2.64*.

It is sometimes necessary to indicate machining processes and surface finish. The machining symbol together with examples of process notes and the surface finishes in micrometres (mm) is shown in *Figure 2.65*.

Activity 2.32

Figure 2.66 shows a component drawn in isometric projection. Use a CAD package to redraw it in first angle orthographic projection and add the dimensions using the following techniques:

(a) Absolute dimensioning using parallel dimension lines.
(b) Absolute dimensioning using superimposed running dimensions.

Present your work in the form of printed drawings using appropriate drawing sheets.

Sectioning

Sectioning is used to show the hidden detail inside hollow objects more clearly than can be achieved using dashed thin (type E) lines. *Figure 2.67a* shows an example of a simple sectioned drawing. The cutting plane is the

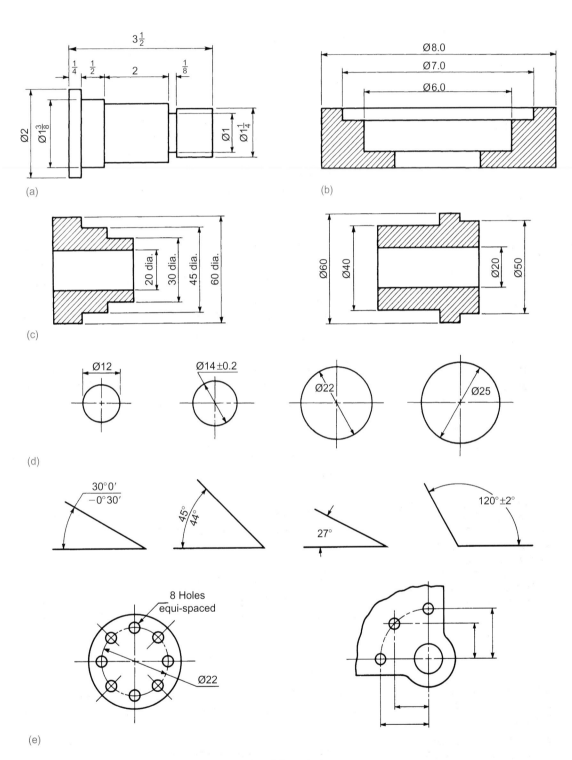

Figure 2.64 More examples of dimensioning

line A–A. In your imagination you remove everything to the left of the cutting plane, so that you see only what remains to the right of the cutting plane looking in the direction of the arrowheads. Another example is shown in *Figure 2.67b*.

Figure 2.67c shows how to section an assembly. Note how solid shafts and the key are not sectioned. Also note that thin webs that lie on the section

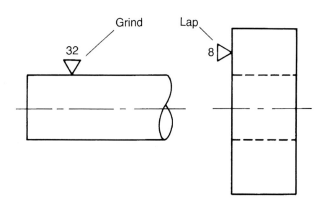

Figure 2.65 Indicating surface finishes

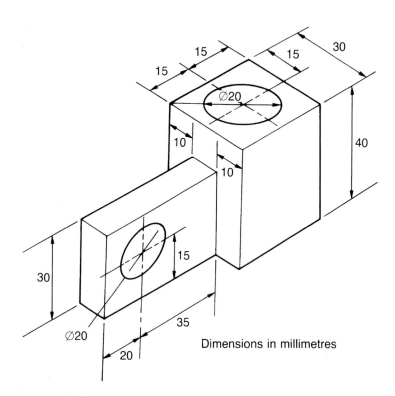

Figure 2.66 See Activity 2.32

plane are not sectioned. When interpreting sectioned drawings, some care is required. It is easy to confuse the terms sectional view and section:

Sectional view	In a sectional view you see the outline of the object at the cutting plane. You also see all the visible outlines seen beyond the cutting plane in the direction of viewing. Therefore, *Figure 2.67a* is a sectional view.
Section	A section shows only the outline of the object at the cutting plane. Visible outlines beyond the cutting plane in the direction of viewing are not shown. Therefore, a section has no thickness.

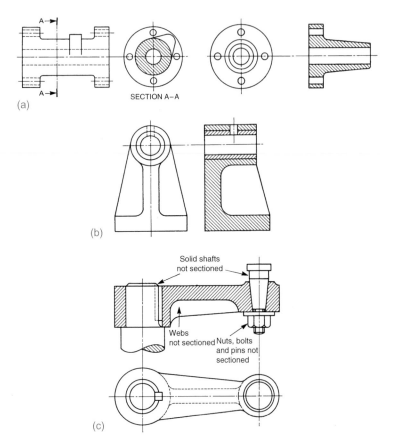

Figure 2.67 Sectioning – see also Activity 2.33

Activity 2.33

(a) **Use a CAD package to redraw *Figure 2.67a* as a section. (Remember that my drawing is a sectional view.)**
(b) **Explain why *Figure 2.67b* can be a section or a sectional view.**

Present your work in the form of a printed drawing (using an appropriate drawing sheet) together with a brief hand-written explanation.

Cutting planes You have already been introduced to cutting planes in the previous examples. They consist of type G lines. That is a thin chain line that is thick at the ends and at changes of direction. The direction of viewing is shown by arrows with large heads. The points of the arrowheads touch the thick portion of the cutting plane. The cutting plane is labelled by placing a capital letter close to the stems of the arrows. The same letters are used to identify the corresponding section or sectional view.

Hatching You will have noticed that the shading of sections and sectional views consists of sloping, thin (type B) lines. This is called hatching. The lines are equally spaced, slope at 45° and are not usually less than 4 mm apart. However, when hatching very small areas the hatching can be reduced, but by never less than 1 mm. The drawings in this book may look as though they do not obey these rules. Remember that they have been reduced from much bigger drawings to fit onto the pages.

UNIT 2

Figure 2.68 shows the basic rules of hatching. The hatching of separated areas is shown in *Figure 2.68a*. Separate sectioned areas of the same component should be hatched in the same direction and with the same spacing.

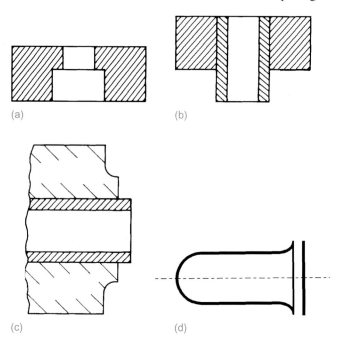

(a) (b)

(c) (d)

Figure 2.68 Hatching

Figure 2.68b shows how to hatch assembled parts. Where the different parts meet on assembly drawings, the direction of hatching should be reversed. The hatching lines should also be staggered. The spacing may also be changed.

Figure 2.68c shows how to hatch large areas. This saves time and avoids clutter. The hatching is limited to that part of the area that touches adjacent hatched parts or just to the outline of a large parts.

Figure 2.68d shows how sections through thin materials can be blocked in solid rather than hatched. There should be a gap of not less than 1 mm between adjacent parts even when these are a tight fit in practice.

Finally, we have included some further examples of sectioning in *Figure 2.69*. These include assemblies, half-sections, part sections and revolved sections. Then it will be your turn to produce some engineering drawings including some or all of the features outlined in this section.

Activity 2.34

Use a CAD package to:

(a) Redraw *Figure 2.70* in third angle projection. Include an end view looking in the direction of arrow 'A', and section the elevation on the cutting plane XX.

(b) *Figure 2.71* shows a cast iron pipe bend.
   (i) Redraw, adding an end view looking in the direction of arrow A.
   (ii) Section the elevation on the centre line.
   (iii) Draw an auxiliary view of flange B.

Present your work in the form of printed drawings using appropriate drawing sheets.

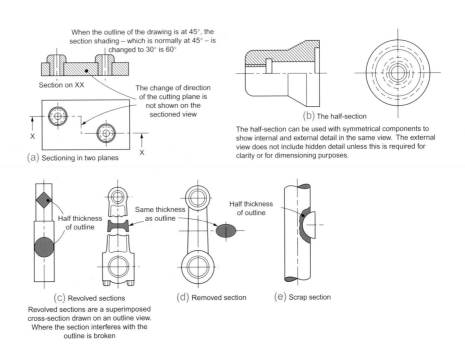

When the outline of the drawing is at 45°, the section shading – which is normally at 45° – is changed to 30° is 60°

Section on XX

The change of direction of the cutting plane is not shown on the sectioned view

(a) Sectioning in two planes

(b) The half-section

The half-section can be used with symmetrical components to show internal and external detail in the same view. The external view does not include hidden detail unless this is required for clarity or for dimensioning purposes.

Half thickness of outline

Same thickness as outline

Half thickness of outline

Half thickness of outline

(c) Revolved sections

Revolved sections are a superimposed cross-section drawn on an outline view. Where the section interferes with the outline is broken

(d) Removed section

(e) Scrap section

Figure 2.69 Examples of sectioning

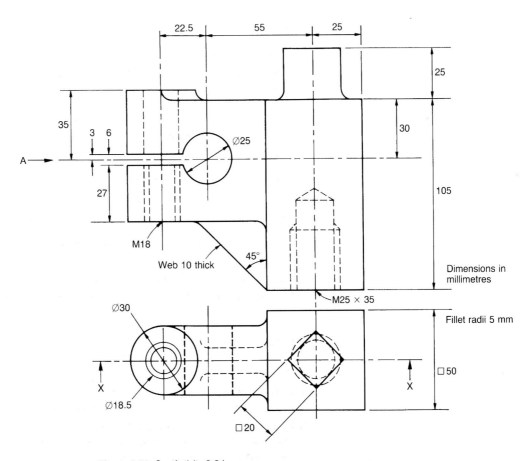

Figure 2.70 See Activity 2.34

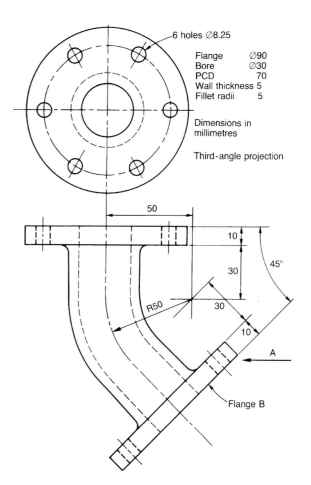

6 holes ⌀8.25

Flange   ⌀90
Bore   ⌀30
PCD   70
Wall thickness 5
Fillet radii   5

Dimensions in
millimetres

Third-angle projection

Figure 2.71 See Activity 2.34

Fluid power schematic diagrams

These diagrams cover both pneumatic and hydraulic circuits. The symbols that we shall use do not illustrate the physical make-up, construction or shape of the components. Neither are the symbols to scale or orientated in any particular position. They are only intended to show the 'function' of the component they portray, the connections and the fluid flow path.

Complete symbols are made up from one or more basic symbols and from one or more functional symbols. Examples of some basic symbols are shown in *Figure 2.72* and some functional symbols are shown in *Figure 2.73*.

Energy converters

Let us now see how we can combine some of these basic and functional symbols to produce a complete symbol representing a component. For example, let us start with a motor. The complete symbol is shown in *Figure 2.74*.

The large circle indicates that we have an energy conversion unit such as a motor or pump. Notice that the fluid flow is into the device and that it is

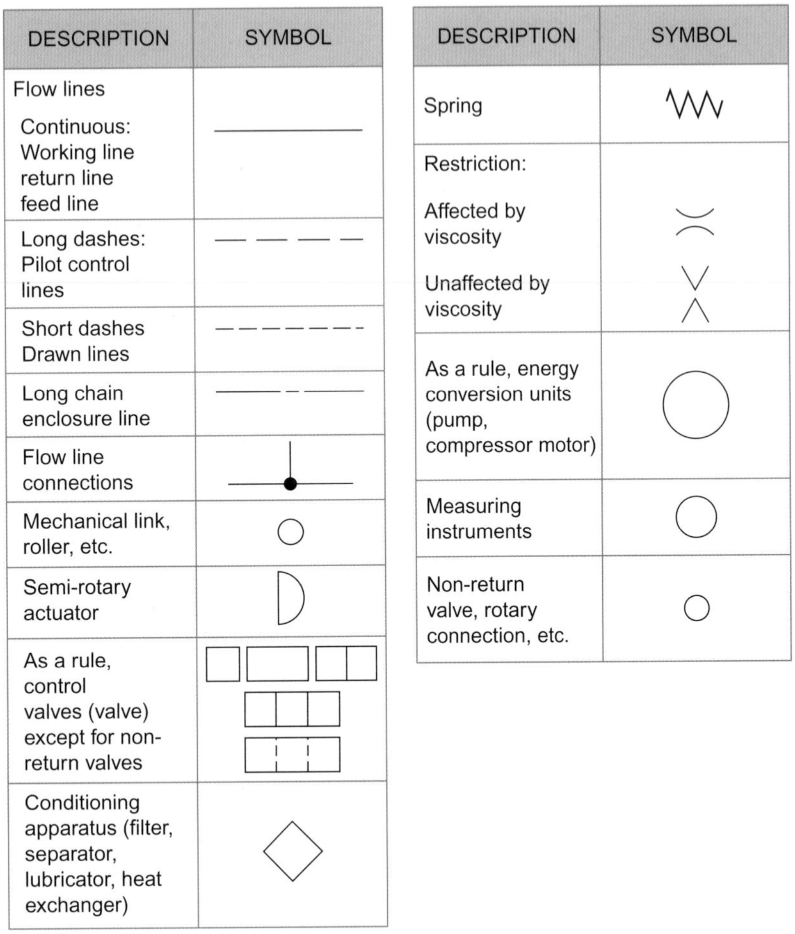

DESCRIPTION	SYMBOL	DESCRIPTION	SYMBOL
Flow lines Continuous: Working line return line feed line	————	Spring	(symbol)
Long dashes: Pilot control lines	— — — —	Restriction: Affected by viscosity	(symbol)
Short dashes Drawn lines	– – – – – –	Unaffected by viscosity	(symbol)
Long chain enclosure line	—— — ——	As a rule, energy conversion units (pump, compressor motor)	(circle symbol)
Flow line connections	(symbol)		
Mechanical link, roller, etc.	(circle)	Measuring instruments	(circle symbol)
Semi-rotary actuator	(symbol)	Non-return valve, rotary connection, etc.	(circle symbol)
As a rule, control valves (valve) except for non- return valves	(symbols)		
Conditioning apparatus (filter, separator, lubricator, heat exchanger)	(diamond symbol)		

Figure 2.72 Basic symbols used in fluid power diagrams

pneumatic. The direction of the arrowhead indicates the direction of flow. The fact that the arrowhead is clear (open) indicates that the fluid is air. Therefore, the device must be a motor. If it were a pump the fluid flow would be out of the circle. The single line at the bottom of the circle is the outlet (exhaust) from the motor and the double line is the mechanical output from the motor.

Now let us analyse the symbol shown in *Figure 2.75*.

- The circle tells us that it is an energy conversion unit.
- The arrowheads show that the flow is from the unit so it must be a pump.
- The arrowheads are solid so it must be a hydraulic pump.
- The arrowheads point in opposite directions so the pump can deliver the hydraulic fluid in either direction depending upon its direction of rotation.
- The arrow slanting across the pump is the variability symbol, so the pump has variable displacement.
- The double lines indicate the mechanical input to the pump from some engine or motor.

Summing up, we have a variable displacement, hydraulic pump that is bi-directional.

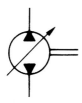

Figure 2.74 Basic symbol for a motor

Figure 2.75 Energy converter symbol

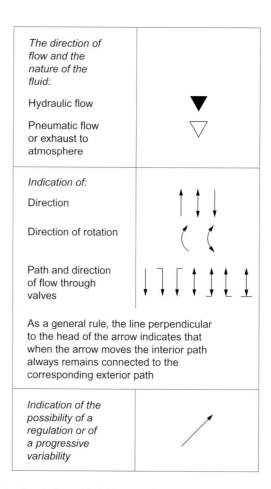

The direction of flow and the nature of the fluid: Hydraulic flow Pneumatic flow or exhaust to atmosphere	
Indication of: Direction Direction of rotation Path and direction of flow through valves	
As a general rule, the line perpendicular to the head of the arrow indicates that when the arrow moves the interior path always remains connected to the corresponding exterior path	
Indication of the possibility of a regulation or of a progressive variability	

Figure 2.73 Functional symbols used in fluid power diagrams

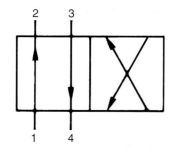

TYK *2.25*

Draw the symbol for:
(a) a unidirectional, fixed displacement pneumatic pump (compressor)
(b) a fixed capacity hydraulic motor.

Directional control valves

The function of a directional control valve is to open or close flow lines in a system. Control valve symbols are always drawn in square boxes or groups of square boxes to form a rectangle. This is how you recognize them. Each box indicates a discrete position for the control valve. Flow paths through a valve are known as 'ways'. Thus, a 4-way valve has four flow paths through the valve. This will be the same as the number of connections. We can, therefore, use a number code to describe the function of a valve. *Figure 2.76* shows a 4/2 directional control valve (DCV). This valve has four flow paths, ports or connections and two positions. The two boxes indicate the two positions. The appropriate box is shunted from side to side so that, in your imagination, the internal flow paths line up with the connections. Connections are shown by the lines that extend 'outside' the perimeters of the boxes.

Figure 2.76 4/2 directional control valve

TYK 2.26

A valve symbol is shown in *Figure 2.77*.
(a) State the numerical code that describes the valve.
(b) Describe the flow path drawn.
(c) Sketch and describe the flow path when the valve is in its alternative position.

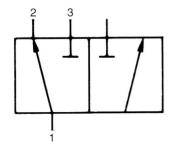

Figure 2.77 See TYK 2.26

As drawn, the fluid can flow into port 1 and out of port 2. Fluid can also flow into port 3 and out of port 4. In the second position, the fluid flows into port 3 and out of port 1. Fluid can also flow into port 4 and out of port 2.

Valve control methods

Before we look at other examples of directional control valves, let us see how we can control the positions of a valve. There are five basic methods of control, these are:

- Manual control of the valve position.
- Mechanical control of the valve position.
- Electromagnetic control of the valve position.
- Pressure control of the valve positions (direct and indirect).
- Combined control methods.

The methods of control are shown in *Figure 2.78*. With simple electrical or pressure control, it is possible only to move the valve to one, two or three discrete positions. The valve spool may be located in such positions by a spring loaded detent.

Combinations of the above control methods are possible. For example, a single solenoid with spring return for a two-position valve. Let us now look at some further DCVs.

- *Figure 2.79a* shows a 4/2 DCV controlled by a single solenoid with a spring return.
- *Figure 2.79b* shows a 4/3 DCV. That is, a directional control valve with four ports (connections) and three positions. It is operated manually by a lever with spring return to the centre. The service ports are isolated in the centre position. An application of this valve will be shown later.
- *Figure 2.79c* shows a 4/2 DCV controlled by pneumatic pressure by means of a pilot valve. The pilot valve is actuated by a single solenoid and a return spring.

TYK 2.27

Describe the DCV whose symbol is shown in *Figure 2.80*.

DESCRIPTION	SYMBOL
Manual control:	
general symbol	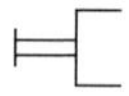
by push-button	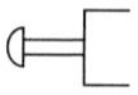
by lever	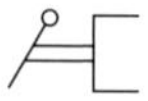
by pedal	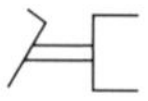
Mechanical control:	
by plunger or tracer	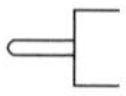
by spring	
by roller	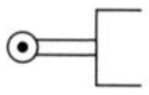
by roller, operating in one direction only	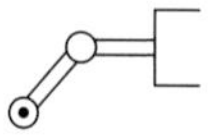
Electrical control:	
by solenoid (one winding)	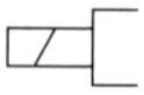
(two windings operating in opposite directions)	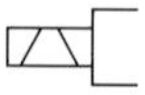
by electric motor	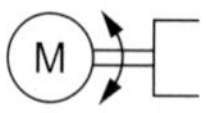

DESCRIPTION	SYMBOL
Control by application or release of pressure	
Direct acting control:	
by application of pressure	
by release of pressure	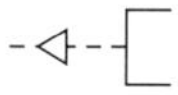
by different control areas	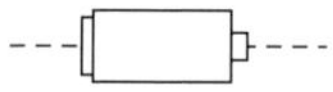
Indirect control, pilot actuated:	
by application of pressure	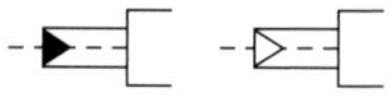
by release of pressure	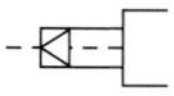
Interior control paths (paths are inside the unit)	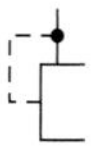
Combined control:	
by solenoid and pilot directional valve (pilot directional valve is actuated by the solenoid)	
by solenoid or pilot direction valve (either may actuate the control independently)	

Figure 2.78 Methods of control

Linear actuators

A linear actuator is a device for converting fluid pressure into a mechanical force capable of doing useful work and combining this force with limited linear movement. Put more simply, a piston in a cylinder. The symbols for linear actuators (also known as 'jacks' and 'rams') are simple to understand and some examples are shown in *Figure 2.81*.

- *Figure 2.81a* shows a single-ended, double-acting actuator. That is, the piston is connected by a piston rod to some external mechanism through

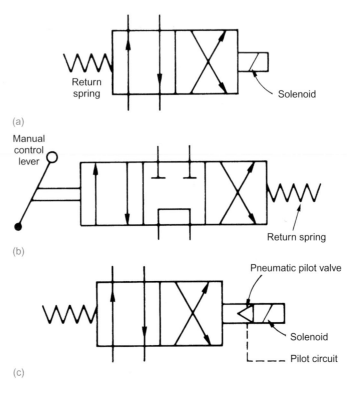

(a)

(b)

(c)

Figure 2.79 Various types of DCV

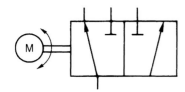

Figure 2.80 See TYK 2.27

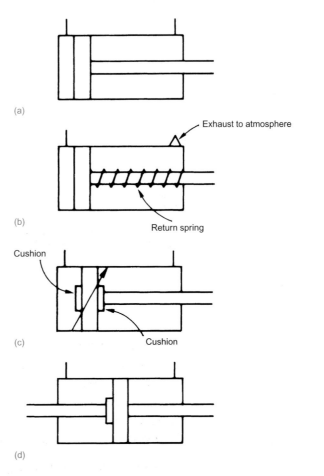

(a)

(b)

(c)

(d)

Figure 2.81 Various types of linear actuator

UNIT 2

one end of the cylinder only. It is double acting because fluid pressure can be applied to either side of the piston.

- *Figure 2.81b* shows a single-ended, single-acting actuator with spring return. Here, the fluid pressure is applied only to one side of the piston. Note, the pneumatic exhaust to atmosphere so that the air behind the piston will not cause a fluid lock.
- *Figure 2.81c* shows a single-ended, single-acting actuator, with double variable cushion damping. The cushion damping prevents the piston impacting on the ends of the cylinder and causing damage.
- *Figure 2.81d* shows a double-ended, double-acting actuator fitted with single, fixed cushion damping.

We are now in a position to use the previous component symbols to produce some simple fluid power circuits.

Figure 2.82 shows a single-ended, double-acting actuator controlled by a 4/3 tandem centre, manually operated DCV. Note that in the neutral position both sides of the actuator piston are blocked off, forming a hydraulic lock. In this position the pump flow is being returned directly to the tank. Note the tank symbol.

This system is being supplied by a single direction fixed displacement hydraulic pump.

Figure 2.83 shows a simple pneumatic hoist capable of raising a load. The circuit uses two 2-port manually operated push-button valves connected to a single-ended, single-acting actuator. Supply pressure is indicated by the circular symbol with a black dot in its centre. Valve 'b' has a threaded exhaust port indicated by the extended arrow. When valve 'a' is operated, compressed air from the air line is admitted to the underside of the piston in the cylinder. This causes the piston to rise and

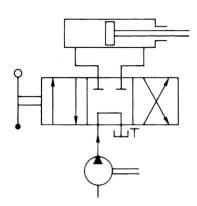

Figure 2.82 Actuator controlled by a DCV

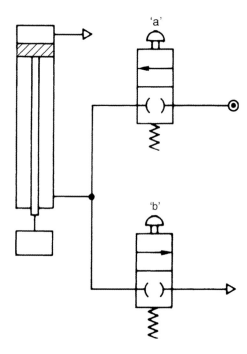

Figure 2.83 A simple pneumatic hoist

to raise the load. Any air above the piston is exhausted to the atmosphere through the threaded exhaust port at the top of the cylinder. Again this is indicated by a long arrow. When valve 'b' is operated, it connects the cylinder to the exhaust and the actuator is vented to the atmosphere. The load is lowered by gravity.

Both these circuits are functional, but they do not have protection against overpressurization, neither do they have any other safety devices fitted. Therefore, we need to increase our vocabulary of components before we can design a safe, practical circuit. We will now consider the function and use of pressure and flow control valves.

Pressure relief and sequence valves

Figure 2.84 shows an example of a pressure relief (safety) valve. In *Figure 2.84a* the valve is being used in a hydraulic circuit. Pressure is controlled, by opening the exhaust port to the reservoir tank against an opposing force such as a spring. In *Figure 2.84b*, the valve is being used in a pneumatic circuit so it exhausts to the atmosphere.

Figure 2.84c and d shows the same valves except that this time the relief pressure is variable, as indicated by the arrow drawn across the spring. If the relief valve setting is used to control the normal system pressure as well as acting as an emergency safety valve, the adjustment mechanism for the valve must be designed so that the maximum safe working pressure for the circuit cannot be exceeded.

Figure 2.84e,f shows the same valves with the addition of pilot control. This time the pressure at the inlet port is not only limited by the spring

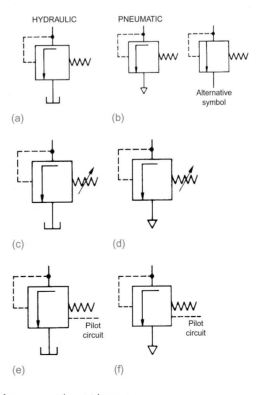

Figure 2.84 Use of a pressure release valve

but also by the pressure of the pilot circuit superimposed on the spring. The spring offers a minimum pressure setting and this can be increased by increasing the pilot circuit pressure up to some predetermined safe maximum. Sometimes the spring is omitted and only pilot pressure is used to control the valve.

Sequence valves are closely related to relief valves in both design and function, and are represented by very similar symbols. They permit the hydraulic fluid to flow into a subcircuit, instead of back to the reservoir, when the main circuit pressure reaches the setting of the sequence valve. You can see that *Figure 2.85* is very similar to a pressure relief valve (PRV) except that, when it opens, the fluid is directed to the next circuit in the sequence instead of being exhausted to the reservoir tank or allowed to escape to the atmosphere.

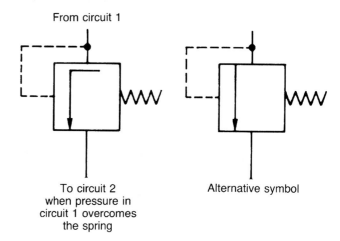

Figure 2.85 Sequence valve

Flow control valves

Flow control valves, as their name implies, are used in systems to control the rate of flow of fluid from one part of the system to another. The simplest valve is merely a fixed restrictor. For operational reasons this type of flow control valve is inefficient, so the restriction is made variable as shown in *Figure 2.86a*. This is a throttling valve. The full symbol is shown in *Figure 2.86b*. In this example the valve setting is being adjusted mechanically. The valve rod ends in a roller follower in contact with a cam plate.

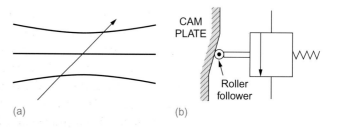

Figure 2.86 Fluid control valves

Sometimes it is necessary to ensure that the variation in inlet pressure to the valve does not affect the flow rate from the valve. Under these circumstances we use a pressure compensated flow control valve (PCFCV). The symbol for this type of valve is shown in *Figure 2.87*. This symbol suggests that the valve is a combination of a variable restrictor and a pilot operated relief valve. The enclosing box is drawn using a long-chain line. This signifies that the components making up the valve are assembled as a single unit.

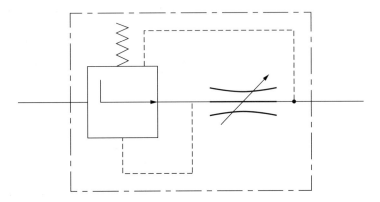

Figure 2.87 Pressure compensated flow control valve

Non-return valves and shuttle valves

The non-return valve (NRV), or check valve as it is sometimes known, is a special type of directional control valve. It allows the fluid to flow in one direction only and it blocks the flow in the reverse direction. These valves may be operated directly or by a pilot circuit. Some examples are shown in *Figure 2.88*.

- *Figure 2.88a* shows a valve that opens (is free) when the inlet pressure is higher than the outlet pressure (back pressure).
- *Figure 2.88b* shows a spring-loaded valve that only opens when the inlet pressure can overcome the combined effects of the outlet pressure and the force exerted by the spring.
- *Figure 2.88c* shows a pilot controlled NRV. It opens only if the inlet pressure is greater than the outlet pressure. However, these pressures can be augmented by the pilot circuit pressure.
  - (i) The pilot pressure is applied to the inlet side of the NRV. We now have the combined pressures of the main (primary) circuit and the pilot circuit acting against the outlet pressure. This enables the valve to open at a lower main circuit pressure than would normally be possible.
  - (ii) The pilot pressure is applied to the outlet side of the NRV. This assists the outlet or back pressure in holding the valve closed. Therefore, it requires a greater main circuit pressure to open the valve. By adjusting the pilot pressure in these two examples we can control the circumstances under which the NRV opens.
- *Figure 2.88d* shows a valve that allows normal full flow in the forward direction, but restricted flow in the reverse direction. The valves previously discussed did not allow any flow in the reverse direction.

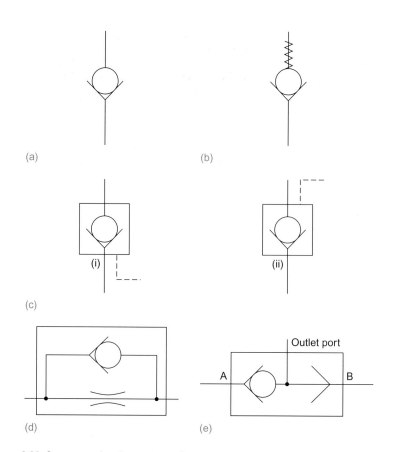

Figure 2.88 Some examples of non-return valves

- *Figure 2.88e* shows a simple shuttle valve. As its name implies, the valve is able to shuttle backwards and forwards. There are two inlet ports and one outlet port. Imagine that inlet port A has the higher pressure. This pressure overcomes the inlet pressure at B and moves the shuttle valve to the right. The valve closes inlet port B and connects inlet port A to the outlet port. If the pressure at inlet port B rises, or that at A falls, the shuttle will move back to the left. This will close inlet port A and connect inlet port B to the outlet. Thus, the inlet port with the higher pressure is automatically connected to the outlet port.

Conditioning equipment

The working fluid, be it oil or air, has to operate in a variety of environments and it can become overheated and/or contaminated. As its name implies, conditioning equipment is used to maintain the fluid in its most efficient operating condition. A selection of conditioning equipment symbols is shown in *Figure 2.89*. Note that all conditioning device symbols are diamond shaped.

Filters and *strainers* have the same symbol. They are normally identified within the system by their position. The filter element (dashed line) is always positioned at 90° to the fluid path.

Water traps are easily distinguished from filters since they have a drain connection and an indication of trapped water. Water traps are

Filters, water traps, lubricators and miscellaneous apparatus

DESCRIPTION	SYMBOL
Filter or strainer	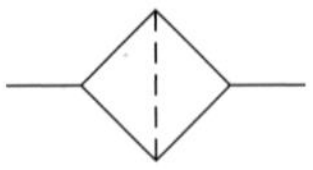
Water trap: with manual control	
automatically drained	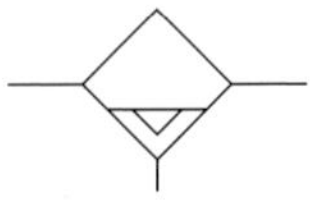
Filter with water trap: with manual control	
automatically drained	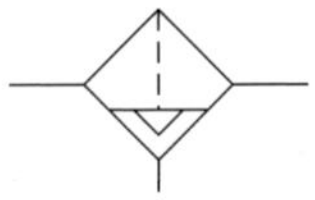
Air dryer	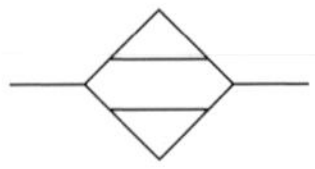
Lubricator	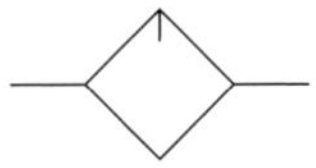
Conditioning unit detailed symbol	
simplified symbol	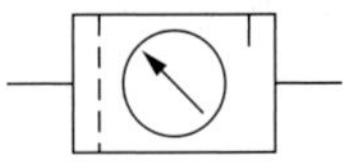

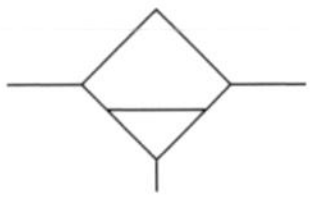

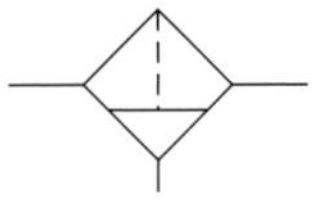

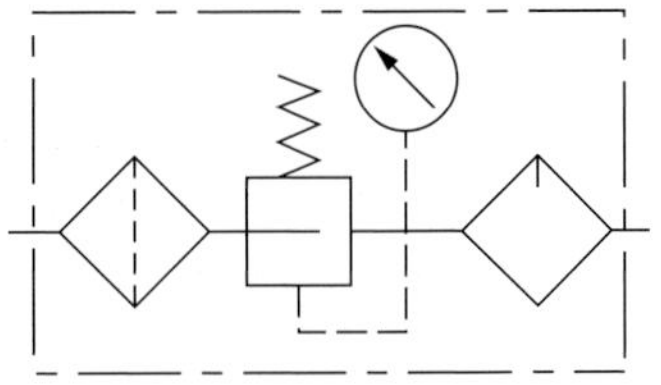

Heat exchangers

DESCRIPTION	SYMBOL
Temperature controller (arrows indicate that heat may be either introduced or dissipated)	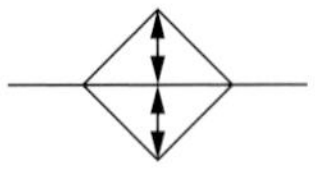
Cooler (arrows indicate the extraction of heat) without representation of the flow lines of the coolant	
with representation of the flow lines of the coolant	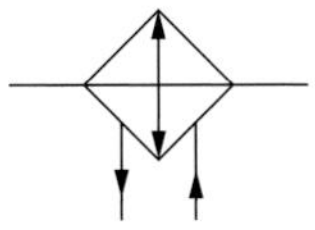
Heater (arrows indicate the introduction of heat)	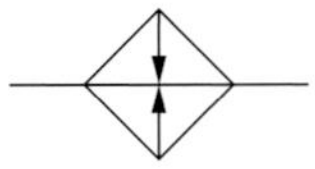

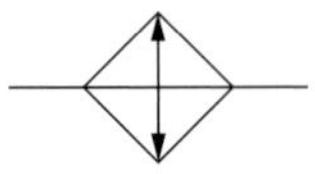

Figure 2.89 Symbols for air conditioning devices

particularly important in pneumatic systems because of the humidity of the air being compressed.

Lubricators are particularly important in pneumatic systems. Hydraulic systems using oil are self-lubricating. Pneumatic systems use air, which has no lubricating properties so oil, in the form of a mist, has to be added to the compressed air line.

Heat exchangers can be either heaters or coolers. If the hydraulic oil becomes too cool it becomes thicker (more viscous) and the system becomes sluggish. If the oil becomes too hot it will become too thin (less viscous) and not function properly. The direction of the arrows in the symbol indicates whether

TYK 2.28

Figure 2.90 shows a selection of fluid circuit symbols. Name the symbols and briefly explain what they do.

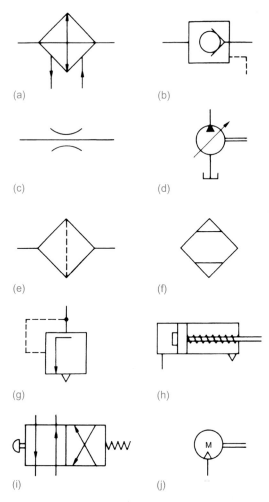

(a)

(b)

(c)

(d)

(e)

(f)

(g)

(h)

(i)

(j)

Figure 2.90 See TYK 2.28

heat energy is taken from the fluid (cooler) or given to the fluid (heater). Notice that the cooler can show the flow lines of the coolant.

There is one final matter to be considered before you can try your hand at designing a circuit, and that is the pipework circuit to connect the various components together. The correct way of representing pipelines is shown in *Figure 2.91*.

- *Figure 2.91a* shows pipelines that are crossing each other but are not connected.
- *Figure 2.91b* shows three pipes connected at a junction. The junction (connection) is indicated by the solid circle (or large dot, if you prefer).

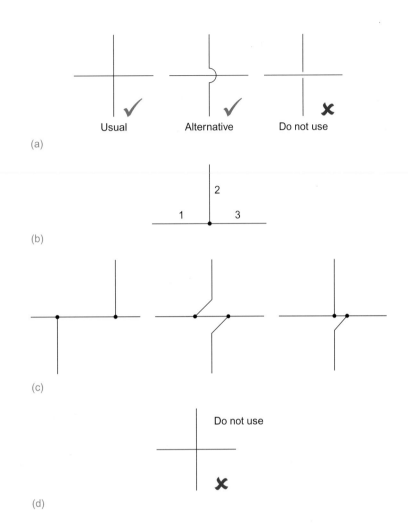

Figure 2.91 Representing pipelines

- *Figure 2.91c* shows four pipes connected at a junction. On no account can the connection be drawn as shown in *Figure 2.91d*. This is because there is always a chance of the ink running where lines cross on a drawing. The resulting 'blob' could then be misinterpreted as a connection symbol with disastrous results.

Activity 2.35

Figure 2.92 shows the general principles for the hydraulic drive to the ram of a shaping machine. The ram is moved backwards and forwards by a double-acting single-ended hydraulic actuator. The drawing was made many years ago and it uses outdated symbols. Use CAD or a technical illustration package to draw a schematic hydraulic diagram for the machine using current symbols and practices (as set out in BS PP 7307).

Present your work in the form of a printed diagram.

Electrical and electronic circuit schematics

Electrical and electronic circuits can also be drawn using schematic symbols to represent the various components. The full range of symbols

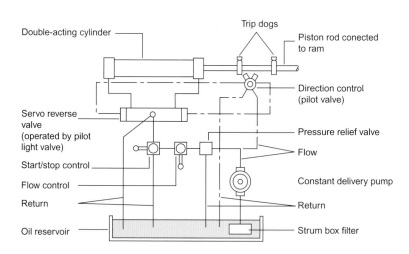

Figure 2.92 See Activity 2.35

and their usage can be found in BS 3939. This is a very extensive standard and well beyond the needs of this book. For our immediate requirements you should refer to PP 7307 Graphical Symbols for Use in Schools and Colleges. *Figure 2.93* shows a selection of symbols that will be used in the following examples:

- A *cell* is a source of direct current (DC) electrical energy. Primary cells have a nominal potential of 1.5 V each. They cannot be recharged and are disposable. Secondary cells are rechargeable. Lead–acid cells have a nominal potential of 2 V and nickel cadmium (NiCd) cells have a nominal potential of 1.2 V. Cells are often connected in series to form a battery.
- *Batteries* consist of a number of cells connected in series to increase the overall potential. A 12 V car battery consists of six lead–acid secondary cells of 2 V each.
- *Fuses* protect the circuit in which they are connected from excess current flow. This can result from a fault in the circuit, from a fault in an appliance connected to the circuit or from too many appliances being connected to the same circuit. The current flowing in the circuit tends to heat up the fuse wire. When the current reaches some predetermined value the fuse wire melts and breaks the circuit so the current can no longer flow. Without a fuse the circuit wiring could overheat and cause a fire.
- *Resistors* are used to control the magnitude of the current flowing in a circuit. The resistance value of the resistor may be fixed or it may be variable. Variable resistors may be preset or they may be adjustable by the user. The electric current does work in flowing through the resistor and this heats up the resistor. The resistor must be chosen so that it can withstand this heating effect and sited so that it has adequate ventilation.
- *Capacitors*, like resistors, may be fixed in value or they may be preset or variable. Capacitors store electrical energy but, unlike secondary cells, they may be charged or discharged almost instantaneously.

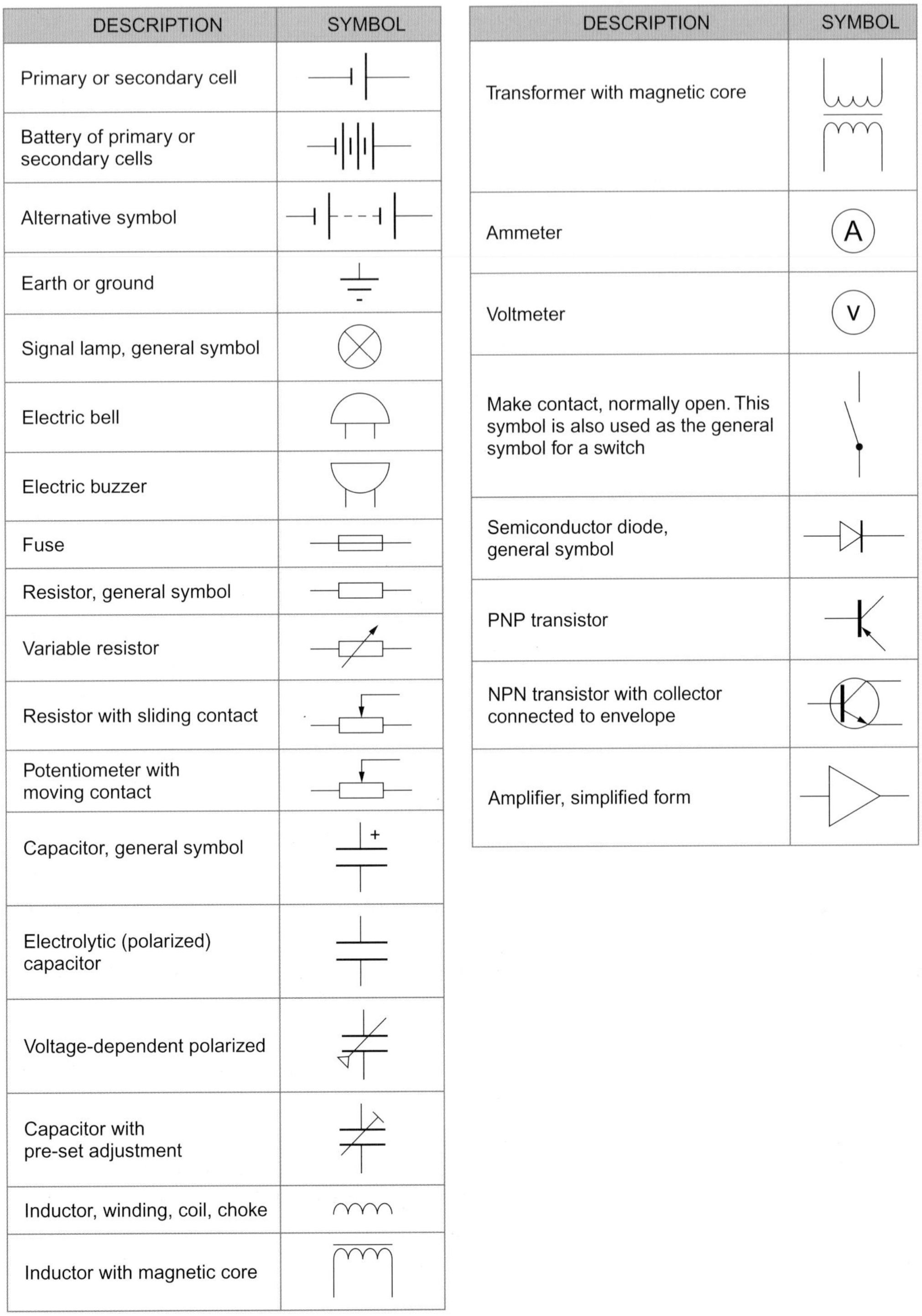

UNIT 2

DESCRIPTION	SYMBOL		DESCRIPTION	SYMBOL
Primary or secondary cell			Transformer with magnetic core	
Battery of primary or secondary cells			Ammeter	
Alternative symbol			Voltmeter	
Earth or ground			Make contact, normally open. This symbol is also used as the general symbol for a switch	
Signal lamp, general symbol				
Electric bell			Semiconductor diode, general symbol	
Electric buzzer			PNP transistor	
Fuse			NPN transistor with collector connected to envelope	
Resistor, general symbol			Amplifier, simplified form	
Variable resistor				
Resistor with sliding contact				
Potentiometer with moving contact				
Capacitor, general symbol				
Electrolytic (polarized) capacitor				
Voltage-dependent polarized				
Capacitor with pre-set adjustment				
Inductor, winding, coil, choke				
Inductor with magnetic core				

Figure 2.93 Electronic symbols

The stored charge is much smaller than the charge stored by a secondary cell. Large value capacitors are used to smooth the residual ripple from the rectifier in a power pack. Medium value capacitors are used for coupling and decoupling the stages of audio frequency amplifiers. Small value capacitors are used for coupling and decoupling radio frequency signals and they are also used in tuned (resonant) circuits.

- *Inductors* act like electrical 'flywheels'. They limit the build-up of current in a circuit and try to keep the circuit running by putting energy back into it when the supply is turned off. They are used as current limiting devices in fluorescent lamp units, as chokes in telecommunications equipment and, together with capacitors, to make up resonant (tuned) circuits in telecommunications equipment.
- *Transformers* are used to raise or lower the voltage of alternating currents. Inductors and transformers cannot be used in direct current circuits. You cannot get something for nothing, so if you increase the voltage you decrease the current accordingly so that (neglecting losses), $V \times I = k$, where k is a constant for the primary and secondary circuits of any given transformer.
- *Ammeters* measure the current flowing in a circuit. They are always wired in series with the circuit so that the current being measured can flow through the meter.
- *Voltmeters* measure the potential difference (voltage) between two points in a circuit. To do this they are always wired in parallel across that part of the circuit where the potential is to be measured.
- *Switches* are used to control the flow of current in a circuit. They can only open or close the circuit. So the current either flows or it does not.
- *Diodes* are like the non-return valves in hydraulic circuits. They allow the current to flow in one direction only as indicated by the arrowhead of the symbol. They are used to rectify AC and convert it into DC.
- *Transistors* are used in high speed switching circuits and to magnify radio and audio frequency signals.
- *Integrated circuits* consist of all the components necessary to produce amplifiers, oscillators, central processor units, computer memories and a host of other devices fabricated onto a single slice of silicon; each chip being housed in a single compact package.

Let us look at some examples of schematic circuit diagrams using these symbols. All electric circuits consist of:

- A source of electrical energy (e.g. a battery or a generator).
- A means of controlling the flow of electric current (e.g. a switch or a variable resistor).
- An appliance to convert the electrical energy into useful work (e.g. a heater, a lamp or a motor).
- Except for low power battery operated circuits, an overcurrent protection device (fuse or circuit breaker).
- Conductors (wires) to connect these various circuit elements together. Note that the rules for drawing conductors that are connected and conductors that are crossing but not connected are the same as for drawing pipework as previously described in *Figure 2.74*.

Figure 2.94 shows a very simple circuit that satisfies the above requirements. In *Figure 2.94a*, the switch is 'closed' therefore the circuit as a whole is also a closed loop. This enables the electrons that make up the electric current to flow from the source of electrical energy through the

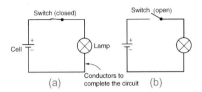

Figure 2.94 A simple electronic circuit

appliance (lamp) and back to the source of energy ready to circulate again – rather like the fluid in our earlier hydraulic circuits. In *Figure 2.94b*, the switch is 'open' and the circuit is no longer a closed loop. The circuit is broken. The electrons can no longer circulate. The circuit ceases to function. We normally draw our circuits with the switches in the 'open' position so that the circuit is not functioning and is 'safe'.

Figure 2.95 shows a simple battery operated circuit for determining the resistance of a fixed value resistor. The resistance value is obtained by substituting the values of current and potential into the formula, $R = V/I$. The current in amperes is read from the ammeter and the potential in volts is read from the voltmeter. Note that the ammeter is wired in series with the resistor so that the current can flow through it. The voltmeter is wired in parallel with the resistor so that the potential can be read across it. This is always the way these instruments are connected.

UNIT 2

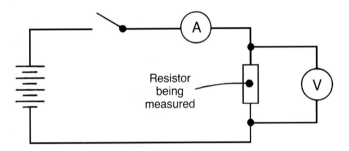

Figure 2.95 Circuit for determining resistance

Figure 2.96 shows a circuit for operating the light over the stairs in a house. The light can be operated either by the switch at the bottom of the stairs or by the switch at the top of the stairs. Can you work out how this is achieved? The switches are of a type called 'two-way, single-pole'. The circuit is connected to the mains supply. It is protected by a fuse in the 'consumer

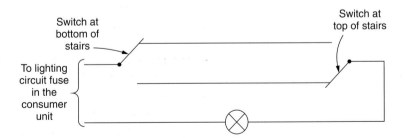

Figure 2.96 Two-way lighting switch

unit'. This unit contains the main switch and all the fuses for the house and is situated adjacent to the supply company's meter and main fuse.

Figure 2.97 shows a two-stage transistorized amplifier. It also show a suitable power supply. Table 2.3 lists and names the components.

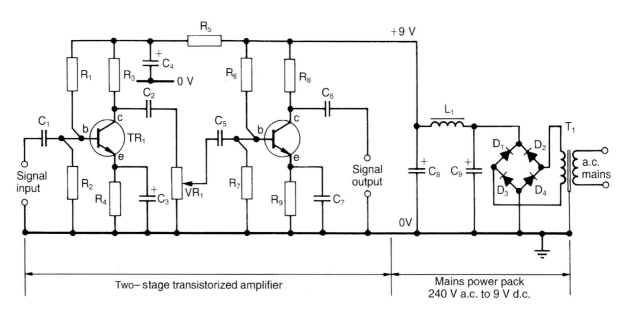

Figure 2.97 A two-stage transistor amplifier

Table 2.3 Components used in a two-stage transistorized amplifier

Component	Description
R_1–R_9	Fixed resistors
VR_1	Variable resistor
C_1–C_9	Capacitors
D_1–D_4	Diodes
TR_1, TR_2	Transistors
T_1	Mains transformer
L_1	Inductor (choke)

Figure 2.98 shows a similar amplifier using a single chip. Such an amplifier would have the same performance, but fewer components are required. Therefore, it is cheaper and quicker to make.

Activity 2.36

Draw a schematic circuit diagram for a battery charger having the following features:

- the primary circuit of the transformer (i.e. the side that is connected to the AC mains supply) is to have an on/off switch, a fuse and an indicator lamp
- the secondary circuit of the transformer is to have a bridge rectifier, a variable resistor to control the charging current, a fuse and an ammeter to indicate the charging current.

Use CAD or a technical illustration package to produce the schematic circuit diagram and present your work in the form of a printed diagram.

TYK 2.29

Figure 2.99 shows a selection of electrical and electronic symbols. Name the symbols and briefly explain what they do.

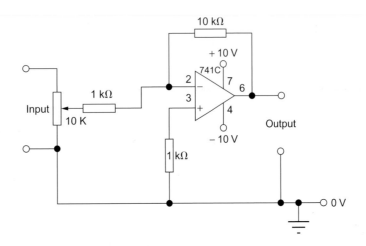

Figure 2.98 A single-chip amplifier circuit

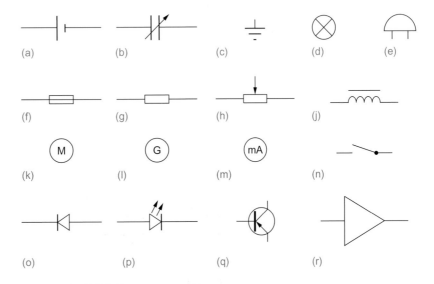

Figure 2.99 See TYK 2.29

Activity 2.37

Figure 2.100 shows an electronic circuit.

(a) Draw up a component list that numbers and names each of the components (include values where given).

(b) Suggest what the circuit might be used for. (Hint: The circuit only has an output!)

Present your work in the form of a single A4 printed page of word-processed text.

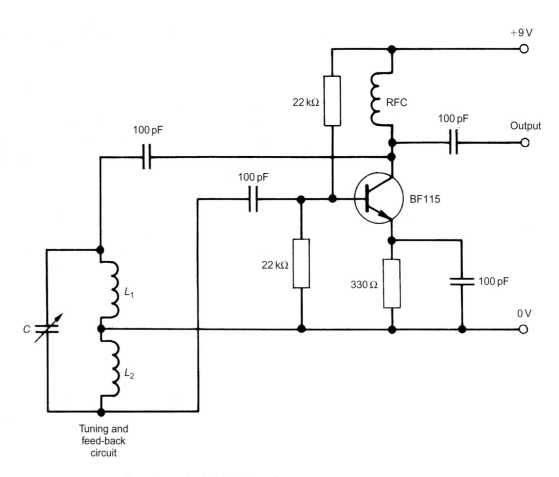

Figure 2.100 See Activity 2.37

Activity 2.38

Your tutor will supply you with a selection of (at least five) engineering components (mechanical, hydraulic, pneumatic, electrical and electronic). Identify each component and make a freehand sketch of it.

Present your work in the form of a portfolio of sketches.

Verbal and Written Communication

Verbal communication (i.e. speaking and listening) is widely used in everyday situations, including:

- Informal discussions either on the telephone or face to face.
- Formal presentations to groups of persons who all require the same information.

Where a group of persons all require the same information, a formal presentation must be used. On no account should information be 'passed down the line' from person to person because errors are bound to creep in. There is a story that during World War I the message 'send reinforcements, we are going to advance' arrived at headquarters, by word of mouth, as 'send three and four pence (old money), we are going to a dance'. We will let you decide on the truth behind this story, but we feel it makes the point.

In any event, it is important to remember that the spoken word is easily forgotten and oral communication should be reinforced by:

- Notes taken at the time.
- Tape recording the conversation.
- A written summary. For example, the published 'proceedings' of formal lectures and presentations. Another example is a 'press release' that is provided to journalists and reporters in order to ensure the factual accuracy of information intended for the public.

Oral communication must be presented in a manner appropriate to the audience. It must be brief and to the point. The key facts must be emphasized so that they can be easily remembered. The presentation must be interesting so that the attention of the audience does not wander.

When communicating by the spoken word, it is as equally important to be a good listener as it is to be a good speaker. This applies to conversations between two or three people as well as to formal presentations.

> **KEY POINT**
>
> When speaking to people or giving a verbal presentation it is important to check that everybody has understood what you have been saying.
>
> There are various ways to do this including, in a small group, asking each person if they have any questions and if necessary, testing their understanding by asking *them* questions. Your tutor should be doing this all the time!

Activity 2.39

Use presentation software to prepare a 5-minute presentation to the rest of the class (using appropriate visual aids) on any one of the following topics:

- How to choose a digital camera.
- How to connect to the Internet.
- What to look for when purchasing a second-hand car.

You should prepare a set of brief printed notes summarizing the key points for your audience. Also include printed copies of any screens or overhead projector transparencies that you use. At the end of your talk you should invite questions from your audience and provide appropriate answers.

Activity 2.40

Conduct a brief interview (lasting no longer than 15 minutes) with another student and take notes to summarize the outcome. Do not forget to allow time for questions at the end of the interview. Your interview should be based around the following questions:

- Why did you decide to take a course in Engineering?
- Why did you choose the BTEC Level 3 course?
- What made you choose this school/college?
- What subjects/topics have you enjoyed the most?
- What subjects/topics have you enjoyed the least?
- What plans have you got for the future?
- Where would you hope to be and what would you hope to be doing in 10 years' time?

You should add further questions to clarify the above. Do not forget to thank your interviewee! Present your findings in the form of hand-written interview notes.

Written communication

This is a more reliable method of communication since it usually provides a permanent written record of the key information. The same information is available for all those who require it.

Anyone who has ever marked an English comprehension test will know that the same written passage can mean very different things to different people. Therefore, care must be taken in preparing written information. To avoid confusion, the normal conventions of grammar and punctuation must be used. Words must be correctly spelt. Use a dictionary if you are uncertain. If you are using a word-processing package use the spell checker. However, take care, many software packages originate in the USA and the spell checker may reflect this.

Never use jargon terms and acronyms unless you are sure that those reading the message are as equally familiar with them as is the writer.

An engineer often has to write notes, memoranda and reports. He/she often has to maintain logbooks and complete service sheets. An engineer may also have to communicate with other engineers, suppliers and customers by letter. Being able to express yourself clearly and concisely is of great importance.

Activity 2.41

Prepare a brief article for the local press (using not more than 1000 words) on any one of the following topics:

- A sporting event that you took part in.
- A recent school or college activity.
- A newly available product or technology.

Include contact or other details for further information. Present your work in word-processed form and include relevant photographs, diagrams or sketches.

Activity 2.42

Prepare:

(a) a word-processed letter
(b) an e-mail message

to an engineering supplier requesting details of a product or service. This may simply take the form of a request for a short-form catalogue or for the supply of a data sheet or application note. Present your work in the form of printed copies of correspondence and e-mail messages.

End of Unit Review

1. Describe the main features of:

(a) a detail drawing

(b) a general arrangement drawing

(c) an assembly drawing.

2. Explain how a search engine is used to locate information on the World Wide Web.

3. What information is typically held in each of the following types of database?

(a) a product database

(b) a manufacturing database

(c) a customer database

(d) a spare parts database.

4. List the main headings used in a technical report.

5. Draw a graph showing how the following current varies with time:

Time, t (s)	0	0.1	0.2	0.3	0.4	0.5	0.6	0.7
Current, i (V)	0	0.22	0.45	0.65	0.89	1.09	1.32	1.54

What does the shape of the graph suggest? Use the graph to determine the value of i when $\tau = 0.25$ s.

6. The address of a website is:
`http://www.daviesengineering.co.uk`.
Explain how this address is constructed.

7. The following data relates to a 12-month operating period for an engineering company:

Total income:	£19.2 million
Material costs:	£6.8 million
Overhead costs:	£2.2 million
Labour costs:	£5.9 million

Use a pie chart to illustrate the income, expenditure and profit made by the company.

8. Explain, with the aid of an example, how a scatter diagram is used.

9. The following data refers to the number of TV sets produced by a particular manufacturer:

1997	2500
1998	4000
1999	5500
2000	7250
2001	4500

Illustrate this information using an ideograph.

10. Sketch typical line styles used to illustrate:

(a) a centre line

(b) the limit of a partial or interrupted view.

11. How many A2 drawing sheets can be cut from an A0 drawing sheet? Explain your answer with a sketch.

12. State FOUR items that should be included within the title block of an engineering drawing.

13. Sketch engineering drawing symbols that are used to indicate the following components:

(a) a 4/2 directional control valve

(b) a non-return valve

(c) a battery

(d) a variable resistor

(e) a semiconductor diode

(f) an iron-cored transformer.

14. Draw, using appropriate symbols, a two-way lighting circuit. Label your drawing clearly.

15. List four advantages of using CAD in the preparation of engineering drawings compared with purely manual methods.

16. Identify the projection used in *Figures 2.101 and 2.102*.

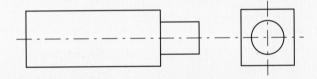

Figure 2.101 See End of unit reviews question 16

Figure 2.102 See End of unit review question 16

17. *Figure 2.103* shows a pie chart. If the labour costs amount to £2.5 million, what profit is made by the company?

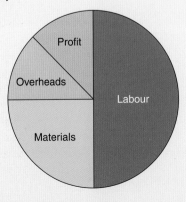

Figure 2.103 See End of unit review question 17

18. Convert the pie chart in End of unit review question 2.17 into a bar chart.

19. Identify the lines marked A and B in *Figure 2.104*.

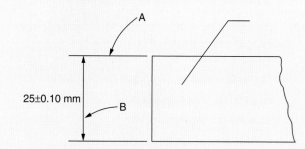

Figure 2.104 See End of unit review question 19

20. Identify each of the components shown in *Figure 2.105*.

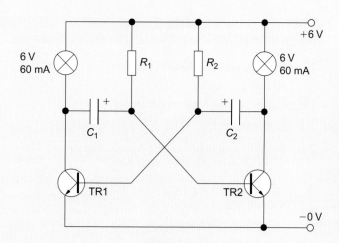

Figure 2.105 See End of unit review question 20

6

The Millau Viaduct in France has the tallest road deck in the world at 270 metres, with one of the pier's summits reaching 343 metres. This engineering project involved a very large multi-disciplinary team of architects, designers, construction engineers and technicians to see the project through from concept to successful completion. The project spanned many years, with the bridge being finally opened to traffic in December 2004.

Engineering Project

Throughout the world, almost everything that is man made has been engineered in some way, everything from large civil engineering projects, such as bridges, hydroelectric power stations and highly sophisticated transport systems, to the humble office paper clip. It is this amazing diversity that makes engineering such an interesting and rewarding profession to follow. Each and everyone of these engineering achievements requires engineers and technicians to project manage them. In fact engineers and technicians spend the majority of their time working on, or managing projects, no matter how large or how small. Large complex projects will involve a multidisciplinary engineering team to see it through from start to finish, while smaller projects may be managed or worked on by individual engineers or technicians.

Engineers and technicians need to acquire the skills, knowledge and techniques to be able to see a project through from the ideas stage to the commissioning and maintenance stage. The successful completion of this unit should greatly assist you in acquiring and applying the knowledge and skills you have learnt, to a realistic engineering project, that is within your own chosen engineering discipline.

You will be encouraged to select an appropriate engineering project for yourself or you may be given a project by your tutor, dependent on your circumstances. The end result of your project, no matter whether chosen by you or selected for you, is likely to be an engineered product, device or service, or the production of a modification to an existing process or product, or something similar. Whatever your choice of project, you will have to produce an engineered solution, in a specified time, to an appropriate standard. In order for you to do this, you will need not only to draw upon and apply the knowledge and skills you have learnt from other units within the programme, but also undertake a substantial amount of your own research in order to achieve your project aims.

The unit outcomes have been designed to provide you with the process skills necessary for you to produce a project specification, plan a course of action, implement your plan, produce a worthwhile project solution and present your findings verbally and in the form of a written report. All these process skills will prove to be of great benefit to you when you start to practice as an engineer or technician.

BTEC National Engineering. DOI: 10.1016/B978-0-12-382202-4.00003-9

Introduction

Before getting involved with the detail in the outcomes, there is a need for you to have an overview of the whole of the project process so that you are clear about what is required of you and the direction you must take in order to achieve a successful project solution. You will be given a lot of preliminary information by your project tutor, prior to the start of your project, and you will need to sift through it and separate out what is vital from what may be useful or what has little relevance in your case. Here are three key questions that you may well need answer in order to help you decide what is important and how to go about successfully completing this unit.

1. What do I need to do and produce?
2. When do I need to do it?
3. How will the project be assessed?

What you need to do and produce?

Let us try and answer these questions, starting with what you need to do and produce.

You first need to set up a system to **administer your project**. This will require you to prepare and keep a record of all events related to your project, right from the time you have your initial ideas until you present your final solution. To enable you to do this, you will need to **prepare and keep a logbook** to record all events on a day-to-day basis throughout the life of your project. The sort of events you will need to record in your logbook and the use and contents of the logbook will be explained in detail in outcome 1 of this unit. You will need to submit your logbook on a regular basis for scrutiny by your project tutor and hand it in for final assessment when asked for.

At the same time as setting up your logbook, you also need to **choose a suitable project**. There are a number of factors that you should think about when making your choice. You will first need to consider the type of project that you would like to undertake, and whether it is to be based on design, production, maintenance or modification of an engineering product or service. In practice, projects concerned with engineering services such as production, systems of works, schedules, cost benefit analysis or new procedures and processes are likely to be more challenging because of difficulties with research and the production of meaningful data and evidence to substantiate your decisions.

When making your choice you should realize that the subject matter of your project is the vehicle through which you are to be assessed and for this reason you would be well advised to choose a subject that will allow you to maximize your potential for gaining marks against the full range of requirements stated in the assessment criteria. For example, it is worth remembering that the best projects are those that arise out of a real requirement, possibly one identified by a customer, or through an industrial or commercial contact. Your project should not be over ambitious or over demanding in terms of the knowledge, skills, materials, resources or time available for the unit.

Finally, you should choose a project that will extend your understanding of a subject that already interests you and that you wish to further develop. For example, if you are interested in motor vehicle technology you might like to consider a project that will allow you to explore a particular aspect of the design of performance cars. This is one of the reasons why it is better to choose your own project subject, rather than having one chosen for you in which you have no particular interest!

You need to *produce a project specification*. This is essentially a document that details the feasibility, aims, objectives and deliverables of your project and indicates how you intend to achieve them. The project specification provides an overview or plan of the technical content of your report, and includes the tasks you will need to undertake to meet your project aims and objectives, and details the nature of what you intend to produce and present as your project solution. The project specification normally forms part of *a preliminary report* that you may need to submit (depending on your centre's policy and procedures) which apart from the specification itself, will also include detail on the background and need for your project together with the methods you have adopted for evaluating your final solution.

Apart from the day-to-day administration and progress of your project using your logbook, you will also have to *produce a long-term plan and a method for monitoring your progress* in accordance with your plan. Examples of appropriate planning techniques and methods for monitoring progress and achievement are detailed in outcome 2 of this unit.

You will for the majority of your time be involved in *implementing your project plan* in order to *produce your solution (your deliverable)*. This may involve you initially in carrying out research, manipulating data, performing calculations, producing a design, building and testing a product or other activities directly associated with your project solution (deliverable). You may also be involved with analysing and discussing your results or checking and proving the worth and financial viability of your design, product, process or procedure, depending on the type of project you will have chosen. Examples of the kind of work you will need to undertake in order to implement your plan and produce your deliverable are given in outcome 3 of this unit.

You also need to *prepare and deliver a verbal presentation and present your logbook and your final written report for assessment*. You may also be able to provide other deliverables, such as technical drawings, photographic evidence, working models and engineered products, that all contribute to your project solution. Examples of presentation methods, software and equipment are provided in outcome 4 of this unit together with typical report writing and presentation methods.

So, hopefully you now have some idea of what you need to do and produce! Do not worry if it looks a little daunting. By following the above process and producing what is required, you will be able to successfully complete the project unit. The order in which you should complete the above tasks is now detailed.

When you need to do it?

The timings given here are based on the assumption that you are enrolled on the new National Level 3 Diploma or Extended Diploma that is being delivered conventionally: over 2 years part time or 2 years full time, respectively. In your first year you will have covered Unit 1 Health and Safety in the Engineering Workplace and Unit 2 Communications. If you are enrolled on a certificate or subsidiary diploma programme you are likely to start your project in the second term and the suggested timings given here will need to be adjusted accordingly. Based on this assumption, you need to think about your project towards the end of your first year of studies so that, acting on advice from your tutor, you have the summer to consider an appropriate project subject. You will also be given the project brief at this stage, so you should record all your actions, by starting your logbook, at this early stage. If you are enrolled on a two term course, you should be given all the necessary information and be in a similar position to start your project at the beginning of the second term.

Figure 3.1 shows a flow diagram indicating the order in which you will need to complete the tasks indicated. Note that there is a lot of organizational work required towards the beginning of the project; this is why it is advisable to make an early start.

As a guide, a rough *timeline* has been including down the left side of *Figure 3.1* that indicates when these major tasks should be completed, so that your project solution is completed and delivered on time.

How will the project be assessed?

In answering this final question, all we need do is turn to the grading criteria given in your project brief. Thus, *in order to pass this unit* you will need to produce information (evidence) when completing the above tasks that shows you able to:

- Prepare and maintain project records (your logbook and possible use of software)
- Prepare and produce a project specification (production of your specification/preliminary report)
- Evaluate and outline potential project solutions (carried out during the production of your project specification/preliminary report)
- Monitor and record achievement during the life of your project (primarily logbook, also possible additional software)
- Implement the plan and produce and check the project solution (evidence from logbook, research, analysis and the final report and other deliverables)
- Prepare and deliver a presentation (evidence from marking guide prepared by your tutor)
- Present a written report (quality dependent on your report writing knowledge and skills)

In order to achieve a *merit grade*, your record keeping must be detailed and show how changes to your plan have been identified and dealt with. In addition, you should use a wide range of techniques to justify your chosen

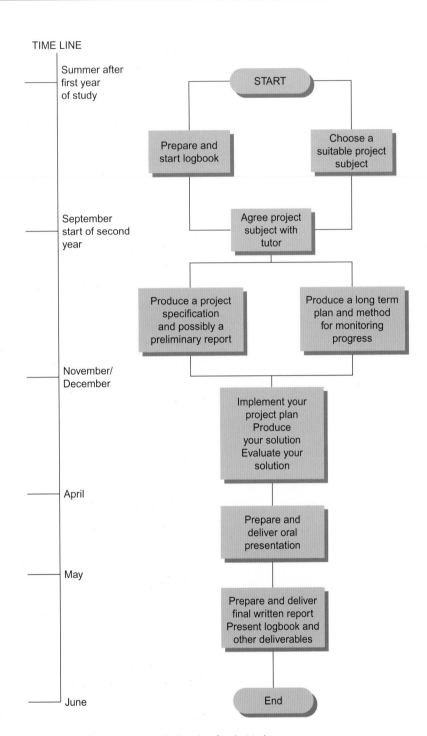

TIME LINE

Summer after
first year
of study

September
start of second
year

November/
December

April

May

June

START

Prepare and
start logbook

Choose a
suitable project
subject

Agree project
subject with
tutor

Produce a project
specification
and possibly a
preliminary report

Produce a long term
plan and method
for monitoring
progress

Implement your
project plan
Produce
your solution
Evaluate your
solution

Prepare and
deliver oral
presentation

Prepare and deliver
final written report
Present logbook and
other deliverables

End

UNIT 3

Figure 3.1 Flow diagram indicating likely order of project tasks

project option and present well-structured development records and final report. To achieve a ***distinction grade*** you should in addition to the above, have managed your project independently only seeking guidance when necessary and produced evidence that shows you have critically evaluated your project solution and proposed methods for improvement.

Having presented you with an overview of what you need to do, the time in which you have to do it and how you are to be assessed, we now turn our attention to a more detailed study of the project process and project solution by looking at the module outcomes.

UNIT 3

Activity 3.1

Produce a flow chart that shows the tasks you need to undertake and the order in which they need to be carried out, so that you meet all of the criteria for a *merit grade* for your project.

Specifying the Project

As mentioned in the introduction, in this outcome we will look in more detail at the nature and way in which we produce a specification or more globally produce a preliminary report that encompasses a specification. We will also look again at the way in which we choose a project and in particular look at the concepts and techniques that not only influence the type of project we choose but also offer us a way of deciding on a solution for our selected project. We start by looking at some of the techniques we can use to prepare and maintain records throughout the life of the project. In particular we look at the use and form of the logbook, leaving plans and planning until we study outcome 2.

Preparing and maintaining project records

In order to successfully complete your project you are required to keep and maintain comprehensive records of your progress and planning. These records may be in both written and electronic form. So, for example, you may have written records in the form of notes, sketches and drawings, with written time and resource plans and a manual method for modifying these plans when things change during the time of your project. In addition, you may want to use electronically available planning tools for setting and monitoring your progress. Here, we will only be looking at manual ways of record keeping, where this can be done with the use of a *logbook*. In fact, as you already know, a mandatory part of the assessment process for this module requires you to keep and maintain a logbook. Long-term planning aids that form part of the record keeping and monitoring process will be covered in outcome 2, where both manual and electronic forms of these planning tools will be covered.

The Logbook

KEY POINT

You must keep a logbook and record all events and actions associated with your project

Your logbook, if properly organized and maintained, will become your primary source of information and should prove to be an invaluable source of reference when it comes to you preparing your verbal presentation and writing up your final report. It should be a living history of events and actions that take place in chronological order over the whole life of your project. For this reason your logbook needs to be regularly maintained, preferably on a daily basis, or at any time when ideas come into your head or when useful information is at hand from which you can make notes. All entries need to be dated and referenced at the time you make them; doing this immediately saves you the time and effort of trying to remember where you got the information from at some later date! Your logbook will also be the place where you will keep hard copies of your planning schedules (see outcome 2) that will show evidence of change as the project progresses.

Your logbook should also contain a communications log that is generally started in the back of the book by turning the book upside down. Your communications log will reference and list all verbal and written correspondence (*Figure 3.2*), such as e-mails, letters, website visits and important interviews or conversations with appropriate professionals.

Log Book

E-mail 221	06 May 06 1938	I have consolidated all the projects into one big group project and started up a daisy chain editing process.
E-mail 222	06 May 06 1939	Sent to Mat to help give him ideas of how to mail Neil
E-mail 223	06 May 06 2123	From Mat ref e-mail 221, agreeing with the editing process.
Tel 182	06 May 06 2200	Spoke to Mark about the editing process for the project
Tel 183	07 May 06 2114	Spoke to Rodger about how to distribute the project for consolidation and arranging a date for a final group meeting.
E-mail 224	08 May 06 1517	From Mark with his first edited version of the group project. He was forwarding it onto Neil
Tel 184	12 May 06 1045	Neil phoned to discuss the editing process
E-mail 225	12 May 06 2200	From Mark telling Mat that Neil had edited his parts and passed them onto him, as Rodger's inbox was full and not accepting any mail.
E-mail 226	14 May 06 1912	From Frank saying that he thinks the project looks good. He suggests that we should look at combining the conclusion with an epilogue to help finish the project better
E-mail 227	16 May 06 1114	From Mat with his edited version of the group project.
Tel 185	17 May 06 1400	Frank phoned up to discuss the editing process
E-mail 228	18 May 06 1308	From Rodger with his edited part of the project.
E-mail 229	18 May 06 1730	Sent to everyone to see if everyone can make a meeting on 30th May to finish off the project together.
E-mail 230	18 May 06 1947	From Mat ref e-mail 229, saying that he can make the meeting on the 30th. He also asked Rodger to pass his project on to Frank
Tel 186	19 May 06 1512	Neil phoned to ask for the assessment form GP2
E-mail 231	19 May 06 1641	Ref tel 186, sent marking assessment form to Neil
Tel 187	19 May 06 1830	Neil phoned to discuss the peer assessment form and how to allocate the marks.
E-mail 232	21 May 06 1211	From Rodger saying that he is having trouble sending the project because it is such a large file. He suggests we wait

Figure 3.2 Example of e-mail communications log for a group project

The actual detailed information gained as a result of these communications should not appear in your logbook: only a brief statement of the purpose and outcome of such communication as shown in *Figure 3.2*. All primary source material, such as photocopied articles, website downloads, magazine articles, large amounts of technical data, etc., you obtain may be kept in a separate A4 binder for your future reference and viewing by the project tutor if desired. In addition, if you take notes or short passages from articles that you intend to use or quote in your final report, then these should be listed separately in your logbook, under the heading ***references*** (see outcome 4, report writing).

To assist you in finding your way around your logbook and for ease of reference, you should consider sectioning your logbook in an appropriate manner. These sections may include information on your initial project ideas and selection (memory maps, brainstorming diagrams, etc., see *Figure 3.3*), the production and form of your specification (including your project aims, objectives and proposed solution/deliverables), your project planning documentation and monitoring methods, and your project implementation (that may have subsections such as research, calculations, findings, data collection and analysis, recommendations/conclusions). Drawings, sketches and illustrations would be scattered throughout your log under these various sections.

Your logbook is best presented as an A4 bound notebook with sufficient pages to last throughout the whole lifetime of your project. Important

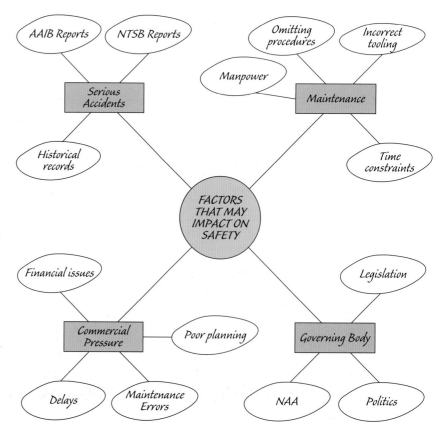

Figure 3.3 A mind map of factors that may affect aircraft safety

information that cannot be transcribed, such as complex drawings, detailed tables and client briefs may be pasted into your logbook, but you should try and keep this process down to a minimum: you do not want your logbook turning into a scrapbook! *Figure 3.4* illustrates a typical complex diagram for an electronic circuit that you would not be expected to transcribe!

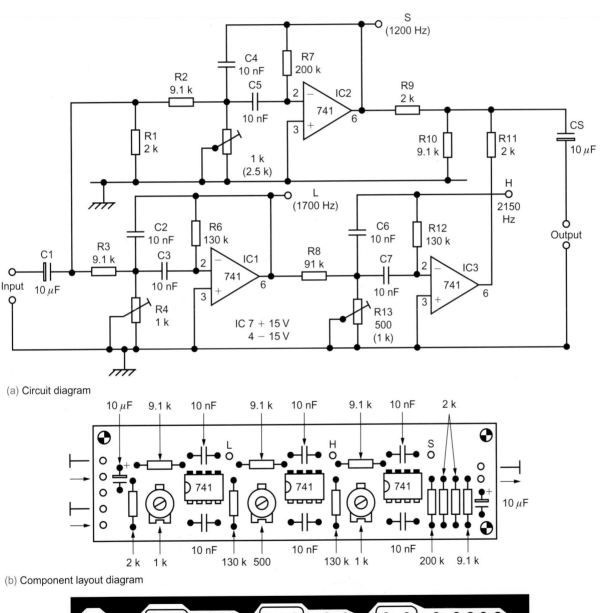

(a) Circuit diagram

(b) Component layout diagram

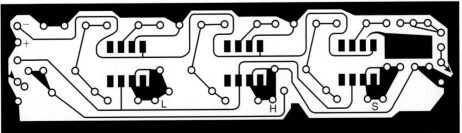

(c) PCB copper track layout

Figure 3.4 A typical electronic circuit diagram with corresponding component layout and PCB track diagrams

You should not use an A4 diary as your logbook because there is no flexibility for including abnormally large entries on any one particular day and diaries contain irrelevant material, both front and back, that leave no room for a communications log or allow for any effective way of sectioning them when required. You can, however, use a diary in tandem with your logbook to book meetings and interviews and as a way of reminding yourself of important milestones, like submission dates, presentation dates, etc.

TYK

TYK *3.1*

Write a list of the different types of activity and entries your tutor may expect to see in your logbook

KEY POINT

Your chosen project must have an outcome, deliverable or solution

Project specification

We now consider the design and production of the specification and see how, to some extent, the ease of producing your specification can depend on your appropriate choice of project. Project specifications will vary dependent on the type of project being undertaken: for example whether it is concerned with design, production, processes, maintenance, etc. No matter what the variations in the specifications, you should ensure that it contains all the essential elements needed to fully meet the grading criteria for this unit! These will all be fully explained in this section. Lets start by revisiting the subject of choosing a project.

Choosing an appropriate project

You have already been given some advice on choosing your project in the introduction. Here we define in more detail the types of project that are open to you, including the advantages and disadvantages of considering a group project which up till now has not been mentioned.

You have been told that the engineered solution to your project may lead to a form of product or service. In the context of this unit (and its assessment) an ***engineered product or service*** can be taken as meaning:

- a physical product, for example an item of mechanical, electrical, electronic or fluidic equipment;
- a service product, for example the delivery of an engineering service such as the routine testing, inspection and maintenance of portable electrical equipment;
- a 'system' product, for example remote monitoring of the performance of the engine of a rally car using appropriate hardware, software and telecommunications equipment.

KEY POINT

You should carefully consider the advantages and disadvantages of a group project before making it your choice

What you might choose in terms of the product, service or system, will very much depend on your own engineering specialization, but what we can do with each of the areas above in engineering terms is common to all engineering fields. So, for physical, service and system products we can: *specify*, *design*, *build*, *test*, *modify* and *evaluate* all of them and no matter what your choice of project you will need to at least 'specify' and 'evaluate' your solution.

Here is a short list of possible project subjects that fit the criteria for selection given above:

1. The specification, design and evaluation of light emitting diodes (LEDs) to replace conventional aircraft navigation lights.
2. The test and evaluation of steel wire reinforced composites for use in motor vehicle bodies.
3. The specification, design and evaluation of a low cost, low weight, ultra portable electric drill.
4. The modification of a car handbrake system to give better driver access.
5. The specification, production and evaluation of a modified maintenance procedure for railway train electrical generators.
6. The novel design of a solid-state thermionic device for converting heat directly to electricity.
7. A novel design for testing the strength of toothpaste tubes.
8. A novel process for the production of metal pressings for mobile phone bodies.
9. The evaluation of a novel design for binary metal, fluid transfer refrigerants.
10. The test and evaluation of heat sensitive colour changing cable insulation for ease of electrical fault finding.
11. The production and evaluation of a modified maintenance schedule for the Airbus A 300 series cabin air-conditioning system.
12. A modified method to improve availability of spare parts for the garage maintenance of motor vehicles.

Although not directly apparent in the above project titles, they all involve the production of a specification, although you will note that this looks easier for some titles than others. For example, producing a specification and evaluating the solution for the last two (11 and 12) projects does not look quite as easy as the others. Thus, projects involving the redesign or modification of procedures, particularly to engineering services, will be a little more difficult to evaluate! This is not to say that you should not choose these kinds of project, just that you need to be aware of the difficulties that may arise if you do.

Your circumstances may be such that a ***group project*** would suit you best. There are sound advantages associated with working in a group including the ability to tackle a much more challenging project that would normally be beyond the reach of an individual, together with being able to develop your team work skills, something that industry puts a high premium on when you are seeking work. Group projects also offer companionship and moral support and encouragement, and more opportunity to generate the free flow of ideas through such methods as brainstorming and group discussion, putting into practice the old adage that 'many brains are better than one'! There is also the down side to group projects: the first hurdle being the need to hold and to attend scheduled meetings; then there is the possibility of problems with the dynamics of the group, for example every individual contributing equally and the need for the group to prove that every individual has made the contributions necessary to meet the grading criteria. There are also likely to be differences in the quality of the contributions made by individual members of the group and problems associated with individuals in the group who continually fail to meet time goals. When deciding on whether or not to embark on a group or individual project, you need to take the above points into account and make your decision accordingly; your project tutor will also offer advice on the 'pros and cons' of group work.

When choosing a possible project remember that the best projects will be those offered by industry where you are trying to find a solution to a justified need and that any project you may think of yourself should be sufficiently self-contained to ensure completion in the specified time. Do not attempt too much. Designing and building even the simplest of products takes a great deal of time to achieve a meaningful solution!

Activity 3.2

Based on the selection criteria given above produce a list of five possible alternative subjects for an individual project in your chosen specialization and two alternatives for a specialized group project.

Production of a project specification

We now need to look at what must be covered when preparing and producing your project specification and whether other factors need to be included as part of a preliminary report – the nature of these additional factors being dependent on the type of specification you need to produce. We will base our 'ideal project specification' on a typical model for an **engineering design** that is to be generated from a customers design brief and then look at what needs to be added to meet the grading criteria when other project subjects have been chosen. To start, you must carry out a **needs analysis**, where you will be involved in a fair amount of initial research to canvas opinion and discover whether or not there is a technical or commercial market for your product or service. Even if your project is not involved with design or production, you will still have to provide evidence of the **need** for your chosen project. You must resist the temptation to rush into the first bright idea you have! Instead, you should ensure that you devote sufficient time to clarifying the problem and collecting together information from a variety of sources to test the viability of your ideas and establish which one of them would make a good subject for a project. You should discuss your ideas with your project tutor before committing too much time to initial research. The result of your research should enable you to produce a ***design brief*** that details the requirements upon which your design specification will be based. If you are lucky enough to undertake a design or production project for industry then you are likely to be given the design brief as your starting point. No matter how the design brief has been obtained, from it you will produce your ***design specification***.

A typical project design brief and specification

Your project design specification will therefore contain key points identified from the *client brief*, if your project is industrially based, or from your own brief, if you are pursuing your own design or producing your own product. Your specification should be as comprehensive as possible and should contain measurable parameters against which it can be evaluated. You should develop your ***project design specification*** from the brief in consultation with industry or with your project tutor. It should typically contain information on:

- the function of the product
- user requirements
- performance requirements
- material and component requirements
- quality, safety and environmental issues
- required conformance standards
- scale of production and cost.

The example that follows illustrates how, as a result of market research, a need was established for a product. A list of basic requirements was established (the client brief) and from this brief a technical design specification was produced.

Market research indicated that a market existed for a small adjustable DC power supply that could be used in a school or college science laboratory. The research was based on discussions with a large number of potential users, where opinion was gathered on the type of construction, display features and performance required for the power supply. These user requirements are detailed below in the client brief.

Client Brief

The power supply should:

1. be mechanically and electrically rugged;
2. use proven and reliable technology;
3. be low cost and easy to maintain;
4. comply with appropriate European legislation (e.g. 'the electromagnetic compatibility' (EMC) and low-voltage directives);
5. have one variable (3–15 V) output and one fixed (+5 V) accessory output and that both outputs should be protected against a short circuit;

UNIT 3

6. have (with the exception of the accessory output and the AC mains input) all controls, connectors and switches made available on the front panel;

7. have colour coded output terminals that will accept standard 4 mm plugs and the terminals should also allow wires that have not been fitted with plugs to be clamped directly using a screw action;

8. operate from a standard 220 V AC mains supply that should be connected using a standard 3-pin connector;

9. be 'tamper-proof' (it should not be possible to remove the knobs or the enclosure without having to resort to the use of special tools);

10. be lightweight and portable;

11. have LED indicators to show that the power supply is switched on and that the outputs are present.

Having determined a need for the product and obtained a detailed list of user requirements, the next stage was that of firming up on the client brief and producing a detailed specification for the power supply. The design specification, in this case, simply provided further detail on the numerical performance parameters of the power supply outputs and inputs. These have been incorporated into the client brief to provide the technical specification (see below) upon which the design can eventually be carried forward to the prototype build and test stage.

Technical specification

The laboratory DC power supply will:

1. be mechanically and electrically rugged;

2. use proven and reliable technology;

3. be low cost and easy to maintain;

4. comply with appropriate European legislation (e.g. EMC and low-voltage directives);

5. *Variable output*: voltage adjustable from 3 to 15 V, current adjustable from 50 mA to 1 A max, 4 mm binding post connectors (red and black);

6. *fixed output*: voltage fixed at +5 V ±5%, current 1 A max, 4 mm binding post connectors (yellow and black);

7. have (with the exception of the accessory output and the AC mains input) all controls, connectors and switches made available on the front panel;

8. have colour coded output terminals that will accept standard 4 mm plugs and the terminals should also allow wires that have not been fitted with plugs to be clamped directly using a screw action;

9. *Input* 200–240 V, AC via an integrated electrical connector, 1 A fuse and EMC filter;

10. be 'tamper-proof' (it should not be possible to remove the knobs or the enclosure without having to resort to the use of special tools);

11. be lightweight and portable;

12. have LED indicators to show that the power supply is switched on and that the outputs are present.

TYK 3.2

Look carefully at the technical specification produced for the power supply example and identify and write down any omissions or points that you think need further clarification in order to move to the build and test stages.

We will return to the power supply example later because there is still the need to examine a range of solutions for its design. I would first like you to attempt Activity 3.3 either on your own or with help from your project tutor.

Activity 3.3

A potential customer approaches you, as a design engineer working for A.B. Brown Engineering, with a brief for the design of an electric drill. In order to secure the contract, you need on behalf of your company to produce a comprehensive specification that takes into account all of this customer's requirements that are detailed below.

Design requirements for electric drill

Performance	Capable of taking drill bits up to 0.75 in. in diameter; operate from a 220 V 50 Hz power supply; capable of two speed operation; have a hammer action; operate continuously for long periods of time; be suitable for drilling into soft and hard materials; the eccentricity of the drill action must be limited to ±0.01 in., with drill bits up to 12 in.; have a minimum cable reach of 5 m.
Environment	Able to operate internally and externally within a temperature range of −20 to +40°C. Have no adverse effects from dirt, dust or, ingress of oil or grease. Capable of operation in wet conditions and where combustible dusts are present.
Maintenance	Capable of being dismantled into component parts, for ease of maintenance. Requires no special tools for dismantling and assembly operations. Component parts need to last a minimum of 2 years before requiring replacement or rectification.
Costs	To have a maximum unit cost of £30.00.
Quantity	2000 required on first production run.
Aesthetics	Polymer body shell to have a two colour finish.
Ergonomics	Drill to be fitted with pistol grip handle and upper body steady handle.
Size/weight	Overall length of drill must not exceed 30 cm and have a maximum weight of 3 kg.

Note: the electric drill is to be built for sale in Europe.

UNIT 3

UNIT 3

Based on the information given above, *re-write your customers requirements into the form required for a proper design project specification* for A.B. Brown Engineering, identifying any *important omissions* and rectifying over ambitious and unrealistic requirements, paying particular attention to the problems associated with drill dimensions, component performance, maintenance and costs. Also indicate any areas of the specification that would require further clarification from the customer in order to prepare your specification.

Selecting and evaluating possible project solutions

There is now just one more important part of the project specification we need to consider and that is the selection and evaluation of alternative solutions. There are several techniques available to help you do this; here we will consider only one, the **evaluation matrix**. An evaluation matrix sets each possible solution against a list of selection criteria (*Figure 3.5*). For each criterion some kind of scoring system is used to indicate how the individual design concept agrees with what is judged to be average.

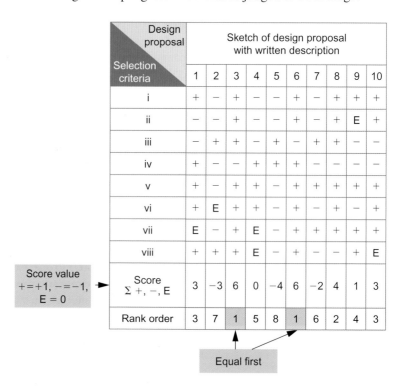

Figure 3.5 The evaluation matrix

The concept proposals are best presented in the form of sketches (that will be in your logbook) together with a short written explanation. One form of scoring system is to use the letter E for equal to the average (norm), then +1 for a criterion considered better than average and −1 for a criterion that is considered worse than average. Other more sophisticated scoring systems can be used, such as 1–5 for poor, below average, average, good and excellent, respectively. The scoring system you choose is entirely up to you.

The use of an evaluation matrix is illustrated in the following example, where a company has been tasked with producing a large flywheel for a heavy pressing machine.

Example 3.1

Assume that you are to manufacture a large diameter flywheel for a heavy pressing machine at minimum cost. Write a short list of selection criteria against which the given project solutions can be evaluated and produce an evaluation matrix using the casting method of manufacture as your average (norm).

Figure 3.6 shows the proposed design project solutions. We now need to produce our evaluation criteria.

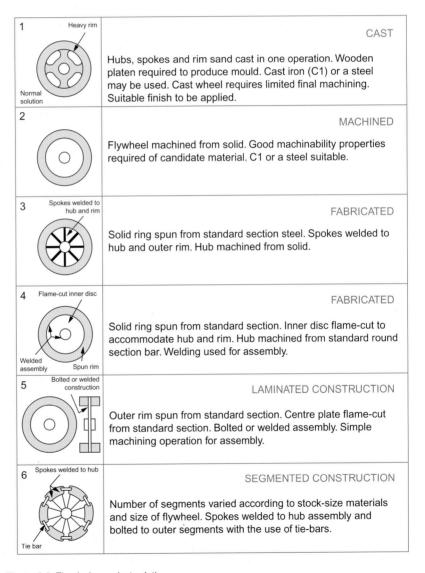

| 1 | Heavy rim / Normal solution | CAST |
| | | Hubs, spokes and rim sand cast in one operation. Wooden platen required to produce mould. Cast iron (C1) or a steel may be used. Cast wheel requires limited final machining. Suitable finish to be applied. |

| 2 | | MACHINED |
| | | Flywheel machined from solid. Good machinability properties required of candidate material. C1 or a steel suitable. |

| 3 | Spokes welded to hub and rim | FABRICATED |
| | | Solid ring spun from standard section steel. Spokes welded to hub and outer rim. Hub machined from solid. |

| 4 | Flame-cut inner disc / Welded assembly / Spun rim | FABRICATED |
| | | Solid ring spun from standard section. Inner disc flame-cut to accommodate hub and rim. Hub machined from standard round section bar. Welding used for assembly. |

| 5 | Bolted or welded construction | LAMINATED CONSTRUCTION |
| | | Outer rim spun from standard section. Centre plate flame-cut from standard section. Bolted or welded assembly. Simple machining operation for assembly. |

| 6 | Spokes welded to hub / Tie bar | SEGMENTED CONSTRUCTION |
| | | Number of segments varied according to stock-size materials and size of flywheel. Spokes welded to hub assembly and bolted to outer segments with the use of tie-bars. |

Figure 3.6 The design project solutions

Since in this example costs appear to be very important, we will just look closely at the manufacturing methods that minimize costs in order to produce our evaluation criteria. We will assume that the requirements of the specification have been met by all the design alternatives and that all the designs are compatible for use with the pressing machine under all conditions.

Then for each of the possible design solutions shown in *Figure 3.6*, we need to consider:

- material costs
- skill and amount of labour required
- complexity of construction
- tooling costs
- machine and finishing costs
- safety (this will be related to the integrity of the design solution assuming it is chosen)
- amount of waste generated
- company preference – knowledge, skills and equipment.

The above list of criteria is not exhaustive, but should enable us to select one or two design alternatives. *Figure 3.7* shows the completed evaluation matrix for this problem.

Selection criteria / Design proposal	See Figure 3.6 for design proposal sketch and written decription					
	1	2	3	4	5	6
Materials	+	E	−	−	E	+
Labour	−	−	+	+	+	−
Complexity of constraint	+	+	E	E	E	−
Tooling	−	−	E	+	+	+
Machining/finish	+	−	E	−	+	+
Safety	+	+	E	E	E	−
Waste generation	+	−	+	−	E	−
Company preference	−	+	+	+	+	+
Score Σ +, −, E	2	−1	2	0	4	3
Rank order	NORM	5	3	3	1	2

Figure 3.7 The completed evaluation matrix showing the rank order of the alternative design solutions

You will note that the company preference immediately skews the scores. The company does not have or does not wish to use foundry facilities, hence the reduction in the score for the casting method (design 1). This is probably due to costs, where the production of a mould would be prohibitively expensive for a 'one off' job. Obviously you as the design technician or engineer would be aware of these facts before evaluating the options.

Note that proposals 2–6 all involve some form of fabrication, assembly or machining, which we will assume is the company preference, so there is no need to consider proposal 1.

Proposal 2:   Machining parts for a heavy flywheel requires several machining operations and the use of elaborate fixtures, not to mention operator skill for an object of such size; so labour tooling and machining costs are relatively high. This process also generates a large amount of material waste.

Proposal 3:   The major advantage of this method is that standard stock materials can be used. Difficulties include the use of jigs and fixtures, weld decay, and the possibility of complicated heat treatments.

Proposal 4:   Similar advantages and disadvantages to option 3.

Proposal 5:   Advantages include the use of standard stock materials, little machining required after assembly, relatively easy to assemble. Disadvantages include necessity for positive locking of bolts after assembly, and outer ring would require skimming after spinning.

Proposal 6:   Labour intensive fabrication and assembly, complex assembly and integrity of construction would raise a safety issue. No specialist tooling required, finish relatively easy and cheap, minimal waste from each machining operation, and a company preference.
Note: If company preference had been for casting, then *proposal 1* would probably have been preferable, provided it met the cost requirements. Options 5 and 6 appear next to favourite, although option 3 might also be worth looking at again, dependent on the skills of the labour force. If there was insufficient evidence on which to make a decision, then more selection criteria would need to be considered. For example, do the options just meet or exceed the design specification? Also, bursting speeds and other safety criteria might need to be taken into account.

To finish this section, let us return to the ***laboratory power supply*** example. The next stage in the development of this power supply was that of examining a range of solutions. These were split into the electrical/electronic and mechanical aspects of the design. This required essentially three major areas of the specification to be evaluated to establish the best solution, these being:

- the selection of the best electronic circuit,
- the best mounting for the electrical components,
- the most appropriate material and design for the casing.

Research was carried out to find a suitable circuit and to identify and evaluate the components needed to build it. The evaluation of the circuit was based on three low cost readily available integrated circuits devices that met requirements 2 and 3 of the technical specification. The circuit choice was made based on these criteria and is illustrated in *Figure 3.8*.

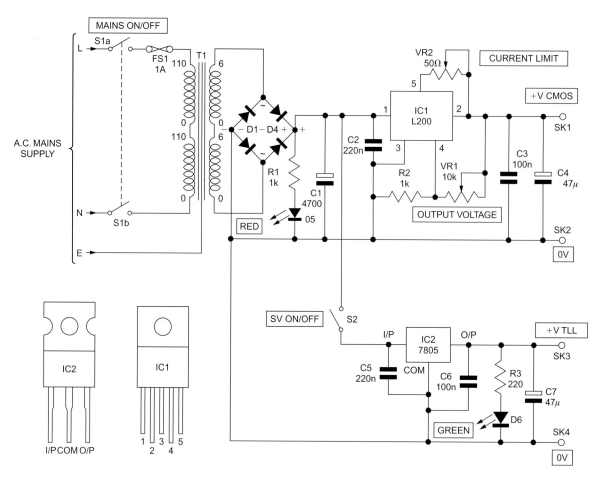

Figure 3.8 Circuit diagram for a variable DC power supply

The data sheet for the voltage regulator used in the circuit is shown in *Figure 3.9*.

Three alternative mountings were used to construct the prototypes and each would be evaluated for functionality and accuracy to determine which best meet criteria 1, 5, 6, 9, 11 and 12 of the technical specification. These prototypes are illustrated in *Figures 3.10–3.12*.

Finally, several casings were evaluated that satisfied or part satisfied requirements 3, 4, 7, 8, 9 and 10 of the technical specification and as a result of this evaluation a fully screened metal enclosure was used for the final prototype (*Figure 3.13*).

For those individuals or groups whose project is much more concerned with process, rather than product, especially if this process is for a service industry such as a procedure, a management plan, a critical investigation, developing a physical resource, work planning, commercial viability, modifying/improving procedures, etc., then the design orientated project

Howard Associates

DATA SHEET	L200 Adjustable Voltage Regulator

MAIN FEATURES

- Adjustable output voltage down to 2.85 V
- Adjustable output current up to 2 A
- Input overload protection (up to 60 V for 10 ms)
- Thermal overload protection
- 5-pin Pentawatt® package
- Low bias current on regulation pin
- Low standby current drain
- Low cost

DESCRIPTION

The L200 is a monolithic integrated circuit voltage regulator which features variable voltage and variable current adjustment. The device is supplied in a 5-pin Pentawatt® package (a TO-3 packaged version is also available to special order). Current limiting, power limiting, thermal shutdown and input over-voltage protection (up to 60 V for 10 ms) make the L200 virtually blow-out proof. The L200 can be used in a wide range of applications wherever high-performance and adjustment of output voltage and current is required.

DIMENSIONS

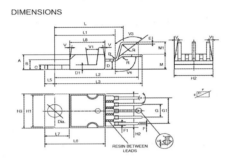

DIM.	mm			inch		
	MIN.	TYP.	MAX.	MIN.	TYP.	MAX.
A			4.8			0.189
C			1.37			0.054
D	2.4		2.8	0.094		0.110
D1	1.2		1.35	0.047		0.053
E	0.35		0.55	0.014		0.022
E1	0.76		1.19	0.030		0.047
F	0.8		1.05	0.031		0.041
F1	1		1.4	0.039		0.055
G	3.2	3.4	3.6	0.126	0.134	0.142
G1	6.6	6.8	7	0.260	0.268	0.276
H2			10.4			0.409
H3	10.05		10.4	0.396		0.409
L	17.55	17.85	18.15	0.691	0.703	0.715
L1	15.55	15.75	15.95	0.612	0.620	0.628
L2	21.2	21.4	21.6	0.831	0.843	0.850
L3	22.3	22.5	22.7	0.878	0.886	0.894
L4			1.29			0.051
L5	2.6		3	0.102		0.118
L6	15.1		15.8	0.594		0.622
L7	6		6.6	0.236		0.260
L9		0.2			0.008	
M	4.23	4.5	4.75	0.167	0.177	0.187
M1	3.75	4	4.25	0.148	0.157	0.167
V4			40° (typ.)			

Figure 3.9 Howard Associates voltage regulator data sheet

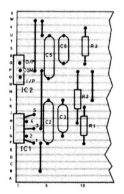

Figure 3.10 Stripboard prototype layout

Figure 3.11 Assembled stripboard prototype

report may only fit in part and a more global report needs to be produced for submission. This may still be referred to as the project specification or the ***preliminary report***. In addition to a technical, commercial or business orientated specification, it is likely to contain an executive summary; some introduction/background to the project that establishes the need, which will not necessarily require any product design; the aims, objectives and proposed solutions (deliverables) of your project and the methods you intend to use to deliver them. Your proposed solution and the methods you intend to adopt to achieve your solution will still be contained within your specification but the evidence you require to back up your solution may not come from test data, calculations, prototype build or any hard facts. It may have to come from canvassing opinions from experts in the field, surveys, literature searches and other confirmed methods of research. It is the execution of your project plan that may prove difficult with this kind of project, but evidence to back up your solution is still required. One word of warning, it is notoriously difficult to produce survey questions that provoke a meaningful response and provide you with the answers you are looking for. Conducting surveys is also extremely time consuming and the response is often very disappointing! If you intend to use survey evidence, for example, to justify the need for your project, then make sure you carry out such surveys at a very early stage and have a back-up method of some

UNIT 3

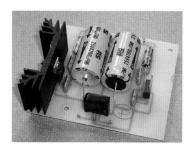

Figure 3.12 Printed circuit board prototype

Figure 3.13 Final prototype showing casing and controls

sort for gaining the same information in another way. The one advantage of undertaking projects that are likely to produce a 'paper solution' is that you are unlikely to need the same amount of physical resource provision as that required for a design orientated project. Physical resource requirements is something you should consider, right back at your ideas stage. Your college is unlikely to fund you for the purchase of expensive equipment, so you need to think about this when preparing your project plan. This will be discussed further in outcome 2 of this unit.

The actions you need to take and the processes you need to follow in order to produce a project specification/preliminary report for a product or process orientated project are summarized in the flow diagram shown in *Figure 3.14*. With what you now know, you should be clear as to exactly what each of these actions means in terms of time and commitment.

As a final exercise before you leave this outcome, consider the list of possible project titles I gave you earlier, in particular title (4) *The modification of a car handbrake system to give better driver access*, and try and answer the questions given in TYK 3.3.

TYK

TYK 3.3

A ratchet type handbrake mechanism is so positioned in a particular model of car that short drivers are required to reach back an uncomfortable distance to operate it. You are required to come up with a modification to remedy the situation. By applying the techniques learnt in this outcome, answer the following questions:
(a) Think of and write down three possible ideas for a suitable solution
(b) Generate five relevant selection criteria and so produce an evaluation matrix
(c) Evaluate your three possible solutions using your matrix and so justify your choice for a solution.

Planning the Project

Introduction

As part of this unit you need to be able to apply basic project planning and scheduling methods, including establishing time scales and resource requirements and the relationships that exist between the various activities that make up a project.

Project planning is different from other forms of planning and scheduling simply because the set of activities that constitute a project are unique and only occur once. Whereas, for example, production planning and scheduling relate to a set of activities that may be performed a large number of times.

In this outcome we are going to look at one or two of the techniques that engineers use when they plan a project. Planning is an important task and it

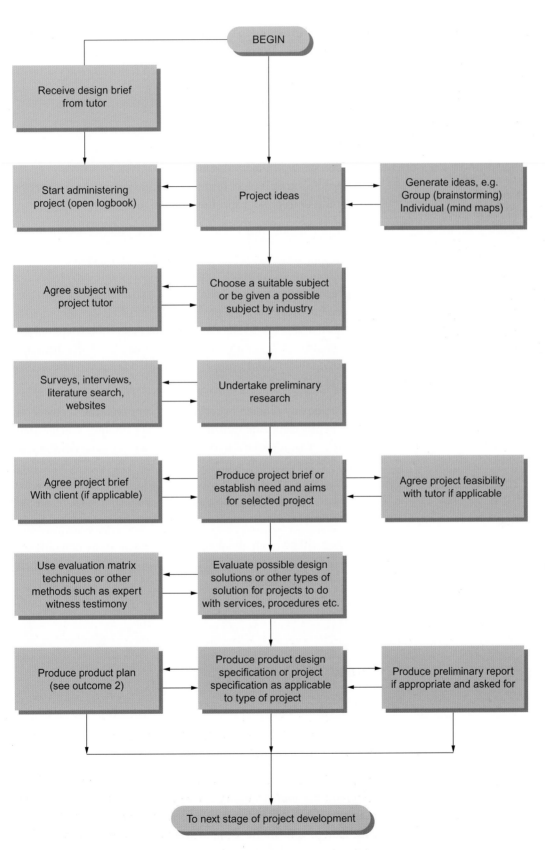

Figure 3.14 Flow diagram leading to production of project specification

needs to be carried out in detail at an early stage in the life of a project. In fact, you have already been involved in planning in order to produce your project specification, where you have been following your short-term plan to:

1. Generate ideas
2. Administer your project
3. Choose or be given a suitable subject for a project
4. Undertake preliminary research
5. Determine your project feasibility and/or produce a project brief
6. Produce your project specification.

The above planning process is in fact part of your long-term project plan!

An effective plan saves both time and money and it will ensure that resources are not wasted. A plan defines a **strategy** (i.e. a strategic or management overview) of a project in terms of all the individual tasks and activities that take place and the resources that will be required to carry out the whole project. It is important to share this plan with others involved and obtain their agreement to supply the required input and resources within the timescale defined in the plan.

In producing any product or service one very important consideration is **cost** and costs can be minimized with good planning! When establishing the feasibility of your project, in other words when establishing whether or not your product will sell or your service will be of benefit, then cost is the primary driving factor. So carrying out some kind of cost benefit analysis, to justify your project solution at an early stage, is all part of **project planning**.

Note: You will learn all about costing when you study outcome 4 in Unit 7 Business Systems, and you will need to apply what you learn there to your project planning process.

TYK 3.4

Explain why project planning is different from other forms of planning.

When implementing your plans you will also need to find some method of **monitoring and recording achievement** to ensure that you are making progress so that, for example, if you get stuck at any time trying to obtain information, you will have planned an alternative method of gaining this information; in other words you plan a 'what if' route that enables you to change your plan should the need arise.

In this outcome, we consider one or two of the most common methods for monitoring and recording achievement, leaving the **implementation** of your plan, that is the way in which you carry out your tasks and make appropriate changes, until you study outcome 3.

Most engineering projects tend to follow a similar process and, therefore, have similar activities that are carried out in a set sequence to achieve the project goals (solutions). You have in fact already been introduced to a rather crude version of your overall project plan when we produced a flow diagram (*Figure 3.1*) in the introduction to this unit that showed the process you needed to follow in order to meet your project goals and deliver a worthwhile solution. Listed below, in a little more detail, are the typical activities you would need to consider in order to produce the plan for the 'whole of your project'.

Typical activities for the whole project process
1. Generate ideas
2. Administer the project (monitor and record progress)
3. Identify need for the product or service
4. Undertake preliminary research
5. Determine project feasibility
6. Evaluate possible alternative solutions
7. Carry out a cost benefit analysis
8. Produce the project specification/preliminary report
9. Produce a detailed project plan
10. Implement the plan
11. Produce project solution/deliverable
12. Evaluate project solution (product or service)
13. Modify solution as necessary
14. Produce final solution
15. Present project solution (i.e. sell service/product to target audience)
16. Produce written report and other deliverables for target audience

Looking at the activities listed above, it can be seen that as part of the overall project process there is a need to produce a detailed plan that must be followed in order to produce and report on the final solution. The planning techniques (tools) we are about to consider enable us to incorporate this detail within the whole strategic plan. These planning tools can be produced manually or generated using computer software. After studying these techniques we will go on to discuss, very briefly, the methods used for recording progress, including further discussion on the use of the logbook for this purpose.

The program evaluation and review technique

Introduction

Programme evaluation and review technique (PERT) was developed for the US Navy in 1958 for planning and control of the Polaris nuclear submarine project. This project involved around 3000 contractors and the use of PERT was instrumental in reducing the project completion time by 2 years. PERT is widely used today as a fundamental project management tool in many areas of commerce and industry. When very large projects of this nature are planned the strategic planning activities that need to be considered may differ slightly and you as a project engineer might only be involved with planning and carrying out one or two particular aspects of such a project. A typical set of activities for a very large engineering project might include:

- Appointing consultants
- Appointing suppliers
- Forming a team to be responsible for carrying out the project
- Preparing a budget for the project
- Preparing a detailed costing
- Producing drawings
- Producing itemized parts lists
- Producing specifications
- Obtaining management approval
- Complying with international and national legislation and safety requirements
- Obtaining planning permission
- Planning and scheduling the phases of the project.

PERT requires that project activities should be discrete and have definite start and end points. The technique provides most benefit when projects have a very large number of interrelated activities where it can be very effective in helping to identify the most effective sequence of activities from a variety of possibilities.

One important aspect of PERT is that it allows us to identify the path through the network for which the total activity times are the greatest. This is the **critical path**.

Critical path method

Critical path method (CPM) is also widely used in the engineering industry. CPM and PERT are very similar and the critical path is important for several reasons:

1. Since it represents the most time-critical set of activities, the total time to reach the project goal can *only* be reduced by reducing time spent in one or more of the activities along the critical path. In other words, the critical path highlights those activities that should be critically reviewed to see whether they can be shortened in any way. Putting extra resources into one or more of the critical path activities can usually be instrumental in reducing the overall project time.
2. The critical path is unique for a particular set of activities and timings. If any of these are changed (e.g. by directing extra resources into them) a new critical path will be revealed. We can then apply PERT evaluation to this new critical path, critically reviewing the activities that it points us to. This process is iterative (may be frequently repeated), so that in a large project we can continue to reduce overall project time making changes as the project develops.
3. The critical path shows us where the most risky and potentially time threatening activities occur. Since any problems or delays with activities on the critical path may jeopardize the entire project it is in our interests to focus particular attention on these tasks.

KEY POINT

PERT and CPM allow us to analyse the activities that make up a project using a network diagram. This, in turn, allows us to identify the critical path.

There are two reasons for this:

- problems or delays with the activities on a project's critical path can potentially jeopardize the entire project
- the overall time spent on the project can only be reduced by reviewing (and hopefully reducing) the time spent on the activities that appear on the critical path

KEY POINT

The critical path is the path through the network diagram for which the activity time is greatest

TYK 3.5

Explain what is meant by programme evaluation and review technique (PERT).

TYK 3.6

Explain the term 'critical path' in relation to PERT and CPM. Give two reasons why the critical path is important.

UNIT 3

The PERT process

PERT is a straightforward process. We have to:

1. Identify all of the activities that make up the project.
2. Identify the sequence of the activities in step 1 above and, in particular, the order of precedence of these activities.
3. Estimate the timing of the activities.
4. Construct a diagram that illustrates steps 1, 2 and 3 above.
5. Evaluate the network and, in particular, identify and clearly mark the critical path.
6. Monitor actual performance as the project is carried out against the schedule produced, revise and re-evaluate the network as appropriate.

PERT network diagrams

The network diagram used in PERT consists of a series of ***events*** which form the nodes in the network. Events are linked together by arrows that denote the ***activities***. *Figure 3.15* shows two events, 0 and 1 linked by a single activity, A, which would normally be described in words. The two events (0 and 1) could be stated on the diagram or referred to elsewhere in a table. The ***expected time for the activity*** (here shown as 3 units) would normally be specified in hours, days, weeks, months, etc.

Figure 3.16 shows a slightly more complicated network diagram where there are three events (0, 1 and 2) linked by two activities (A and B). Note that activity A must be completed before activity B (another way of saying this is that 'activity A precedes activity B'. The expected times for activities A and B are respectively 3 and 2 units of time and so it takes 5 units of time to reach event 2 from event 0, the start point.

The network diagram shown in *Figure 3.17* shows four events, 0, 1, 2 and 3, linked by four activities, A, B, C and D. In this network, activity A precedes activity C, whilst activity B precedes activity D. Note that event 3 is not reached until activities C and D have both been completed.

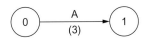

Figure 3.15 A network diagram showing two events linked by a single activity

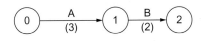

Figure 3.16 A network diagram with three events and two activities

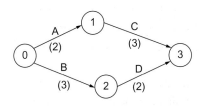

Figure 3.17 A network diagram with four activities

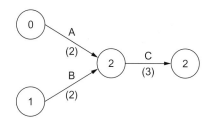

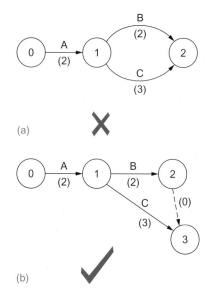

Figure 3.18 A network diagram in which activity C cannot start until both activities A and B have been completed

Figure 3.19 A network diagram with a dummy activity

KEY POINT

Reducing the time for activities along the critical path reduces overall time for the project

The expected times for activities A, B, C and D are 2, 3, 3 and 2 units of time, respectively. There are several other things to note about this network:

- the events occur in the following order: 0, 1, 2 and 3;
- activity B is performed at the same time as activity A;
- activity D is performed at the same time as activity C;
- the total time to reach event 3, whichever route is chosen, amounts to 5 units of time.

Another network diagram is shown in *Figure 3.18*. In this network, activities A and B both precede activity C (in other words, activity C cannot start until both activities A and B have been completed). The total time to reach event 3, whichever route is chosen, amounts to 5 units of time.

Within a network diagram, the activities that link two events must be unique. Consider *Figure 3.19*, which shows that event 2 can be reached via activities B and C, where the expected time for activity B is 2 units whilst the expected time for activity C is 3 units.

To avoid potential confusion, we have introduced a ***dummy activity*** between event 2 and event 3. This activity requires no time for completion and thus its expected time is 0 (note that we have adopted the convention that dummy activities are shown as a dashed line).

Finally, you should see that event 2 is reached before event 3 and that the total expected time through the network amounts to 5 units and that there is ***slack time*** associated with activity B amounting to 1 unit of time (in other words, activity B can be performed up to 1 unit of time late without affecting the expected time through the network).

The critical path

Within a network diagram, the **critical path** is the path that links the activities that have the greatest expected time. In other words, it is the longest route through the network in terms of time.

Consider the network diagram shown in *Figure 3.20*. This diagram shows five events linked by five real activities plus one dummy activity. The relationship between the activities and their expected times can be illustrated in the form of a table:

Activity	Preceding activity	Expected time
A	None	7
B	None	3
C	B	5
D	B	2
E	A, C, D	1

The critical path (shown as a thick blue line) constitutes activities B, C and E that produce a total expected time, between event 0 (the start of the project) and event 4 (the completion of the project) of 9 units.

The critical path allows us to identify those activities that are critical. Remember that, by reducing the time spent on activities along the critical path, we can reduce the expected time for the complete project. If, for

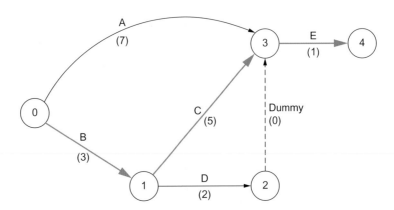

Figure 3.20 A network diagram with a dummy activity

example, we can direct extra resources into the project we would only be able to reduce the overall time by concentrating our efforts on activity B, C or E. Reducing the time spent on activities A or D would have no effect on the overall time spent on the project!

Now let us put this into context by taking a practical example of using network diagrams and applying critical path analysis. Let us assume that Audio Power Systems (APS) a manufacture of loudspeakers for hi-fi sound systems are about to produce a prototype for one of their new loudspeaker designs. This particular design is to use two drivers, a bass unit (or 'woofer') for low-frequency signals and a treble unit (or 'tweeter') for high-frequency signals. In order to feed the correct frequency range of sounds to the appropriate driver, the signal applied to the loudspeaker is to be split into two bands by means of a *cross-over* unit.

The loudspeaker enclosure will house the drivers together with the cross-over unit. The design of the enclosure and cross-over unit (both designed using CAD), will depend upon the specification and the drivers used.

Before they proceed with the project, APS has produced a network diagram so that everyone involved with the project knows the timescale and where any potential problems are likely to arise. The network diagram is based on a total of nine activities and is shown in *Figure 3.21*. The events associated with the project are listed below:

Event number	Event
0	Commencement of project
1	Specifications and budget agreed
2	Materials and parts specified
3	Design of cross-over completed
4	Design of enclosure completed
5	Components (i.e. enclosure, cross-over and driver) ready for assembly
6	Loudspeaker assembled
7	Testing complete

From *Figure 3.21* it is easy to see that the design and construction of the cross-over network takes place at the same time as designing and building the enclosure. These activities are quite separate and can be performed by different people using different resources. Note, however, that it is not

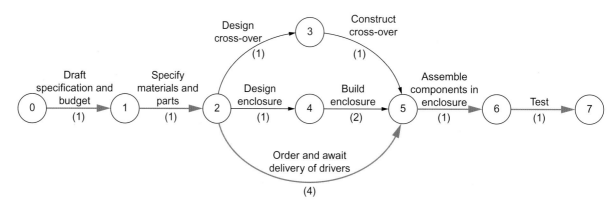

Figure 3.21 Network diagram with the APS prototype loudspeaker

possible to start assembling the loudspeaker until *both* the cross-over and the enclosure have been designed and built, and the drivers have been delivered.

The critical path shows that APS should pay particular attention to the ordering and delivery of the driver units. It also shows that there is slack time amounting to 2 days associated with the design and construction of the cross-over unit and 1 day associated with the design and building of the enclosure.

As with most projects, there is a trade-off between time and cost. By deploying additional resources (assuming, of course, that APS has the money available to do this!) along the critical path it can make significant reductions in the overall time taken. We will look at this next.

TYK *3.7*

In relation to the network diagram shown in *Figure 3.21*:
(a) How many days are required to:
    (i) construct the cross-over?
    (ii) design the enclosure?
    (iii) assemble the components?
(b) What is the total time for the project?
(c) If the time taken to order and deliver the drivers can be reduced to 2 days, what happens to the critical path?

TYK *3.8*

The following data refers to the activities that make up a project. Use this information to construct a network diagram and identify the critical path. Also determine the expected time to complete the project.

Activity	Preceding activity	Expected time
A	None	1
B	A	2
C	None	2
D	C	3
E	D	2
F	E	1

TYK 3.9

The following data refers to the activities that make up a project. Use this information to construct a network diagram and identify the critical path. Also determine the expected time to complete the project.

Activity	Preceding activity	Expected time
A	None	1
B	None	1
C	B	1
D	A, C	2
E	D	2
F	A, C	5
G	B	2

Activity 3.4

As an engineering student you have been asked by the Head of Engineering in your college to produce a portable display stand to advertise engineering courses and the BTEC National Extended Diploma in Engineering in particular. You have been asked to include photographs of students' work, details of college resources and some examples of students' projects. The display stand is to fold up for transport into a container that will fit in the boot of a car and it should be equipped with leaflet storage and display lighting.

(a) List the activities that will make up this project and estimate the expected time for each activity.
(b) Construct a network diagram showing how the activities are related.
(c) Identify the critical path on the network diagram.
(d) Determine the expected time to complete the project.

Present your answer in the form of a word-processed report with a printed chart to be submitted to the Head of Engineering for approval. Use a computer-aided drawing package to produce the network diagram.

Project costs

As mentioned earlier you will be studying costing techniques in some detail in Unit 7 Business Systems. Presented below is just a few words of clarification concerning costs that you will need to follow the project planning process.

Projects involve two types of costs: *indirect costs* and *direct costs*. Indirect costs include items such as administrative overheads and facilities costs (heating, lighting, etc.). Direct costs are concerned with additional labour costs, equipment leasing, etc. We can spend extra money to reduce the time taken on the project; however, this only makes sense up to the point where further direct cost expenditure (such as the cost of employing additional contract staff) becomes equal to the savings in indirect project costs (such as heating, lighting and other overheads).

To examine the trade-off between project time and costs, we need to have the following information:

1. A network diagram for the project showing expected times and indicating the initial critical path. We also need to know the minimum time for each activity when there are no resource constraints (this is known as the **crash time**).
2. Cost estimates for each project activity expressed in terms of indirect expenditure per unit time.
3. The costs of providing additional resources for each project activity and the consequent time saving expressed in terms of expenditure per unit time reduction.

With the above information we can reduce the critical path activity times, beginning with the activity that offers the least expenditure per unit time reduction. We can then continue with the second least costly, and continue until we are left with the most costly until either we reach the target minimum time for the project or the additional direct cost expenditure becomes equal to the savings in direct costs. Having briefly discussed costs, we now return to look at time analysis in a little more detail.

Note: Refer to Unit 7 for further details on costing techniques.

Project time analysis using PERT

PERT defines a number of important times in the project life cycle. These are as follows:

Expected time, t_e	The expected time for an activity is simply the average time for the activity.
Optimistic time, t_o	This is the fastest time for the completion of the activity. This time will rarely be achieved and will only be bettered under exceptionally favourable circumstances.
Pessimistic time, t_p	This is slowest time for the completion of the activity. This time will nearly always be bettered and will only be exceeded under exceptionally unfavourable circumstances.
Most likely time, t_m	This time represents the 'best guess' time for the completion of the activity. This time is the statistical mode of the distribution of the times for the activity.

Estimates of project times are often based on previous experience of performing similar tasks and activities. The expected time, t_e, is usually calculated from the formula:

$$t_e = \frac{(t_o + 4t_m + t_p)}{6}$$

The expected time to reach a particular event, T_E, can be found by adding up all of the values of expected time, t_e, that lead up to the event in question. To explain how this works take a look at *Figure 3.22*. This shows a network diagram with five activities: A, B, C, D and E. The critical path is through A, B, C and E.

Activity	Preceding activity	Time (weeks)			
		t_o	t_m	t_p	t_e
A	None	1	2	4	2.2
B	A	1	2	4	2.2
C	B	2	3	5	3.2
D	A	2	4	8	4.3
E	C, D	1	2	4	2.2

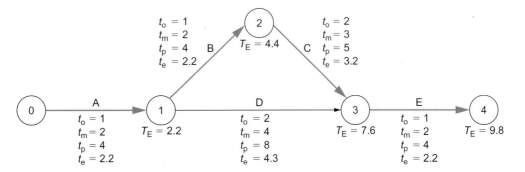

Figure 3.22 Project time analysis using a network diagram

Adding together the values of t_e for the critical path activities (A, B, C and E) gives a total time for the project of 2.2 + 2.2 + 3.2 + 2.2 = 9.8 weeks.

The values of T_E can be found by adding together the values of t_e for the preceding activities. For example, the time taken to reach event 2 will be 2.2 + 2.2 = 4.4 weeks whilst that to reach event 3 will be 2.2 + 2.2 + 3.2 = 7.6 weeks. You should also see that the slack time for activity D is 7.6 − 4.3 = 3.3 weeks.

Activity 3.5

Skylane Aircraft Manufacturing (SAM) an aircraft manufacturer has engaged a consultant to advise on the modification of its new twin-engined aircraft for traffic spotting and aerial surveillance. The consultant has suggested that the aircraft should be fitted with a gyro-stabilized camera platform and has identified the following sequence of activities prior to supplying a prototype to a major client:

Ref.	Activity	Preceding activity	Time estimate (days)		
			t_o	t_m	t_p
A	Design and costing	None	3	5	10
B	Client review	A	2	7	14
C	Specify camera and optical equipment	B	1	2	4
D	Order camera and await delivery	C	7	14	21
E	Detailed design of camera platform	D	1	3	6
F	Construct camera platform	E	3	7	11
G	Assemble platform and camera	D, F	1	2	4
H	Design camera controller	B	3	4	7
I	Construct camera controller	H	3	7	11
J	Fit platform and controller to aircraft	G, I	1	2	4
K	Client acceptance testing	J	3	5	10

(i) Draw the network diagram for the project.

(ii) Determine the expected time for each activity, t_e, and mark this on the network diagram.

(iii) Determine the critical path and mark this in the network diagram.

(iv) Determine the expected time to each event, T_E, and the time for the complete project. Mark these on the network diagram.

Present your work in the form of a network diagram produced using a computer-aided drawing package.

Gantt charts

Another time planning tool that is often used for less complex projects is the **Gantt chart**.

A Gantt chart is simply a bar chart that shows the relationship of activities over a period of time. When constructing a Gantt chart, activities are listed down the page whilst time runs along the horizontal axis. The standard symbols used to denote the start and end of activities and the progress towards their completion are shown in *Figure 3.23*.

A simple Gantt chart is shown in *Figure 3.24*. This chart depicts the relationship between four activities A to D that make up a project. The horizontal scale is marked off in intervals of 1 day with the whole project completed by day 14. At the start of the sixth day (see time now) the following situation is evident:

- Activity A has been completed
- Activity B has been partly completed and is on schedule
- Activity C has not yet started and is behind schedule
- Activity D is yet to start.

Symbol	Meaning
[	Start of an activity
]	End of an activity
[—————]	Actual progress of an activity
☐	(alternative representation)
V	Time now

Figure 3.23 Symbols used in Gantt

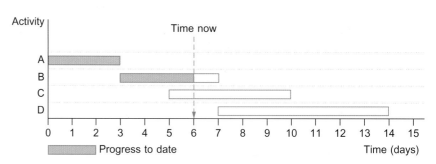

Figure 3.24 A simple Gantt chart

Another example is shown in *Figure 3.25*. This chart depicts the relationship between six activities A to F that make up a project. The horizontal scale is marked off in intervals of 1 day, with the whole project completed by

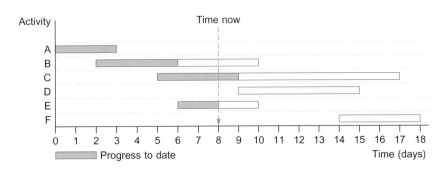

Figure 3.25 Another simple Gantt chart

day 18. At the start of the eighth day (again marked time now) the following situation is evident:

- Activity A has been completed
- Activity B has been partly completed but is running behind schedule by 2 days
- Activity C has been partly completed and is running ahead of schedule by 1 day
- Activity D is yet to start
- Activity E has started and is on schedule
- Activity F is yet to start.

As an example of how a Gantt chart is used in practical situation take a look at *Figure 3.26*. This shows how APS (the loudspeaker manufacturer that you met earlier) could use this technique to track the progress of its project to produce a new loudspeaker design. The chart shows the situation at the beginning of day 4 with all activities running to schedule. You might like to compare this with the network diagram for the same project.

Activity	Days									
	1	2	3	4	5	6	7	8	9	10
Draft spec. and budget	⊢———⊣									
Specify materials/parts		⊢———⊣								
Design cross-over			⊢———⊣							
Construct cross-over				[——]						
Design enclosure			⊢———⊣							
Build enclosure				[————]						
Order/await drivers			⊢————]							
Assemble components							[—]			
Test								[——]		

Figure 3.26 Gantt chart for the new APS loudspeaker (see also *Figure 3.21*)

UNIT 3

TYK 3.10

Figure 3.27 shows the Gantt chart for the APS loudspeaker design project with the situation at the beginning of day 6.

Which activities are:
(a) On schedule
(b) Behind schedule by 1 day, or less
(c) Behind schedule by more than 1 day
(d) Ahead of schedule.

Activity	Days									
	1	2	3	4	5	6	7	8	9	10
Draft spec. and budget										
Specify materials/parts										
Design cross-over										
Construct cross-over										
Design enclosure										
Build enclosure										
Order/await drivers										
Assemble components										
Test										

Figure 3.27 Gantt chart for TYK 3.10

Activity 3.6

Use a computer-aided drawing package to create a Gantt chart for the construction of the Engineering Department display stand in Activity 3.4. Present your work in the form of a printed chart to be submitted to the Head of Engineering for approval. The chart should show each of the activities plotted against a common timescale.

Recording Progress

Having spent some time looking at the ways in which we produce project plans and monitor our progress, we now very briefly discuss the methods you should adopt to **record your progress** and maintain your records. The primary vehicle for the day-to-day recording of events is your **logbook**. You have already been introduced to the nature and use of the logbook, when we discussed the preparation and maintenance of records. Look back now at what was said. We not only discussed the actual form of the logbook but also mentioned the kind of information you needed to keep in it. In order to record your progress it is this area that we expand on a little here.

Within your communications log or separately you need to keep a record of the names, addresses and telephone numbers of all personal contacts which can then be added to the contact list in your logbook as and when they arise. You may also wish to keep contact details in a diary as well. However, do remember what was said earlier about diaries: these should not be used as a logbook. By recording all relevant information concerning the progress of your project in one place, information will be more readily available when you need to access it or amend it.

Another place you may choose for recording and sorting personal contact details is of course your mobile phone directory. This is fine providing your mobile phone is not lost or stolen! Remember also that you will need to provide written evidence of your contacts and sources of reference, for assessment purposes, so if you use a mobile phone for this purpose it should have a means of downloading this information.

Hard copy transcripts could also be produced if you keep a computer diary and this is fine for the recording and maintenance of contacts, but is a little less convenient for jotting down notes or sketching a situation when you are out and about information gathering or conducting interviews. Asking your interviewee to wait while you set up your laptop and recording equipment might not go down too well! Having said that, it may on occasions be appropriate to ask permission to record a conversation, especially if you are being provided with important facts on a subject. Asking permission will very much depend on circumstances and how well you know the person you are interviewing, so do use discretion!

Returning to logbooks for a moment, these should be used to store hard copies of your Gantt charts or PERT printouts or other planning documents, so that you have them readily to hand as a permanent reminder of your current state of progress. You should also amend these planning documents from time to time as your plans change over the life of your project. Keeping these documents in your logbook, suitably amended, enables you to provide evidence of project planning, record keeping and monitoring, all from one source!

Your logbook should not only contain administrative information concerning project progress, but also provide details of any technical data you may have obtained and the ways in which you have processed this data. Thus, results and data obtained from tests, experiments, research or expert witness testimony will be included together with any analysis you may have carried out on this data.

Having spent some time on planning, planning techniques and recording we will now look at ways to implement our project.

Implementing the Project

In this outcome we will concentrate on the ways in which you implement your project, that is how you go about achieving your project aims and producing your solution. The type of information you need and the type of decisions you have to make in order to achieve your project aims will very much depend on the type of project you have chosen. For example, if you have chosen a design and build style of project then you may very well be engaged in producing your design alternatives, making choices about component selection, building a prototype, testing your solution, evaluating your solution and producing the final design.

It will be an altogether different set of tasks you will need to complete if you are, for example, producing an improved maintenance procedure for a particular model of motor car. In this case, you will be spending most of your research time trying to justify how your modified procedure will be better than the existing procedure and why. For example, you would need to justify your suggested procedure on both technical and financial grounds. Producing evidence of technical improvements and financial savings may be difficult, because unless you are undertaking an industrial or commercially based project obtaining meaningful data will prove very difficult.

An integral part of the project implementation process is to ensure that you check your solutions. If you have a clear design brief against which to work and a well-structured technical specification, then checking your solution becomes a relatively simple process that may involve verification of calculations and testing your product against known parameters. While again, if you are involved with improving a procedure or schedule for a service, you may have to check your solution by gathering data, based on the opinion of those who are using your trial solution in the field and a lot of this data will have to be based on opinion rather than fact!

No matter what the nature of your project or your chosen solution, when implementing your plan and producing your outcome you will need to adhere to all related safety and legislative issues. With design-related projects for example, you will need to be aware of all the related British Standards and ISO standards that govern your design and/or product, as well as being fully conversant with all safety procedures related to the build and test of your product. If modifying or trying to devise a new procedure for a service, you will need to be fully conversant with the legislation and documentation issued by the appropriate governing body. For example, imagine you are trying to devise an improved procedure for the maintenance of the Airbus A380 landing gear. For this project you would need to be fully conversant with the maintenance documentation for the aircraft and the maintenance directives issued by the European Aviation Safety Agency (EASA) and the National Aviation Authority (NAA) which in the United Kingdom is the Civil Aviation Authority the CAA.

To try and illustrate the ways in which you might implement your project we start by first considering a *case study*, where we illustrate the implementation of a design and build project from the design brief stage to finished product. I have made a number of comments at various stages throughout the case study to encourage you to think about what you might do or should do at a particular stage in the development of *your* project solution.

Project implementation case study

James Taylor is an engineering student at North Downs College. His chosen project involved the development of an engineered product/system that will solve a particular problem identified by a local garage, car dealership and service agent, Portmore Motors.

Client brief

Portmore Motors asked James to find a way of extending the coverage of their existing wireless local area network (LAN) to include a PC that is installed in the Service Manager's Office. Unfortunately, the coverage of the wireless LAN (specified by the manufacturer) is only 30 m but the line-of-sight (LOS) distance between the wireless access point (WAP) and the PC in the Service Manager's Office is 100 m. The client brief, therefore, was to extend the wireless LAN in order that the Service Manager can access the company's client records and database and also to allow him to place orders for spare parts using the Internet.

Note 1: Is the client brief clear or is further clarification needed? If you are working to a brief or you are attempting to justify the need for your chosen project that has not got a brief, make sure you are fully aware of the problem and you are able to fully justify the need for your project topic.

In order to clarify the client brief (and before attempting to develop a technical specification) James arranged to carry out an initial visit to Portmore Motors in order to talk to the Service Manager, Greg Smith. James was able to view the existing network and assess the layout of the site and, using sketches and rough measurements, he was later able to produce the drawing shown in *Figure 3.28*.

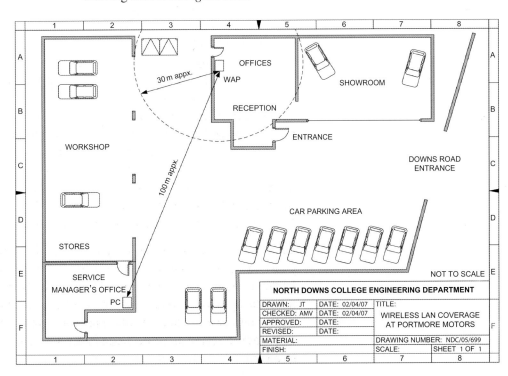

Figure 3.28 Portmore Motors and the location of the two horn antennas

This drawing provided useful clarification of the client brief and was included in James's logbook. James also took along his digital camera and obtained some photographs of the existing wireless LAN, including the wireless adapters (shown in *Figure 3.29*) and the WAP and cable MODEM (shown in *Figure 3.30*).

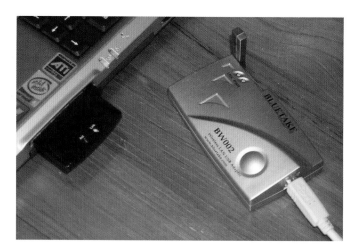

Figure 3.29 Wireless network adapters used at Portmore Motors: the PCMCIA adapter (left) has a low-gain internal antenna while the USB wireless adapter (right) has a small quarter-wave vertical antenna

Figure 3.30 Cable MODEM and wireless network adapter

Note 2: You may not understand the depth of technical detail in the text, but you would be expected to include the same amount of detail for a similar project in your own chosen area of engineering.

James also spent some time discussing alternative solutions (e.g. those based on cables and fixed LAN access points), but these were felt to be unsuitable on the basis of the likely cost and lack of flexibility. The company had found the wireless network to be very satisfactory in the office and showroom areas, but unfortunately it was inaccessible from the workshop and the Service Manager's Office.

Initial research

Following his initial visit to Portmore Motors, James decided to carry out some research using the North Downs College Library and the Internet. James searched specifically for information on wireless networks and on ways in which the range (effective distance) of this equipment could be improved, James found that the easiest way of increasing the range of the wireless LAN would be to make improvements to the antenna system used. By replacing one of the low-gain omni-directional antennas with a directional antenna he would be able to increase the coverage by adding 'gain' into the system (the gain being associated with the directivity of the antenna compared with an omni-directional component). James also found that the required minimum fourfold increase in power gain would be equivalent to a power gain of 6 dB. James was also pleased to find that this increase in gain could be achieved by simply replacing the current antennas and that no additional power supply or amplifying equipment would be required. James kept a record of the information sources that he used so that he could add these to his logbook.

Note 3: Any Internet search you may make should be targeted towards the specific information you are looking for. Do not forget to write down the web addresses you visit in your logbook and your visit date, for use later as references.

Technical specification

James was now in a position to formulate a full draft technical specification for the project:

Project aim	Extension of existing wireless LAN to provide access from the Service Manager's Office
Product	Directional antenna system
Required range	100 m approx. (current range = 30 m)
Antenna gain	12 dBi to 14 dBi (min. additional gain = 2.6 dB)
Frequency	2.450 GHz
Data rate	100 Kbps min.
Connector	N-type female coaxial connector
Impedance	50 Ω
Weight	Less than 600 g
Size	Less than 300 mm square
Standards	IEEE 802.11b, IEEE 802.11 g
Features	Rugged, weatherproof, simple mounting, easily adjustable
Production	2 units

He included this specification in his logbook and returned to it on numerous occasions to confirm that the specification was being met. The specification became particularly important at the evaluation stage.

Note 4: This is a detailed specification and at this stage an outline solution to the problem has been found. However, the exact nature of the solution has yet to be decided.

Generation of alternative ideas

Having completed the technical specification and having arrived at a means of solving the problem that Portmore Motors had identified in their brief, James was confronted with the problem of arriving at a suitable antenna design. This required him to carry out some focused library and Internet research from which he was able to identify the five different antennas (or 'candidate solutions') shown in *Figure 3.31*.

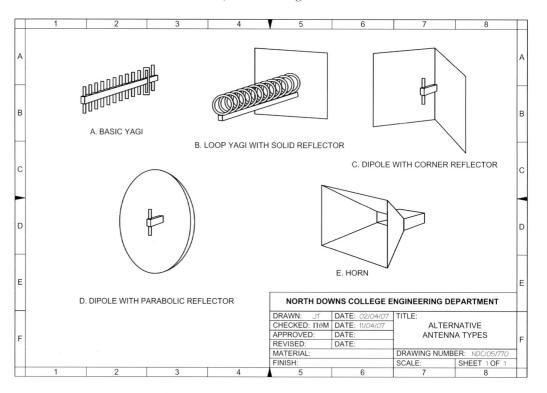

Figure 3.31 Alternative antenna types

In order to evaluate these potential solutions, James decided to make use of an evaluation matrix (see *Figure 3.32*), that he produced from a rough copy he had made in his logbook. He chose six different features to take into consideration and arrived at an overall 'suitability rating' for each of them. The result of the evaluation was that the horn antenna was found to be most suitable followed closely by the corner reflector and basic Yagi antennas (both of which were slightly more complex). In order to confirm his choice, James consulted several technical references, including the VHF/UHF Manual by George Jessop, from which he found that '... *horns are an attractive form of antenna, particularly for use at the higher microwave frequencies. They are fundamentally broadband devices which show a virtually perfect match over a wide range of frequencies. They are simple to design, tolerant of dimensional inaccuracies in construction, and they need no adjustment.*' James incorporated this quotation in his final report, having recorded the full reference for its origin in his logbook.

Note 5: The use of an evaluation matrix and the criteria used in this case to make decisions. For your project you should ensure that any evaluation matrix you use has realistic and appropriate criteria upon which to base your selection.

KEY POINT

The evaluation matrix is a very good way to select a project solution

Antenna	A	B	C	D	E
Type	Basic Yagi	Loop Yagi	Corner reflector	Parabolic reflector	Horn
Gain (typical)	12 dBi to 15 dBi	15 dBi to 18 dBi	12 dBi to 15 dBi	15 dBi to 30 dBi	10 dBi to 20 dBi
Construction	Fairly simple	Complex	Fairly simple	Complex	Simple
Size	Medium	Medium	Large	Large	Medium
Mounting	Simple	Fairly simple	Fairly simple	Could be difficult	Fairly simple
Signal feed	Fairly simple	Could be difficult	Fairly simple	Must be at focal point	Easy
Materials	Aluminium or brass rods	Brass or copper strip	Tinplate or aluminium sheet	Aluminium sheet	Tinplate or brass sheet
Suitability	Possible?	Too complex	Possible?	Too complex	Good

Figure 3.32 James's evaluation matrix used by him to evaluate possible design solutions

Final design solution

James's final design solution was the horn antenna shown in *Figure 3.33*. His next problem was that of determining the dimensions of the antenna and the specification of suitable materials and production processes.

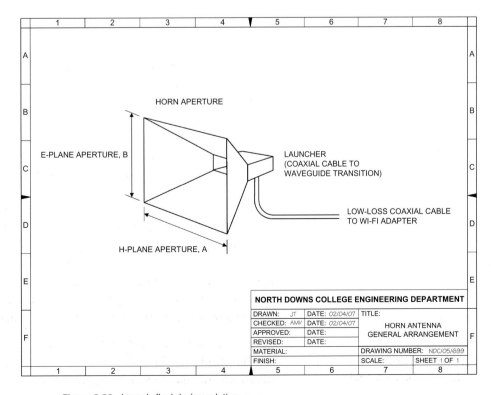

Figure 3.33 James's final design solution

A further Internet search provided James with formulae for calculating the dimensions of the horn aperture and launcher, and he was also able to locate some CAD software that simplified the design process (see *Figure 3.34*). James compared the calculated values with those from the CAD software and found that they were in close agreement. James recorded these calculations and screen dumps in his logbook and reproduced them for inclusion in an appendix in his final report.

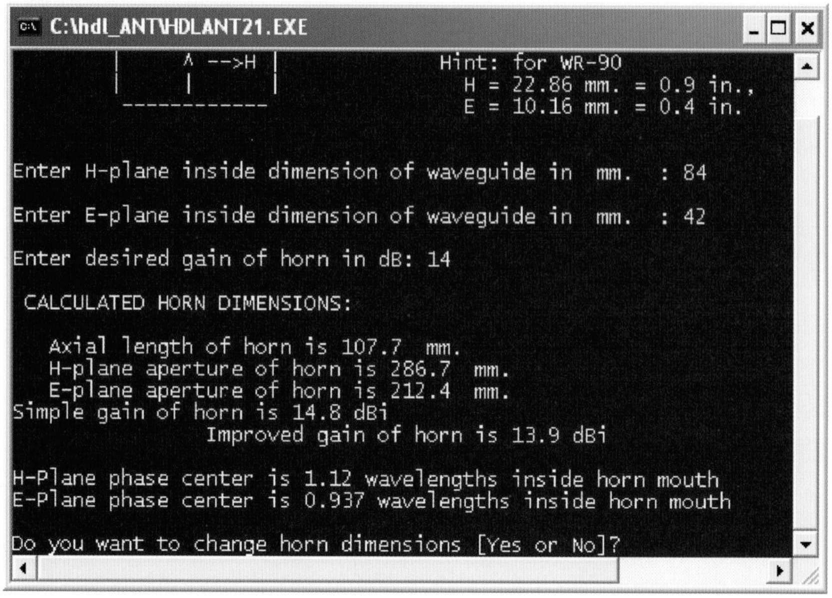

Figure 3.34 The horn antenna design software

Note 6: The Internet was used to establish the type of calculations necessary to ensure compliance with the specification. Not only were the dimensions and the materials for the antenna verified, but also the dimensions for the horn aperture and launcher. All draft calculations were recorded in James's logbook.

Other avenues of research could well have been pursued at this stage in order to make choices concerning the final production and build of the antenna.

Planning for production

James was now ready to put his ideas into practice and plan the production of his prototype. First, however, he needed to give some thought as to how he would actually construct the horn and the launcher and join them together. *Figure 3.35* shows the different techniques that James considered for constructing the horn antenna.

By using a further evaluation matrix, James arrived at method D (fully folded construction), but he also decided to build an earlier prototype for comparison using method A, that was made from copper laminate, cut and soldered along all four seams (*Figure 3.36*).

Note 7: The use again of an evaluation matrix together with the results of an earlier evaluation of a prototype, resulted in James's choice of final design for the antenna and launcher.

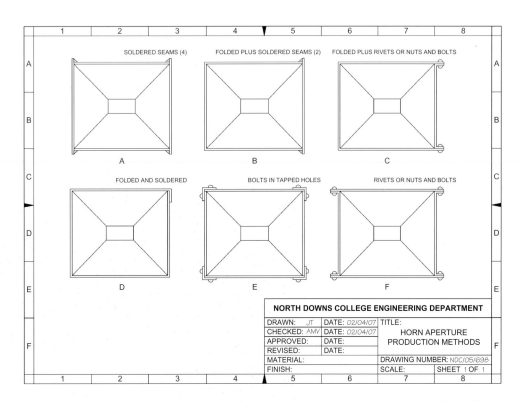

Figure 3.35 The six different methods considered for constructing the horn antenna. Method D (fully folded construction) was chosen

Figure 3.36 Earlier prototype horn antenna using method A

This was tested in order to 'prove' the design before moving on to production of the first two antennas using folded construction (as shown in *Figure 3.37*) and the launcher (as shown in *Figure 3.38*). All of these drawings and photographs were placed in James's logbook ready to be incorporated into James's final report.

Final evaluation

Having completed the production and assembly of his two horn antennas James was ready to carry out a full site test using the wireless LAN at

UNIT 3

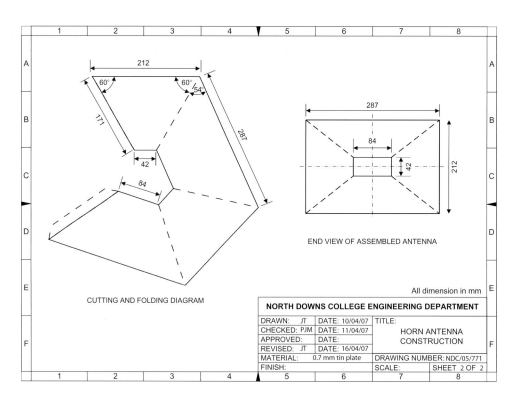

Figure 3.37 Horn antenna construction showing the arrangements for cutting and folding the brass or tinplate sheet

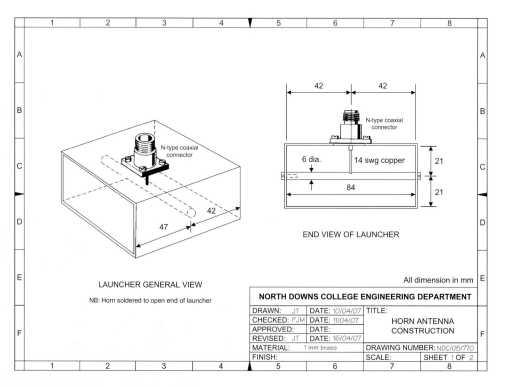

Figure 3.38 Horn antenna construction showing the launcher (coaxial cable to waveguide transition). The launcher is to be soldered to the base of the horn aperture

UNIT 3

Portmore Motors (see *Figure 3.39*). He obtained two low-loss coaxial connectors to link the horn antennas to the wireless network adapter (H2) and to one of the two antenna ports provided on the WAP (H1).

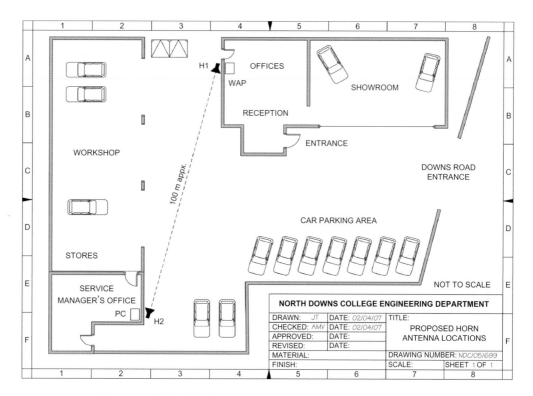

Figure 3.39 The location of the two horn antennas (H1 and H2) at the Portmore Motors site

James used the wireless LAN configuration utility to set up the network (see *Figure 3.40*) and also to check the signal strength and overall quality of the link (see *Figure 3.41*).

Figure 3.40 Wireless network configuration screen

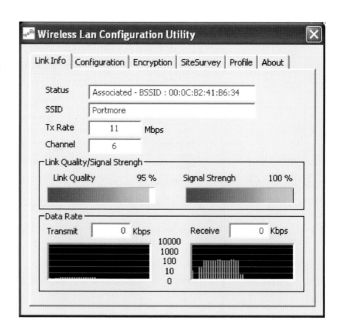

Figure 3.41 Wireless network performance information. Screen evidence used to show that horn antenna had complied with its original technical specification

Note 8: Final project solution evaluated both 'on-site' by functional testing and verified using computer-based wireless LAN configuration utility software.

The indicated signal strength was '100%' and the signal quality '95%' (showing some errors in data transmission). The data rate indicated reached a maximum of about 100 Kbps (in line with the original technical specification). Greg Smith was invited to test the network connection and report any problems to James. After a period of approximately 2 weeks, Greg reported that the system had worked without fault but he expressed concern about the need to seal the horn aperture against the ingress of water and airborne dust and dirt. James included this valuable feedback in his final report and modified his original design by fitting two Perspex covers to the front aperture of the two horn antennas. He included full details of this modification in the 'evaluation and modification' section of his report. James also included several photographs of the completed system including close-ups of the antennas showing their construction and how they were mounted.

Notice that throughout the implementation of this project design solution, constant use was made of the logbook to record results and findings and these were then easily available for inclusion into the final Report.

Notice also the methods that James, in the case study, used to select and check his solutions. He used an evaluation matrix to evaluate his initial overall designs and again when considering both the horn antenna and the launcher. In addition, he checked his final solution by comparing his calculated values with those laid down in the antenna design software. He also checked his project solution by field testing – perhaps the most effective and useful check of them all!

Remember that for some project topics it will be impossible or impractical to field test, except perhaps by undertaking a pilot study, where in general, this type of study yields statistical (rather than factual) evidence that must be sorted and analysed. This is particularly appropriate when considering projects that are concerned with the production or modification of procedures, training programmes, work practices, resources, etc.

Finally with respect to the case study, it cannot have escaped your notice that for a fairly simple solution a considerable amount of work had to be done. Please be guided by your project tutor or industrial link as to the size and scope of your project and try not to be too ambitious!

Before you leave this outcome I would like you to attempt Activity 3.7, which requires you to put into practice much of what you have learnt up till now.

Activity 3.7

As a consultant engineer with expertise in hydraulic system design and maintenance, you receive the following customer brief.

*We currently produce portable hydraulic power packs used on a range of agricultural machinery. Our current design of flexible hose fittings used with the pack have caused difficulties when removing/fitting the hydraulic power pack from/to the machines during maintenance. Hose end connectors are required **that will allow the attachment and removal of hoses to be carried out quickly and safely, without undue leakage of hydraulic fluid**. The fittings will need to accommodate 10, 15, 20 and 30 mm flexible wire reinforced rubber hoses, capable of withstanding pressures up to a maximum of 70 bar. The end fittings/connectors must be robust enough to cope with harsh agricultural environment conditions.*

Study the brief and then use the Internet, your library and any other information sources you think appropriate to answer the following questions:

1. What British/European Standards govern the design and use of both flexible hydraulic hoses and hose connectors/end fittings?
2. By studying the brief think of a set of questions you would want to ask in order to ensure sufficient technical information for the production of a technical specification.
3. Give as many reasons as you can think of as to why a site visit to view the current tractor fittings would be necessary.
4. Identify at least five criteria against which you can evaluate three alternative hose fitting assemblies and using your evaluation matrix, choose the most appropriate proprietary (readily available) set of hose assemblies that meet your customer's requirements.
5. What would you now need to do in order to check the validity of your project solution?

Presenting the Project

In this outcome we first look at what written material you need to present that is appropriate to your project solution. In other words, your written deliverables and the form in which these should be presented. Then as the

UNIT 3

final section of this unit we look at what you need to prepare and how to deliver your oral presentation.

When presenting your project deliverables you will need to draw on the knowledge and techniques you have acquired during your study of Unit 2 Communications for Engineering Technicians. Only one or two of the more important written and oral presentation skills that you were introduced to in that unit will be reviewed here. The more familiar you are with, for example, the production and form of technical drawings, writing reports, computer packages and other forms of presentation aids, then the easier you will find the delivery of your project outcomes.

Formal written report

You are already aware that you need to present your logbook for assessment and we have spent some time learning about its style, format and content. In addition to your logbook and your source material files, etc., the most important submission document needed as a permanent record of your project is the *formal written report*. Most of the information required for your report should be readily available in your logbook or in other electronic records that you have kept over the life of your project. It is at this time, when you are ready to produce your report, that accurate recording of data, results, analysis and communications history becomes most useful.

Typical report content

The following general information is given for guidance only. The report content and layout may differ slightly from that given here; it often depends on the house style of the company/institution who will read your report. Specific information on technical report writing may be found in BS 4811, *The presentation of research and development reports* (British Standards Institute 1972). Two useful web addresses for information on report writing standards and academic writing in general are:

- `www.standardsdirect.org/standards`
- `http://en.wikipedia.org` (academic writing).

The project report normally contains:

Title page: This will include a clear and concise unambiguous title for your report and will also contain details on the academic institute, company or business being targeted.

Acknowledgements: These should always appear at the front of a report. They should include individuals, companies or any associated bodies who have provided you with help and advice. This may include assistance with literature, materials, information, finance or any form of resource.

Summary: This should provide a brief statement of the problem, design, its solution, and any further recommendations with respect to modification or development testing. It should be aimed at an executive readership (the decision makers) that do not necessarily have time to read the whole report. In an academic thesis and in research articles, the summary is often replaced by an **abstract** that serves the same purpose as the summary, explaining the main arguments, topics or findings in the article, appearing before the table of contents.

List of contents: This list should provide the page number of all the main headings as they appear in the report. A separate list of all figures, diagrams, sketches, drawings, illustrations and photographs should also be provided giving figure numbers, page numbers, plate numbers and drawing numbers, as appropriate.

Introduction: This should provide all background detail to the project and clearly indicate to the reader the need for the project and whether the author is acting on a given brief or otherwise.

Specification/aim: This section should include the technical detail or proposed deliverable (aim) of the project and give a clear indication of what is required.

Project implementation: This is not a section heading. It merely indicates that this is the section where the project execution is discussed (what you did and the way you went about it). If presenting a design project then details of the design parameters, design solution, production and design evaluation will be given here. If other than a design/build project is being attempted, then the methods adopted to gather information, validate this information, justify your solution and achieve your project aims and objectives will be discussed in this section. Detailed calculations and other technical content that have been used to confirm a design or substantiate the technical/commercial worth of a process would not appear here; these should appear in the appendices.

Results and findings: These would be presented in this section in the form of tables, graphs, statistical diagrams and charts, where their meaning and significance would be explained. The raw data used to formulate these illustrations would not appear here; it would be placed in an appendix.

Conclusions/recommendations/evaluation: It is here that the most significant findings/results of your project should be discussed, their implications identified and interpreted and related to your design or process. In the case of a design project, it is at this stage in your report that you would detail the recommendations and conclusions reached as a result of your final evaluation process. The special features of your project solution should be outlined, its limitations and discrepancies addressed and possible future improvements/modifications/recommendations suggested, for example, the fitting of James's plastic antenna covers!

References: The reference list should contain only those references that you have quoted from and mentioned in the main body of your report. They are normally numbered and quoted in brief at the point of use (see references in the section on report presentation and writing standards that follows).

Appendices: These contain all supporting material necessary for the report which is not essential for inclusion into the main body of the report (e.g. everything that appears before the list of references). The material contained in the appendices should be referred to in the text of the main report. If this cannot be done then that particular piece of information should not appear as an appendix!

Appendices are often identified using a capital Roman numeral (I, II, III, IV, etc.). The following topics are typical of appendix material found in a formal written project report:

- Evaluation of alternative design solutions, including sketches and description of alternatives.
- Details of decision making processes, such as evaluation matrices, decision trees and survey questionnaires.
- Theoretical calculations, mathematical derivations, formulae and repetitive calculations.
- Evaluation of materials and manufacturing processes.
- Detailed costs and pricing policy.
- Details of correspondence associated with the project solution.
- Descriptions of specialist test and development equipment.
- Details of experimentation and record of associated data (this does not include the detail of building a prototype of your design product, this will appear in the main body of your report).
- Computer programmes and evaluation of computer printouts.

Report presentation writing standards

Your final report should always be word processed using 10, 11 or 12 point font for everything except headings. Single line spacing is normally used within paragraphs, although your tutor or industrial project supervisor may prefer you to use 1.5 spacing for your draft report, so that comments and amendments can be made prior to the production of the final version.

Paragraph, line and sentence spacing

- Left and right hand margins are normally indented by 2.5 cm.
- Headings must occur on the same page as the start of the first paragraph.
- Do not leave the odd line of text floating on its own.
- Each full stop, question mark or exclamation mark should be followed by one space.
- Each comma, and other punctuation within the sentence, should be followed by a space.
- Never put spaces before punctuation marks (your grammar checker on your computer will show the green error line if you do this).

Section numbering and headings

Below is an illustration of the standard decimal numbering system used in a report. This has the advantage of easy and accurate cross-referencing when required. Note that conventions for the size and use of emboldening for headings may vary. That given below is quite commonly used, although dependent on the nature and complexity of the report subheadings may not be emboldened. Your project tutor or communications tutor will remind you of the convention expected.

2. FIRST LEVEL HEADING (normally in bold capitals, 14 point font)

2.1 Second level heading (sub-heading, in lower case bold 12 point font)

2.1.1 Third level heading (for paragraphs, not emboldened)

Writing style

You should write in a style to suit your audience. Your choice of language should be such that it is able too be read comfortably by a secondary school aged child. It should be grammatically correct, with no spelling errors. If the report is for industry or is aimed at tendering for engineering business (trying to sell your idea), then it should be written in *third person*, with emphasis being placed on the achievement of the solution. The final report is not a place to air your personal problems or try to defend yourself against under achievement! You are, however, expected to talk about the short falls with your project solution (if there are any) in the discussion/conclusion section of your report and how you would do things differently in the future, given the opportunity.

Referencing

References must be given to all information sources used and listed at the end of the report. Examples of the most common form of referencing 'The Harvard' system are given below.

A single author book reference:

Buch, A., *Fatigue Strength Calculation*, Trans Technical Publishing, 1988

A multiauthor report reference:

Schijve, J.; Broek, D., et al, *Fatigue tests with random and programme loading sequences with and without GAG*, A comparative study on full-scale wing centre sections, NLR Report S-613, 1965

Within your report at the point of use of these references, they will be referred to as (Buch, 1988) and (Schijve and Broek, 1965). Your text may alternatively be identified with a superscript number and the abbreviated reference placed at the bottom of the appropriate page.

Further details on this system may be found by visiting numerous websites. One useful site is: `http://library.curtin.edu.au/referencing/harvard.pdf`

Illustrating the written report

There are many methods that you may use to good effect to present statistical data, illustrate your findings or convey complex ideas in your project report. These include: computer-aided design packages, spread sheets and word processing software. The use and nature of this information and communications technology (ICT) will be taught to you when you study Communications in Unit 2 of your programme. For this reason we will not be covering this topic here. However, to remind you of what is available, there follow a few examples of the types of drawing and charts you may wish to use to illustrate important aspects of your written report. Then, in the final section of this outcome you will find a few tips on preparing and presenting your verbal presentation.

Types of drawing

Engineers use many different types of drawing and diagram to communicate their ideas. These drawings include: block diagrams, flow diagrams, exploded diagrams, circuit diagrams and schematic layout diagrams. It is very useful to be able to understand these different type of drawing and be aware of the circumstances in which they can be used to illustrate important aspects of your project. To avoid confusion (and to ensure that drawings are correctly interpreted) nationally and internationally recognized symbols, conventions and abbreviations are used. These are listed (and their use explained) in British Standard BS 8888. This specification superceded BS 308 in September 2000.

Block diagrams show the relationship between the various elements of a complex system. They can be used to simplify such systems by dividing it into a number of much smaller functional elements. *Figure 3.42* shows a typical example in which the links and dependencies between various elements (inputs and outputs) can be clearly seen.

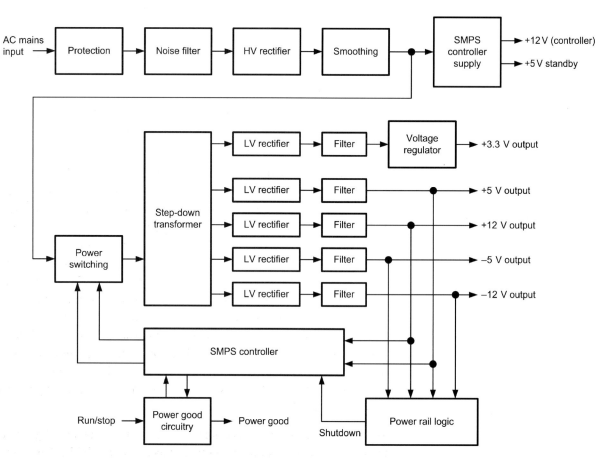

Figure 3.42 Block diagram of a computer power supply

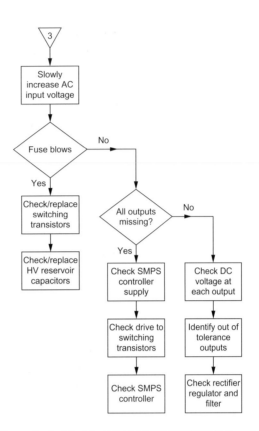

Figure 3.43 Flow diagram for fault finding on the power supply shown in *Figure 3.42*

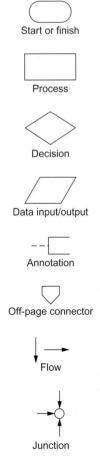

Figure 3.44 Flow diagram symbols

Flow diagrams (or *flowcharts*) are used to illustrate the logic of a sequence of events. They are frequently used in fault finding, computer programming (software engineering) and in process control. They are also used by production engineers when working out the best sequence of operations in which to manufacture a product or component. *Figure 3.43* shows a flowchart for fault location on the computer power supply shown in *Figure 3.42*. The symbols used for flow diagrams are shown in *Figure 3.44*.

Circuit diagrams are used to show the functional relationships between the components in a circuit. The components are represented by symbols and their position in the circuit diagram does not represent their actual position in the final assembly. Pictorial circuit diagrams are referred to as *schematic diagrams* or circuit schematics. If you look back at *Figure 3.4*, you will see that *Figure 3.4a* shows the circuit diagram for an electronic filter unit using standard component symbols in accordance with BS 3939. *Figure 3.4b* shows a layout diagram. This pictorial diagram is a circuit schematic, as is *Figure 3.4c* that shows in schematic form the printed circuit board. The overarching standard governing these schematic diagrams is BS EN 60617 and depending on what electrical application is being represented will depend on which part of this standard is applicable. For guidance on the use of BS 3939 and BS EN60617 (Graphical symbols for diagrams) you should refer to BS 7845:1996.

Apart from electrical and electronic circuits an understanding of general engineering circuits and fluid power circuits is useful. Fluid power circuits such as air-operated pneumatic circuits and oil-operated hydraulic circuits

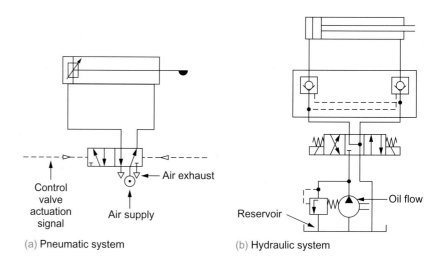

Figure 3.45 Typical pneumatic and hydraulic circuit diagrams using ISO 1219 symbols

share the same symbols using BS 2917 and ISO 1219 standards. To tell which circuit is which pneumatic circuits have open triangles to indicate air exhausts (*Figure 3.45a*), while hydraulic circuits have solid black triangles to indicate oil flow from pump, etc. (*Figure 3.45b*). Also, pneumatic circuits exhaust to the atmosphere, whilst hydraulic circuits have return lines that direct the oil back to the reservoir.

A plumbing circuit (that you first met in Unit 2) is illustrated again in *Figure 3.46*. The symbols shown in the circuit diagram (*Figure 3.46a*) are taken from BS 1553-1. *Figure 3.46b* shows the pictorial circuit schematic of the piping.

Apart from block diagrams, flow diagrams and circuit diagrams that all have a systems focus, you may wish to use general arrangement drawings, detail drawings or exploded views to illustrate your project report. *General arrangement* (GA) drawings (as you will already have learnt from your Communications unit) can be used to show the overall arrangement of an engineering assembly such as a pump, gearbox, motor drive, clutch, etc. GA drawings are often supported by a number of *detail drawings* that provide additional information and clarity on the component parts of the assembly. Detail drawings will include: the dimensions, tolerances, materials specification and finish of the individual component.

You may remember from the time you studied the core Mathematics unit how you handled statistical data and were able to represent this data in the form of statistical charts and graphs. The way in which you present information from your project research, such as data obtained from experimental results, survey results, mathematical functions and other statistics, can have a powerful effect on your target audience, if it is done well. When presenting your project both in writing and orally, you can hold your readers and listeners attention by presenting (sometimes dry facts) in an appealing and eye-catching way that stimulates interest and wins over your audience! Some of the statistical charts and graphs that you might find of most use when compiling your report and/or oral presentation are shown in the figures that follow.

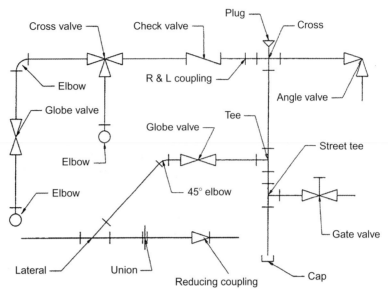

(a) Circuit diagram

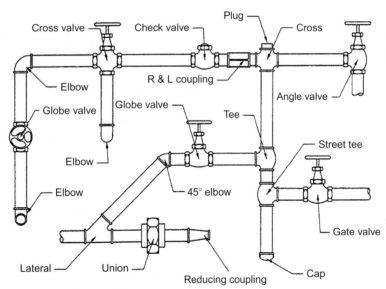

(b) Circuit schematic diagram showing piping

Figure 3.46 A typical plumbing circuit

Figure 3.47 (as seen in Unit 2) illustrates the bond strength against curing time of a particular adhesive joint. The diagram is useful in that it quickly establishes a trend. In this particular case it shows that maximum bond strength of the adhesive is achieved some 25–30 hours after application. Thus in the recommendations for use given to customers, it could be stated that the device 'should not be used for at least 36 hours after joining'.

Test your knowledge

?

TYK

TYK *3.11*

With reference to *Figure 3.47* and considering a line of best fit, estimate the bond strength for curing times of:

(a) 10 hours   (b) 20 hours

UNIT 3

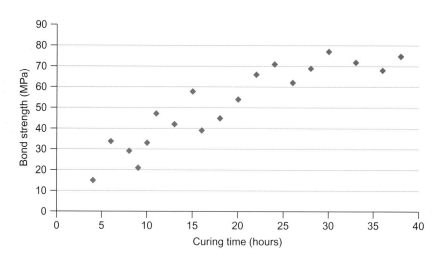

Figure 3.47 Engineering data represented in the form of a scatter diagram

Figure 3.48 shows a ***pictogram***, this type of diagram is particularly useful for illustrating company, monthly and yearly turnover and/or production statistics.

TYK *3.12*

With reference to *Figure 3.48*, how may TTL logic chips were produced in:

(a) 2003?   (b) 2006?

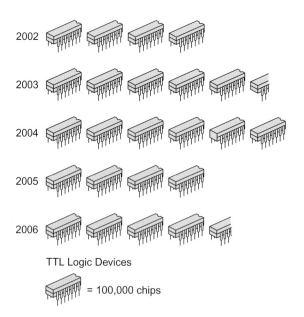

2002

2003

2004

2005

2006

TTL Logic Devices

= 100,000 chips

Figure 3.48 Engineering data represented in the form of a pictogram

Figure 3.49 shows two ***Pi charts*** that illustrate the results of a survey concerning night shift workers at a company.

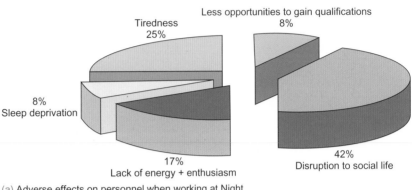

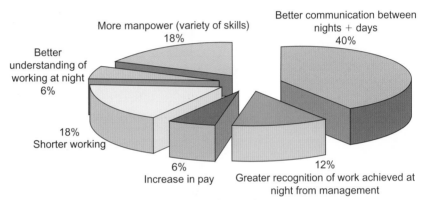

(b) Suggested improvements for the night shift operation

Figure 3.49 Pi charts illustrating the results of a survey

Figure 3.49a shows the five most important adverse effects on night shift workers, while *Figure 3.49b* shows the six top suggestions for improvement. These charts could also be used as part of the oral presentation, to talk about (in this case) the plight of the night shift workers!

Another very simple, yet very effective way of presenting information is the humble ***bar chart***. *Figure 3.50* shows a bar chart being used to indicate the range in nautical miles of a number of different types and marks of aircraft.

KEY POINT

Always use a diagram to present complex results and data

TYK *3.13*

Figure 3.50 shows that the aircraft with the longest range is the Boeing 777-200 LR, at around 9300 nautical miles. If you were an airline operator intending to purchase a new fleet of aircraft, detail four important selection criteria you would use in addition to range.

Finally, here are a few do's and don'ts you should remember, before submitting your final written report.

UNIT 3

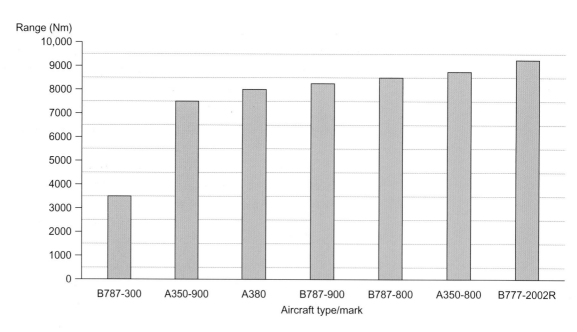

Figure 3.50 Bar chart illustrating range of aircraft by type/mark

Don'ts:
- Write your report on lined paper
- Use plastic sleeves
- Present your report in a loose-leaf ring binder
- fill your report with source material down-loaded from the Internet
- Overload your report with appendix material
- Leave the production of your report to the last minute.

Do's:
- Use your computer spelling and grammar checker *before* printing off your hard copy
- Get someone to proof read your report *before* you print off your hard copy
- Make sure you include your centres cover sheet and your title page, before binding your report
- Make sure you have included your list of references
- Make sure you have a disk back-up copy of your report
- Spiral-bind or fully bind your report for submission
- Hand-in your report on time.

Oral presentation

As you are aware you are required to give an oral presentation of your project to a group that is likely to be made up of your peers, project tutor, visiting lecturers and industrial contacts if applicable. Presenting in front of a group of people can be quite intimidating but it can be made much easier with **good preparation**. Set out in this final section of this outcome are a few tips and hints concerning the preparation and delivery of your oral presentation. Remember that although this part of the project process is being discussed last in this unit, you will in fact have to present your project before you submit your written report.

Preparing your presentation

Students often underestimate the amount of time that needs to be set aside to prepare for a presentation. The more effort you put in, the more likely you are to succeed. Feeling confident in what you are going to say and the way in which you are going to say it helps a great deal with presentation nerves!

You will be told by your project tutor sometime in advance how long you are to be given to talk (this is typically 10–15 minutes) and how long you have to answer questions (typically 5–10 minutes). There will also be time allowed to set up the equipment and props you need to deliver your presentation, such as overhead projectors, computers, viewing screens, flip charts, marker pens, video recorders, white boards, smart boards and any other technology you may or may not need!

The tendency when preparing to give any talk is to write down verbatim everything that you want to say. You must try and avoid this at all costs! You will most certainly not be able to remember everything you have written and will end up reading your project to the audience. This is both embarrassing for them and for you and is a trap that some people fall into when they have not adequately prepared their talk.

So the first important point to remember is that you must ***know your own project*** by the time you come to present it. Then you can feel quite confident in having the minimum amount of information as a memory jogger. This brings us on to what you ought to prepare for your talk.

It should not be necessary to prepare much new or novel material for your talk. Your primary source of information should be your project plan and your logbook. If you look at your project plan it shows you the major 'milestones' (targets) you have set yourself and the way in which you intended to achieve them in order to produce your project solution (deliverable). It is these areas that you need to emphasize in your talk and the detail on each of these areas will be recorded in your logbook.

So an overall plan for your talk should include something on the following topics:

1. Project title
2. Background/need and overview of the project
3. Project aims and objectives (what you hope to achieve)
4. How you went about meeting your aims/achieving your solution (implementing)
5. Your achievements (solution/deliverable, what you finally achieved)
6. Strengths and weaknesses of your achievement
7. Recommendations/Conclusions/Further work
8. Summary
9. Any questions.

Now we need to prepare our presentation centred on the topics (1–9) given above. For the purpose of your presentation these topics can be collected into the major areas of your presentation. These are:

- Introduction/background
- Proposed outcomes/aims

UNIT 3

- Achieving the outcomes
- Evaluating outcomes
- Conclusions and summary.

Topics 1 and 2 (the project title, background and need for the project) can all be grouped under your introduction. Topic 3 is only concerned with your proposed outcome/aim (this is your proposed solution). Topics 4 and 5 cover the areas concerned with the outcomes you achieved, as opposed to those you proposed. Topic 6 is concerned with assessing the worth (evaluating) your outcomes. Finally, topics 7 and 8 are concerned with making recommendations and summarizing your achievement.

The way you present these areas of your project is up to you. For example, you could use a white board or flip chart to write down these areas of your project and then talk about each in turn, although this is not the most interesting method of presentation I can think of! It is more likely that you will either produce one or two overhead transparencies on each area, or even better produce a power point presentation that covers each area. The method(s) you choose will depend on your knowledge of the available technology. It is really worth mastering power point or similar computer-aided presentation packages to make an impression and give you the props around which you can talk.

KEY POINT

Keep the number of presentation slides down to a workable number

The amount of information you put on each overhead slide or page of your power point presentation should be limited, as should the number of slides you use for your presentation. A maximum of 8–10 slides/pages should be your aim. Using power point does allow you the flexibility of having other information readily to hand should you need it. For example, if a particular question has been raised and you need to justify your answer you may well be able to refer back to that information, which had been prepared previously in the form of a power point table, graph or bar chart. Remember also that talking around an illustration is far easier than talking around words because again, without being adequately prepared, you can end up reading what you are presenting! To avoid this, apart from really knowing your project, you can always have a few flip cards that you keep in your hand and refer to as an aide-mémoire like 'the autocue' used by television readers!

In order to get an idea of exactly what you might need to include on any one individual slide/page and how you might present your talk on a particular subject, let us consider an example.

Example 3.2

You have been approached to give a short presentation on a very topical issue where you are required to try and answer the question:

What effect will the increase in air travel have on the environment, what are the future implications for airline operators and what can be done to limit emissions?

This is not only a very topical issue but is also a very contentious one! Our speaker introduces the subject using the following two slides:

Increased Air Travel:
Implications for the environment and airline operators
What can be done?

Overview:

- The engine emission problem and the environment
- Growth of aviation
- Future global policy on engine emissions
- Airline environmental operating costs

Our speaker has produced about the right amount of text in each slide, there is a clear indication of what the talk is about and an overview has been given as part of the introduction to the subject.

This sets up the next set of slides for the talk where each one of the topic areas in the overview can be talked about in turn. So, for example, considering the first bullet, let us look at a sample slide that addresses the possible impact on the environment, leaving aside the engine emissions problem.

Impact on the Environment:

- Sea level rise, damage to coastal areas
- Changes in growing seasons and vegetation patterns
- Dramatic shifts in weather patterns
- Increase in tropical diseases

The subject is gradually being developed through the use of the slides/power point presentation.

For example, the growth of aviation would be considered and statistical tables and charts presented to indicate the actual growth to date compared with the predicted growth of air travel and the effects for example, this would have on CO_2 emissions. Here is a slide that emphasizes this point.

Forecast Growth in UK Aviation Emissions

Estimated total CO_2 including domestic flights in:

1990   15.4 million tonnes
2000   33.4 million tonnes
2030   67 to 75 million tonnes

By 2030 UK aviation could account for 25% of the UK total contribution to global warming

UNIT 3

To add visual impact to the presentation, photographs or short video clips could be used when considering the bullet points within the slides where, for example, to emphasize the effect of rising sea levels and disturbed weather patterns, a picture or video could be shown of a recent natural disaster.

If time is pressing then you may need to 'home in' on a particular aspect of your talk that you wish to emphasize. In this case, for example, it might be the effect that increased growth and emissions have on airline operator costs, so you would expand the topic concerned with airline environmental operating costs from the overview slide.

I hope that the above example has given you some idea of what you need to produce in order to present your oral presentation.

Do remember that there is a whole range of computer software designed to help you in the preparation of both written and oral presentation material and you will have the opportunity of using and practicing with some of this software while studying Unit 2, your Communications unit. In this unit the emphasis has been placed on just one or two tried and tested methods for preparing and presenting your written report and oral presentation.

Giving your presentation

Having prepared your oral presentation carefully you will eventually be confronted with having to deliver it! Good preparation and knowing your project are a strong foundation from which to make a start, but there is still more you can do to ensure you are successful with your delivery.

Timing is a crucial issue and the only real way to get an idea of how long your talk is likely to last is to practice it. If you are really self-conscious then try practicing what you are going to say in front of a mirror. Alternatively, a better way would be to talk through your presentation in front of friends or family. *Do not rush it, do not mumble, and do not speak to the floor or the ceiling!* Do also ensure that you have the project presentation times given to you by your project tutor well in advance of the day. Based on the feedback you get from practicing your talk, you will know whether you should add or leave out content and you will have the time to make the amendments, as necessary.

You next need to make sure that the ***audiovisual aids*** you have selected for your talk will be available on the day and that you know intimately how to operate them! In this way you are not relying on your friends or technicians to help you out and it will also help to stop you panicking if things go wrong! Do not over complicate your presentation by trying to use too wide a range of audiovisual aids. If you are giving a demonstration of some sort of equipment as part of your presentation make absolutely certain that the equipment is reliable and is fully functional on the day. All demonstrations should be set up ready for use before the start of the first presentation. You should also have rehearsed your demonstration to assess its effect on your delivery time. If you are showing a model or your final design product make sure it is sound and safe to handle. If those assessing your talk get oil or grease all over them, they will not be best pleased! If your product is dangerous or dirty to handle, just set it up so that it may be viewed.

KEY POINT

Practice your presentation and become familiar with the audio visual aids you intend to use

Remember to check the criteria against which your presentation is to be assessed. For example, do you have to prepare and present a handout? If so, would a copy of your presentation slides be sufficient? What proportion of the marks go towards your presentation skills, your use of audiovisual aids, your ability to answer questions and your overall manner and bearing? Make sure that you plan to cover all aspects of the *assessment* in your presentation. Also make sure that you are appropriately dressed. Treat your presentation rather like a job interview and dress accordingly. Remember that you are out to 'sell' your project idea/deliverable!

TYK *3.14*

List five key points you intend to follow prior to and during the delivery of your presentation

If you do get nervous when having to speak in public have something in your hand to act as a prop. This might be your cue cards, mentioned earlier, or a short pointer to pick out important facts on the screen. Another tip is to take deep breaths and speak slowly. Finally, do not hide behind the overhead project screen or computer monitor. Be seen and try to act confident even if you are not!

UNIT 3

End of Unit Review

The review problems set out below are designed to reinforce your understanding of the **processes** you need to follow in order to achieve a successful project outcome.

1.   (a) Write down in chronological (time) order a list of activities you need to carry out and evidence you need to produce in order to obtain a pass grade for your project.

     (b) Detail the additional requirements you need to meet in order to gain a merit grade and a distinction grade for your project.

2.   You have been given the task of investigating the feasibility of replacing conventional filament type aircraft navigation lights with LEDs. Carry out an Internet search and/or use other information sources to find five very sound reasons why it would be advantageous to switch to LED illuminated aircraft navigation lights.

3.   Study the following technical specification for an electric kettle and then answer the questions set out below it.

Aquaspeed Kettle Technical Specification

Performance	Electrical rating 1 kW
	Boiling time 90 seconds
Environment	For domestic use only
Maintenance	Replaceable fuse, element and seal
	(no other replaceable parts)
Cost	£14 to £20
Quantity	10,000 (initial production run)
Aesthetics/ ergonomics	Choice of colour: red, green, blue, brown, white
	Easy clean plastic surface
	Good pouring characteristics
	Water level indicator
	Easy grip handle
Physical features	Basic unit 120 mm diameter
	Max jug diameter 140 mm
	Overall height 250 mm
	Capacity 1 litre
	Weight (empty) 0.25 kg
	Weight (full) 0.95 kg
Safety factors	Mains on/off switch
	Shrouded mains connectors between base and jug
	Double insulated element
	Moulded plug
	3 A mains fuse

(a) Use the Internet to determine the British (BS) and International (ISO) standards for electrical domestic appliances.

(b) The technical specification contains some serious errors. Identify the errors and provide details of what is wrong.

(c) The technical specification contains some serious omissions. Identify these and state what is missing.

(d) The technical specification contains insufficient detail. Identify at least three items that need improvement and state the nature of these improvements.

(e) Re-write the extract from the technical specification in relation to the answers you have provided for b, c and d of this question.

4.   The following data refers to the activities that make up a project. Use this information to construct a network diagram and identify the critical path.

Activity	Preceding activity	Expect time
A1	None	1
A2	A1	2
B1	None	2
B2	B1	3
C	A2, B2	2
D	C	5
E	C	3
F	D, E	1

5.   KarKare (an engineering company specializing in automobile accessories) is developing a roof box, mounted on a metal rack. This new product will go into production at the end of the year. The Gantt chart for this project is shown in *Figure 3.51*. Use the Gantt chart to identify:

(a) the number of weeks required to complete the project

(b) activities that are due to be completed at the end of the current month (July)

(c) activities that are ahead of schedule but not yet completed

(d) activities that are more than 2 weeks behind schedule.

6.   The following questions refer to the example presentation we considered (page 224) on the implications for the environment and airline operators of increases in air travel, and what if anything could be done about it. Look back at this sample presentation now and determine by research the answers to the following questions.

(a) What is the current UK policy on aircraft engine emissions and what effect does it have on airline operating costs?

(b) What can be done, and what is being done, by aircraft engine manufacturers and airline operators to help the situation?

(c) Produce two or three slides that you would use to present your answers for part (a) and part (b) of this question.

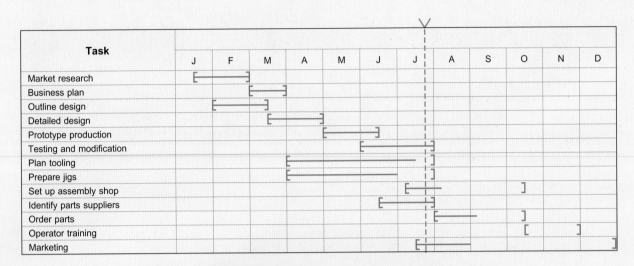

Figure 3.51 Gantt chart for End of unit review question 5

Mathematics is an essential tool for all aspects of
engineering design, manufacture and maintenance.
Engineers for instance use differential calculus to determine
performance characteristics such as speed and acceleration
and apply trigonometry in vehicle chassis design and
steering geometry set up of high performance racing cars.
In order to apply these and other mathematical concepts,
technicians and engineers also require familiarity with
algebraic formulae and their manipulation.

Unit 4

Mathematics for Engineering Technicians

In all branches and at all levels of engineering there is a need for the engineer and engineering technician to be able to solve problems quickly, efficiently and accurately. Most problems in engineering are expressed mathematically. This is done so that important information is presented succinctly and without ambiguity. It is therefore vital that all those who purport to be engineers or engineering technicians have a firm grasp of fundamental mathematical principles and are able to apply these principles to engineering problems.

This unit has been designed to provide the essential mathematics knowledge and understanding necessary for those who wish to practice as engineering technicians and extends and deepens the mathematical concepts already acquired from GCSE or BTEC First Diploma study. It has also been designed to provide you with the numerical skills necessary to successfully complete all other scientific units in the National Engineering programme. In addition, it will act as a foundation for the study of further mathematics for those wishing to progress their studies within this programme and then, coupled with Further Mathematics (Unit 28), enable progression onto engineering programmes offered by Higher Education establishments.

The unit is split into four outcomes covering elements of algebra, trigonometry, statistical data handling and display methods and elementary calculus techniques. Those who feel they have some possible gaps in their prior mathematical knowledge may wish to attempt the introductory material that has been provided on the accompanying website at http://www.key2study/btecnat before tackling the unit outcomes. The material covered includes essential revision on arithmetic, decimal powers, estimation techniques and the use of Pythagoras theorem.

The first of the unit outcomes covers the algebraic techniques concerned with factors, products, powers and indices and their application to the manipulation and factorization of simple algebraic expressions, formulae and linear and quadratic equations. Transposition of formulae although not strictly part of the unit syllabus has also been covered in view of its importance to prospective engineers. The graphical representation and solution of problems concerning linear and simultaneous linear equations is also covered. Then in the final part of outcome one, the nature and use of logarithms is considered, particularly with respect to the linearization of experimental and engineering data. In the second outcome of the unit we start with revision of the trigonometric ratios and then cover aspects of triangular measurement and the graphical representation of the trigonometric functions and see how these may be applied to such topics as vector force diagrams and rotating electrical phasors. Circular measure is then covered and this together with triangular measurement is applied to the solution of problems concerned with finding the surface area and volume of a number of regular solids. Outcome three is concerned with gathering, manipulating and displaying statistical data and finding the mean, median and modal values of such data. Finally in outcome four, the elementary rules of calculus arithmetic are used to solve problems involving simple differentiation and integration of algebraic and trigonometric functions and the application of these methods to determine rates of change and areas under curves.

In order to aid your understanding of technician mathematics, you will find numerous fully worked examples and TYK exercises spread throughout this unit.

BTEC National Engineering. DOI: 10.1016/B978-0-12-382202-4.00004-0

UNIT 4

Algebraic Methods

Introduction

In this section, the first of four outcomes that make up the Mathematics for Technicians unit, you will be introduced to some of the basic rules, laws and methods needed to manipulate algebraic expressions, functions and equations.

Mastery of algebra is vitally important, because without the ability to manipulate equations and functions quickly and efficiently, the remainder of your mathematical and scientific topics become that much more difficult.

When students tell me that they are 'unable to differentiate' or 'unable to solve a problem involving trigonometry', the chances are that what they really mean is that they are unable to simplify or transpose the equations associated with these subjects, which require fluency in using algebra. You will, therefore, not be surprised to discover that we will spend some time studying the underpinning algebra necessary for the satisfactory resolution of mathematical, scientific and engineering problems.

We start by looking at factors, products, powers and indices (exponents) and use these algebraic tools to manipulate and simplify algebraic expressions and formulae. Then, although not strictly part of the outcome, we consider the methods necessary to transpose and evaluate formulae, these methods being of vital importance to those wishing to practice as technicians and engineers. The skills learnt in transposing and evaluating formulae may then be applied to the analytical solution of equations. Indeed, after the introduction of some graphical methods, we consider the analytical and graphical solution of linear and simultaneous equations together with the analytical solution of quadratic equations, using factorization and the quadratic formula. Finally, the nature and use of logarithms is considered, particularly with respect to the linearization of experimental and engineering data.

Factorization, products, powers and indices

Factors

When two or more numbers are multiplied together, each of them, or the product of any number of them (apart from them all), is a factor of the product. This applies to explicit arithmetic numbers and to literal numbers.

So, for example, if we multiply the numbers 2 and 6, we get that $2 \times 6 = 12$, thus 2 and 6 are factors of the number 12. However, the number 12 has more than one set of factors, $3 \times 4 = 12$ so 3 and 4 are also factors of the number 12. We can also multiple $2 \times 2 \times 3$ to get 12. So, the numbers 2, 2 and 3 are yet another set of factors of the number 12. Finally, you will remember that any number n multiplied by 1 is itself, or $n \times 1 = n$. So, every number has itself and 1 as factors. 1 and n are considered *trivial factors* and when asked to find the factors of an explicit or literal number, we will exclude the number itself and 1.

Example 4.1

Find the factors of: (a) 8   (b) *xy*   **(c) 24   (d)** *abc*   **and   (e)** −*n*.

(a) Apart from the trivial factors 1 and 8, which we agreed to ignore, the number 8 has the factors $2 \times 2 \times 2$ and 2×4 since $2 \times 2 \times 2 = 8$ and $2 \times 4 = 8$. Remember that this last set of factors can be presented in reverse order, $4 \times 2 = 8$, but although the factor 2 is repeated the factors 2 and 4 are still the only factors.

(b) Similarly, the literal number *xy* can only have the factors *x* and *y*, if we ignore the trivial factors. Thus the numbers *x* and *y* multiplied together to form the product *xy* are factors of that product.

(c) The number 24 has several sets of factors with varying numbers in each set. First we find the number of sets with two factors. These are:

$24 = 6 \times 4$
$24 = 8 \times 3$
$24 = 12 \times 2$

More than two factors:

$24 = 2 \times 2 \times 6$
$24 = 4 \times 3 \times 2$
$24 = 2 \times 2 \times 2 \times 3$

However if we look closely we see that the number 24 has only six *different* factors: 12, 8, 6, 4, 3 and 2.

(d) So, what about the factors in the number *abc*? Well I hope you can see that the product of each individual factor *a*, *b* and *c* constitute one set of factors. Also *ab* and *c*; *a* and *bc*; and *b* and *ac*, form a further three sets. So, extracting the different factors from these sets we have: *a*, *b*, *c*, *ab*, *ac* and *bc* as the six factors of the number *abc*.

(e) We have two sets of factors here 1 and (e) −*n*, which are the trivial factors, but also the set *n* and −1. Notice the subtlety with the sign change. When dealing with minus numbers, any two factors must have opposite signs.

Products and factorization

There are many occasions when we are required to determine the factors and products of algebraic expressions. Literal numbers are used in expressions and formulae to provide a precise, technically accurate way of generalizing laws and statements associated with mathematics, science and engineering, as mentioned previously. When manipulating such expressions, we are often required to multiply them together (determine their **product**) or carry out the reverse process, that of **factorization**. You will see, in your later studies, that these techniques are very useful when it comes to changing the subject of a particular algebraic formula, in other words when you are required to **transpose a formula**, in terms of a particular variable.

We begin by considering the products of some algebraic expressions. Once we are familiar with the way in which these expressions are 'built-up' we can then look at the rather more difficult inverse process, that of factorization.

Consider the two factors $(1 + a)$ and $(1 + b)$, noting that each factor consists of a **natural number** and a **literal number**. Suppose we are required

UNIT 4

to find $(1 + a)(1 + b)$, in other words, their product. Providing we follow a set sequence, obeying the laws of multiplication of arithmetic, then the process is really quite simple!

In order to describe the process accurately, I need to remind you of some basic terminology. In the factor $(1 + a)$ the natural number *1* is considered to be a **constant** because it has no other value. On the other hand the literal number *a* can be assigned any number of values; therefore, it is referred to as a **variable**. Any number or group of numbers, whether natural or literal, separated by a $+$, $-$ or $=$ sign is referred to as a **term**. So, for example, the expression $(1 + a)$ has *two* terms.

When multiplying $(1 + a)$ by $(1 + b)$ we start the multiplication process from the left and work to the right, in the same manner as reading a book. We multiply each term in the left hand bracket by each of the terms in the right hand bracket, as follows:

$$(1 + a)(1 + b) = (1 \cdot 1) + (1 \cdot b) + (a \cdot 1) + (a \cdot b) = 1 + b + a + ab$$
$$= 1 + a + b + ab$$

Note: 1. The 'dot' notation for multiplication has been used to avoid confusion with the variable *x*.
2. It does not matter in which order the factors are multiplied. Refer back to the commutative law of arithmetic if you do not understand this fact.

Example 4.2

Determine the product of the following algebraic factors:

(a) $(a + b)(a − b)$ **and**   **(b)** $(2a − 3)(a − 1)$.

(a) In this example we proceed in the same manner as we did above, that is, $(a + b)$ $(a − b) = (a \cdot a) + (a)(−b) + (b \cdot a) + (b)(−b) = a^2 + (−ab) + (ba) + (−b^2)$, which by the laws of signs $= a^2 − ab + ba − b^2$ and by the commutative law this can be written as $a^2 − ab + ab − b^2$ or $(a + b)(a − b) = a^2 − b^2$. I hope you have followed this process and recognize the notation for multiplying two bracketed terms.

The product $a^2 − b^2$ is a special case and is known as the *product of two squares*. This enables you to write down the product of any two factors that take the form $(x + y)$ $(x − y)$ as equal to $x^2 − y^2$, where *x* and *y* are any two variables.

(b) Again for these factors, we follow the process where we get:

$$(2a − 3)(a − 1) = 2a \cdot a + (2a)(−1) + (−3)(a) + (−3)(−1) = 2a^2 − 2a − 3a + 3$$

and so,

$$(2a − 3)(a − 1) = 2a^2 − 5a + 3$$

I hope you are getting the idea of how to multiply factors to produce products. So far we have restricted ourselves to just two factors. Can we adopt the process for three or more factors? Well, if you did not know already, you will be pleased to know that we can!

Example 4.3

Simplify the following:

(a) $(x + y)(x + y)(x - y)$ and   **(b)** $(a - b)(a^2 + ab + b^2)$.

(a) These expressions may be simplified by multiplying out the brackets and collecting like terms. I hope you recognize the fact that the product of $(x + y)(x - y)$ is $x^2 - y^2$, then all we need do is multiply this product by the remaining factor to get:

$$(x + y)(x^2 - y^2) = x^3 - xy^2 + x^2y - y^3.$$

Note the convention of putting the variables in alphabetical order and the fact that it does not matter in what order we multiply the factors, the result will be the same.

(b) This is a straightforward product where:

$$(a - b)(a^2 + ab + b^2) = a^3 + a^2b + ab^2 - a^2b - ab^2 - b^3 = a^3 + b^3.$$

Notice that there are six terms resulting from the necessary six multiplications. When we collect like terms and add we are left with the product known as the *difference between cubes*.

Factorization and quadratic expressions

Factorizing is the process of finding two or more factors which when multiplied together will result in the given expression. Therefore, factorizing is really the opposite of multiplication or finding the product. It was for this reason that we first considered the simpler process of finding the product.

Thus, for example, $x(y + z) = xy + xz$. This product resulted from the multiplication of the two factors x and $(y + z)$. If we now unpick the product you should be able to see that x is a **common factor** that appears in **both terms** of the product.

What about the expression $x^2 - 16$? I hope you are able to recognize the fact that this expression is an example of the **difference between two squares**. Therefore, we can write down the factors immediately as $(x + 4)$ and $(x - 4)$. Look back at Example 4.2 if you are unsure. We can check the validity of our factors by multiplying and checking that the product we get is identical to the original expression that we were required to factorize, that is $(x + 4)(x - 4) = x^2 - 4x + 4x - 16 = x^2 - 16$, as required.

Here is an illustration of the technique for finding common factors that is very useful when simplifying algebraic expressions, Example 4.4 illustrates the method that was explained earlier.

Example 4.4

Simplify the following expressions by extracting common factors:

(a) $abc + bcd$ and   **(b)** $wx^2yz - w^2xy^2z + wxyz^2$.

(a) Looking at the two terms in the expression, it can be seen that bc is common to both; therefore, it can be removed from both the terms. After removal, it is necessary to bracket what is left of the original expression, so that the factors are held together by

UNIT 4

the arithmetic operation of multiplication. So, we get $bc(a + d)$ and we know that these are factors because if we multiply these factors back together we get $bca + bcd$ and following convention this may be written in alphabetical order as $abc + bcd$, as before.

(b) A similar argument follows for this example, it is just a question of recognizing what is common to all three terms in the expression. I hope you can see that *wxyz* is common to all three terms and because $x^2 = x \cdot x$ (*x* multiplied by *x*) and $y^2 = y \cdot y$, etc., then on removal of the common factor we are left with the factors **wxyz(x − wy + z)**. Again if these two sets of factors are multiplied up we get back to our original expression.

Quadratic expressions involve at least one term that has an element in it raised to the power two, that is squared. Suppose you are asked to factorize the expression $a^2 − 6a + 9$, how do we go about it? Well a good place to start is with the term involving the highest power of the variable, that is a^2. Remember that **convention dictates we lay out our expression in descending powers of the unknown**, starting with the highest power positioned at the extreme left side of the expression. a^2 can only have factors of itself and 1 or *a* and *a*, therefore, ignoring the trivial factors, $a^2 = a \cdot a$. At the other end of the expression we have the natural number 9, this has the trivial factors *1* and *9* or the factors *3* and *3* or *−3* and *−3*. Note the importance of considering the **negative** case, where from the laws of signs $(−3)(−3) = 9$. So now we have several sets of factors we can try. These are:

(i) $(a + 3)(a + 3)$ or   (ii) $(a − 3)(a − 3)$ or   (iii) $(a + 3)(a − 3)$.

Now, we could try multiplying up each set of factors until we obtained the required result: that is, determine the factors by **trial and error**. This does become rather tedious when there are a significant number of possibilities. So, before resorting to this method, we need to see if we can eliminate some combinations of factors by applying one or two simple rules.

I hope you can see why the factors $(a + 3)(a − 3)$ can be immediately excluded. These are the factors for the difference between two squares, which is not the original expression we need to factorize.

What about the factors $(a + 3)(a + 3)$? Both factors contain only positive terms, therefore any of their products must also be positive by the laws of signs! In our expression $a^2 − 6a + 9$ there is a *minus sign*, so again this set of factors may be eliminated. This leaves us with the factors $(a − 3)(a − 3)$ and on multiplication we get: $(a − 3)(a − 3) = a^2 − 3a − 3a + 9 = a^2 − 6a + 9$, giving us the correct result.

You may have noticed that we left out the sets of factors $(a − 1)(a − 9)$, $(a − 1)(a + 9)$, $(a + 1)(a − 9)$ and $(a + 1)(a + 9)$ from our original group of possibles! Well in the case of $(a + 1)(a + 9)$, this would be eliminated using the laws of signs, but what about the rest?

There is one more very useful technique we can employ when considering just two factors. This technique enables us to check the accuracy of our factors by determining the middle term of the expression we are required to factorize. So, in our case for the expression $2a − 6a + 9$, $−6a$ is the middle term.

The middle term is derived from our chosen factors by multiplying the outer terms, multiplying the inner terms and adding.

So, in the case of the correct factors $(a - 3)(a - 3)$, the outer terms are a and -3, which on multiplication $(a)(-3) = -3a$ and similarly the inner terms $(-3)(a) = -3a$ and so their sum $= -3a + (-3a) = -6a$, as required.

If we apply this technique to any of the above factors involving 1 and 9, we will see that they can be quickly eliminated. For example $(a - 1)(a - 9)$ has an outer product of $(a)(-9) = -9a$ and an inner product of $(-1)(a) = -a$, which when added $= -9a - a = -10a$. This of course is incorrect.

Example 4.5

Factorize the expressions: (a) $x^2 + 2x - 8$ and   (b) $12x^2 - 10x - 12$.

(a) To determine the factors for this expression, we follow the same procedure as before.

First we consider the factors for the outer term x^2 (apart from the trivial factors). We have $x^2 = x \cdot x$ and the factors of the outer term -8 are $(2)(-4)$ or $(-2)(4)$ or $(-1)(8)$ or $(8)(-1)$. So, by considering only outer and inner terms we have the following possible combination of factors:

$$(x + 2)(x + 4),\ (x + 2)(x - 4),\ (x - 2)(x - 4)\ \text{and}$$
$$(x + 1)(x + 8),\ (x + 1)(x - 8),\ (x - 1)(x + 8).$$

Now we eliminate the sets of factors that only have positive terms (by the law of signs). This leaves $(x + 2)(x - 4)$, $(x - 2)(x + 4)$, $(x + 1)(x - 8)$ and $(x - 1)(x + 8)$. The last two sets of factors can be eliminated by applying the outer and inner term rule. If you apply this rule, neither of these sets gives the correct middle term. We are therefore left with the two sets of factors:

$$(x + 2)(x - 4)\quad \text{or}\quad (x - 2)(x + 4).$$

So let us try $(x + 2)(x - 4)$. Applying the outer and inner term rule we get $(x)(-4) = -4x$ and $(2)(x) = 2x$, which on addition gives $-2x$. But we require $+2x$, so these are not the factors. So, finally we try $(x - 2)(x + 4)$, where on application of the rule we get $(x)(4) = 4x$ and $(-2)(x) = -2x$, which on addition gives $4x - 2x = 2x$, as required, where the factors of the expression $x^2 + 2x - 8$ are **$(x - 2)(x + 4)$**.

(b) For the expression $12x^2 - 10x - 12$, we have the added complication of several possibilities for the term involving the square of the variable x, that is $12x^2$. This term could be the product of the factors: $(x)(12x)$ or $(2x)(6x)$ or $(3x)(4x)$ and the right-hand term could be the product of the factors $(-1)(12)$ or $(1)(-12)$ or $(-2)(6)$ or $(2)(-6)$ or $(-3)(4)$ or $(3)(-4)$. By the **rule of signs**, no set of factors can have all positive terms, so these can be eliminated from the possible solutions. This leaves us with:

set 1: $(3x + 1)(x - 12),\ (3x - 1)(x + 12),\ (x - 1)(3x + 12)$ or $(x + 1)(3x - 12)$

set 2: $(3x + 2)(4x - 6),\ (3x - 2)(4x + 6),\ (4x + 2)(3x - 6)$ or $(4x - 2)(3x + 6)$

set 3: $(3x + 3)(x - 4),\ (3x - 3)(x + 4),\ (x + 3)(3x - 4)$ or $(x - 3)(3x + 4)$.

The choice of possible solution does seem to be getting complicated! However, if we apply the **multiplication of outer and inner terms rule** to sets 1 and 3, they are quickly eliminated leaving us with just set 2. Application of the rule, once more, to the factors in set 2 gives us our required solution, where the factors of the expression $12x^2 - 10x - 12$ are **$(3x + 2)(4x - 6)$**.

UNIT 4

Example 4.6

Factorize the expression $3x^3 - 18x^2 + 27x$.

We are now dealing with an unknown variable x raised to the third power! Do not worry, in this particular case the trick is to recognize the **common factor**. If we first consider the integers that multiply the variable we have $3x^3 - 18x^2 + 27x$. All these numbers are divisible by 3, therefore *3* is a common factor.

Also, in a similar manner, the variable itself has a common factor, since all are divisible by x. So, all we need do is remove these common factors to produce the expression:

$$3x(x^2 - 6x + 9)$$

Note that on multiplication you will obtain the original expression, so that $3x$ and $x^2 - 6x + 9$ must be factors.

This expression now has one factor where the greatest power of the unknown is 2. This factor can itself be broken down into two **linear factors** (i.e., where the unknown is raised to the power 1) using the techniques described before, where the factors of the expression $3x^3 - 18x^2 + 27x$ are **$(3x)(x - 3)(x - 3)$**.

Example 4.7

Factorize the following algebraic expressions: (a) $-x^2 + x + 6$, (b) $5x^2y^3 - 40z^3x^2$ and (c) $x^2 - 4x - 165$.

(a) This is a straightforward example of factorizing a **quadratic** expression (an algebraic expression where the highest power of the unknown is 2).

We simply follow the rules you have used in the previous examples. I will go through the procedure once more to remind you.

We first consider the left-hand term $-x^2$, which obeying the rules of multiplication must have factors $-x$ and x (ignoring trivial factors). Also, for the right-hand term we have 2 and 3 or the trivial factors 1 and 6. Again ignoring the trivial factors, we first try 2 and 3.

So, we have the following sets of factors:

$$(-x + 2)(x + 3), (x + 2)(-x + 3), (-x - 2)(x - 3), (x - 2)(-x - 3)$$

Now remembering the rule for determining the middle term (addition of outer and inner products), then by applying the rule and trial and error, we eliminate all sets of factors, except the correct solution, which is: **$(x + 2)(-x + 3)$**.

(b) Here the trick is to recognize the common factor(s) and pull them out behind a bracket. In this case, I hope you can see that x^2 is common to both terms as is the number 5. Then we can write the factors as: **$5x^2(y^3 - 8z^3)$**.

Your answer can always be checked by multiplying up the factors. You should then, of course, obtain the original expression providing your factors are correct and your subsequent multiplication is also correct!

(c) With this example, the only difficulty is in recognizing possible factors for the rather large number 165. Well this is where it is useful to know your fifteen times table! With trial and error and use of your calculator you should eventually find that apart from the trivial factors, the numbers 15 and 11 are factors of 165. Also recognizing that $15 - 11 = 4$, we know that some combination of these numbers will produce the required result. Then by obeying the rules of signs you should eventually find the correct factors as: **$(x - 15)(x + 11)$**.

Finally, before we leave our study of factorization, some common algebraic expressions are shown in *Table 4.1* with their factors.

Table 4.1 Common expressions and their factors

Expression	Factors
$xy + xz$	$x(y + z)$
$x^2 - y^2$	$(x + y)(x - y)$
$x^2 + 2xy + y^2$	$(x + y)^2$
$x^2 - 2xy + y^2$	$(x - y)^2$

TYK 4.1

1. Multiply out the following brackets:
   (a) $(2 - x)(1 + x)$
   (b) $(3 - a)(1 - 2a)$
   (c) $(a - 3)(b - 4)$
   (d) $(3x - 2)(x + 3)$.

2. Multiply out the following brackets and simplify where possible:
   (a) $(x - y)(x - 2y)$
   (b) $(3 + 2p)(p - q)(p)$
   (c) $(y)(2x - z)(y - z)$.

3. Find the common factors in the expression $ab^2c^2 + a^3b^2c^2 + ab^2c$.

4. Find the factors (other than the trivial factors) of:
   (a) 16
   (b) n^2
   (c) $wxyz$.

5. Determine the product of the following:
   (a) $(3a - 1)(2a + 2)$
   (b) $(2 - x^2)(2 + x^2)$
   (c) $ab(3a - 2b)(a + b)$.

6. Factorize the following expressions:
   (a) $x^2 + 2x - 3$
   (b) $a^2 - 3a - 18$
   (c) $4p^2 + 14p + 12$.

UNIT 4

Powers and indices

As you have already seen, when a number is the product of the same factor multiplied by itself this number is called a **power** of the factor. For example, we know that $3 \times 3 = 9$. Therefore, we can say that 9 is a power of 3. To be precise, it is the second power of 3, because two 3s are multiplied together to produce 9. Similarly, 16 is the second power of 4. We may use literal terminology to generalize the relationship between powers and factors.

So, the second power of *a* means $a \times a$ (or $a \cdot a$), this is written as a^2, where *a* is known as the *base* (factor) and 2 is the *index* (or exponent). Thus writing the number 9 in index form we get $9 = 3^2$, where **9 is the second power of 3, 3 is the base (factor) and 2 is the index (exponent)**.

The above ideal can be extended to write arithmetic numbers in index or exponent form, for example, $5^2 = 25$, $9^2 = 81$ and $3^3 = 27$. Notice that the second power of 5 gives the number 25 or $5 \times 5 = 25$. Similarly, 3^3 means the third power of 3, literally $3 \times 3 \times 3 = 27$. The idea of powers and indices (exponents) can be extended to literal numbers. For example:

$a \cdot a \cdot a \cdot a \cdot a$ or a^5 or in general a^m, where a is the base (factor) and the index (or exponent) m is any positive integer. a^m means a used as a factor m times and is read as the 'mth power of a'.

Note that since any number used as a factor once would simply be the number itself, the index (exponent) is not usually written. In other words 'a' means a^1.

Now, providing the base of two or more numbers expressed in index (exponent) form are the same, we can perform multiplication and division on these numbers by adding or subtracting the ***indices*** accordingly.

We will from now on only refer to the index of a number rather than to the exponent of a number as its index so that we avoid confusion with other particular functions, such as the exponential function, which we will study later.

Consider the following literal numbers in index form:

$$x^2 \cdot x^2 = (x \cdot x)(x \cdot x) = x \cdot x \cdot x \cdot x = x^4$$

$$x^2 \cdot x^4 = (x \cdot x)(x \cdot x \cdot x \cdot x) = x \cdot x \cdot x \cdot x \cdot x \cdot x = x^6$$

$$\frac{x^2}{x^2} = \frac{x \cdot x}{x \cdot x} = x^0 = 1$$

$$\frac{x^2}{x^4} = \frac{x \cdot x}{x \cdot x \cdot x \cdot x} = \frac{1}{x \cdot x} = x^{-2}$$

What you are looking for is a pattern between the first two literal numbers that involve multiplication and the second two that involve division.

For multiplication of numbers with the same base we add the indices, and for division of numbers with the same base we subtract the indices in the **denominator** (below the line) from those in the **numerator** (above the line). Remember also that the base number $x = x^1$.

We will now generalize our observations and so formulate the ***laws of indices***.

The laws of indices

In the following laws ***a*** is the common ***base***, ***m*** and ***n*** are the ***indices***. Each law has an example of its use alongside.

1. $a^m \times a^n = a^{m+n}$ $\qquad$ $2^2 \times 2^4 = 2^{2+4} = 2^6 = 64$

2. $\dfrac{a^m}{a^n} = a^{m-n}$ $\qquad$ $\dfrac{3^4}{3^2} = 3^{4-2} = 3^2 = 9$

3. $(a^m)^n = a^{mn}$ $\qquad$ $(2^2)^3 = 2^{2\times3} = 2^6 = 64$

4. $a^0 = 1$ Any number raised to the power 0 is always 1
5. $a^{\frac{m}{n}} = \sqrt[n]{a^m}$ $27^{\frac{4}{3}} = \sqrt[3]{27^4} = 3^4 = 81$
6. $a^{-n} = \dfrac{1}{a^n}$ $6^{-2} = \dfrac{1}{6^2} = \dfrac{1}{36}$

We need to study these laws carefully in order to understand the significance of each.

Law 1 you have already met. It enables us to *multiple numbers* given in index form that have a common base. In the example the common base is 2, the first number raises this base (factor) to the power 2 and the second raises the same base to the power 4. In order to find the result we simply *add* the indices.

Law 2 we have again used when *dividing numbers* with a common base; in this case the base is 3. Note that since division is the opposite arithmetic operation to multiplication. It follows that we should perform the opposite arithmetic operation on the indices, that of *subtraction*. Remember we always subtract the index in the denominator from the index in the numerator.

Law 3 is concerned with raising the powers of numbers. Do not mix this law up with law 1. When *raising powers of numbers* in index form, we *multiple* the indices.

Law 4 you have also met. This law simply states that *any number raised to the power 0 is always 1*. Knowing that any number divided by itself is also 1, we can use this fact to show that a number raised to the power 0 is also 1. What we need to do is use the second law concerning the division of numbers in index form.

We know that $^9/_9 = 1$ or $3^2/3^2 = 3^{2-2} = 3^0 = 1$ which shows that $3^0 = 1$ and in fact because we have used the second law of indices, this must be true in all cases.

Law 5 This rather complicated looking law simply enables us to find the decimal equivalent of a number in index form, where the index is a fraction. All that you need to remember is that the index number above the fraction line is raised to that power and the index number below the fraction line has that number root.

So, for the number $8^{\frac{2}{3}}$, we raise 8 to the power 2 and then take the cube root of the result. It does not matter in which order we perform these operations. So we could have just as easily taken the cube root of 8 and then raised it to the power 2.

Law 6 is very useful when you wish to convert the division of a number to multiplication. In other words, bring, a number from underneath the division line to the top of the division line. *As the number crosses the line we change the sign of its index.* This is illustrated in the example that accompanies this law.

The following examples further illustrate the use of the above laws when evaluating or simplifying expressions that involve numbers and symbols.

Example 4.8

Evaluate the following expressions:

(a) $\dfrac{3^2 \times 3^3 \times 3}{3^4}$ (b) $(6)(2x^0)$ (c) $36^{-\frac{1}{2}}$ (d) $16^{-\frac{3}{4}}$ (e) $\dfrac{(2^3)^2(3^2)^3}{(3^4)}$

(a) $\dfrac{3^2 \times 3^3 \times 3}{3^4} = \dfrac{3^{2+3+1}}{3^4}$ (law 1) $= \dfrac{3^6}{3^4} = 3^{6-4}$ (law 2) $= 3^2 =$ **9**

(b) $(6)(2x^0) = (6)(2) =$ **12** remembering that $x^0 = 1$ (law 4)

(c) $36^{-\frac{1}{2}} = \dfrac{1}{36^{\frac{1}{2}}} =$ (law 6) $= \dfrac{1}{\sqrt{36}}$ (law 5) $= \pm\dfrac{1}{6}$ (note $\pm$ square root)

(d) $16^{-\frac{3}{4}} = \dfrac{1}{16^{\frac{3}{4}}}$ (law 6) $= \dfrac{1}{\sqrt[4]{16^3}}$ (law 5) $= \dfrac{1}{2^3} = \dfrac{1}{8}$

(e) $\dfrac{(2^3)^2(3^2)^3}{3^4} = \dfrac{(2^{3\times2})(3^{2\times3})}{3^4}$ (law 3) $= \dfrac{2^6 \times 3^6}{3^4} = 2^6 \times 3^{6-4}$ (law 2)

$= 2^6 \times 3^2 = 64 \times 9$ (law 6) $=$ **576**

Example 4.9

Simplify the following expressions

(a) $\dfrac{12x^3y^2}{4x^2y}$ (b) $\left(\dfrac{a^3b^2c^4}{a^4bc}\right)\left(\dfrac{a^2}{c^2}\right)$

(c) $[(b^3c^2)(ab^3c^2)(a^0)]^2$

(a) $\dfrac{12x^3y^2}{4x^2y} = 3x^{3-2}y^{2-1}$ (rule 2 and simple division of integers) $=$ **3xy**.

(b) $\left(\dfrac{a^3b^2c^4}{a^4bc}\right)\left(\dfrac{a^2}{c^2}\right) = a^{3+2-4}b^{2-1}c^{4-1-2}$ (rule 2 and operating on like bases) $=$ **abc**.

Note also in the above problem that there was no real need for the second set of brackets, since all numbers were multiplied together.

(c) $[(b^3c^2)(ab^3c^2)(a^0)]^2 = [(b^3c^2)(ab^3c^2)(1)]^2$ (rule 4)

$= [ab^{3+3}c^{2+2}]^2$ (rule 1) $= [ab^6c^4]^2 =$ **$a^2b^{12}c^8$** (rule 3).

To solve the problems in the next example (Example 4.10), you will need to recall your previous knowledge on collecting and adding like terms and multiplying out brackets, as well as the use of indices.

KEY POINT

Remember the laws of precedence by the acronym B.O.D.M.A.S., brackets, of, division, multiplication, addition, subtraction

Example 4.10

Carry out the appropriate arithmetic operations on the following algebraic expressions:
(a) $3ab + 2ac - 3c + 5ab - 2ac - 4ab + 2c - b$ (b) $3x - 2y \times 4z - 2x$
(c) $(3a^2b^2c^2 + 2abc)(2a^{-1}b^{-1}c^{-1})$ (d) $(3x + 2y)(2x - 3y + 6z)$.

(a) All that is required here is to add or subtract like terms, so we get:

$$3ab + 5ab - 4ab + 2ac - 2ac - 3c + 2c - b = \mathbf{4ab - b - c}.$$

(b) Here you need to remember the laws of precedence that you will have studied previously, remembering them by the acronym B.O.D.M.A.S. From this law we carry out multiplication before addition or subtraction. So we get:

$$3x - 8yz - 2x = \mathbf{x - 8yz}.$$

(c) With this expression, when multiplying up the brackets, we need to remember the law of indices for multiplication. Using this law we get:

$$6a^{2-1}b^{2-1}c^{2-1} + 4a^{1-1}b^{1-1}c^{1-1} = 6a^1b^1c^1 + 4a^0b^0c^0 = \mathbf{6abc + 4}.$$

(Do not forget the 4. Remember that any number raised to the power zero is 1 and $4 \times 1 \times 1 \times 1 = 4$.)

(d) This is just the multiplication of brackets where we multiply all terms in the right-hand bracket by both terms in the left-hand bracket so that before any simplification we should end up with $2 \times 3 = 6$ terms.

$$(3x + 2y)(2x - 3y + 6z) = 6x^2 - 9xy + 18xz + 4xy - 6y^2 + 12yz \text{ (6 terms)}$$

and so after simplification, which involves only two terms in this case, we get:

$$\mathbf{6x^2 - 5xy + 18xz - 6y^2 - 12yz}.$$

TYK 4.2

1. Simplify:
   (a) $(a^2b^3c)(a^3b^{-4}c^2d)$
   (b) $(12x^2 - 2)(2xy^2)$.

2. Simplify:
   (a) $\dfrac{1}{2^3} \times 2^7 \times \dfrac{1}{2^{-5}} \times 2^{-4}$

   (b) $\left(\dfrac{16}{81}\right)^{\frac{3}{4}}$

   (c) $\dfrac{b^3b^{-8}b^2}{b^0b^{-5}}$

3. Simplify:
   (a) $(2^2)^3 - 6 \times 3 + 24$
   (b) $\dfrac{1}{2^{-2}} + \dfrac{1}{3^{-2}} - \dfrac{1}{3^{-1}}$

4. Reduce the following fractions to their lowest terms:
   (a) $\dfrac{21a^3b^4}{28a^9b^2}$

   (b) $\dfrac{abc}{d} \div \dfrac{abc}{d^2}$

   (hint turn the last fraction upside down and multiply)

5. Simplify the following expression: $\left[\left(\dfrac{a^3b^2c^4d^2}{a^4bcd}\right)\left(\dfrac{a^2}{c^2}\right)\right]^2$

Transposition and evaluation of formulae

Transposition of formulae

We now consider the transposition of formulae and although this subject does not form a mandatory part of this outcome, it is considered a vital tool for all prospective technicians and engineers. You will see how some of the mathematical methods we have learnt up to now may be used to manipulate these formulae. These same techniques may also be used to simplify and solve linear, simultaneous and quadratic equations analytically, as you will see when we look at these equations later on.

As mentioned earlier, formulae provide engineers with a method of writing down some rather complex relationships and ideas in a very precise and elegant way. For example, the formula $v = u + at$ tells us that the final velocity (v) of, say, a car is equal to its initial velocity (u) plus its acceleration (a) multiplied by the time (t) the car is accelerating. If the car is neither accelerating nor decelerating, then $v = u$ because the acceleration $a = 0$ and $0 \times t = 0$, as you already know. I think you are already beginning to realize that to explain the meaning of one simple formula requires rather a lot of words! It is for this reason that formulae are used rather than just words to convey engineering concepts. If you have met the transposition of formulae before, then this section will serve as good revision and consolidation and enable you to solve the equations you will meet later.

Before considering the techniques needed to manipulate or transpose formulae, we first need to define some important terminology. We will use our equation of motion $v = u + at$ for this purpose.

Term This is defined as any variable or combination of variables separated by a +, a − or an = sign. You will have already met this definition, if you studied the introductory material on the laws of arithmetic. Therefore, in our formula, according to the definition, there are three (3) terms, they are: v, u and at.

Variable These are represented by literal numbers, which may be assigned various values. In our case, v, u, a and t are ***all variables***. We say that v is a ***dependent variable*** because its value is determined by the values given to the ***independent variables***, u, a and t.

Subject The subject of a formula ***sits on its own on one side of the equals sign***. Convention suggests that the subject is placed to the left of the equals sign. In our case, v is the subject of our formula. However, the position of the subject whether to the left or to the right of the equals sign makes no difference to the sense of a formula. So, $v = u + at$ is identical to $u + at = v$, the subject is simply pivoted about the equals sign.

In the following examples we will ***transpose simple formulae***, using the basic arithmetic operations of addition, subtraction, multiplication and division, to rearrange the subject of a formula.

> **KEY POINT**
>
> Formulae enable engineers to write down complex ideas in a very precise way

> **KEY POINT**
>
> A term in an algebraic formula or expression is always separated by a plus (+), a minus (−) or an equals (=) sign

Example 4.11

Transpose the following formula to make the letter in brackets the subject.

(1) $a + b = c$ (b)　(2) $y − c = z$ (c)　(3) $x = yz$ (y)　and　(4) $y = \dfrac{a}{b}$ (b).

(1) In this formula we are required to make b the subject; therefore, b needs to sit on its own on the left-hand side (LHS) of the equals sign. To achieve this we need to remove a term from the LHS. We ask the question, how is a attached to the LHS? It is in fact added, so to remove it to the right-hand side (RHS) of the equals sign we apply the inverse arithmetic operation, that is, we subtract it. To maintain the equality in the formula we need in effect to subtract it from both sides, that is, $a - a + b = c - a$, which of course gives

$$b = c - a$$

You will remember this operation as: *whatever we do to the LHS of a formula or equation, we must do to the other or, when we take any term over the equals sign we change its sign.*

(2) Applying the procedure we used in our first example to $y - c = z$, then we subtract y from both sides to give $y - y - c = z - y$, which again gives $-c = z - y$. Now unfortunately in this case we are left with $-c$ on the LHS and we require $+c$, or just c as we normally write it when on its own. Remembering from your study of fundamentals that a minus multiplied by a minus gives a plus and that any number multiplied by one is itself then:

$$(-1)(-c) = (-1)(z) - (y)(-1) \text{ or } c = -z + y$$

and exchanging the letters on the RHS gives

$$c = y - z$$

Now all that we have done, in this rather long-winded procedure, is to multiply every term in the formula by (-1) or as you may remember it, we have changed the sign of every term in order to eliminate the negative sign from the subject of our formula.

(3) Now with the formula, $x = yz$ we have just two terms and our subject z is attached to y by *multiplication*. So, all we need to do is *divide* it out. In other words, apply the inverse arithmetic operation, then we get:

$$\frac{x}{y} = \frac{yz}{y} \quad \text{or} \quad \frac{x}{y} = z$$

and reversing the formula about the equals sign gives

$$z = \frac{x}{y}$$

KEY POINT

When transposing a formula for a variable, you are making that variable the subject of the formula

(4) With the formula $y = \dfrac{a}{b}$, then b is attached to a by *division*, so we *multiply* it out to give:

$$by = \frac{ab}{b} \quad \text{or} \quad by = a$$

This leaves us with y attached to b by *multiplication* so to eliminate y we *divide* it out, then:

$$\frac{by}{y} = \frac{a}{y} \quad \text{or} \quad b = \frac{a}{y} \text{ as required.}$$

In the above examples I have shown every step in full. We often leave out the intermediate steps, so for example if $p = (q - m)/r$ and we wish to make q the subject of the formula, then multiplying both sides by r gives $pr = q - m$. Adding m to both sides we get $pr + m = q$, and reversing the formula, $q = pr + m$.

What about *transposing simple formula with common factors*? You have already learnt to factorize, so now we can put that knowledge to good use.

KEY POINT

Always change the sign of a term, variable or number when crossing the equals (=) sign

UNIT 4

Example 4.12

Transpose the following formulae to make c the subject:

(a) $a = c + bc$ (b) $2c = pq + cs$ (c) $x = \dfrac{ab + c}{a + c}$

(a) All we need do here is to take out c as a common factor, then:

$$a = c(1 + b)$$

now dividing through by the *whole* of the bracketed expression, we get:

$$\frac{a}{1 + b} = c$$

and reversing the formula we get:

$$c = \frac{a}{1 + b}$$

(b) Transposition of this formula is essentially the same as in (a), except that we first need to collect all the terms involving the common factor onto one side of the formula, so subtracting cs from both sides gives $2c - cs = pq$ and after taking out the common factor we get $c(2 - s) = pq$ and after division by the *whole of the bracketed expression* we get $c = \dfrac{pq}{(2 - s)}$ or $\mathbf{c = \dfrac{pq}{2 - s}}$ since there is no longer any need for the bracket.

(c) Now on multiplication of both sides by $a + c$ we get $x(a + c) = ab + c$. Notice that we have placed $a + c$ in brackets. This is very important because x is multiplied by both a and c. When transferring complicated expressions from one side of the formula to the other, a convenient way of doing it is to *place the expression in a bracket*, then move it.

Now we can remove the brackets by multiplying out. Having transferred the whole expression, we get

$$ax + cx = ab + c$$

and collecting the terms containing c onto one side gives $cx - c = ab - ax$ and taking out c as a common factor we get $c(x - 1) = ab - ax$ and again, after dividing out the bracketed expression, we get

$$c = \frac{ab - ax}{x - 1} \quad \text{or} \quad \mathbf{c = \frac{a(b - x)}{x - 1}}$$

Transposing of formulae involving powers and roots. You may remember from your previous study that when we write a number, say, 25 in index form we get $5^2 = 25$, where the 5 is the base and the 2 is the *index* or *power*. Look back at the work we did on indices, in particular, on *powers* and the *laws of indices*. We are going to use this knowledge to transpose formulae that involve terms with powers they may be *positive, negative or fractional*, for example p^2, p^{-3} or $p^{\frac{1}{2}} = \sqrt{p}$, respectively.

If $x^2 = yz$ and we wish to make x the subject of the formula, all we need to do is to take the *square root of both sides*, that is, $\sqrt{x^2} = \sqrt{yz}$ or $x = \sqrt{yz}$. In index form this is the equivalent to $x^{(2)\left(\frac{1}{2}\right)} = y^{(1)\left(\frac{1}{2}\right)}z^{(1)\left(\frac{1}{2}\right)}$ or $x^1 = y^{\frac{1}{2}}z^{\frac{1}{2}}$. Similarly, if we are required to make x the subject of the formula $\sqrt{x} = yz$, then all that we need to do is to square both sides, that is, $\left(\sqrt{x}\right)^2 = (yz)^2$ or $x = y^2z^2$. Suppose we wish to make p the subject in the formula $\left(\sqrt[3]{p}\right)^2 = abc$. Then writing this formula in index form we have, $p^{\frac{2}{3}} = a^1 b^1 c^1$ and to get p^1 we need to multiply both sides of the formula by the power $\frac{3}{2}$ so $p^{\left(\frac{2}{3}\right)\left(\frac{3}{2}\right)} = (a^1 b^1 c^1)^{\frac{3}{2}}$ or $p = (abc)^{\frac{3}{2}}$ or $p = \left(\sqrt{abc}\right)^3$. What the above working shows is that *if we wish to find the subject of a formula, that itself has been raised to a power, we multiply it by its inverse power*. It does not matter whether this power is greater than one (>1) or less than one (<1); in other words whether it is a power or a root, respectively.

KEY POINT

If we wish to find the subject of a formula, that itself has been raised to a power, we multiply it by its inverse power

---

Example 4.13

(a) If $a = b\sqrt{c}$, make c the subject of the formula.

(b) If $Z = \sqrt{R^2 + X^2}$, transpose the formula for X.

(c) If $a^{\frac{3}{4}} + b^2 = \dfrac{c - d}{f}$, make a the subject of the formula.

(a) Our subject c is under the square root sign, so our first operation must be to square both sides and release it!

Squaring both sides:

$$a^2 = \left(b\sqrt{c}\right)^2 \quad \text{or} \quad a^2 = b^2\left(\sqrt{c}\right)^2 \text{ then } a^2 = b^2 c$$

Dividing through by b^2:

$$\frac{a^2}{b^2} = c$$

and reversing:

$$c = \frac{a^2}{b^2}$$

so that:

$$c = \left(\frac{a}{b}\right)^2$$

(b) Again we need to release X from underneath the square root sign.

Squaring both sides: $Z^2 = R^2 + X^2$

Subtracting R^2 from both sides: $Z^2 = R^2 + X^2$ and reversing $X^2 = Z^2 - R^2$

Then taking the square root of both sides we get $x = \sqrt{Z^2 - R^2}$

Note that we square root the *whole* of both sides!

(c) Isolating the term involving a by subtracting b^2 from both sides, we get:

$$a^{\frac{3}{4}} = \left[\frac{c - d}{f}\right] - b^2.$$

Now, multiplying *all* of both sides by the inverse power, that is by:

$$\left(\frac{4}{3}\right),$$

we get:

$$a^{\left(\frac{3}{4}\right)\left(\frac{4}{3}\right)} = \left[\left(\frac{c - d}{f}\right) - b^2\right]^{\frac{4}{3}}$$

and so,

$$a = \left[\left(\frac{c - d}{f}\right) - b^2\right]^{\frac{4}{3}}$$

KEY POINT

When carrying out any transposition, remember that the object of the transposition is to isolate the term involving the subject then obtain the subject from the term by using multiplication or division

KEY POINT

Multiplying every term by (-1) is the same as changing the sign of every term

---

So far we have been transposing relatively simple formulae where the order in which we carried out the operations was reasonably obvious.

With more complex formulae and equations, you may have doubts about the order of operations. If you are in doubt, then try using this sequence:

1. Remove root signs, fractions and brackets (in an order which suits the particular problem)
2. Rearrange the formula for the subject, following the arithmetic operations
3. Collect all terms on one side of the equation that contain the subject
4. Take out the subject as a common factor, if necessary
5. Divide through by the coefficient of the subject
6. Take roots, as necessary.

Note that the **coefficient** is a decimal number multiplying a literal number in a formula. So, for example, in the simple formula $3b = cde$, the number 3 is the *coefficient of b* and on division by 3, we get $b = (cde)/3$.

The above procedure is best illustrated by the following example.

Example 4.14

(a) If $s = ut + \frac{1}{2}at^2$, transpose the formula for a.

(b) If $\dfrac{D}{d} = \sqrt{\dfrac{f + p}{f - p}}$, transpose the formula for f.

(a) Following our procedure, there is really only one fraction that we can eliminate: it is $\frac{1}{2}$. If we multiply every term by the inverse of $\frac{1}{2}$, that is $\frac{2}{1}$, we get $2s = 2ut + at^2$ Subtracting $2ut$ from both sides gives $2s - 2ut = at^2$ and dividing both sides by t^2 gives $((2s - 2ut)/t^2) = a$. Reversing the formula and pulling out the common factor gives $a = [(2(s - ut))/t^2]$. Alternatively, remembering your laws of indices, we can bring up the t^2 term and write the formula for a as $\boldsymbol{a = 2t^{-2}(s - ut)}$.

(b) We again follow the procedure, first clearing roots, then fractions, in the following manner.

Squaring:

$$\left(\frac{D}{d}\right)^2 = \frac{f + p}{f - p} \quad \text{or} \quad \frac{D^2}{d^2} = \frac{f + p}{f - p}$$

and multiplying both sides by the terms in the denominator, or *cross-multiplying d^2* and $(f - p)$ gives:

$$D^2(f - p) = d^2(f + p)$$

and $D^2f - D^2p = d^2f + d^2p$. So collecting terms on one side containing the subject, we get $D^2f - d^2f = d^2p - D^2p$. After pulling out common factors, we have $f(D^2 - d^2) = (d^2 - D^2)p$ and dividing both sides by $(D^2 - d^2)$ yields the result: $\boldsymbol{f = \dfrac{(d^2 - D^2)p}{D^2 - d^2}}$.

Evaluation of formulae

We have spent some time transposing formulae and equations. We will now combine transposition with the substitution of numerical values into formulae and equations. In example 4.14 we transposed the formula $s = ut + \frac{1}{2}at^2$ for a. This equation relates the *distance s* with the *initial velocity u, acceleration a* and the *time t*. Suppose we want to find a numerical value for the acceleration a. We may first transpose the formula in terms of a as we did earlier to give

$$a = \frac{2(s - ut)}{t^2}.$$

Then, if we are told that distance $s = 125\,\text{m}$, initial velocity $u = 20\,\text{ms}^{-1}$ and time $t = 5s$, we may substitute these directly into our rearranged formula for a to give

$$a = \frac{[2(125 - (20)(5)]}{5^2} = \frac{(250 - 200)}{25}$$

$$= \frac{50}{25} = 2\,\text{ms}^{-2}$$

In this procedure we have found a numerical value for one of the variable, a. In other words we have *evaluated* the formula.

In this fairly simple formula it was not strictly necessary to transpose the formula for a before substituting in the given numerical values. However, we need to be careful with this direct substitution method, especially if the formula is complex. It is easy to assign the incorrect numerical value to a variable or experience difficulties in simplifying numbers and letters for the desired unknown.

In the next example we will combine the idea of substitution with that for solving a simple equation where the power of the unknown is one.

If you are unsure what this means, look back at your work on powers and exponents where you will find numbers written in index form. As a brief reminder, 5^2 is the number 5 raised to the power 2; in other words, five squared. If the literal number z is an unknown it is, in index form, z^1 or z raised to the power one. We normally ignore writing the power of a number when raised to the power one *unless* we are simplifying expressions where numbers are given in index form and we need to use the *laws of indices*, which you learnt in the fundamentals section.

Example 4.15

If $a^2x + bc = ax$, find x, given $a = -3$, $b = -4$, $c = -1$.

In this case we will substitute the numerical values before we simplify the formula. Then

$$(-3)^2x + (-4)(-1) = (-3)x$$
$$9x + 4 = -3x$$
$$9x + 3x = -4$$
$$12x = -4$$
$$x = \frac{-4}{12}$$
$$\text{then } x = -\frac{1}{3}$$

Notice the important use of brackets on the first line; this prevents us from making mistakes with signs!

In the next example where we use the formula for centripetal force, we will solve for the unknown m, using both direct substitution and by transposing first and then substituting for the values.

Example 4.16

If $F = \dfrac{mV^2}{r}$, find m when $F = 2560$, $V = 20$ and $r = 5$.

Then, by direct substitution:

$$2560 = \frac{m(20)^2}{5} \quad \text{so} \quad (2560)(5) = m(400)$$

$$400\,m = 12800$$

$$m = \frac{12800}{400} \quad \text{then} \quad \boldsymbol{m = 32}$$

Alternatively, we can transpose the formula for m and then substitute for the given values:

$$F = \frac{mV^2}{r} \quad \text{and} \quad Fr = mV^2 \quad \text{so} \quad \frac{Fr}{V^2} = m$$

$$\text{then } m = \frac{Fr}{V^2} \quad \text{and} \quad m = \frac{(2560)(5)}{(20)^2} = \frac{12800}{400} = \boldsymbol{32}$$

giving the same result as before.

In our final example on substitution, we use a formula that relates electric charge Q, resistance R, inductance L and capacitance C.

Example 4.17

Find C if $Q = \dfrac{1}{R}\sqrt{\dfrac{L}{C}}$ where $Q = 10$, $R = 40\Omega$, $L = 0.1$.

$$QR = \sqrt{\frac{L}{C}} \quad \text{and squaring both sides gives}$$

$$(QR)^2 = \frac{L}{C} \quad \text{or} \quad Q^2R^2 = \frac{L}{C}$$

$$C(Q^2R^2) = L \text{ then, } C = \frac{L}{Q^2R^2}$$

Substituting for the given values, we get:

$$C = \frac{0.1}{10^2 40^2} = \boldsymbol{6.25 \times 10^{-7}\, Farads}$$

TYK 4.3

1. Transpose the formula $v = u + at$ for u.

2. Transpose the formula $v^2 = u^2 + 2as$ for v.

3. Transpose the formula $F = mg + ma$ for m.

4. Transpose the formula $\rho_1 A_1 v_1 = \rho_2 A_2 v_2$ for A_1.

5. If $F = \dfrac{mv^2}{r}$, find F when $m = 40$, $v = 10$ and $r = 0.2$.

6. Transpose the formula $v = \pi r^2 h$ for r.

7. If the value of the resistance to balance a Wheatstone bridge is given by the formula:

   $R_1 = \dfrac{R_2 R_3}{R_4}$, find R_2 if $R_1 = 3$, $R_3 = 8$ and $R_4 = 6$.

8. If $Q = A_2 \sqrt{\dfrac{2gh}{1 - \left(\dfrac{A_2}{A_1}\right)^2}}$ find Q

   when $A_1 = 0.0201$, $A_2 = 0.005$, $g = 9.81$ and $h = 0.554$.

9. Make a the subject of the formula $S = \dfrac{n}{2}[2a + (n - 1)d]$.

10. Transpose the equation, $\dfrac{x - a}{b} + \dfrac{x - b}{c} = 1$, for x.

11. If $X = \dfrac{1}{2\pi f C}$, calculate the value of C when $X = 405.72$ and $f = 81.144$.

12. Simplify $\dfrac{1}{x - 1} - \dfrac{1}{x + 1} - \dfrac{3}{2(x^2 - 1)}$.

Solution of equations

In the following section we will be looking at the analytical and graphical solution of linear and simultaneous equations, together with the analytical solution of quadratic equations by factorization and by use of the quadratic formula. These equations appear a great deal in engineering problems and require a certain amount of algebraic rigour for their solution, as well as an understanding of graphical methods.

Solution of linear equations

Although you may not have realized, you have already solved linear *equations analytically*. However, before we start our study of the *graphical solution* of equations, here is an example which shows that in order to solve simple equations analytically, all we need to do is apply the techniques you have learnt when transposing and manipulating formula. The important point about equations is that the *equality sign must always be present*!

Example 4.18

Solve the following equations:

(a) $3x - 4 = 6 - 2x$.

(b) $8 + 4(x - 1) - (5x - 3) = 2(5 + 2x)$.

(c) $\dfrac{1}{2x + 3} + \dfrac{1}{4x + 3} = 0$.

(a) For this equation, all we need do is to collect all terms involving the unknown x on to the left-hand side of the equation simply by using our rules for transposition of formula.

Then $3x + 2x - 4 = 6$; so, $3x + 2x = 6 + 4$ or $5x = 10$ and so $\boldsymbol{x = 2}$.

(b) In this equation we need first to multiply out the brackets, then collect all terms involving the unknown x onto one side of the equation and the numbers onto the other side, then divide out to obtain the solution.

$$8 + 4(x - 1) - (5x - 3) = 2(5 + 2x)$$
$$8 + 4x - 4 - 5x + 3 = 10 + 4x$$
$$4x - 5x - 4x = 10 - 3 - 8 + 4$$

So:

$$-5x = 3$$
$$x = \frac{3}{-5} \quad \text{or} \quad x = -\frac{3}{5}$$

Note the care taken with the signs!

(c) To solve this rather complex looking equation, we need to manipulate fractions, or apply the inverse arithmetic operation, to every term! The simplification to obtain x using the rules for transposition is laid out in full below.

$$\frac{1}{2x + 3} + \frac{1}{4x + 3} = 0$$

$$\frac{1 \ (2x + 3)}{2x + 3} + \frac{1 \ (2x + 3)}{4x + 3} = 0 \ (2x + 3)$$

$$1 + \frac{2x + 3}{4x + 3} = 0$$

and

$$1 \ (4x + 3) + \frac{(2x + 3)(4x + 3)}{4x + 3} = 0 \ (4x + 3)$$

$$(4x + 3) + (2x + 3) = 0$$

$$\text{or } 4x + 3 + 2x + 3 = 0$$

$$6x = -6$$

$$x = -1.$$

KEY POINT

For all linear equations the highest power of the unknown is 1

We could have carried out the multiplication by the terms in the denominator in just one operation simply by multiplying every term by the product $(2x + 3)(4x + 3)$. Notice also that when multiplying any term by zero, the product is always **zero**.

Graphical axes, scales and coordinates

To plot a graph, you know that we take two lines at right angles to each other (*Figure 4.1*). These lines, being the axes of reference, are known as **Cartesian coordinates** or *rectangular coordinates* where their intersection at the point zero is called the *origin*. When plotting a graph a suitable scale must be chosen; this scale need not be the same for both axes. In order to plot points on a graph we need to identify them by their coordinates. The coordinate points (2, 4) and (5, 3) are shown in *Figure 4.1b*. Note that the *x-ordinate* or *independent variable* is always quoted first. Also remember that when we use the expression plot *s against t*, then all the values of the *dependent* variable *s* are plotted up the *vertical axis* and the other *independent variable* (in this case *t*) are plotted along the *horizontal axis*.

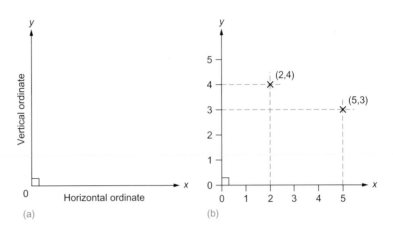

Figure 4.1 Axes and coordinates of graphs

You will have met the concept of dependent and independent variables during your study on fundamentals, just remember that the values of the dependent variable are determined by the values assigned to the independent variable. So, for example, in the simple equation $y = 3x + 2$, if $x = 2$ then $y = 8$ and if $x = -2$ then $y = -4$, and so on. So, to plot a graph all we need to do is:

1. Draw the two axis of reference at right angles to each other.
2. Select a suitable scale for the dependent and independent variable, or both.
3. Ensure that values of the dependent variable are plotted up the vertical axis.
4. Produce a table of values as necessary to aid your plot.

If the graph is either a straight line or a smooth curve, then it is possible to use the graph to determine other values of the variables, apart from those given.

Example 4.19

Plot the graph of y against x given the following coordinates:

x (m)	0	1	2	3	4	5	6	7	8	9	10
y (m)	2	5	8	11	14	17	20	23	26	29	32

and find the corresponding value of y when x = 5.5 and the value x when y = 38.

The graph is plotted below (*Figure 4.2*). Note that when we join the coordinate points we get a straight line. The x axis scale is 1 cm = 1 m and the y axis scale is 1 cm = 2 m.

To find the value of y corresponding to x = 5.5, we find 5.5 on the horizontal axis and draw a vertical line up until it meets the graph at point P, then draw a horizontal line until it meets the vertical y ordinate and read off the value which is *18.5*.

Should we wish to find a value of x given y, we reverse this procedure. So, to find the value of x corresponding to y = 38, we first find 38 on the y axis and draw a horizontal line across to meet the line. However, using the tabulated values, in this case the line does not extend this far. It is therefore necessary to *extend or extrapolate* the line. In this particular case it is possible to do this, as shown above, where reading vertically down we see that the intercept is at x = 12. This process involves extending the graph, without data being available to verify the accuracy of our extended line. Great care must be taken

UNIT 4

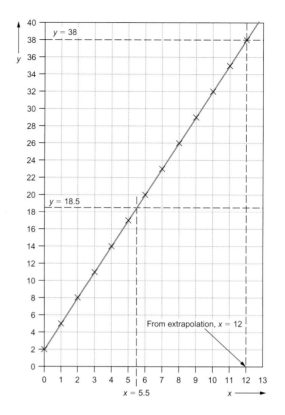

Figure 4.2 A straight line graph

when using this process to prevent excessive errors. In the case of a straight-line graph or linear graph, this is acceptable practice. This process is commonly known as graphical extrapolation.

Graphs of linear equations

In the above example all values of the coordinates are positive. This is not always the case and to accommodate negative numbers, we need to extend the axes to form a cross (*Figure 4.3*), where both positive and negative values can be plotted on both axes.

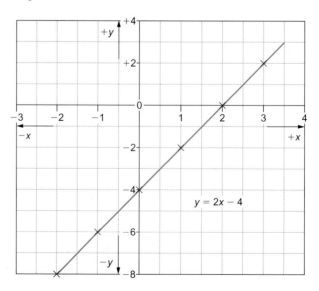

Figure 4.3 Graph of the equation $y = 2x - 4$

Figure 4.3 not only shows the positive and negative axes, but also the plot of the equation $y = 2x - 4$. To determine the corresponding y ordinates shown for values of x between -2 and 3, we use a table.

x	-2	-1	0	1	2	3
$2x$	-4	-2	0	2	4	6
-4	-4	-4	-4	-4	-4	-4
$y = 2x - 4$	-8	-6	-4	-2	0	2

So, for example, when $x = -2$, $y = 2(-2) - 4 = -4 - 4 = -8$.

The scale used on the *y-axis* is $1\,\text{cm} = 1$ unit and on the *x-axis* $2\,\text{cm} = 1$ unit.

This equation, where the highest power of the variable x, y is 1.0, is known as an *equation of the first degree* or a *linear equation*.

Now every linear equation may be written in *standard form*, that is:

$$y = mx + c$$

So, for our equation $y = 2x - 4$, which is in the standard form, $m = 2$ and $c = -4$.

Also, every linear equation may be re-arranged so that it is in standard form. For example:

$$4y + 2 = 2x - 6 \text{ then re-arranging for } y$$

$$4y = 2x - 6 - 2 \quad \text{or} \quad 4y = 2x - 8 \text{ and on division by 4}$$

$$y = \frac{2}{4}x - \frac{8}{4} \quad \text{or} \quad y = \frac{1}{2}x - 2 \text{ where } m = \frac{1}{2} \text{ and } c = -2.$$

Determining m and c for the equation of a straight line

In *Figure 4.4*, point A is where the straight line cuts the *y-axis* and has coordinates $x = 0$ and $y = c$. Thus c in the equation $y = mx + c$ is the point

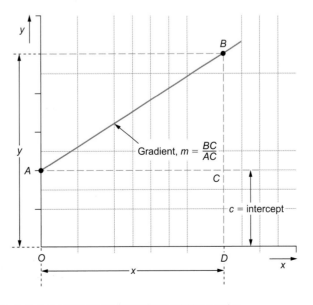

Figure 4.4 Graph demonstrating the meaning of the variables c and m

where the line meets the y-axis, when the value of $x = 0$ or *the variable*
$c =$ the y intercept when $x = 0$.

Also, from *Figure 4.4*, the value $\dfrac{BC}{AC}$ is called the *gradient* of the line. Now the length

$$BC = \left(\frac{BC}{AC}\right)AC = AC \times \text{gradient of the line}$$
$$\begin{aligned} y &= BC + CD = BC + AO \\ &= AC \times \text{the gradient of the line} + AO \\ &= x \times \text{the gradient of the line} + c \end{aligned}$$

But $y = mx + c$. So, it can be seen that:

$m =$ the gradient of the line
$c =$ the intercept on the y-axis

Example 4.20

(a) Find the law of the straight line illustrated in *Figure 4.5*.

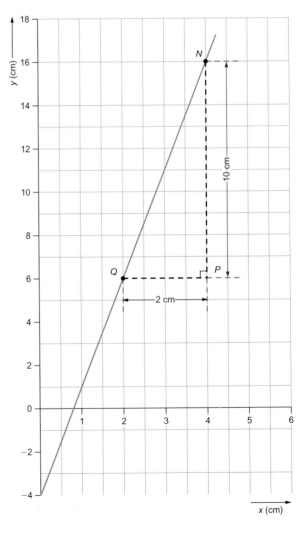

Figure 4.5 Figure for Example 4.20

(b) If a straight line graph passes through the point (−1,3) and has a gradient of 4 find the values of *m* and *c* and then write down the equation of the line.

(a) Since the intercept *c* is at the origin, it can be read-off the graph as −4. The value of *m*, the gradient of the line, is found by taking convenient values of *x* and *y*, then the gradient *m* from the graph $= \frac{NP}{QP} = \frac{10}{2} = 5$. So, the equation of the line $y = mx + c$ is **$y = 5x + 4$**.

(b) We are given the gradient $m = 4$, therefore $y = 4x + c$ and this line passes through the point (−1,3). So we know that $y = 3$ when $x = −1$ and substituting these values into the equation of the straight line gives $3 = 4(−1) + c$ and so $c = 7$. Then the equation of the line is **$y = 4x + 7$**.

Note that in the questions given in examples 4.21 and 4.22, the gradient (or slope) of the straight lines were both *positive*. Straight-line graphs can also have a *negative gradient*, this will occur when the graph of the line slopes downwards to the right of the *y*-axis. Under these circumstances a negative value of *y* results, so that $m = −y/x$ and so the gradient *m* is negative.

We leave our study of linear equations and their straight line graphs with an example of the application of the law of a straight line, $y = mx + c$, to experimental data.

Example 4.21

During an experiment to verify Ohm's law, the following experimental results were obtained.

E (volts)	0	1.1	2.3	3.4	4.5	5.65	6.8	7.9	9.1
I (amperes)	0	0.25	0.5	0.75	1.0	1.25	1.5	1.75	2.0

Plot voltage against current and so determine the equation connecting *E* and *I*.
The resulting plot is shown in *Figure 4.6*.

From the plot, it can be seen that the experimental data produce a straight line. Therefore, the equation connecting *E* and *I* is of the form $y = mx + c$. Since the graph goes *directly through the origin*, the variable $c = 0$. Also, from the graph taking suitable values of *E* and *I*, the gradient $m = 4.57$ is correct to three significant figures. So, the equation connecting *E* and *I* is **$E = 4.57I$**.

Quadratic equations

A quadratic equation is the one in which the unknown variable is raised to the second power. For example, the equation $x^2 = 4$ is perhaps one of the simplest of quadratic equations. We can solve this equation by taking the square root of both sides, something that you are familiar with when transposing a formula. Then: $\sqrt{x^2} = \sqrt{4}$ or $x = ±2$. Note that even for this simple equation there are two possible solutions, either $x = +2$ or $x = −2$, remembering your laws of signs! When we square a positive number we get a positive number $(+2)(+2) = +4$ or simply 4, also $(−2)(−2) = 4$ from the laws of signs.

UNIT 4

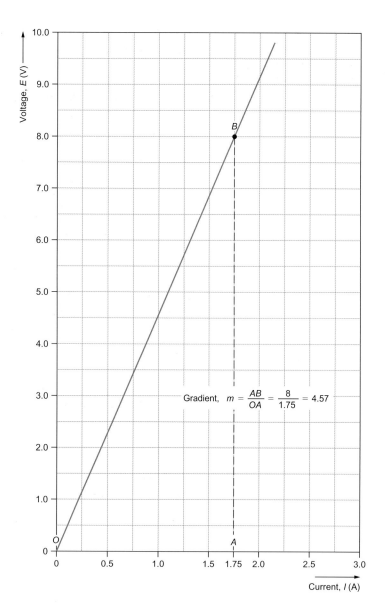

Figure 4.6 Graph of *E* against I

In general, a quadratic equation is of the type $ax^2 + bx + c = 0$, where the constants *a*, *b*, *c* can take *any* numerical value, positive or negative, decimal or fraction. Like linear equations, quadratic equations do not always appear in *standard form*, that is, they are not always arranged in exactly the same order as their qualifying equation, $ax^2 + bx + c = 0$.

How is our simple equation $x^2 = 4$ related to its qualifying equation? Well the coefficient of x^2, that is, the number multiplying the x^2 term $a = 1$. What about the constant *b*? Well there is no *x* term in our equation so $b = 0$. What about the constant *c*? Our equation is not in standard form, because the equation should be equated to zero. Then in standard form our equation becomes $x^2 - 4 = 0$ by simple transposition! So, now we know that for our equation the constant term $c = -4$. A quadratic equation may contain only the square of the unknown variable, as in our simple equation, or it may contain the square and the first power of the variable, for example, $x^2 - 2x + 1 = 0$. Also, the unknown variable may have ***up to two*** possible

KEY POINT

For all quadratic equations, the highest power of the independent variable is 2 (two)

real solutions. The equations we deal with in this course will always have at least one real solution.

There are several ways in which quadratic equations may be *solved*, that is finding the values of the *unknown* variable. We shall concentrate on just two methods of solution: the first is by *factorization* and the second involves the use of the *quadratic formula*.

We concentrate first on the solution by *factorization*. Take the equation $x^2 - 2x + 1 = 0$. If we ignore for the moment the fact that this is an equation and concentrate on the expression $x^2 - 2x + 1$, then you may remember how to find *the factors* of this expression! Look back now at your work on factors to remind yourself.

I hope you were able to identify the factors of this expression as $(x - 1)(x - 1)$. Now all we need to do is to equate these factors to *zero* to solve our equation. Thus $(x - 1)(x - 1) = 0$. Then for the equation to *balance* either the first bracket $(x - 1) = 0$ or the second bracket (the same in this case), $(x - 1) = 0$. Solving this very simple linear equation gives $x = 1$, no matter which bracket is chosen. So in this case our equation only has one solution $x = 1$. Note that if any one of the bracketed expressions $(x - 1) = 0$, then the other bracket is multiplied by zero, that is, $0(x - 1) = 0$. This is obviously true, because any quantity multiplied by zero is zero.

Example 4.22

Solve the equation $3x^2 - 5 = -2x - 4$ by factorization.

The first thing to note before we attempt a solution is that this equation **is not in standard form**. All we need do is transpose the equation to get it into standard form. You should, by now, be able to do the transposition with ease, so make sure you obtain:

$$3x^2 + 2x - 1 = 0.$$

Now using the techniques for factorization that you learnt earlier, after trial and error, you should find that:

$$(3x - 1)(x + 1) = 0$$

then either $3x - 1 = 0$ giving $\boldsymbol{x = \frac{1}{3}}$ or $(x + 1) = 0$ giving $\boldsymbol{x = -1}$.

Note that in this case the equation has two different solutions. Both can be checked for accuracy by substituting them into the **original** equation.

Then either

$$3\left(\frac{1}{3}\right)^2 - 5 = -2\left(\frac{1}{3}\right) - 4 \quad \text{or} \quad \frac{3}{9} - 5 = -\frac{2}{3} - 4 \text{ therefore } -4\frac{2}{3} = -4\frac{2}{3}$$

which is correct, or $3(-1)^2 - 5 = -4 - 2(-1)$ or $3 - 5 = -4 + 2$ therefore $-2 = -2$, which is also correct. Note the need to manipulate fractions and be aware of the laws of signs, skills I hope you have acquired at this stage in your learning.

It is not always possible to solve quadratic equations by factorization. When we cannot factorize a quadratic expression, we may resort to use of the *standard quadratic formula*. Now we know that *the standard form of the quadratic equation is*:

$$ax^2 + bx + c = 0$$

and it can be shown that the solution of this equation is:

$$x = \frac{-b \pm \sqrt{b^2 - 4ac}}{2a}$$

The quadratic formula

This equation may look complicated but it is relatively simple to use. The coefficients a, b, c are the same coefficients as in the standard form of the quadratic. So in finding a solution for the variable x, all we need to do is to substitute the coefficients into the above formulae for the quadratic equation we are considering. All you need to remember is that *before using the quadratic formula, always put the equation to be solved into standard form*. Also note that in the above formula, the whole of the numerator, including the $-b$, is divided by $2a$.

Example 4.23

Solve the equation $5x(x + 1) - 2x(2x - 1) = 20$.

The above equation is not in standard form, in fact until we simplify it, we may not be aware that it is a quadratic equation. So, simplifying by multiplying out the brackets and collecting like terms gives:

$$5x^2 + 5x - 4x^2 + 2x = 20$$

and so,

$$x^2 + 7x - 20 = 0.$$

This equation is now in standard form and may be solved using the formula. You may have attempted to try a solution by factorization first. If you cannot find the factors reasonably quickly, then you can always resort to the formula, unless told otherwise!

Then from $x = \frac{-b \pm \sqrt{b^2 - 4ac}}{2a}$ we get $x = \frac{-7 \pm \sqrt{7^2 - (4)(1)(-20)}}{2(1)}$ and simplifying gives

$$x = \frac{-7 \pm \sqrt{129}}{2} \quad \text{or} \quad x = \frac{-7 \pm 11.358}{2} \quad \text{and so,}$$

$$x = \frac{-7 + 11.358}{2} \quad \text{or} \quad x = \frac{-7 \pm 11.358}{2}$$

giving the values of the unknown x, correct to 3 significant figures as:

$x = 2.18$ *or* $x = -9.18$.

KEY POINT

Quadratic equations may have up to two real solutions

Simultaneous equations

Simultaneous equations involve more than one variable or unknown. We can solve a simple linear equation with one unknown using the laws of algebra, which you have already learnt. It is often required to represent an engineering problem that involves more than one unknown.

For example, if an engineering problem involves the solution of an equation such as $3x + 2y = 12$. How do we go about solving it? Well the answer is that a single equation with two unknowns is unsolvable unless we know the value of one of the variables. However, if we have *two equations* with *two unknowns*, it is possible to solve these equations *simultaneously*, that is, at the same time. Three linear equations with three variables can also be solved simultaneously. In fact, any number of linear equations with a corresponding number of unknowns (variables) can be solved simultaneously. However, when the number of variables is greater than three, it is better to solve the system of equations using a computer!

These systems of equations occur in many aspects of engineering, particularly when we model static and dynamic behaviour of solids and liquids. You will be pleased to know that we will only be considering *two equations* simultaneously, involving *two unknowns*! Even so, the distribution of currents and voltages, for example, in electrical networks sometimes involves the solution of such equations with just two unknowns.

Analytical solution of simultaneous equations. Consider the pair of equations:

$$3x + 2y = 12 \qquad (1)$$
$$4x - 3y = -1 \qquad (2)$$

Now to solve these equations, all we need to do is to use *elimination* and *substitution* techniques, working on both equations simultaneously.

Let us try to eliminate the variable x from both equations. This can be achieved by multiplying each equation by a constant. When we do this, we do not alter the nature of the equations. If we multiply equation (1) by the constant 4, and equation (2) by the constant 3, we get:

$$12x + 8y = 48 \qquad (3)$$
$$12x - 9y = -3 \qquad (4)$$

Note that we have multiplied *every term* in the equations by the constant! Now, how does this help us to eliminate x? Well if we now add both equations together we end up with the first term being $24x$. This is not very helpful. However, if we subtract equation (4) from equation (3), we get:

$$\begin{aligned} 12x + 8y &= 48 \\ -(12x - 9y &= -3) \\ \hline 0 + 17y &= 51 \end{aligned}$$

from which we see that $y = 3$. Now having found one of the unknown variables, we can substitute its value into *either one of the original equations* in order to find the other unknown. Choosing equation (1), from

$3x + 2y = 12$ we get $3x + (2)(3) = 12$ or $3x = 6$ and therefore $x = 2$. So the required solution is $y = 3$ and $x = 2$.

KEY POINT

To solve equations simultaneously, we require the same number of equations as there are unknowns

UNIT 4

When solving any equation, the solution can always be checked by substituting values into the original equation, so substituting the values into equation (2) gives:

$4(2) - (3)(3) = -1$, which is correct.

Example 4.24

Applying Kirchoff's voltage law to an electrical circuit produced the following simultaneous equation involving the two currents in each branch of the circuit.

$8I_1 + 2I_2 = 12$ (1)

$-6I_1 + 10I_2 = 0$ (2)

By solving these equations, find the value of these two currents.

Multiplying equation (1) by 5 and subtracting equation (2), we are able to eliminate I_2. So,

$$40I_1 + 10I_2 = 60$$
$$-(-6I_1 + 10I_2 = 0)$$
$$46I_1 = 60 \quad \text{and} \quad I_1 = \textbf{1.304A}$$

and substituting the value of I_1 into equation (1) gives

$$(8)(1.304) + 2I_2 = 12 \text{ and so } I_2 = \textbf{0.784A}$$

Graphical solution of two simultaneous equations This method of solution is illustrated in Example 4.25. For each of the linear equations, we plot their straight-line graphs and where the plots intersect we find the unique solution for both equations.

Example 4.25

Solve the following simultaneous equations, graphically:

$\dfrac{x}{2} + \dfrac{y}{3} = \dfrac{13}{6}; \dfrac{2x}{7} - \dfrac{y}{4} = \dfrac{5}{14}$

Now we first need to simplify these equations and rearrange them in terms of the independent variable y. I hope you can remember how to simplify fractions! Make sure that you are able to rearrange the equations and obtain:

$$2y = 13 - 3x$$
$$-7y = 10 - 8x$$

Now transposing in terms of y, we get:

$$y = \frac{13}{2} - \frac{3}{2}x$$
$$y = -\frac{10}{7} + \frac{8}{7}x$$

Now we can find the corresponding values of y, for our chosen values of x. Using just four values of x, say, 0, 1, 2 and 3, will enable us to plot the straight lines. Then

x	0	1	2	3
$y = \dfrac{13}{2} - \dfrac{3}{2}x$	$\dfrac{13}{2}$	5	$\dfrac{7}{2}$	2
$y = -\dfrac{10}{7} + \dfrac{8}{7}x$	$-\dfrac{10}{7}$	$-\dfrac{2}{7}$	$\dfrac{6}{7}$	2

From the plot (*Figure 4.7*), the intersection of the two straight lines yields the required result, that is $x = 3$ and $y = 2$.

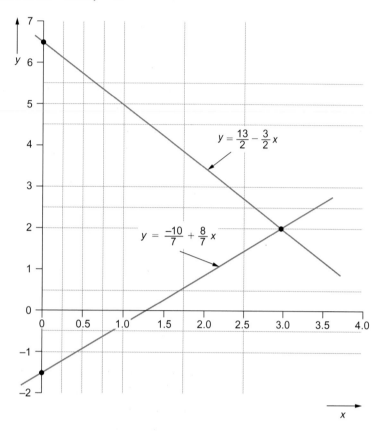

Figure 4.7 Simultaneous graphs of the functions $y = \dfrac{13}{2} - \dfrac{3}{2}x$ and $y = -\dfrac{10}{7} + \dfrac{8}{7}x$

In this particular example, it would have be easier to solve these equations using an algebraic method.

UNIT 4

Test your knowledge

TYK

TYK *4.4*

1. Solve the following linear equations:
   (a) $5x - 1 = 4$ (b) $6p - 7 = 17$ (c) $4(x - 5) = 7 - 5(3 - 2x)$
   (d) $3(x - 2) = 2(x - 1)$ (e) $\dfrac{a}{5} - \dfrac{a}{3} = 2$
   (f) $\dfrac{1}{p} + \dfrac{1}{p + 1} = \dfrac{2}{p - 1}$

2. The values in the table below show how instantaneous current i varies with voltage v. Plot a graph of v against i and so find the value of v when $i = 3.0$.

v	15	25	35	50	70
i	1.1	2.0	2.5	3.2	3.9

3. Solve the following quadratic equations by factors or otherwise:
   (a) $x^2 - 2x + 1 = 0$ (b) $3p^2 - 48 = 0$
   (c) $x^2 - x - 6 = 0$ (d) $p^2 + 4p - 32 = 0$
   (e) $4x^2 - 4x = 3$ (f) $6b^2 - 2b - 8 = 0$.

4. Solve the following quadratic equations by formula:
   (a) $6x^2 + x - 2 = 0$ (b) $7x^2 + 8x - 2 = 0$
   (c) $-2x^2 - 20x = 32$ (d) $f + \dfrac{1}{f} = 3$
   (e) $(2x - 3)^2 = 13$ (f) $\dfrac{1}{a+1} + \dfrac{1}{a+2} - \dfrac{2}{3} = 0$

5. Solve the following simultaneous equations analytically:
   (a) $3x + 2y = 7$ (b) $2x + 3y = 8$ (c) $5x + 4y = 22$
         $x + y = 3$ $2x - 3y = 2$ $3x + 5y = 21$

   (d) $\dfrac{p}{2} + \dfrac{q}{3} = 2$
       $2p + 3q = 13.$

6. If $y = ax + b$ find the value of y, when $x = 4$, given that $y = 4$ when $x = 1$ and that $y = 7$ when $x = 2$.

7. Solve graphically the following simultaneous equations:
   (a) $x + 3y = 7$ (b) $7x - 4y = 37$
       $2x - 2y = 6$ $6x + 3y = 51$

Logarithms and logarithmic functions

We have already studied the laws and use of indices. It is now time to briefly consider logarithms. In fact logarithms are really indices. The logarithm of a number is in fact its index. For example $10^3 = 1000$. The left-hand side of this equation 10^3 is the number 1000 written in index form. The index 3 is in fact the *logarithm* of 1000. Check this by pressing your *log* button (which is the logarithm to the base ten) then key in the number 1000 and press the $=$ button. You will obtain the number 3.

Manipulation of numbers, expressions and formulae that are in index form may be simplified by using logarithms. Another use for logarithms is being able to reduce the sometimes more difficult arithmetic operations of multiplication and division to those of addition and subtraction. This is often necessary when manipulating more complex algebraic expressions.

We start by considering the laws of logarithms in a similar manner to the way in which we dealt with the laws of indices earlier.

The laws of logarithms

The **laws of logarithms** are tabulated below and they are followed by simple examples of their use. In all these examples, we use *common logarithms*, that is, logarithms to the *base 10*. Later, we will look at one other type, the *Napierian logarithm*, or natural logarithm, where the *base* is the number e.

KEY POINT

The power or index of a number, when in index form, is also its logarithm, when taken to the base of the number

Number	Logarithmic law
1	If $a = b^c$, then $c = \log_b a$
2	$\log_a MN = \log_a M + \log_a N$
3	$\log_a \dfrac{M}{N} = \log_a M - \log_a N$
4	$\log_a (M^n) = n \log_a M$
5	$\log_b M = \dfrac{\log_a M}{\log_a b}$

Law 1

All these laws look complicated, but you have already used law 1 when you carried out the calculator exercise above. So again, we know that $1000 = 10^3$. Now if we wish to put this number into *linear* form (decimal form), then we may do this by *taking logarithms*. Following law 1, where in this case, $a = 1000$, $b = 10$ and $c = 3$, then $3 = \log_{10} 1000$. You have already proved this fact on your calculator! So, you are probably wondering why we need to bother with logarithms? Well in this case we are dealing with **common logarithms**, that is, numbers in index form where the **base of the logarithm is 10**. We can also consider numbers in index form that are not to the base ten, as you will see later. We may also be faced with a problem where the index (power) is not known.

Suppose we are confronted with this problem: *find the value of x where* $750 = 10^x$. The answer is not quite so obvious, but it can easily be solved using our first law of logarithms. So, again following the law, that is taking logarithms to the appropriate base, we get $x = \log_{10} 750$ and *now* using our calculator we get $x = 2.875$, correct to 4 significant figures.

Law 2

One pair of factors for the number 1000 is 10 and 100. Therefore, according to the second law, $\log_a (10)(100) = \log_a 10 + \log_a 100$. If we choose logarithms to the base 10 then we already know that the $\log_{10} 1000 = 3$, then using our calculator again, we see that $\log_{10} 10 = 1$ and $\log_{10} 100 = 2$. **What this law enables us to do is to convert the multiplication of numbers in index form into that of addition.** Compare this law with the first law of indices you studied earlier! Remember also that we are at liberty to choose any base we wish, providing we are able to work in this base. Your calculator only gives you logarithms to two bases, 10 and e.

Law 3

This law allow us to convert the division of number in index form into that of subtraction. When dealing with the *transposition* of more complex formulae, these conversions can be particularly useful and help us with the transposition.

So, using the law directly then, for example:

$$\log_{10} \frac{1000}{10} = \log_{10} 100 = \log_{10} 1000 - \log_{10} 10$$

or from your calculator $2 = 3 - 1$.

Law 4

This laws states that if we take the logarithm of a number in index form M^n then this equals to the logarithm of the base of the number $\log_a M$ multiplied by the index of the number, $n \log_a M$, so for example, $\log_{10}(100^2) = \log_{10}$

$10000 = 2\log_{10} 100$. This is easily confirmed on your calculator as $4 = (2)(2)$.

Law 5

This law is rather different from the others, in that it enables us to change the base of a logarithm. This of course is very useful if we have to deal with logarithms or formulae involving logarithms that have a base not found on our calculator!

For example, suppose we wish to known the numerical value of $\log_2 64$, then using law 5, we have

$$\log_2 64 = \frac{\log_{10} 64}{\log_{10} 2} = \frac{1.806179974}{0.301029995} = 6, \text{ interesting!}$$

If we *use law 1* in reverse, then $\log_2 64$ is equivalent to the number $64 = 2^6$, which of course is now easily verified by your calculator! This example, again demonstrates that given a number in *index form*, the *index* of that number is also its *logarithm*, providing the logarithm has the same base.

We will now consider, through example, one or two engineering uses for the laws of *common* and *natural* logarithms.

KEY POINT

Common logarithms have the base 10

Example 4.26

An equation connecting the final velocity *v* of a machine with the machine variables, *w*, *p* and *z*, is given by the formula $v = 20^{\left(\frac{w}{pz}\right)}$. Transpose the formula for *w*, and find its numerical value when $v = 15$, $p = 1.24$ and $z = 34.65$.

This formula may be treated as a number in *index form*. Therefore to find w, as the subject of the formula, we need to apply the laws of logarithms. The first step in this type of problem is to *take logarithms of both sides*. The base of the logarithm chosen is not important provided we are able to find the numerical values of these logarithms when required. Thus, we generally take logarithms to the base 10 or to the base e. As yet, we have not considered logarithms to the base e, so we will take *common logarithms* of both sides. However, if the number or expression is not to a base of logarithms we can manipulate, then we are at liberty to change this base using law 5.

So, $\log_{10} v = \log_{10} 20^{\left(\frac{w}{pz}\right)}$. At this stage, taking logarithms seems to be of little help! However, if we now apply the appropriate logarithmic laws, we will be able to make w the subject of the formula.

Applying law 4 to the right-hand side of the expression we get

$$\log_{10} v = \left(\frac{w}{pz}\right)\log_{10} 20$$

Then, finding the numerical value of $\log_{10} 20 = 1.30103$, we can now continue with the transposition.

$$\log_{10} v = \left(\frac{w}{pz}\right)1.30103 \quad \text{or} \quad \frac{\log_{10} v}{1.30103} = \frac{w}{pz} \quad \text{and so, } w = \frac{(pz)(\log_{10} v)}{1.30103}$$

Having transposed the formula for w, we can substitute the appropriate values for the variables and find the numerical value of w.

Then

$$w = \frac{(1.24)(34.65)(\log_{10} 15)}{1.30103} = \frac{(1.24)(34.65)(1.17609)}{1.30103} = \mathbf{38.84}$$

Napierian (natural) logarithms and the exponential function

If you look at your calculator you will see the *ln* or *Napierian logarithm* button. The *inverse* of the Napierian logarithm function is e^x or exp x, the exponential function. This logarithm is sometime known as the *natural logarithm*, because it is often used to model naturally occurring phenomena, such as the way things grow or decay. In engineering, for example, the decay of charge from a capacitor may be modelled using the natural logarithm. It is therefore a very useful function, and both the natural logarithm and its inverse, the exponential function, are very important within engineering.

We will now consider the transposition of a formula that involves the use of natural logarithms and the logarithmic laws.

KEY POINT

Napierian or natural logarithms have the base e where $e \simeq 2.718281828$ correct to nine decimal places

Example 4.27

Transpose the formula, $b = \log_e t - a \log_e D$ to make t the subject.

First, note that the natural or Napierian logarithm may be expressed as $\log_e$ or ln, as on your calculator. **Do not mix-up** the expression $\log_e$, or its inverse e^x or exp x with the exponent function (EXP) on your calculator, which multiplies a number by powers of ten!

We first use the laws of logarithms as follows:

$b = \log_e t - \log_e D^a$ from law 4, then

$$b = \log_e \left(\frac{t}{D^a} \right) \text{ from law 3.}$$

Now, for the first time we take the inverse of the natural logarithm or **anti-logarithm noting that any function multiplied by its inverse is 1 (one)**. Then multiplying both sides of our equation by e the inverse of ln ($\log_e$), we get:

$e^b = \dfrac{t}{D^a}$ (since e is the inverse or antilogarithm of $\log_e$ = ln or (e)($\log_e$) = 1), then

$\boldsymbol{t = e^b D^a}$ as required.

KEY POINT

The inverse function of the natural logarithm is the exponential function, which in symbols is expressed as exp x or e^x

As mentioned before, the ***exponential function* e^x *or exp x and its inverse* *ln***, the natural logarithm, have many uses in engineering, because they can be used to ***model growth and decay***. So, the way solids expand, electrical resistance changes with temperature, substances cool pressure changes with altitude or capacitors discharge can all be modelled by the *exponential* function.

UNIT 4

Here are just two engineering examples of the use of the exponential function.

Example 4.28

If the pressure p at height h (in metres) above the ground is given by the relationship $p = p_0 e^{\frac{h}{k}}$, where p_0 is the sea-level pressure of $101325\,\text{Nm}^{-2}$ determine the value of the height h, when the pressure at altitude p is $70129\,\text{Nm}^{-2}$ and $k = -8152$.

We need first to transpose the formula for h. This will involve taking *natural logarithms*, the inverse function of $e^{\frac{h}{k}}$. Before we do so we will first isolate the exponential term, then $\dfrac{p}{p_0} = e^{\frac{h}{k}}$ and taking logarithms gives $\log_e\left(\dfrac{p}{p_0}\right) = \dfrac{h}{k}$, then $k\log_e\left(\dfrac{p}{p_0}\right) = h$, then substituting the given values,

$$h = -8152\log_e\left(\frac{70129}{101325}\right) = (-8152)\log_e(0.692) = (-8152)(-0.368) = 3000\text{m},$$

correct to 4 significant figures. Thus the altitude $\boldsymbol{h = 3000\,m}$.

This final example is concerned with the information contained in a radio communications message. It is not necessary to understand the background physics in order to solve the problem, as you will see.

Example 4.29

It can be shown that the information content of a message is given by $I = \log_2\left(\dfrac{1}{p}\right)$. Show using the laws of logarithms that the information content may be expressed as $I = -\log_2(p)$ and find the information content of the message if the chances of receiving the code (p) is $\frac{1}{16}$.

So, if we are being asked to show that $I = \log_2\left(\dfrac{1}{p}\right) = -\log_2(p)$ the left-hand side of this expression may be written as $\log_2(p^{-1})$. I hope you remember the laws of indices! Now if we compare this expression with law 4 where $\log_a(M^n) = n\log_a M$, then in this case $M = p$ and $n = -1$ so $\log_2(p^{-1}) = -1\log_2, p = -\log_2 p$, as required.

Now to find the information content of the message, we need to substitute the given value of $p = \frac{1}{16}$ into the equation $\log_2(p^{-1}) = \log_2\left(\dfrac{1}{p}\right) = \log_2(16)$. Now our problem is that we cannot easily find the value of logarithms to the base 2. However, if we use logarithmic law 5, then we get $\log_2 16 = \dfrac{\log_{10}16}{\log_{10}2} = 4$. **Then the information content of the message = 4.**

I hope you were able to following the reasoning in the above two quite difficult examples. There is just one more application of the laws of logarithms that we need to cover. It is sometimes very useful when considering experimental data to determine if such data can be related to a particular law. If we can relate this data to the law of a straight line $y = mx + c$, then we can easily determine useful results. Unfortunately the data is not always related in this form. However, a lot of engineering data follows the general form $y = ax^n$, where as before x is the *independent variable*, y *is the dependent variable* and in this case, a and n are *constants* for the particular experimental data being considered.

We can use a technique involving logarithms to reduce equations of the form $y = ax^n$ to a linear form, following the law of the straight line, $y = mx + c$. The technique is best illustrated by example.

Example 4.30

The pressure p and volume v of a gas at constant temperature are related by Boyle's law, which can be expressed as $p = cv^{-0.7}$, where c is a constant. Show that the experimental values given in the table follow this law and from an appropriate graph of the results determine the value of the constant c.

Volume v (m³)	1.5	2.0	2.5	3.0	3.5
Pressure p (10^5 Nm^{-2})	7.5	6.2	5.26	4.63	4.16

The law is of the form $p = ax^n$. So, taking common logarithms of both side of the law $p = cv^{-0.7}$ we get $\log_{10} p = \log_{10} (cv^{-0.7})$ and applying law 2 and law 4 to the right-hand side of this equation gives $\log_{10} p = -0.7 \log_{10} v + \log_{10} c$. Make sure you can see how to get this result. Then comparing this equation with the equation of a straight line $y = mx + c$, we see that $y = \log_{10} p$, $m = -0.7$, $x = \log_{10} v$ and $c = \log_{10} c$.

So, we need to plot $\log_{10} p$ against $\log_{10} v$. A table of values and the resulting plot is shown below.

Volume v (m³)	1.5	2.0	2.5	3.0	3.5
$\log_{10} v$	0.176	0.301	0.398	0.447	0.544
Pressure p (10^5 Nm^{-2})	7.5	6.2	5.26	4.63	4.16
$\log_{10} p$	0.875	0.792	0.721	0.666	0.619

Then from the plot (*Figure 4.8*) it can be seen that the slope of the graph is -0.7, and the y intercept at $\log_{10} v = 0$ is given as 1.0 or $\log_{10} c = 1.0$ and so $c = 10$. Therefore, the plotted results do follow the law, $p = 10v^{-0.7}$.

This use of logarithms to manipulate experimental data is very useful.

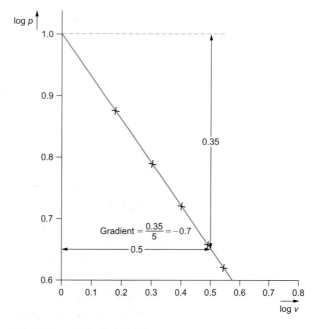

Figure 4.8 Plot of $\log_{10} p$ against $\log_{10} v$

Hopefully, you are now in a position to tackle TYK 4.5. You may find these questions rather difficult but persevere because the algebraic technique involved with logarithmically reducing data to a linear form is so useful for your experimental work!

TYK

TYK *4.5*

1. Transpose the formulae $q = rx^{\frac{s}{t}}$ for t and then find its value when $q = 30\pi$, $r = 3\pi$, $x = 7.5$ and $s = 16$.

2. The formula $P = T(1 - e^{-\mu\theta})v$ relates the power (P), belt tension (T), angle of lap (θ), linear velocity (v) and coefficient of friction (μ) for a belt drive system. Transpose the formula for (μ) and find its value when $P = 2500$, $T = 1200$, $V = 3$ and $\theta = 2.94$.

3. In an experiment, values of current I and resistance R were measured, producing the results tabulated below.

R	0.1	0.3	0.5	0.7	0.9	1.1	1.3
I	0.00017	0.0015	0.0043	0.0083	0.014	0.021	0.029

Show that the law connecting I and R has the form $I = aR^b$, where a and b are constants and determine this law.

Trigonometric Methods and Formulae

In this section the second of the outcomes for the unit is discussed. We first revise the trigonometric ratios and the solution of right-angled triangles using these ratios and Pythagoras theorem. The trigonometric ratios are then applied to the resolution of forces. Next the general solution of non-right-angled triangles using the sine and cosine rules is covered.

A separate section then follows on the circle and circular measure, introducing the concept of the radian; this being necessary preparation for the section on the trigonometric functions that follows. The oscillatory natures of the sine, cosine and tangent functions are covered with emphasis being placed on the properties and uses of these functions.

Finally, some of the methods and formulae learnt in the earlier sections are applied to the solution of problems associated with trigonometric measurement, in particular, the determination of surface areas and volumes of regular solids, such as cones, cylinders and spheres.

Solution of right-angled triangles and the trigonometric ratios

We start this section by reminding you of the basic trigonometric ratios and use them to solve triangles and apply them to the resolution of forces. What we mean by solving triangles is to find their missing angles and/or sides.

UNIT 4

Fundamental trigonometric ratios

I am sure that in your previous studies you have met the fundamental trigonometric ratios. However, they are an essential part of this outcome and for those with this gap in their knowledge and to serve as a reminder to others, they are repeated here.

For any right-angled triangle (*Figure 4.9*),

1. $\dfrac{\textbf{The side opposite the angle}}{\textbf{the hypotenus}}$ is called the **sine** of the reference angle and is often abbreviated to *sin*. Therefore,

$$\sin A = \frac{opposite}{hypotenuse} = \frac{a}{c}.$$

Note that from now on we will use only the capital letter to represent angles, dropping the $\hat{A}$ (hat) sign above the letter. Also note that *lower case letters represents the sides of the triangle.*

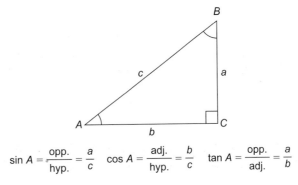

$$\sin A = \frac{\text{opp.}}{\text{hyp.}} = \frac{a}{c} \qquad \cos A = \frac{\text{adj.}}{\text{hyp.}} = \frac{b}{c} \qquad \tan A = \frac{\text{opp.}}{\text{adj.}} = \frac{a}{b}$$

Note: 1. The hypotenuse is opposite the right angle

2. $\tan A = \dfrac{\sin A}{\cos A}$

Figure 4.9 The right-angled triangle

2. $\dfrac{\textbf{The side adjacent to the angle}}{\textbf{the hypotenuse}}$ is called the **cosine** of the reference angle

and is often abbreviated to *cos*. Therefore,

$$\cos A = \frac{adjacent}{hypotenuse} = \frac{b}{c}.$$

3. $\dfrac{\textbf{The side opposite the angle}}{\textbf{the side adjacent to the angle}}$ is called the **tangent** of the angle and is often abbreviated to *tan*. Therefore,

$$\tan A = \frac{opposite}{adjacent} = \frac{a}{b}.$$

These fundamental ratios are very important and should be remembered. One aid to memory is to use the acronym **SOHCAHTOA** (pronounced *sock-ca-tow-ah*) where the letters mean: sine, opposite, hypotenuse; cosine, adjacent, hypotenuse; tangent, opposite, adjacent. There are other aids to memory – stick with what you know!

One final very important point concerning these ratios: remember that the **sine, cosine and tangent function must be followed by the angle they refer to**. Thus, for example, *sin A* makes sense, but *sin* on its own is nonsense. Remember that the **sin** and the **A cannot be separated**!

Before we look at an example of the use of these ratios, we need to ensure that you can use your calculator to evaluate them.

Suppose, for example, that we wish to find the numerical value of sin 51°. That is, find the value of the sine of 51°. Then the following is a typical sequence you need to follow:

(a) ensure your calculator is selected in the degree mode (shown in the viewing window)
(b) press the sine key
(c) key-in 51
(d) press the = button
(e) Read-off the result as 0.77715 (correct to 5 decimal points).

Note that with some calculators operations (b) and (c) may be reversed and on pressing the sine key the result is automatically shown in the viewing window.

Also, note that fractions of a degree are best entered into your calculator in decimal form. If you are required to find the trigonometric ratio of an angle given in degrees and minutes, the following procedure may be used.

Find, using your calculator, the numerical value of cos 26° 45′ (45 minutes).

We first convert the 45 minutes into degrees by dividing it by 60, then $45'/60 = 0.75°$

Now we follow the procedure as before:

(a) ensure calculator is in degree mode
(b) press the cosine key
(c) key in 26.75
(d) press the = button
(e) read-off the result as 0.89298 (correct to 5 decimal points).

Example 4.31

For the triangle shown in *Figure 4.10*, find the sine, cosine and tangent for ∠A (angle A) and ∠B.

For angle A, the opposite $= \sqrt{3}$, adjacent $= 1$ and the hypotenuse $= 2$. So, the respective trigonometric ratios are

$\sin A = \dfrac{\sqrt{3}}{2}$, $\cos A = \dfrac{1}{2}$ and $\tan A = \dfrac{\sqrt{3}}{1}$. In fractional form, these are exact values for

these ratios, so we will leave them in this form for the moment. Similarly, for ∠B we get

$\sin B = \dfrac{1}{2}$, $\cos B = \dfrac{\sqrt{3}}{2}$ and $\tan B = \dfrac{1}{\sqrt{3}}$.

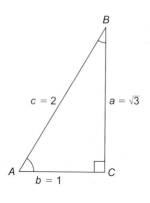

Figure 4.10 Figure for Example 4.31

One very useful relationship that can be seen from Example 4.31 is as follows:

$$\frac{\sin A = \dfrac{\sqrt{3}}{2}}{\cos A = \dfrac{1}{2}} = \frac{\dfrac{\sqrt{3}}{2}}{\dfrac{1}{2}} = \left(\frac{\sqrt{3}}{2}\right)\left(\frac{2}{1}\right) = \frac{\sqrt{3}}{1} = \tan A.$$

This relationship is true in all cases, that is, for the general angle θ,

$$\frac{\sin \theta}{\cos \theta} = \tan \theta.$$

Now, if we wish to evaluate these ratios and *find the angle* in degrees that they represent, we need to consider the **inverse trigonometric ratio**. Thus, if we consider the trigonometric ratio $\sin A = \dfrac{\sqrt{3}}{2}$ from the previous example, we need to find the *angle* represented by this ratio. In general terms, the inverse ratio is, in symbols, represented as $\mathbf{sin^{-1}}$ **A**, or in words as *the angle A whose sign is*, in our particular example: $\sin A = 0.8660254$ and so the angle whose sign is 0.8660254, or $\sin^{-1} A = 0.8660254$, is the angle we are required to find! This inverse function of sine (in this case) is easily found on our calculator as follows:

(a) ensure calculator is in degree mode
(b) press the *shift* key
(c) press the sin key; ($\sin^{-1}$) will appear in the display window
(d) key-in desired value of trigonometric ratio (0.8660254 in our case)
(e) press = button and read-off angle in degrees (correct to 4 decimal places: we get $A = 60°$).

Following this procedure for any of the trigonometric ratios we found in Example 4.31 will yield the following results:

Trigonometric ratio	Angle in degrees
$\sin A = 0.8660$	60
$\cos A = 0.5$	60
$\tan A = 1.732$	60
$\sin B = 0.5$	30
$\cos B = 0.866$	30
$\tan B = 0.577$	30

Solving right-angled triangles

So far we have used the trigonometric ratios sine, cosine and tangent to find angles, given the 3 sides of a triangle. We can in fact solve any right-angled triangle, given any side and two angles using the trigonometric ratios, and where necessary, combining these ratios with Pythagoras.

Consider the triangle shown in *Figure 4.11*, with sides a, b, c and angles, $\angle A$, $\angle B$ and $\angle C$.

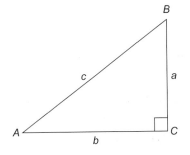

Figure 4.11 Labelled right-angled triangle

UNIT 4

Then from the figure:

$$\sin\theta = \frac{a}{c}, \text{ therefore } a = c\sin\theta$$

$$\cos\theta = \frac{b}{c}, \text{ therefore } b = c\cos\theta$$

$$\tan\theta = \frac{a}{b}, \text{ therefore } a = b\tan\theta$$

Also, from Pythagoras: $c^2 = a^2 + b^2$.

Example 4.32

In the right-angled triangle shown in *Figure 4.12* find $\angle A$, $\angle B$ and side c.

We find angle A by noting that we are given the opposite side and the adjacent side; therefore, I hope you can see that we should use the tangent ratio to find $\angle A$. Then $\tan A = \frac{opp}{adj} = \frac{12}{9}$ and using our calculator we find **A = 53.13°** and **B** = (90 − 53.13°) = **36.87°**.

To find side c, we could again use $\angle A$ with the sine ratio. Unfortunately, if we have made a mistake in our calculation of $\angle A$, we will create yet another error. So, it is wise to use given information when we can. One way to do this is to find side c using Pythagoras. Then:

$$c^2 = 12^2 + 9^2 = 144 + 81 = 225 \text{ and so, side } c = 15.$$

Now in this example for the sake of completeness, we will check this result using:

$$\sin A = \frac{12}{h} \quad \text{or} \quad h = \frac{12}{\sin 53.13°} = \frac{12}{0.79999} = \frac{12}{0.8} = 15.$$

Note that the calculator value for sin 53.13° was 0.79999, rather than the exact value 0.8. This is because, the original value for $\tan A = \frac{12}{9} = 1.3333$. If you enter this recurring decimal value into your calculator, then the resulting value of $\angle A$ is not quite exact and this non-exact value was keyed in again when finding sin A. Yet another reason to use Pythagoras and exact fractions!

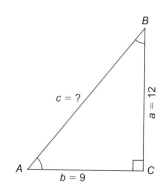

Figure 4.12 Figure for Example 4.32

Example 4.33

From two points A and B, on level ground and on opposite sides of a vertical pylon of height 26 m, the angle of elevation to the top of the pylon are 25° and 48°, respectively. Find the horizontal distance between the two points.

This type of problem requires us to draw a space diagram of the situation. The angle of elevation is that angle which starts at zero degrees along the horizontal and increases as we look upwards. In our case the angle from the horizontal to the top of the pylon viewed from both sides is shown in *Figure 4.13*.

It can be seen that the vertical pylon forms a common side between the two triangles *ACD* and *BCD*. We are required to find the distance between the two points *A* and *B*. This

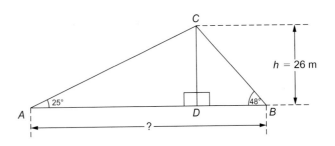

Figure 4.13 Figure for Example 4.33

distance can be found by first finding the distance *AD* and then adding this distance to the side *DB*.

Then with respect to $\angle A = 25°$, we need to select the trigonometric ratio that involves the height of the pylon and the unknown adjacent side *AD*. I hope you can see that this will be the tangent ratio, where:

$$\tan A = \frac{CD}{AD},\quad \text{and on substituting given values we get}$$

$$\tan 25° = \frac{26}{AD} \text{ so that } AD = \frac{26}{\tan 25} = \frac{26}{0.4663} = \textbf{55.76 m}.$$

Similarly for $\angle B$ we have

$$\tan B = \frac{CD}{DB} \text{ or } \tan 48° = \frac{26}{DB}, CB = \frac{26}{\tan 48} = \frac{26}{1.1106}$$

$$\text{and } DB = \textbf{23.41 m}.$$

Then the distance between the points *A* and *B* = *AD* + *DB* = 55.76 + 23.41 = **79.17 m**.

In order to be able to solve some problems involving right-angled triangles, it is necessary to understand the relationship between complementary angles. This really means that we need to memorize two facts.

The first is that there are 180° in a semicircle, which I am sure you know. So, for example, if one angle in a semi-circle is 40°, then its complementary angle is 140°.

The second fact is that there are also 180° in a triangle. So that if a right-angled triangle has two angles, say, 90° and 55°, then the complementary angle is easily found as 180 − 90 − 55 = 35°.

Example 4.34

Calculate the length of *AB*, in *Figure 4.14*. Given that length *BC* = 200 mm.

To solve this problem, we need to remember the facts concerning complementary angles.

Then $\angle ADC = 180 - 32 - 90 = 58°$ and with respect to the angle $CBD = 57°$, we can use the tangent ratio to find *CD*, then

$$\tan 57 = \frac{CD}{200} \text{ or } CD = 200 \tan 57 \text{ and so } CD = (200)(1.5399) = 308 \text{ mm}$$

to the nearest mm.

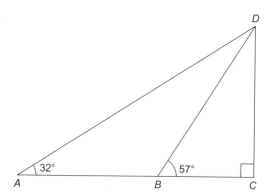

Figure 4.14 Figure for Example 4.34

Now, with respect to angle $CAD = 32°$, again using the tangent ratio, we get

$$\tan 32 = \frac{CD}{AC} \text{ or } AC = \frac{CD}{\tan 32} = \frac{308}{0.62486} = 493\,mm, \text{ again to the nearest mm.}$$

Then, required side $AB = AC - BC = 493 - 200 = \textbf{293\,mm}$.

In the above example, we could go on to find the remaining sides for both triangles and so solve them completely. This is left as an exercise for you later.

Using the trigonometric ratios to resolve forces

An application of the solution of triangles is used in your mechanical principles where you need to be able to **resolve forces** when they are represented as vectors (having both magnitude and direction). What we mean by resolving forces is to split each force into its equivalent vertical and horizontal component. At this point in your learning you need not worry about the science behind the problems; just concentrate on the use of trigonometry to resolve these forces.

Consider a force F acting on a bolt A (*Figure 4.15*). The force F may be replaced by two forces P and Q, acting at right angles to each other, which together have the same effect on the bolt.

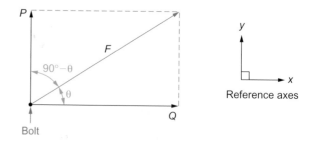

Figure 4.15 Resolving force F into its components

From your knowledge of the trigonometric ratios you will know that

$$\frac{adjacent}{hypotenuse} = \cos\theta$$

therefore, applying this ratio to the situation shown in *Figure 4.15*,

we get $\dfrac{Q}{F} = \cos\theta$ and so $Q = F\cos\theta$.

Also, $\dfrac{P}{F} = \cos(90 - \theta)$ and we know that $\cos(90 - \theta) = \sin\theta$ therefore, $P = F\sin\theta$.

So, from *Figure 4.15*, $P = F\sin\theta$ and $Q = F\cos\theta$.

So, the single force F has been resolved or split into two equivalent forces of magnitude $F\cos\theta$ and $F\sin\theta$, that act at right angles to each other. **$F\cos\theta$** is known as the ***horizontal component of F*** and **$F\sin\theta$** is known as the ***vertical component* of F**.

Another useful thing to do with forces is to find their resultants; that is, applying the reverse process to the resolution of forces. To do this we use Pythagoras theorem, as you will see in Example 4.35.

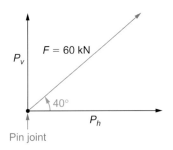

F = 60 kN

P_v

40°

P_h

Pin joint

Figure 4.16 Force applied to a pin joint

KEY POINT

Pythagoras theorem is used to find the single resultant force of two forces at right angles

---

Example 4.35

A force *F* is applied to a pin joint (*Figure 4.16*). Determine the equivalent horizontal and vertical components of this force that act on the pin joint.

Using the trigonometric ratios $\dfrac{P_h}{60} = \cos 40$ and so $P_h = 60\cos 40 = 45.96$ kN, and

$\dfrac{P_v}{60} = \sin 40$ and so $P_v = 60\sin 40 = 38.57$ kN, the horizontal and vertical components of the force F are **$P_h = 45.96\,kN$** and **$P_v = 38.57\,kN$**, respectively.

Also from Pythagoras, $F = \sqrt{(45.96)^2 + (38.57)^2} = 3600 = 60$kN and

$\sin^{-1}\theta = \dfrac{38.57}{60} = 0.64283 = 40°$. So we have found the original resultant force and the angle through which it acts using Pythagoras.

Area of the right-angled triangle

In order to completely solve right-angled triangles, we also need to find their area. ***The formula for the area of a right-angled triangle is given as***

$$\text{Area, } A = \textbf{half the base} \times \textbf{the perpendicular height}$$

or in symbols, $A = \dfrac{1}{2}bh$.

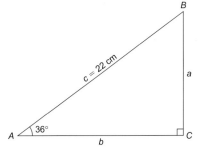

Figure 4.17 Figure for Example 4.36

Example 4.36

Find the area of the triangle shown in *Figure 4.17*.

In order to find the area, we first need to find the two sides, a and b using the sine ratio and $\angle A \sin 36 = \dfrac{a}{22}$ and $a = 22\sin 36 = (22)(0.5878) = 12.93$ cm.

Now, we will use Pythagoras to find the remaining side b, then

$b^2 = c^2 - a^2 = 22^2 - 12.93^2 = 484 - 167.22 = 316.78$; therefore,

side $b = \sqrt{316.78} = 17.8$ cm.

Then the area of the triangle $A = \dfrac{1}{2}ba = \left(\dfrac{1}{2}\right)(17.8)(12.93) = \textbf{115 cm}^2$.

TYK

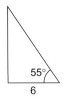

Figure 4.18 Figure for question 4 in TYK 4.6

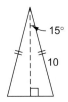

Figure 4.19 Figure for question 5 in TYK 4.6

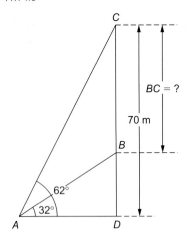

Figure 4.20 Figure for question 8 in TYK 4.6

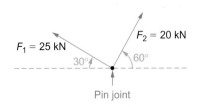

Figure 4.21 Figure for question 10 in TYK 4.6

TYK 4.6

1. Using a suitably labelled right-angled triangle, write down the trigonometric ratios for sin A, cos A and tan A in terms of the sides of the triangle. Also, find the relationship for $\dfrac{\sin A}{\cos A}$ in terms of the sides of your triangle and comment on your answer.

2. Find, using your calculator, the numerical value for (a) sin 45°, (b) cos 45°, (c) tan 45° (d) sin 135°, (e) cos 282.4° and (f) tan 89.9°.

3. Find the angle whose (a) sine is 0.707107, 0.866 and 0.5; (b) cosine is −0.4475, 0.866 and 0.7071; and (c) tangent is 1.000, 3567 and −0.5.

4. Solve the right-angled triangle shown in *Figure 4.18*.

5. Solve the isosceles triangle shown in *Figure 4.19*.

6. In a right-angled triangle, the lengths of the shorter sides are 6 and 9 cm. Calculate the length of the hypotenuse and area of the triangle.

7. All the sides of a triangle are 8 cm in length. What is the vertical height of the triangle and its area?

8. In *Figure 4.20*, the angles of elevation of B and C from A are 32° and 62°, respectively. If DC = 70 m, calculate the length of BC.

9. A vertical radio mast has cable stays of length 64 m, extending from the top of the mast, if each wire makes an angle of 65° with the ground. Find
   (a) the distance each cable is from the base of the mast.
   (b) the vertical height of the mast.

10. Two forces F_1 and F_2 are applied to a pin joint (*Figure 4.21*). Determine the equivalent horizontal and vertical components of each force that act on the pin joint.

General solution of triangles

Angles in any quadrant

If you key into your calculator cos 150, you get the value −0.866. This is the same numerically as cos 30 = 0.866, except that there has been a *sign* change. Whether any one trigonometric ratio is positive or negative depends on whether the projection is on the positive or negative part of the coordinate system. *Figure 4.22* shows the rectangular coordinate system on which two lines have been placed at angles of 30° and 150°, respectively, from the positive horizontal *x ordinate*.

Now, if we consider the sine ratio for both angles, then we get:

$\sin 30 = \dfrac{+ab}{+ob}$ and $\sin 150 = \dfrac{+cd}{+od}$, thus both these ratios are *positive* and, therefore, a positive value for sin 30 and cos 30 will result. In fact from your calculator sin 30 = sin 150 = 0.5.

Now, from the diagram we find that $\cos 30 = \dfrac{+oa}{+ob}$, which will again yield a positive value. In fact cos 30 = 0.866 but $\cos 150 = \dfrac{-oc}{+od}$, a *negative* ratio that yields the negative value −0.866, which you found earlier.

If we continue to rotate our line in an *anti-clockwise* direction, we will find that cos 240 = −0.5 and cos 300 = 0.5. Thus, dependent on which *quadrant* (quarter of a circle, so each 90°) the ratio is placed, depends on whether or not the ratio is positive or negative. This is true for all three of the fundamental trigonometric ratios.

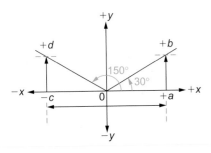

Figure 4.22 Projection of the angles 30° and 150°

Figure 4.23 shows the **signs** for the sine, cosine and tangent functions in any quadrant. A way of remembering when the sign of these ratios is positive is to use the word CAST, starting from the bottom right quadrant and reading anti-clockwise.

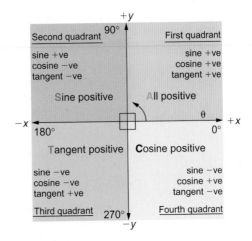

Figure 4.23 Signs of angles of any quadrant

Your calculator automatically shows the correct sign for any ratio of any angle, but it is worth knowing what to expect from your calculator!

UNIT 4

Example 4.37

Find, on your calculator, the value of the following trigonometric ratios and verify that the *sign* is correct by consulting *Figure 4.23*.

(a) sin 57	(b) cos 236	(c) tan 97
(d) sin 320	(e) cos 108	(f) tan 347
(g) sin 137	(h) cos 310	(i) tan 237

The values with their appropriate sign are tabulated below.

(a) 0.8387	*(b)* −0.5592	*(c)* −8.144
(d) −0.6428	*(e)* −0.3090	*(f)* −0.2309
(g) 0.6819	*(h)* 0.6428	*(i)* 1.5397

The sine and cosine rules

We now extend our knowledge to the solution of triangles, which are not right-angled. In order to do this we need to be armed with just two additional formulae. These are tabulated below for reference.

Sine rule	$\dfrac{a}{\sin a} = \dfrac{b}{\sin b} = \dfrac{c}{\sin c}$
Cosine rule	$a^2 = b^2 + c^2 - 2bc \cos A$
	$b^2 = a^2 + c^2 - 2ac \cos B$
	$c^2 = a^2 + b^2 - 2ab \cos C$

The above rules can only be used in specific circumstances.

For the general triangle *ABC* shown in *Figure 4.24* with sides *a*, *b*, *c* and angles $\angle A$, $\angle B$, $\angle C$, the **sine rule may only be used when either**

● *one side and any two angles are known*

or if

● *two sides and an angle (not the angle between the sides) are known.*

The cosine rule may only be used when either

● *three sides are known*

or

● *two sides and the included angle are known*.

Note: When using the sine rule, the equality signs allow us to use any parts of the rule that may be of help. For example, if we have a triangle to solve for which we know the angles $\angle A$ and $\angle C$ and side *a*. We would first use the rule with the terms: $\dfrac{a}{\sin A} = \dfrac{c}{\sin C}$ to find side *c*.

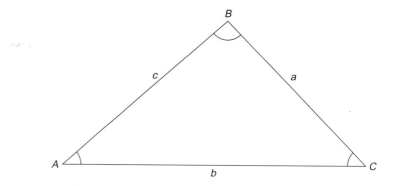

Figure 4.24 The general triangle

Note: When using the cosine rule, the version chosen will also depend on the information given. For example, if you are given sides *a*, *b* and the included angle *C*, then the formula: $c^2 = a^2 + b^2 - 2ab \cos C$ would be selected to find the remaining side *c*.

Example 4.38

In a triangle *ABC*, $\angle A = 48°$, $\angle B = 59°$ and the side $a = 14.5$ cm, find the unknown sides and angle. The triangle *ABC* is shown in *Figure 4.25*.

When the triangle is sketched, it can be seen that we have two angles and one side. So we can use the sine rule. Remembering that the sum of the internal angles of a triangle $= 180°$, we have $\angle C = 180 - 48 - 59 = 73°$. We will use the first two terms of the sine rule, $\dfrac{a}{\sin A} = \dfrac{b}{\sin B}$, to find the unknown side *b*. Then:

$$\frac{14.5}{\sin 48} = \frac{b}{\sin 59} \quad \text{or} \quad b = \frac{(\sin 59)(14.5)}{\sin 48} = \frac{(0.8572)(14.5)}{0.7431} = \textit{16.72 cm}.$$

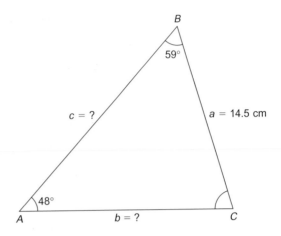

Figure 4.25 Figure for Example 4.38

Similarly, to find side c we use $\dfrac{a}{\sin A} = \dfrac{c}{\sin C}$, which on substitution of the values gives

$$\frac{14.5}{\sin 48} = \frac{c}{\sin 73} \quad \text{or} \quad c = \frac{(\sin 73)(14.5)}{\sin 48} = \frac{(0.9563)(14.5)}{0.7431} = \frac{13.8664}{0.7431} = \textbf{18.66 cm}.$$

When using the cosine rule given three sides, it is necessary to transpose the formula to find the required angles. In the next example, we need to perform this transposition, which you should find relatively simple. If you have difficulties following the steps, you should refer back to the section on transposition of formula in outcome 1.

Example 4.39

A flat steel plate is cut with sides of length 12, 8 and 6 cm. Determine the three angles of the plate. A diagram of the plate, suitably labelled, is shown in *Figure 4.26*, where side $a = 6\,\text{cm}$, $b = 12\,\text{cm}$ and $c = 8\,\text{cm}$.

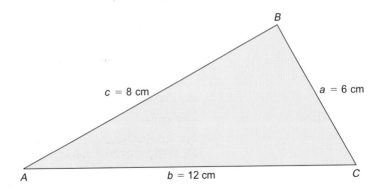

Figure 4.26 Figure for Example 4.39

Now, in this particular case we are free to choose any variant of the formula to find the corresponding angle. We will use:

$$b^2 = a^2 + c^2 - 2ac\cos B.$$

Then transposing for cos B,

$$2ac \cos B = a^2 + c^2 - b^2 \text{ and}$$

$$\cos B = \frac{a^2 + c^2 - b^2}{2ac}$$

then $\cos B = \dfrac{6^2 + 8^2 - 12^2}{(2)(6)(8)} = \dfrac{36 + 64 - 144}{96} = \dfrac{-44}{96} = -0.4583.$

Now $\angle B = 117.28$, using a calculator. Note that cos B is negative. Therefore, $\angle B$ must lie outside the first quadrant (see the next section on angles in any quadrant), that is, it must be greater than 90°. However, since it is the angle of a triangle it must also be less than 180°, thus $\angle B = 117.28°$ is its only possible value.

Now to find another angle, we could again use the cosine rule. However, since we now have an angle and two *non-included* sides, a and b, we are at liberty to use the simpler *sine* rule. Then

$$\frac{a}{\sin A} = \frac{b}{\sin B} \text{ and so, } \frac{6}{\sin A} = \frac{12}{\sin 117.28} \text{ or}$$

$$\sin A = \frac{(6)(\sin 117.28)}{12} = \frac{(6)(0.8887)}{12} = 0.4444$$

and from the calculator, $\qquad\qquad \angle A = 26.38°.$

Finally $\qquad\qquad \angle C = 180 - 117.28 - 26.38 = 36.34°.$

Area of any triangle

Now to complete our study of general triangles, we need to be able to calculate their area. This of course we have already done during our study of areas and volumes in the previous outcome on algebra. Again, as we did for right-angled triangles, let us use one of the formulas we learnt earlier to find the area of any triangle. The formula we will use is

$$\sqrt{s(s-a)(s-b)(s-c)}$$

where a, b and c are the sides and $s = \dfrac{a+b+c}{2}$.

Then in the case of the triangular plate we have just been considering in Example 4.39

$a = 6\,\text{cm}$, $b = 12\,\text{cm}$ and $c = 8\,\text{cm}$, then $s = \dfrac{6+12+8}{2} = \dfrac{26}{2} = 13$

and therefore,

$$\text{the area} = \sqrt{13(13-6)(13-12)(13-8)} = \sqrt{(13)(7)(1)(5)}$$
$$= \sqrt{455} = 21.33\,\text{cm}^2.$$

Now, the area of any triangle ABC can also be found using any of the following formulae.

Area of any triangle

$$ABC = \frac{1}{2}ab\sin C \quad \text{or} \quad = \frac{1}{2}ac\sin B \quad \text{or} \quad = \frac{1}{2}bc\sin A$$

These formulae are quoted here without proof and any variant may be used dependent on the information available. So again, for the triangle in Example 4.39, using the first of the formulae, area of triangle $ABC = \frac{1}{2}ab \sin C = \frac{1}{2}(6)(12)(\sin 36.34) = (0.5)(72)(0.5926) = 21.33\,\text{cm}^2$ as before.

TYK 4.7

1. For the following trigonometric ratios state the quadrant in which the angle θ will sit and then check your answer using your calculator.
   (a) $\sin \theta = 0.5221$, $\sin \theta = 0.1774$, $\sin \theta = -0.3773$
   (b) $\cos \theta = 0.8660$, $\cos \theta = -0.6649$, $\cos \theta = 0.0771$
   (c) $\tan \theta = 2.6641$, $\tan \theta = 0.1174$, $\tan \theta = -0.2243$

2. State the circumstances under which
   (a) the sine rule may be used
   (b) the cosine rule may be used.

3. Use the sine rule to solve the triangle ABC, where side $a = 37.2\,\text{cm}$, side $b = 31.6\,\text{cm}$ and $\angle B = 37°$.

4. Use the cosine rule to solve the triangle ABC, where: $a = 12\,\text{cm}$, $b = 10\,\text{cm}$ and $c = 6\,\text{cm}$ and find its area.

5. Find the area of the following triangles
   (a) $a = 5\,\text{cm}$, $b = 7\,\text{cm}$ and $\angle C = 105°$
   (b) $b = 7.3\,\text{cm}$, $c = 12.2\,\text{cm}$ and $\angle A = 135°$
   (c) $a = 9.6\,\text{cm}$, $c = 11.2\,\text{cm}$ and $\angle B = 163°$

Test your knowledge

?

TYK

UNIT 4

Circular measure

Elements and properties of the circle

You will already be familiar with the way in which we find the circumference and area of a circle, although you will be reminded of this in this section and in the introduction to the final section concerned with finding the surface area and volume of regular solids. Here, however, we extend our knowledge of the circle by identifying and defining certain geometrical and trigonometric elements of the circle, together with the study of radian measure. You will find an understanding of the nature of the circle and radian measure particularly useful when determining areas and volumes and when considering circular motion.

The major elements of the circle are shown in *Figure 4.27*. You will be familiar with most, if not all, of these elements. However, for the sake of completeness, we will formally define them.

A point in a plane whose distance from a fixed point in that plane is constant lies on the *circumference* of a circle. The fixed point is called the centre of the circle and the constant distance is called the *radius*.

A circle may be marked out on the ground by placing a peg or spike at its centre. Then using a length of cord for the radius, we simply walk round with a pointer at the end of the cord and mark out the circumference of the circle.

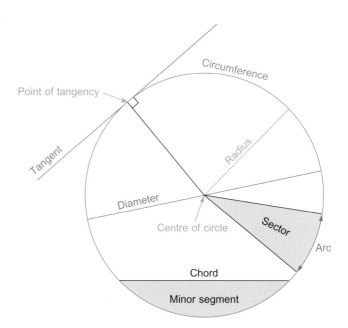

Figure 4.27 Elements of a circle

A **chord** is a straight line that joins two points on the circumference of a circle. A **diameter** is a chord drawn through the centre of a circle.

A **tangent** is a line that just touches the circumference of a circle at one point (the point of tangency). This tangent line lies at right angles to a radius, drawn from the point of tangency.

A chord line cuts a circle into a **minor segment** and **major segment**. A **sector** of a circle is an area enclosed between two radii, and a length of the circumference the **arc** length.

The radian and circular measure

Circular measure using degrees has been with us since the days of the Babylonians, when they divided a circle into 360 equal parts, corresponding to what they believed were the days in the year. An angle in degrees is a measure of rotation and an angle is formed when a line rotates with respect to a fixed line (*Figure 4.28*), when both lines have the same centre of rotation.

The degree may be subdivided into minutes and seconds of arc, where the *minute is $\frac{1}{60}$ of a degree* and a *second is $\frac{1}{60}$ of a minute* or $\frac{1}{3600}$ of a degree of arc. We will restrict ourselves to angular measurement in degrees, and decimal fractions of a degree, as you learnt earlier.

The **degree**, being an arbitrary form of circular measurement dividing the circle into 360 equal parts, has not always proved an appropriate unit for mathematical manipulation. Another less arbitrary unit of measure has been introduced, known as the **radian** (*Figure 4.29*); the advantage of this unit is its relationship with the **arc length** of a circle. You will also need to understand and use radian measure when you consider the graphs of the basic trigonometric functions later in this outcome.

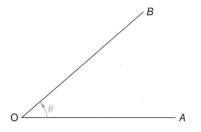

Figure 4.28 The angle as a measure of rotation

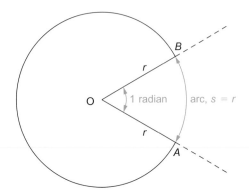

Figure 4.29 Illustration of the radian

> A **radian** is defined as the angle subtended at the centre of a circle by an arc equal in length to the radius of the circle.

Now, we know that the ***circumference of a circle is given by C = 2πr where r is the radius***. Therefore, the circumference contains 2π radii. We have just been told that the arc length for 1 radian is $s = r$. ***Therefore, the whole circle must contain 2π radians*** or approximately 6.28 radians. A circle contains $360°$, so it follows that **2π rad = 360°** or **π rad = 180°**. We can use this relationship to convert from degrees to radians, and radians to degrees.

Example 4.40

(a) Express 60° in radians.

(b) Express $\dfrac{\pi}{4}$ rad in degrees.

(a) Since $180° = \pi$ rad then,

$$1° = \frac{\pi \text{ rad}}{180} \text{ so,}$$

$$60° = 60\left(\frac{\pi \text{ rad}}{180}\right)$$

$$60° = \frac{\pi \text{ rad}}{3} \quad \text{or} \quad \textbf{\textit{1.047 rad}} \text{ (3 decimal points)}$$

Note that if we leave radians in terms of π we have an exact value to use for further mathematical manipulation. For this reason, it is more convenient to leave radians expressed in terms of π.

(b) We follow a similar argument, except we apply the reverse operations.

$$\pi \text{ rad} = 180° \text{ then,}$$

$$1 \text{ rad} = \frac{180°}{\pi} \text{ so,}$$

$$\frac{\pi}{4}\text{rad} = \left(\frac{\pi}{4}\right)\left(\frac{180°}{\pi}\right) \text{ and}$$

$$\frac{\pi}{4}\textbf{rad} = \textbf{45°}$$

To aid your understanding of the relationship between the degree and the radian, *Figure 4.30* shows diagrammatically a comparison between some common angles, using both forms of measure. Note that in the figure, all angles in radian measure are shown in terms of π.

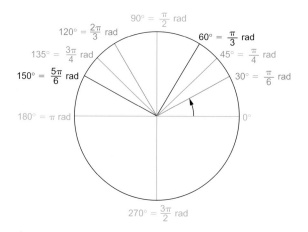

Figure 4.30 Comparison of degree and radian measure

It is often useful to be able to find **the area of the sector of a circle** when considering cross-sectional areas. To determine such areas, we first need to understand the relationship between the *arc length s* and the *angle θ* subtended at the centre of a circle by this arc length.

You have seen that the circumference of a circle subtends 2π rad. So, if we consider the circumference to be an *arc of length $2\pi r$*, we can say that

$$2\pi \text{ rad} = \frac{2\pi r}{r},$$

where r = radius

or, the angle in radians = $\dfrac{arc\ length\ (s)}{radius\ (r)}$ then,

$$\theta \text{ rad} = \frac{s}{r} \quad \text{or} \quad s = r\theta.$$

Always remember that when using this formula, *__the angle θ must be in radian__*.

The area of a sector is now fairly easy to find.

We know that the area of a circle = πr^2. So it follows that when dealing with a portion (sector) of a circle, like that shown in *Figure 4.31*, then the ratio of the angle θ (in radians) of the sector to that of the angle for the whole circle in radians is $\dfrac{\theta}{2\pi}$, remembering that there are 2π radians in a circle (360°).

Then the area of any portion of the circle such as the area of the sector = the area of the circle multiplied by the ratio of the angles, or in symbols:

$$\textbf{Area of sector} = (\pi r^2)\left(\frac{\theta}{2\pi}\right) = \frac{r^2\theta}{2} \ (\theta \text{ in radian}).$$

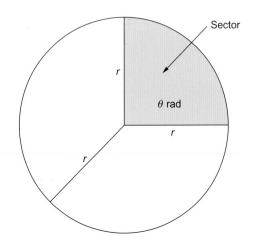

Figure 4.31 Area of sector of a circle

Example 4.41

(a) If the angle subtended at the centre of a circle by an arc length 4.5 cm is 120°, what is the radius of the circle?

(b) Find the angle of a sector of radius 20 cm and area 300 cm².

(a) We must first convert 120° into radians. This we can do very easily, using the conversion factor we found earlier, then

$$120° = \frac{120\pi \, rad}{180} = \frac{2\pi}{3} \, rad.$$ We will leave this angle in terms of π.

Then from $s = r\theta$ we have $r = \frac{s}{\theta} = \frac{4.5}{2/3\pi} = \textbf{\textit{2.149 cm}}$ (correct to 3 decimal points).

(b) To find the angle of the sector we use the area of a sector formula, that is,

$$A = \frac{1}{2}r^2\theta \text{ or } \theta = \frac{2A}{r^2}$$ and on substitution of given values, we get

$$\theta = \frac{(2)(300)}{20^2} = \frac{600}{400} = \textbf{\textit{1.5 rad}}.$$ If we wish to convert this angle to degrees, then

$$1.5 \, rad = (1.5)\frac{180°}{\pi} = \textbf{\textit{85.94°}}$$ (correct to 2 decimal places).

TYK 4.8

1. Express the following angles in degrees:
   (a) 1.26 radian, (b) 5 radian and (c) 1.33 radian.

2. Express the following angles in radian:
   (a) 83°, (b) 189° and (c) 295°.

3. Find the angle in radian subtended by the following arcs:
   (a) arc = 12 cm, radius = 4.4 cm
   (b) arc = 8.0 m, radius = 2.5 m.

4. Find the area of the sector of a circle that subtends an angle of 84° with the centre of the circle that has a radius of 3 m.

5. A chord AB divides a circle of radius 2 m into two segments (*Figure 4.32*). If AB subtends an angle of 60° at the centre of the circle, find the area of the minor segment.

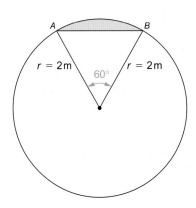

Figure 4.32 Figure for question 5 in TYK 4.8

Trigonometric functions

In this section, we will limit our study of trigonometric functions to the sine, cosine and tangent. In particular, we look at the nature of their graphs and the use to which these may be put. The graphs of these functions are very important, as the sine and cosine curves, in particular, illustrate many kinds of oscillatory motion, which you are likely to meet in your future studies. The sine and cosine functions are used to model the oscillatory motion of currents, voltages, springs, vibration dampers, the rise and fall of the tides and many other forms of vibrating system where the motion is oscillatory.

By *oscillatory, we mean motion that vibrates back and forth about some mean value during even periods of time*. We start by plotting the sine and cosine curves then consider their use for solving sine and cosine functions, in a similar manner to the graphs of algebraic equations we considered earlier. Finally, we look at the tangent function and see its particular properties, introducing the idea of an asymptote.

Graphs of the sine and cosine functions

The basic **sine function** $y = \sin x$ is a wave that lies between the values $+1$ and -1. It is therefore ***bounded***; that is, the value of the dependent variable y reaches a ***maximum value of*** $+1$ and a ***minimum value of*** -1 (*Figure 4.33*). Also the curve is zero at multiples of $180°$ or at multiples of π rad.

The ***x-axis***, in *Figure 4.33*, is marked out in ***degrees and radian***, which measure angular distance. The maximum and minimum values of y are also shown. Other things to note about this graph are the fact that it repeats itself every ***360°*** or 2π rad. Also, this curve reaches it first maximum value at $90°$ or $\frac{\pi}{2}$ rad, and reaches its second maximum $360°$ or 2π rad later at $450°$ or $\frac{5\pi}{2}$ rad. Similarly, it reaches its first minimum value at $270°$ or $\frac{3\pi}{2}$ rad and again $360°$ or 2π rad later, at $630°$ or $\frac{7\pi}{2}$ rad. These ***maximum*** and ***minimum*** values are repeated periodically at $360°$ degree intervals. We therefore say that the sine

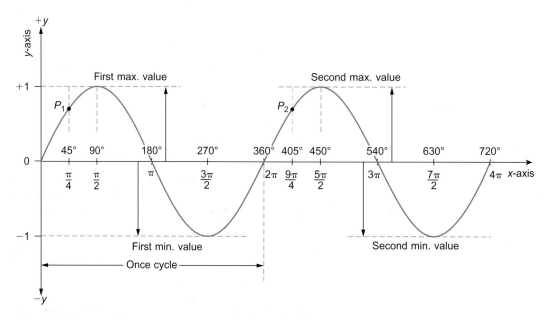

Figure 4.33 Plot of the function $y = \sin x$

wave has periodic motion, where any point on the wave, say, p_1, repeats itself every 360° or 2π rad. These repetitions are known as **cycles**, as shown in *Figure 4.33*, where one complete cycle occurs every 360° or every 2π rad.

Now how do we plot values for sinusoidal functions? Look back at *Figure 4.30* and note how we represented angular measure. In *Figure 4.34*, we can represent angular measure on the set of rectangular coordinates where **the angle** in degrees or radians is measured from the **positive x-axis**, and increases as it rotates in an **anticlockwise** direction, reaching a **positive maximum** value at 90° or $\frac{\pi}{2}$ rad. This maximum value is $+1$ when we make the radius of the circle $r = 1$, as in the diagram.

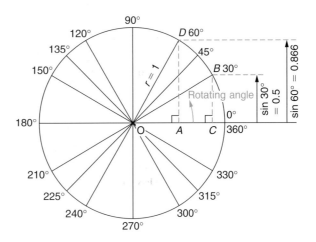

Figure 4.34 The rotating angle

Now, the actual magnitude of this angle (its distance in the y-direction) is found using the *sine* function. For example, the height of the line CB in the triangle OCB can be found by noting that

$$\sin 30° = \frac{opp}{hyp} = \frac{CB}{1} = CB = 0.5.$$ Similarly, as the angle increases, say to 60° or $\frac{\pi}{3}$ rad, then $AD = \sin 60° = 0.866$. It reaches its first maximum value when the rotating angle $= \sin 90° = 1.0 = $ radius r. Compare this value with the value on the curve of the sine function, shown in *Figure 4.33*! Now as the angle continues to increase, it moves into the second quadrant, where the magnitude of the rotating angle gradually reduces until it reaches 180° or π rad, when its value becomes zero once more. As we move into the third quadrant, the magnitude of the rotating angle (vector) once again starts to increase, but in a negative sense, until it reaches its maximum value at 270° or $\frac{3\pi}{2}$ rad, where $\sin 270° = -1$. Finally, in the fourth quadrant, it reduces from the negative maximum (minimum) value, until it once again reaches zero. The behaviour of this point is plotted as the curve shown in *Figure 4.33*, where the curve is produced by connecting the magnitude of this point for many values of the angle, between 0 and 360°, after which the pattern repeats itself every 360°.

A table of values for the magnitude of the rotating angle is given below. Check that these values match the plot of the sine curve shown in *Figure 4.33*.

$x =$ angle θ	$y = \sin\theta$	$x =$ angle θ	$y = \sin\theta$
0	0		
30° or $\frac{\pi}{6}$ rad	0.5	210°	−0.5
45° or $\frac{\pi}{4}$ rad	0.7071	225°	−0.7071
90° or $\frac{\pi}{2}$ rad	1.0	270°	−1.0
120° or $\frac{2\pi}{3}$ rad	0.8660	300°	−0.866
135° or $\frac{3\pi}{4}$ rad	0.7071	315°	−0.7071
150° or $\frac{5\pi}{6}$ rad	0.5	330°	−0.5
180° or π rad	0	360°	0

The above table is similar to that you would need to produce when plotting *any sine function* graphically. For example, suppose you were required to plot the curve for the function $y = 2\sin\theta$. What happens to the values of y in the above table? I hope you can see that every value of y is *doubled*. That means the first *maximum* value for this function will be $y = 2\sin 90° = (2)(1) = 2$. Similarly for all other angles, the y values will be doubled.

I hope you can now appreciate that if $y = 3\sin\theta$, then the magnitude of the y values will all be *trebled*. Then in general, ***the magnitude of the plotted y values is dependent on the value of the constant a, when y = a*** **sin** $\boldsymbol{\theta}$. The magnitude of the y *values* is referred to as their *amplitude*. Then ***the maximum amplitude a will occur when*** $\sin\theta$ ***is a maximum, that is, when*** $\sin\theta = 1.0$. This we know from the table above to first occur at $\theta = 90°$ and then to occur every 360° or 2π rad later. ***The minimum value of the amplitude will first occur when*** **sin** $\boldsymbol{\theta} = -\textbf{1.0}$ This again can be seen to first occur when $\theta = 270°$ and repeat itself every 360° thereafter.

What do you think happens if we plot the graph of $y = \sin 2\theta$? Well if $\theta = \dfrac{\pi}{4}$ rad, then $y = \sin(2)\left(\dfrac{\pi}{4}\right) = \sin\dfrac{\pi}{2} = 1.0$. If we compare this with the plotted values above, then the function $y = \sin 2\theta$ has reached its first maximum, ***twice as fast*** as the function $y = \sin\theta$. The effect of this is to increase the number of oscillations (cycles) in a given angular distance. This is illustrated in *Figure 4.35*.

You should check a few of the plotted values to verify your understanding.

So far we have concentrated our efforts on the sine function. This is because the **cosine function** is very similar to the sine function, except that it reaches its first maximum and minimum values at different angles to that of the sine function. In all other respects it is identical.

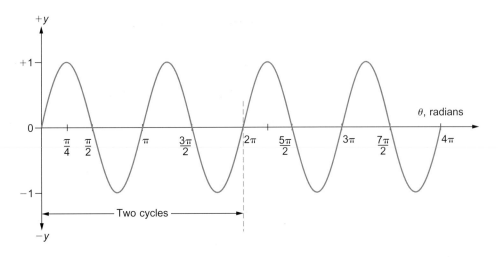

Figure 4.35 Graph of $y = \sin 2\theta$ between 0 and 4π rads

Consider *Figure 4.36*. Here, in the case of the **cosine function, we start** our rotating angle in the **vertical position**. This means that what was 90° for the sine function (*Figure 4.34*) **is now 0° for the cosine function**.

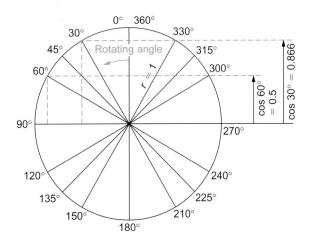

Figure 4.36 Rotating angle and the cosine function

Now, if the cosine of the angle 30° is given by the *height of the y ordinate*, in a similar manner to the sine function, then $y = \cos 30° = 0.866$. Similarly, the cosine of 90° is again the height of the *y ordinate*, which can be seen to be zero, that is, $\cos 90° = 0$. This is easily checked on your calculator! *The net result is that all the cosine function values, for the given angle, are* **90°** *in advance of the sine function*. So, for example, the cosine function starts with its maximum at 0°, which is 90° in advance of the first maximum for the sine function. A plot of the cosine function $y = \cos \theta$ for angles between 0 and 4π rad is shown in *Figure 4.37*.

It can be seen from *Figure 4.37* that apart from the 90° *advance*, the cosine function follows an identical pattern to that of the sine function.

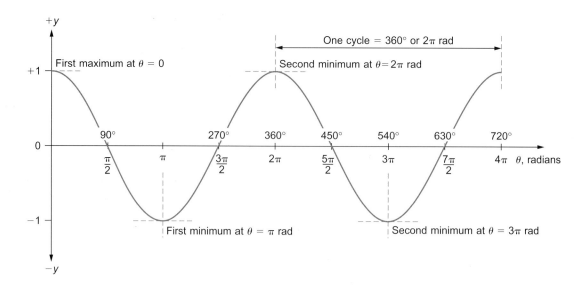

Figure 4.37 Graph of $y = \cos \theta$

Example 4.42

Draw the graph of the function $y = 2\sin \theta + 3\cos \theta$, for values of θ between 0° and 90°. From the graph find
(a) the maximum amplitude of the function
(b) a value of θ which satisfies the equation $2\sin \theta + 3\cos \theta = 3.5$.

(a) Our first task is to set up a table of values and find the corresponding values for θ and y. We will use an interval of 10°.

θ	0	10	20	30	40	50	60	70	80	90
$2\sin \theta$	0	0.35	0.68	1.0	1.29	1.53	1.73	1.88	1.97	2.0
$3\cos \theta$	3	2.95	2.82	2.60	2.30	1.94	1.50	1.03	0.52	0
$y = 2\sin \theta + 3\cos \theta$	3	3.3	3.5	3.6	3.59	3.47	3.23	2.91	2.49	2.0

The table shows only two decimal place accuracy but when undertaking graphical work it is difficult to plot values with any greater accuracy. Note also that we seem to have a maximum value for y when $\theta = 30°$. It is worth plotting a couple of intermediate values either side of $\theta = 30°$ to see if there is an even higher value of y.

I have chosen $\theta = 27°$ and $\theta = 33°$. Then:

when $\theta = 27°$, $y = 3.58$ and when $\theta = 33°$, $y = 3.61$, the latter value is very slightly higher, so this value may be used as the maximum.

The plot is shown in *Figure 4.38*, where it can be seen that within the accuracy of the plot, the maximum value of the amplitude for the function is $y = 3.6$.

(b) Now, the appropriate values for the solution of the equation $2\sin \theta + 3\cos \theta = 3.5$ are read-off from the graph, where the line $y = 3.5$ intersects with the curve $y = 2\sin \theta + 3\cos \theta$. The solutions are that when $y = 3.5$, $\theta = 20°$ and $\theta = 48°$.

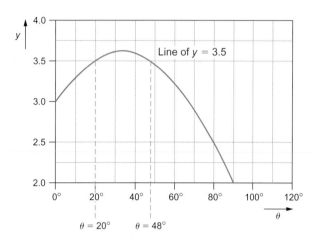

Figure 4.38 Graph of $y = 2\sin\theta + 3\cos\theta$

Example 4.43

For the following trigonometric functions, find the first maximum amplitude and the angular distance it occurs from $\theta = 0°$. Comment on the general form of each function.

(a) $y = 4.2\cos\theta$

(b) $y = 3\sin 2\theta$

(c) $y = \sin\left(\theta - \frac{\pi}{2}\right)$.

(a) The maximum amplitude for all the functions is given when the amplitude a is a maximum ($a \times 1$), in each case.

We know that for $\cos\theta$ this first occurs when $\theta = 0$, so the maximum amplitude is 4.2 at an angular distance of 0°, from the reference angle.

The graph will follow exactly the form of the graph $y = \cos\theta$, except that every value will be amplified by a factor of 4.2.

(b) In this case the maximum amplitude is 3, and it first occurs when $2\theta = 90°$, that is at $+45°$ to the reference angle.

This graph will complete each cycle in half the angular distance, when compared to $y = \sin\theta$.

(c) This function has a maximum amplitude of $a = 1.0$, which first occurs when

$$\theta - \frac{\pi}{2} = \frac{\pi}{2} \text{ rad, therefore, } \theta = \frac{\pi}{2} + \frac{\pi}{2} = \pi \text{ rad}$$

That is, the first maximum occurs 180° after the reference angle. When compared to the function $y = \sin\theta$, each value is found to be lagging by $\frac{\pi}{2}$.

If you are finding it difficult to envisage what is happening, sketch these functions on the same axes and make comparisons.

The properties of the graphs of the basic sine function $y = \sin\theta$ and cosine function $y = \cos\theta$ are summarized below.

Sine function

- It is continuous (i.e., it has no sudden changes in position).
- It lies entirely within the range $-1 \le \sin\theta \le +1$.

- The shape of the graph from $\theta = 0$ to $\theta = 2\pi$ is repeated for each further complete revolution.
- The sine function is said to be *periodic* or *cyclic* and the width of the repeating pattern as measured on the horizontal axis is called the *period*.
- The period for the function $y = \sin \theta$ is 2π.
- The greatest positive value $|\sin \theta|$ is called the amplitude of the sine wave and its value is 1. Its maximum range is between ± 1.

Cosine function

- It is continuous.
- It lies entirely within the range $-1 \le \cos \theta \le +1$.
- It is periodic with a period of 2π.
- It has the same shape as the sine graph but is displaced a distance $\frac{\pi}{2}$ or 90° to the left on the horizontal axis. This displacement is known as *phase shift* and in this case the cosine function is *phase advanced*, when compared with the sine function.
- $y = \cos \theta$ is a cyclic function with period 2π, with a range between ± 1.

The tangent function

When trying to plot a graph of the tangent function it will become very clear that this function $y = \tan \theta$ is different from that of the sine and cosine functions in several respects. If you try to find the value of tan 90° in your calculator it will show up as an error! In other words, it is **not defined**. In actual fact the tangent function has *an infinite value at* $\frac{\pi}{2}$ or 90°. You will find from your calculator that the values of $y = \tan 89 = 57.29$, $y = \tan 89.9 = 572.96$, $y = \tan 89.999 = 57,295.8$! Thus as you approach tan 90, the value of y approaches an infinitely large number. We say that y is approaching positive infinity ($y \rightarrow +\infty$). It is the same story as the function $y = \tan \theta$ approaches -90 or $-\frac{\pi}{2}$ but in this case y approaches negative infinity ($y \rightarrow -\infty$). This behaviour is shown in the plot of the tangent function shown in *Figure 4.39*.

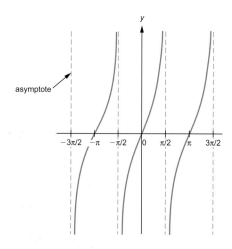

Figure 4.39 Plot of the tangent function $y = \tan \theta$

On the graph (*Figure 4.39*) you will note the dashed vertical lines at the values where the values of y for the tangent function are deemed to be infinite; these lines are known as ***asymptotes***. The function will approach these lines but never meet them because infinity is un-bounded! You will be asked as an exercise to plot the graph of the tangent function and also show that $\dfrac{\sin\theta}{\cos\theta} = \tan\theta$, which you already know.

The properties of the graph of the tangent function $y = \tan\theta$ are summarized below.

Tangent function

- It is not continuous, being undefined when $\theta = -\frac{\pi}{2}, \frac{\pi}{2}, \frac{3\pi}{2}$, etc.
- The range of possible values of $\tan\theta$ is unlimited.
- The tangent function is periodic but the period in this case is π (not 2π as was the case with the sine and cosine functions).

TYK 4.9

1. Explain the essential differences in the properties of the sine function and the tangent function.

2. Sketch the graphs of $y = 2\sin\theta$ and $y = \sin 2\theta$ on the same axis between 0 and 2π radian and describe the differences in the amplitude and time periods of the two functions.

3. Define the term 'phase difference' and explain the phase relationship between the graphs of the basic sine function ($y = \sin\theta$) and cosine function ($y = \cos\theta$).

4. Find the amplitude of the following functions, when $\theta = \frac{\pi}{6}, \frac{\pi}{4}$ and $\frac{\pi}{2}$:
   (a) $y = 2\sin\theta$, (b) $y = \cos 2\theta$, (c) $y = \frac{1}{2}\sin 2\theta$.

5. Plot the graph of the function $y = \dfrac{\sin\theta}{\cos\theta}$ between $-\pi$ and $+\pi$ and comment on the properties of the graph obtained.

UNIT 4

The surface area and volume of regular solids

Before considering the surface area and volume of some regular solids we will use some of the common formulae you will already be aware of to find the area of the triangle, circle and trapezium. You have already dealt with the complete solution of triangles and the circle using trigonometric ratios and radian measure in the previous sections of this outcome. *Table 4.2* provides a list of some of the formulae you will already have met. They are presented here as a revision aid for your convenience.

Table 4.2 Areas of common shapes

Shape	Area
Triangle	Half the base multiplied by the perpendicular height, or $A = \dfrac{1}{2}bh$
	$A = \sqrt{s(s-a)(s-b)(s-c)}$, where a, b, c are the lengths of the sides and $s = \frac{1}{2}(a+b+c)$
Parallelogram	A = base multiplied by the perpendicular height between the parallel sides. The base can be any side of the parallelogram
Circle	$A = \pi r^2$ or $A = \dfrac{\pi d^2}{4}$ where r = radius and d = diameter of circle
Trapezium	Half the sum of the parallel sides (a, b) multiplied by the vertical distance (h) between them, or $A = \left(\dfrac{a+b}{2}\right)h$

Example 4.44

In the triangle *ABC* shown in *Figure 4.40*, side *AB* = 3 cm and side *AC* = 4 cm. Find the area of the triangle using both of the formulae given in *Table 4.2*.

Now, we can see from the diagram that this is a right-angled triangle; therefore, the area A is found simply by using the formula $A = \frac{1}{2}bh$, where the **base (b)** can be taken as either side containing the rectangle. Then $A = \frac{1}{2}(3)(4) = \mathbf{6\,cm^2}$. Note that the other side, not used as the base, is at right angles to the base and is, therefore, the **perpendicular height (h)**. If the triangle were not right-angled, we would need to find the perpendicular height or all of the sides in order to find the area.

To use our second formula for area, we need to know the value of all sides of the triangle, so we need to find side BC. Since this is a right-angled triangle we can find the third side (opposite the right-angle) by using **Pythagoras theorem**, which you have already met. Then using Pythagoras, we know in this case that $(BC)^2 = 3^2 + 4^2 = 9 + 16 = 25$, or $BC = \sqrt{25} = 5$.

We now have three sides and $s = \frac{1}{2}(a+b+c) = \frac{1}{2}(3+4+5) = 6$, therefore the area of the triangle $A = \sqrt{s(s-a)(s-b)(s-c)} = \sqrt{6(6-3)(6-4)(6-5)} = \sqrt{6(3)(2)(1)} = \sqrt{36} = \mathbf{6\,cm^2}$ **as before**.

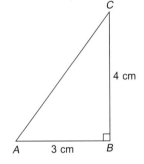

Figure 4.40 Figure for Example 4.44

We will now demonstrate the use of the trapezium formula through another example.

Example 4.45

The cross-section of a metal plate is shown in *Figure 4.41*, find its area.

Using the area rule for a trapezium, where in this case the vertical height is 72.7 mm, then

$$A = \left(\frac{a+b}{2}\right)h = \left(\frac{45.7 + 98.5}{2}\right)72.7 = (72.1)(72.7) = 5241.67 = \mathbf{5242\,mm^2}.$$

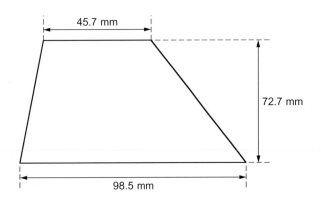

Figure 4.41 Figure for Example 4.45

The rule for the area of a circle I am sure you are familiar with, but we will use it to find the area of an ***annulus***.

Example 4.46

Determine the area of the annulus shown in *Figure 4.42* that has an inner radius of 5 cm and an outer radius of 8 cm.

The shaded area (similar to a doughnut in shape) is the area of the ***annulus*** we require. We know both the inner and outer radii; therefore, we can treat this shape as the *difference* between the *outer* and *inner circles*. We know that the area of a circle $= \pi r^2$. Now, our two circles have two different radii, where $R = 8$ cm and $r = 5$ cm. Then since the area of the *annulus A* is the difference between these two circles we may write:

$$A = \pi R^2 - \pi r^2 \quad \text{or} \quad A = \pi(R^2 - r^2)$$

then substituting the appropriate values of the radii gives $A = \pi(8^2 - 5^2) = \pi(64 - 25) = 39\pi = \textbf{122.5 cm}^2$.

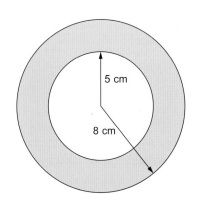

Figure 4.42 Figure for Example 4.46 – the annulus

> **KEY POINT**
>
> The circumference of a circle $= 2\pi r$ or πd

> **KEY POINT**
>
> The area of a circle $= \pi r^2$ or $\dfrac{\pi d^2}{4}$

Table 4.3 Some useful formulae for the volume and surface area of solids

Solid	Volume	Surface area
Right circular cylinder *without* base and top	$V = \pi r^2 h$	$S = 2\pi rh$
Right circular cylinder *with* base and top	$V = \pi r^2 h$	$S = 2\pi rh + 2\pi r^2$ or $S = 2\pi r(h + r)$
Cone *without* base	$V = \dfrac{1}{3}\pi r^2 h$	$S = \pi rl$, where $l = $ the slant height
Cone *with* base	$V = \dfrac{1}{3}\pi r^2 h$	$S = \pi rl + \pi r^2$ or $S = \pi r(l + r)$
Sphere	$\dfrac{4}{3}\pi r^3$	$S = 4\pi r^2$
Hollow pipe of uniform circular cross-section	$V = \pi(R^2 - r^2)l$	$S = 2\pi(R^2 - r^2) + 2\pi(R + r)$
Spherical shell	$V = \dfrac{4}{3}\pi(R^3 - r^3)$	$S = 4\pi(R^2 + r^2)$

UNIT 4

We are now in a position to tabulate some of the more common formulae that we will need to calculate the surface area and volume of regular solids (see *Table 4.3*).

Notes on table

1. For the cylinder the height h is the vertical height. There are two formulae for the surface area of a cylinder dependent on whether or not it has a base and top. The area πr^2 is for the addition of the base *or* top, thus $2\pi r^2$ is for both.
2. The formulae for the surface area of the cone also take into consideration the cone with and without circular base. In the volume formula, the height h is again the vertical height from the base, while the surface area formulae use the slant height l.
3. The hollow pipe takes into account the surface area at the ends of the pipe when the cross-section is cut at right angles to its length. The volume is given by the cross-sectional area of the annulus multiplied by the pipe length.
4. The surface area of the spherical shell includes both the inside and outside surface of the shell.

Example 4.47

Find the volume and total surface area of a right cylinder, with a top and bottom, if the cylinder has a height of 12 cm and a base radius of 3 cm.

In this example it is simply a question of applying the appropriate formula.

Then for the volume, $V = \pi r^2 h = \pi(3)^2 12 = 108\pi = \boldsymbol{339.29\,cm^3}$.

Now the cylinder has a base and a top; therefore the surface area $S = 2\pi r(h + r)$.

Then $S = 2\pi 3(12 + 3) = 90\pi = \boldsymbol{282.74\,cm^2}$.

We finish this short section on areas and volumes with one more example, leaving you to practice use of these formulae by completing the exercises in TYK 4.10.

Example 4.48

Water flows through a circular pipe of internal radius 10 cm at 5 ms^{-1}. If the pipe is always three-quarters full, find the volume of water discharged in 30 minutes.

This problem requires us to find the volume of water in the pipe per unit time, in other words the volume of water in the pipe per second. Note that no length has been given.

The area of the circular cross-section $= \pi r^2 = \pi(10)^2 = 100\pi$; therefore, the area of the cross-section of water $= \left(\dfrac{3}{4}\right)100\pi = 75\pi \text{ cm}^3 = (75\pi)10^{-4}\,m^3$.

Now since water flows at 5 ms^{-1}, the volume of water discharged per second

$= \dfrac{(5)(75\pi)10^{-4}}{1} = (375\pi)10^{-4}\,m^3\,s^{-1}$. Then the number of m^3 discharged in 30 minutes

$= (30)(60)(375\pi)(10^{-4}) = 67.5\pi = \boldsymbol{212\,m^3}$.

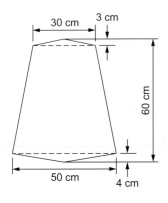

Figure 4.43 Figure for question 2 in TYK 4.10

Test your knowledge

TYK

TYK *4.10*

1. A parallelogram has an area of $60\,cm^2$, if its perpendicular height is $10\,cm$, what is the length of one of the parallel sides?

2. *Figure 4.43* shows the cross-section of a template, what is its area?

3. An annulus has an inside diameter of $0.75\,m$ and an external diameter of $0.9\,m$, determine its area.

4. Find the volume of a circular cone of height $6\,cm$ and base radius $5\,cm$.

5. Find the area of the curved surface of a cone (not including base) whose base radius is $3\,cm$ and whose vertical height is $4\,cm$. *Hint*: you need first to find the slant height.

6. If the area of a circle is $78.54\,mm^2$, find its diameter to 2 significant figures.

7. A cylinder of base radius $5\,cm$ has a volume of $1\,L$ ($1000\,cm^3$), find its height.

8. A pipe of thickness $5\,mm$ has an external diameter of $120\,mm$, find the volume of $2.4\,m$ of pipe material.

9. A batch of 2000 ball bearings are each to have a diameter of $5\,mm$. Determine the volume of metal needed for the manufacture of the whole batch.

10. Determine the volume and total surface area of a spherical shell having an internal diameter of $6\,cm$ and external diameter of $8\,cm$.

UNIT 4

Statistical Methods

Your view of statistics has probably been formed from what you read in the papers, or what you see on the television. Survey use to show which political party is going to win the election, why men grow moustaches, if smoking damages your health, the average cost of housing by area, and all sorts of other interesting data! So statistics is used to analyse the results of such surveys and when used correctly, it attempts to eliminate the bias that often appears when collecting data on controversial issues.

Statistics is concerned with collecting, sorting and analysing numerical facts, which originate from several observations. These facts are collated and summarized, then presented as tables, charts or diagrams, etc.

In this brief introduction to statistics, we look at two specific areas. First, we consider the collection and presentation of data in its various forms. Then we look at how we measure such data, concentrating on finding average values.

If you study statistics beyond this course, you will be introduced to the methods used to make predictions based on numerical data and the probability that your predictions are correct. At this stage in your learning, however, we will only be considering the areas of data handling and measurement of central tendency (averages), mentioned above.

Data manipulation

In almost all scientific, engineering and business journals, newspapers and Government reports, statistical information is presented in the form of charts, tables and diagrams, as mentioned above. We now look at a small selection of these presentation methods, including the necessary manipulation of the data to produce them.

Charts

Suppose, as the result of a survey, we are presented with the following statistical data (*Table 4.4*).

Table 4.4 Results of a survey

Major category of employment	Number employed
Private business	750
Public business	900
Agriculture	200
Engineering	300
Transport	425
Manufacture	325
Leisure Industry	700
Education	775
Health	500
Other	125

Now, ignoring for the moment the accuracy of this data, let us look at typical ways of presenting this information in the form of charts, in particular the **bar chart** and the **pie chart**.

Bar chart

In its simplest form, the bar chart may be used to represent data by drawing individual bars (*Figure 4.44*) using the figures from the raw data (the data in the table).

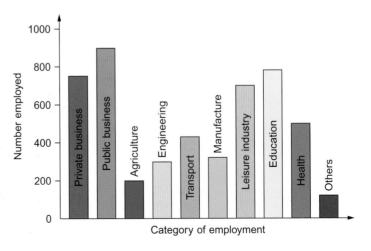

Figure 4.44 Bar chart representing number employed by category

Now, the scale for the vertical axis, the number employed, is easily decided by considering the highest and lowest values in the table, 900 and 125, respectively. Therefore, we use a scale from 0 to 1000 employees. Along the horizontal axis, we represent each category by a bar of even width. We could just as easily have chosen to represent the data using column widths instead of column heights.

Now the simple bar chart above tells us very little that we could not have determined from the table. So, another type of bar chart that enables us to make comparisons, the proportionate bar chart, may be used.

In this type of chart, we use *one bar*, with the same width throughout its height, with horizontal sections marked-off in proportion to the whole. In our example, each section would represent the number of people employed in each category compared with the total number of people surveyed.

In order to draw a proportionate bar chart for our employment survey, we first need to total the number of people who took part in the survey. This total comes to 5000. Now, even with this type of chart we may represent the data either in proportion by height or in proportion by percentage. If we were to choose height, then we need to set our vertical scale at some convenient height, say, 10 cm. Then we would need to carry out 10 simple calculations to determine the height of each individual column.

For example, given that the height of the total 10 cm represents 5000 people, then the height of the column for those employed in private business $= \left(\dfrac{750}{5000}\right)10 = 1.5$ cm. This type of calculation is then repeated for each category of employment. The resulting bar chart is shown in *Figure 4.45*.

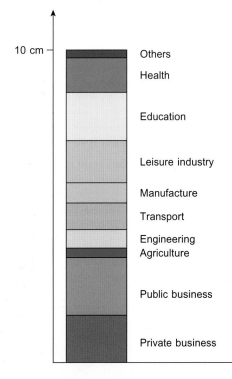

Figure 4.45 Proportionate bar chart graduated by height

Example 4.49

Draw a proportionate bar chart for the employment survey shown in *Table 4.4* using the percentage method.

For this method all that is required is to find the appropriate percentage of the total (5000) for each category of employment. Then, choosing a suitable height of column to represent 100%, mark on the appropriate percentage for each of the 10 employment categories. To save space, only the first five categories of employment have been calculated.

1. private business $= \left(\dfrac{750}{5000} \right) \times 100 = 15\%$

2. public business $= \left(\dfrac{900}{5000} \right) \times 100 = 18\%$

3. agriculture $= \left(\dfrac{200}{5000} \right) \times 100 = 4\%$

4. engineering $= \left(\dfrac{300}{5000} \right) \times 100 = 6\%$

5. transport $= \left(\dfrac{425}{5000} \right) \times 100 = 8.5\%$

Similarly, manufacture $= 6.5\%$, leisure industry $= 14\%$, education $= 15.5\%$, health $= 10\%$ and other categories $= 2.5\%$.

Figure 4.46 shows the completed bar chart.

Other categories of bar chart include ***horizontal bar charts***, where for instance *Figure 4.44* is turned through 90° in a clockwise direction. One last type may be used to depict data given in chronological (time) order. Thus, for example, the horizontal *x*-axis is used to represent, hours, days, years, etc., while the vertical axis shows the variation of the data with time.

Example 4.50

Represent the following data on a chronological bar chart.

Year	Number employed in general engineering (thousands)
2003	800
2004	785
2005	690
2006	670
2007	590

Since we have not been asked to represent the data on any specific bar chart we will use the simplest, involving only the raw data. Then, the only concern is the **scale** we should use for the vertical axis.

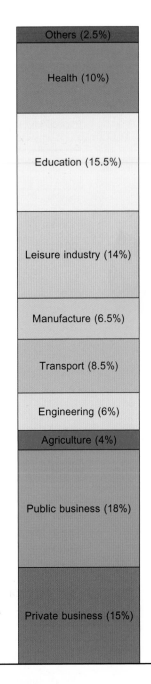

Figure 4.46 Proportionate percentage bar chart

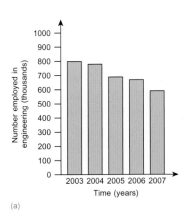

(a)

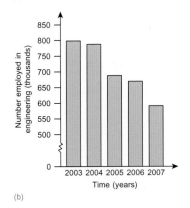

(b)

Figure 4.47 Chronological bar chart: (a) in correct proportion, and (b) with graduated scale

To present a **true** representation, the scale should start from zero and extend to, say, 800 (*Figure 4.47a*). If we wish to emphasize a **trend**, that is, the way the variable is rising or falling with time, we could use a very much exaggerated scale (*Figure 4.47b*). This immediately emphasizes the **downward trend** since 1995. Note that this data is **fictitious** (made-up) and used here merely for emphasis!

Pie chart

In this type of chart the data is presented as a proportion of the total using the angle or area of sectors. The method used to draw a pie chart is best illustrated by example.

Example 4.51

Represent the data given in Example 4.50 on a pie chart.

Remembering that there are 360° in a circle and that the total number employed in general engineering (according to our figures) was 800 + 785 + 690 + 670 + 590 = 3535 (thousands), then we manipulate the data as follows:

Year	Number employed in general engineering (thousands)	Sector angle (to nearest half degree)
2003	800	$\left(\dfrac{800}{3535}\right) \times 360 = 81.5°$
2004	785	$\left(\dfrac{785}{3535}\right) \times 360 = 80°$
2005	690	$\left(\dfrac{690}{3535}\right) \times 360 = 70.5°$
2006	670	$\left(\dfrac{670}{3535}\right) \times 360 = 68°$
2007	590	$\left(\dfrac{590}{3535}\right) \times 360 = 60°$
Total	3535	$= 360°$

The resulting pie chart is shown in *Figure 4.48*.

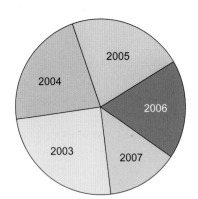

Figure 4.48 Resulting pie chart for Example 4.51: employment in engineering by year

Other methods of visual presentation include pictograms and ideographs. These are diagrams in pictorial form used to present information to those who have a limited interest in the subject matter or who do not wish to deal with data presented in numerical form. They have little or no practical use when interpreting engineering or other scientific data and apart from acknowledging their existence we will not be pursuing them further.

Frequency distributions

One of the most common and most important ways of organizing and presenting raw data is through use of **frequency distributions**.

Consider the data given in *Table 4.5*, which shows the time in hours that it took 50 individual workers to complete a specific assembly line task.

Table 4.5 Data for assembly line task

1.1	1.0	0.6	1.1	0.9	1.1	0.8	0.9	1.2	0.7
1.0	1.5	0.9	1.4	1.0	0.9	1.1	1.0	1.0	1.1
0.8	0.9	1.2	0.7	0.6	1.2	0.9	0.8	0.7	1.0
1.0	1.2	1.0	1.0	1.1	1.4	0.7	1.1	0.9	0.9
0.8	1.1	1.0	1.0	1.3	0.5	0.8	1.3	1.3	0.8

From the data you should be able to see that the shortest time for completion of the task was 0.5 hour, the longest time was 1.5 hours. The **frequency** of appearance of these values is once. On the other hand the number of times the job took 1 hour appears 11 times, or it has a *frequency of 11*. Trying to sort out the data in this *ad hoc* manner is time consuming and may lead to mistakes. To assist with the task we use a *tally chart*. This chart simply shows how many times the *event* of completing the task in a specific time takes place. To record the **frequency of events** we use the number 1 in a tally chart and when the frequency of the event reaches 5, we score through the existing four 1's to show a frequency of 5. The following example illustrates the procedure.

Example 4.52

Use a tally chart to determine the frequency of events, for the data given on the assembly line task in *Table 4.5*.

Time (hours)	Tally	Frequency
0.5	1	1
0.6	11	2
0.7	1111	4
0.8	⊥⊢⊤ 1	6
0.9	⊥⊢⊤ 111	8
1.0	⊥⊢⊤ ⊥⊢⊤ 1	11
1.1	⊥⊢⊤ 111	8
1.2	1111	4
1.3	111	3
1.4	11	2
1.5	1	1
	Total	50

We now have a full numerical representation of the frequency of events. So, for example, 8 people completed the assembly task in 1.1 hours or the time 1.1 hours has a frequency of 8. We will be using the above information later on when we consider measures of central tendency.

The times in hours given in the above data are simply numbers. When data appears in a form where it can be ***individually counted*** we say that it is **discrete** data. It goes up or down in ***countable*** steps. Thus the numbers 1.2, 3.4, 8.6, 9, 11.1, 13.0 are said to be ***discrete***. If, however, data is obtained by measurement, for example, the heights of a group of people, then we say that this data is **continuous**. When dealing with continuous data we tend to quote its limits, that is the limit of accuracy with which we take the measurements. So, for example, a person may be $174 \pm 0.5\,\text{cm}$ in height. When dealing numerically with continuous data or a large

KEY POINT

The grouping of frequency distributions is a means for clearer presentation of the facts

amount of discrete data, it is often useful to **group** this data into **classes or categories**. We can then find out the numbers (frequency) of items within each group.

Table 4.6 shows the height of 200 adults, grouped into 10 classes.

Table 4.6 Height of adults

Height (cm)	Frequency
150–154	4
155–159	9
160–164	15
165–169	21
170–174	32
175–179	45
180–184	41
185–189	22
190–194	9
195–199	2
Total	200

The main advantage of grouping is that it produces a clear overall picture of the frequency distribution. In *Table 4.6*, the first class interval is 150–154. The end number 150 is known as the **lower limit** of the class interval and the number 154 is the **upper limit**. The heights have been measured to the nearest centimetre. That means within ± 0.5 cm. Therefore, in effect, the first class interval includes all heights in the range 149.5–154.5 cm; these numbers are known as the lower and upper class *boundaries*, respectively. The **class width** is always taken as the **difference between the lower and upper class boundaries**, not the upper and lower limits of the class interval.

The histogram and frequency graph

The histogram is a special diagram that is used to represent a frequency distribution, such as that for grouped heights shown above. It consists of a set of rectangles, whose areas represent the frequencies of the various classes. Often when producing these diagrams, the class width is kept the same, so that the varying frequencies are represented by the height of each rectangle. When drawing histograms for grouped data, the **midpoints** of the rectangles represent the midpoints of the class intervals. So, for our data, they will be 152, 157, 162, 167, etc.

An adaptation of the histogram, known as the **frequency polygon**, may also be used to represent a frequency distribution.

Example 4.53

Represent the data shown in *Table 4.6* on a histogram and draw in the frequency polygon for this distribution.

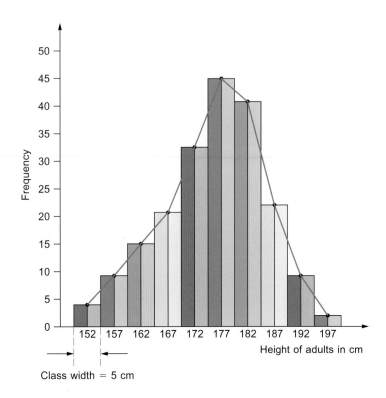

Class width = 5 cm

Figure 4.49 Figure for Example 4.53, histogram showing frequency distribution

All that is required to produce the histogram is to plot frequency against the height intervals, where the intervals are drawn as class widths.

Then, as can been seen from *Figure 4.49*, the area of each part of the histogram is the ***product of frequency × class width***. The frequency polygon is drawn so that it connects the ***midpoint*** of the class widths.

KEY POINT

The frequencies of a distribution may be added consecutively to produce a graph known as a cumulative frequency distribution

Another important method of representation is adding all the frequencies of a distribution consecutively, to produce a graph known as a ***cumulative frequency distribution or Ogive***.

Figure 4.50 shows the cumulative frequency distribution graph for our data given in *Table 4.6*, while *Table 4.7* shows the consecutive addition of the frequencies needed to produce the graph in *Figure 4.50*.

From *Figure 4.50* it is now a simple matter to find, for example, ***the median*** grouped height or as it is more commonly known the 50th-percentile. This occurs at 50% of the cumulative frequency (as shown in *Figure 4.50*), this being, in our case 100 giving an equivalent height of approximately 175 cm. Any percentile can be found: for example, the 75th-percentile, where in our case at a frequency of 150, the height can be seen to be approximately 180 cm.

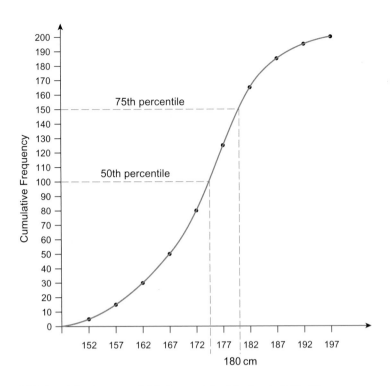

Figure 4.50 Cumulative frequency distribution graph for data given in *Table 4.6*

Table 4.7 Cumulative frequency data

Height (cm)	Frequency	Cumulative frequency
150–154	4	4
155–159	9	13
160–164	15	28
165–169	21	49
170–174	32	81
175–179	45	126
180–184	41	167
185–189	22	189
190–194	9	198
195–199	2	200
Total	200	200

TYK *4.11*

1. In a particular university, the number of students enrolled by a faculty is given in the table below.

Faculty	Number of students
Business and administration	1950
Humanities and social science	2820
Physical and life sciences	1050
Technology	850
Total	**6670**

Illustrate this data on both a bar chart and pie chart.

2. For the group of numbers given below, produced a tally chart and determine their frequency of occurrence.

36	41	42	38	39	40	42	41	37	40
42	44	43	41	40	38	39	39	43	39
36	37	42	38	39	42	35	42	38	39
40	41	42	37	38	39	44	45	37	40

3. Given the following frequency distribution:

Class interval	Frequency (f)
60–64	4
65–69	11
70–74	18
75–79	16
80–84	7
85–90	4

(a) produce a histogram and on it draw the frequency polygon.
(b) produce a cumulative frequency graph and from it determine the value of the 50th-percentile class.

Statistical measurement

When considering statistical data it is often convenient to have one or two values that represent the data as a whole. Average values are often used. You have already found an average value when looking at the median or 50th-percentile of a cumulative frequency distribution. So, for example, we might talk about the average height of females in the United Kingdom being 170 cm, or that the average shoe size of British males is size 9. In statistics, we may represent these average values using the ***mean***, ***median*** or ***mode*** of the data we are considering. We will spend the rest of this short section finding these average values for both discrete and grouped data, starting with the ***arithmetic mean***.

The arithmetic mean

The arithmetic mean or simply the *mean* is probably the average with which you are already familiar. For example, to find the arithmetic mean of the numbers 8, 7, 9, 10, 5, 6, 12, 9, 6, 8, all we need to do is to add them all up and divide by how many there are, or more formally:

$$\text{Arithmetic mean} = \frac{\text{arithmetic total of all the individual values}}{\text{number of values}} = \frac{\sum n}{n}$$

where the Greek symbol $\sum$ = the sum of the individual values, $x_1 + x_2 + x_3 + x_4 + \cdots + x_n$ and n = the number of these values in the data.

So, for the *mean* of our *ten* numbers, we have:

$$\text{mean} = \frac{\sum n}{n} = \frac{8 + 7 + 9 + 10 + 5 + 6 + 12 + 9 + 6 + 8}{10} = \frac{80}{10} = 8.$$

Now, no matter how long or complex the data we are dealing with, *provided* that we are only dealing with individual values (discrete data), the above method will always produce the arithmetic mean. The mean of all the '*x values*' is given the symbol $\bar{x}$, pronounced, '*x bar*'.

Example 4.54

The height of 11 females was measured as follows: 165.6 cm, 171.5 cm, 159.4 cm, 163 cm, 167.5 cm, 181.4 cm, 172.5 cm, 179.6 cm, 162.3 cm, 168.2 cm, 157.3 cm. Find the mean height of these females.

Then, for $n = 11$:

$$\bar{x} = \frac{165.6 + 171.5 + 159.4 + 163 + 167.5 + 181.4 + 172.5 + 179.6 + 162.3 + 168.2 + 157.3}{11}$$

$$\bar{x} = \frac{1848.3}{11} = 168.03 \text{ cm.}$$

Mean for grouped data

What if we are required to find the mean for **grouped data**? Look back at *Table 4.6* showing the height of 200 adults, grouped into *ten* classes. In this case, *the **frequency** of the heights needs to be taken into account.

We select the *class midpoint x* as being the average of that class and then multiply this value by the frequency (f) of the class, so that a value for that particular class is obtained (fx). Then by adding up all class values in the frequency distribution, the total value for the distribution is obtained ($\sum fx$). This total is then divided by the *sum of the frequencies* ($\sum f$) in order to determine the mean. So, for grouped data:

$$\bar{x} = \frac{f_1 x_1 + f_2 x_2 + f_3 x_3 + \cdots + f_n x_n}{f_1 + f_2 + f_3 + \cdots + f_n} = \frac{\sum (f \times \text{midpoint})}{\sum f}$$

This rather complicated looking procedure is best illustrated by example.

Example 4.55

Determine the mean value for the heights of the 200 adults, using the data in *Table 4.6*.

The values for each individual class are best found by producing a table, using the class *midpoints* and *frequencies* and remembering that the *class midpoint* is found by dividing the *sum of the upper and lower class boundaries* by 2. So, for example, the mean value for the first class interval is $\frac{149.5 + 154.5}{2} = 152$. The completed table is shown below.

Midpoint (x) of height (cm)	Frequency (f)	fx
152	4	608
157	9	1413
162	15	2430
167	21	3507
172	32	5504
177	45	7965
182	41	7462
187	22	4114
192	9	1728
197	2	394
Total	$\sum f = 200$	$\sum fx = 35{,}125$

I hope you can see how each of the values was obtained. Now that we have the required totals the mean value of the distribution can be found.

$$mean\,value\ \ \bar{x} = \frac{\sum fx}{\sum f} = \frac{35{,}125}{200} = 175.625 \pm 0.5\ cm$$

Notice that our mean value of heights has the same margin of error as the original measurements. The value of the mean cannot be any more accurate than the measured data from which it was found!

Median

When some values within a set of data vary quite widely, the arithmetic mean gives a rather poor representative average of such data. Under these circumstances another more useful measure of the average is the median.

For example, the mean value of the numbers 3, 2, 6, 5, 4, 93, 7 is 20, which is not representative of any of the numbers given. To find the median value of the same set of numbers, we simply place them in **rank order** that is 2, 3, 4, 5, 6, 7, 93. Then we select the middle (median) value. Since there are seven numbers (items) we choose the fourth item along, the number 5, as our **median value**.

If the number of items in the set of values is **even**, then we add together the value of the **two middle terms** and divide by 2.

Example 4.56

Find the mean and median value for the set of numbers: 9, 7, 8, 7, 12, 70, 68, 6, 5, 8.

The arithmetic mean is found as:

$$mean\ \ \bar{x} = \frac{9 + 7 + 8 + 7 + 12 + 70 + 68 + 6 + 5 + 8}{10} = \frac{200}{10} = \mathbf{20.}$$

This value is not really representative of any of the numbers in the set.

UNIT 4

To find the *median* value, we first put the numbers in rank *order*, that is,

$$5, 6, 7, 7, 8, 8, 9, 12, 68, 70$$

Then from the ten numbers, the two middle values. The 5th and 6th values along are 8 and 8. So, the *median value* $= \dfrac{8 + 8}{2} =$ **8**.

Mode

Yet another measure of central tendency for data containing extreme values is the mode. Now, the **mode** of a set of values containing discrete data is the value that occurs most often. So, for the set of values 4, 4, 4, 5, 5, 5, 5, 6, 6, 6, 7, 7, 7, the mode or **modal value** *is 5* as this value occurs four times. Now, it is possible for a set of data to have more than one mode. For example, the data used in Example 4.62 above has two modes 7 and 8, both of these numbers occurring twice and both occurring more than any of the others. A set of data may not have a modal value at all. For example, the numbers 2, 3, 4, 5, 6, 7, 8 all occur once and there is no mode.

A set of data that has one mode is called **unimodal**, data with two modes is **bimodal** and data with more than two modes is known as **multimodal**.

When considering **frequency distributions** for grouped data, the **modal class** is that group which occurs most frequently. If we wish to find the actual **modal value** of a frequency distribution, we need to draw a histogram.

Example 4.57

Find the modal class and modal value for the frequency distribution of the height of adults given in *Table 4.6*.

Referring back to *Table 4.6*, it is easy to see that the class of heights which occurs most frequently is 175 – 179 cm, which occurs 45 times.

Now, to find the modal value we need to produce a histogram for the data. We did this for Example 4.53. This histogram is shown again here with the modal shown.

From *Figure 4.51* it can be seen that the modal value $= 178.25 \pm 0.5\,cm$.

This value is obtained from the intersection of the two construction lines, *AB* and *CD*. The line *AB* is drawn diagonally from the highest value of the preceding class up to the top right-hand corner of the modal class. The line *CD* is drawn from the top left-hand corner of the modal group to the lowest value of the next class, immediately above the modal group. Then, as can be seen, the *modal value* is read-off where the projection line meets the *x*-axis.

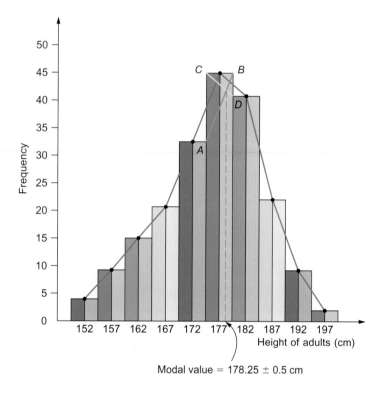

Figure 4.51 Histogram showing frequency distribution and modal value for height of adults

TYK *4.12*

1. Calculate the mean of the numbers 176.5, 98.6, 112.4, 189.8, 95.9 and 88.8.

2. Determine the mean, the median and the mode for the set of numbers 9, 8, 7, 27, 16, 3, 1, 9, 4 and 116.

3. For the set of numbers 8, 12, 11, 9, 16, 14, 12, 13, 10, 9, find the arithmetic mean, the median and the mode.

4. Estimates for the length of wood required for a shelf were as follows:

Length (cm)	35	36	37	38	39	40	41	42
Frequency	1	3	4	8	6	5	3	2

Calculate the arithmetic mean of the data.

5. Calculate the arithmetic mean and median for the data shown in the table.

Length (mm)	167	168	169	170	171
Frequency	2	7	20	8	3

6. Calculate the arithmetic mean for the data shown in the table.

Length of rivet (mm)	9.8	9.9	9.95	10.0	10.05	10.1	10.2
Frequency	3	18	36	62	56	20	5

7. Tests were carried out on 50 occasions to determine the percentage of greenhouse gases in the emissions from an internal combustion engine. The results from the tests showing the percentage of greenhouse gases recorded were as follows:

% greenhouse gases present	3.2	3.3	3.4	3.5	3.6	3.7
Frequency	2	12	20	8	6	2

(a) Determine the arithmetic mean for the greenhouse gases present.
(b) Produce a histogram for the data and from it find an estimate for the modal value.
(c) Produce a cumulative frequency distribution curve and from it determine the median value of greenhouse gases present.

Elementary Calculus Techniques

Introduction

Meeting the calculus for the first time is often a rather daunting business. In order to appreciate the power of this branch of mathematics we must first attempt to define it. So, what is the calculus and what is its function?

Imagine driving a car or riding a motorcycle starting from rest over a measured distance, say 1 km. If your time for the run was 25 seconds, then we can find your average speed over the measured kilometre from the fact that speed = distance/time. Then using consistent units, your average speed would be 1000 m/25 s or $40 \, \text{ms}^{-1}$. This is fine, but suppose you were testing the vehicle and we needed to know its acceleration after you had driven 500 m? In order to find this, we would need to determine how the vehicle speed was changing at this exact point, because the rate at which your vehicle speed changes is its acceleration. To find things, such as rate of change of speed, we can use *calculus techniques*.

The calculus is split into two major areas: the differential calculus and the integral calculus.

The *differential calculus* is a branch of mathematics concerned with finding how things change with respect to variables such as time, distance or speed, especially when these changes are continually varying. In engineering, we are interested in the study of motion and the way this motion in machines, mechanisms and vehicles varies with time, and the way in which pressure, density and temperature change with height or time. Also, how electrical quantities vary with time, such as electrical charge, alternating current, electrical power, etc. *All these areas may be investigated using the differential calculus.*

The *integral calculus* has two primary functions. It can be used to find the length of arcs, surface areas or volumes enclosed by a surface. Its second function is that of *anti-differentiation*. For example, we can use the differential calculus to find the rate of change of distance of our motorcycle

KEY POINT

The differential calculus is concerned with rates of change and we use the differentiation process to find these rates of change

KEY POINT

The integral calculus is anti-differentiation and is concerned with summing things

with respect to time. In other words, we have found its instantaneous speed. We can then use the *inverse process*, the integral calculus to determine the original distance covered by the motorcycle, from its instantaneous speed.

The mathematical process we use when applying the differential calculus is known as **differentiation** and when using the integral calculus, the mathematical process we apply is known as **integration**.

Before we can apply the calculus to some very elementary engineering problems, we need first to understand the notation and ideas that underpin these applications. Thus, at this level we spend the vast majority of our time looking at the basic arithmetic of the calculus and learning the basic techniques that will enable us to *differentiate* and *integrate* a very small number of mathematical functions. If you study further mathematics, you will gain sufficient knowledge to be able to apply the calculus to realistic engineering problems.

We start by first introducing the idea of differentiation by considering the gradient of a curve, in order to find out how functions change primarily with respect to time. Then we introduce the basic rules for the arithmetic of the differentiation process itself and practice differentiating basic mathematic functions. We then introduce the rules and techniques necessary for the anti-differentiation process, that of integration, and finally apply the integration process to that of determining the area bounded by known mathematical functions.

Differentiation

The gradient of a curve and graphical differentiation

As mentioned in the introduction, by finding the gradient of functions we are then able to find out how the variables in the function change with respect to each other. You have already met an example of a change in variables in the introduction, when we considered the way in which a road vehicle's distance and speed varied with time. So, let us start by introducing the simplest of quadratic functions that you met earlier when studying your algebra and see how we find its gradient. Unlike the straight line graphs you have already studied, the gradient of a quadratic function varies, that is, its slope varies.

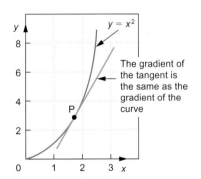

Figure 4.52 Section of the graph for $y = x^2$

Consider the graph of $y = x^2$, a section of which is shown in *Figure 4.52*.

To find the rate of change of the dependent variable y with respect to the independent variable x at a particular point (value) we need to find the gradient of the curve at that point. If we draw a tangent to the curve at a point, then the gradient of the tangent will be the same as the gradient of the curve.

Example 4.58

Plot the graph of the equation $y = x^2$ between the values where $x = \pm 3$ and from the graph find the gradient of the curve at the points where $x = \pm 2$.

To draw the graph we first set up a table of values and show:

x	-3	-2	-1	0	1	2	3
$y = x^2$	9	4	1	0	1	4	9

Notice first that we have included the value zero and the fact that when any negative value is squared we obtain a positive value. There is thus symmetry about zero, that is about the *y*-axis. The required plot, together with the gradient to the tangent at $x = \pm 2$, is shown in *Figure 4.53*.

Note from the graph that the gradient of the two tangents is found in the same way as finding the gradient of a straight line that you learnt earlier when studying simple linear equations in your algebra.

Thus from the graph it can be seen **that when x = 2 the gradient is 4 and when x = −2, the gradient is −4** (the gradient is negative because the tangent slopes down from the left). You should take note of these values of *x* and their corresponding gradient values, where it can be seen **that the gradient values are twice their corresponding x values**.

Also note from the graph the use of the shorthand way of writing the corresponding values of *x* and *y*, using the bracketed expression **(x, y) in that order**. So, the corresponding points at which the tangent line touches the curve are $(-2, 4)$ and $(2, 4)$.

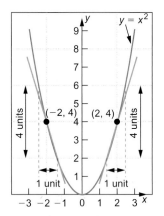

Figure 4.53 Graph of the gradient of the tangents to $y = x^2$ when $x = \pm 2$

Here is another example of finding the gradient to the tangent of the equation $y = x^2$ for different values of the independent variable.

Example 4.59

(a) Draw the graph of the equation $y = x^2$ for values of x from $x = -3$ to $x = 3$.
(b) Find the gradient to the curve at the tangent lines where $x = -1$ and $x = 1$, then comment on your result.

(a) The graph of the function $y = x^2$ is shown in *Figure 4.54*. It is plotted in exactly the same manner as that shown in Example 4.58. It can be seen that it is symmetrical about zero and is parabolic in shape, as before.

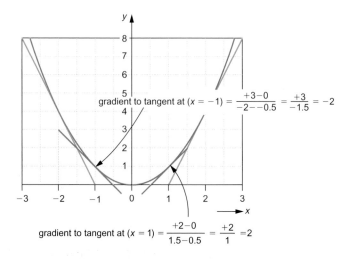

gradient to tangent at $(x = -1) = \dfrac{+3-0}{-2--0.5} = \dfrac{+3}{-1.5} = -2$

gradient to tangent at $(x = 1) = \dfrac{+2-0}{1.5-0.5} = \dfrac{+2}{1} = 2$

Figure 4.54 Finding the gradient of the equation $y = x^2$ at the points ± 1

(b) From the graph it can be seen that the gradient to the tangent lines at the points ±1 are ±2, respectively. ***There seems to be a pattern emerging in that at x = −1 the corresponding gradient = −2, so the gradient is twice as large as the independent variable x.*** This is also true for the gradient at $x = 1$ which again is twice as large. The same pattern was seen in Example 4.58 where for the values ±2 the corresponding gradient values were again twice as large. This pattern is no coincidence, as you are about to see!

For the function $y = x^2$ we have just shown that on two occasions using different independent variables that the gradient of the slope of the tangent line is twice the value of the independent variable or more formally:

the gradient of the tangent at $y = 2x$.

The process of ***finding the gradient of the tangent to the slope at a point*** is known as ***graphical differentiation***. What we have actually done is found the ***differential coefficient*** or ***differentiated*** the function $y = x^2$. In other words, we have found an algebraic expression of how this function varies as we increase or decrease the value of the independent variable.

We can generalize the above procedure for finding the derived function. Consider again part of our function $y = x^2$ (*Figure 4.55*).

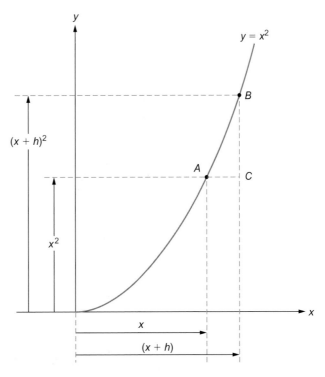

Figure 4.55 Finding the general derived function for $y = x^2$

Suppose that A is the point on our curve $y = x^2$ with ordinate x and that B is another point on the curve with ordinate $(x + h)$. The y-ordinate of A is x^2 and the y-ordinate of B is $(x + h)^2$.

Then:

$$BC = (x + h)^2 - x^2 = 2hx + h^2 \quad \text{and} \quad AC = (x + h) - x = h.$$

Then the gradient of $AB = \dfrac{2hx + h^2}{h} = 2x + h$ (if h is not zero).

Now as h gets smaller and smaller (tends to 0), the gradient tends to (approaches) $2x$.

Therefore, as we found graphically, *the gradient of the tangent is 2x*, or *the derived function (derivative) is 2x*.

In the next example we find the derived function for a different equation.

Example 4.60

Find the derived function (gradient of the curve) for $y = 2x^2 - 2x - 6$ at the point (x, y).

We use the same procedure as before by identifying another point on the curve, say $(x + h, y + k)$. Then we find the slope of the line joining the two points and we bring the two points closer and closer together, until they coincide and become the tangent to the slope of the curve. In other words, the derived function or derivative. We proceed as follows.

For the two points on the x ordinate the y ordinates are:

$$y + k = 2(x + h)^2 - 2(x + h) - 6 \qquad (1)$$

$$y = 2x^2 - 2x - 6 \qquad (2)$$

Then expanding Equation (1) and simplifying, **using your algebra**, we get:

$$y + k = 2(x^2 + 2hx + h^2) - 2x - 2h - 6$$
$$y + k = 2x^2 + 4hx + 2h^2 - 2x - 2h - 6$$
$$y + k = 2x^2 - 2x - 6 + 4hx + 2h^2 - 2h \qquad (3)$$

or

Now subtracting Equation (3) from Equation (1) gives $k = 4hx + 2h^2 - 2h$,

therefore on division by h, the gradient of the chord $\dfrac{k}{h} = 4x + 2h - 2$ and $\dfrac{k}{h}$ tends to $(4x - 2)$ as h tends to 0.

Thus, **$4x - 2$ is the derived function of $y = 2x^2 - 2x - 6$, which is also the gradient of the function at the point (x, y)**. So, for example, at the point $(3, -3)$ the gradient is $(4(3) - 2) = 10$.

You will be pleased to know that we do not need to repeat this rather complicated method of finding the derived function (the tangent to a point of the slope). As you will soon see *all derived functions (derivatives)* of simple algebraic expressions (polynomials) can be found using a simple rule!

Notation for the derivative

There are several ways in which we can represent and describe *the differential coefficient or derived function.* Below are some of the

KEY POINT

To find the gradient of the tangent at a point of a function, we differentiate (or find the derivative of) the function most common methods used to describe the derived function that you will find in textbooks and literature dealing with the **differential calculus**.

These differing terms for finding the derived function include:

- find the derived function of …
- find the derivative of …
- find the differential coefficient for …
- differentiate …
- find the rate of change of …
- find the tangent to the function …
- find the gradient of a function at a point …

This differing terminology is often confusing to beginners. It is further complicated by the fact that different symbols are used for **the differentiation process** (finding the derived function), based on the convention chosen. However, to avoid such confusion we shall only use the more common **Leibniz notation** throughout this unit.

In **Leibniz notation**, the mathematical function is represented conventionally as **y(x)** and its derived function or differential coefficient is represented as $\dfrac{dy}{dx}$. This expression for the derived function can be thought of as finding the slope to the tangent of a point of a particular function, where we take a smaller and smaller bit of x (dx) and divide it into a smaller and smaller bit of y (dy), until we get the slope of a point $\dfrac{dy}{dx}$.

So, in Leibniz notation, for the function $y = x^2$ the differential coefficient is represented as $\dfrac{dy}{dx} = 2x$, which we found earlier. The second derivative in Leibniz notation is represented as $\dfrac{d^2 y}{dx^2}$, the third derivative is $\dfrac{d^3 y}{dx^3}$, and so on.

One other complication arises with all notation in that **the notation differs according to the variable being used**! For example, if our mathematical function is $s(t)$, then in Leibniz notation its first derivative would be $\dfrac{ds}{dt}$ (we are differentiating the variable s with respect to t). In the same way as with $\dfrac{dy}{dx}$ we are differentiating the variable y with respect to x. So, in general, $\dfrac{dy}{dx}, \dfrac{ds}{dt}, \dfrac{du}{dv}$, each represent the first derivative of the functions y, s and u, respectively.

One final type of notation which is often used in mechanics is **dot** notation. For example, $\dot{v}, \ddot{s}$, etc., means that the function is differentiated once ($\dot{v}$) or twice ($\ddot{s}$), and so on. This notation will not be used in this book, but you may meet it in your further studies.

So much for all the hard theory. We are now going to **use one or two rules** to carry out the differentiation process, which once mastered is really quite simple!

KEY POINT

In Leibniz notation $\dfrac{dy}{dx}$ means that we find the first derivative of the function y with respect to x

UNIT 4

TYK 4.13

1. State the function and define (a) the differential calculus, (b) the integral calculus.

2. Explain why finding the slope to the tangent of a function at a point is finding the differential coefficient.

3. Plot the graph of $y = 2x^2 - 5$ for values of x between -2 and 3. From your graph find the gradient of the curve at the points where $x = -1$ and $x = 2$.

4. Plot the graph of $y = x^2 - 3x + 2$ from $x = 2$ to $x = 4$ in 0.5 increments of x and find its gradient at the point where $x = 3$.

5. Using Example 4.60 as a guide, find the derived function (gradient of the curve) for $y = 4x^2 - 3x + 6$ at the point (x, y).

6. Write down the Leibniz notation for the first and third differential of the function $y(x)$.

Differentiating

As you will be aware by now, ***the word differentiate*** is one of many ways of saying that we wish to ***find the derived function***. Again, going back to the simple function $y = x^2$ when we *differentiated* this function, we found that its *derived function* was $\dfrac{dy}{dx} = 2x$. In a similar manner, when we carried out the differentiation process on the function $y = 2x^2 - 2x - 6$ we obtained $\dfrac{dy}{dx} = 4x - 2$. If we were to carry out the rather complex process we used earlier on the following functions $y = 3x^2$, $y = x^3$ and $y = x^3 + 3x^2 - 2$ we would obtain $\dfrac{dy}{dx} = 6x$, $\dfrac{dy}{dx} = 3x^2$ and $\dfrac{dy}{dx} = 6x^2 + 6x$, respectively. I wonder if you can see a pattern in these results?

These results are grouped below for your convenience. Can you spot a pattern?

$$y = x^2 \qquad\qquad \frac{dy}{dx} = 2x$$

$$y = 3x^2 \qquad\qquad \frac{dy}{dx} = 6x$$

$$y = 2x^2 - 2x - 6 \qquad \frac{dy}{dx} = 4x - 2$$

$$y = x^3 \qquad\qquad \frac{dy}{dx} = 3x^2$$

$$y = x^3 + 3x^2 - 2 \qquad \frac{dy}{dx} = 3x^2 + 6x$$

I hope you spotted that we seem to multiply by the index (power) of the unknown, then we subtract (1) one from the index of the unknown. So, for example, with the function $y = 3x^2$, the index is 2 and $(2)(3) = 6$. Also, the original index (power) of x was 2 and on subtracting 1 from this index we get $x^{(2-1)} = x^1 = x$. Thus we finally get $\dfrac{dy}{dx} = 6x$. The use of indices (powers) in this process is one of the reasons you studied them earlier in your algebra!

This technique can be applied to ***any unknown raised to a power***. We can write this rule in general terms.

If $y = x^n$, then $\dfrac{dy}{dx} = nx^{n-1}$. Or in words, ***to find the differential coefficient of the function $y = x^n$ we first multiply the unknown variable by the index and then subtract 1 from the index to form the new index.***

Again, with this rather wordy explanation, you will appreciate the ease with which we can express this rule using a formula!

Note: You may be wondering why the constant (number) in the above functions seems to just disappear. If you remember how we performed the differentiation process graphically by finding the slope at a point on the function, ***then for a constant such as $y = -6$, the graph is simply a straight horizontal line cutting the y-axis at -6. Therefore, its slope is zero, thus its derived function is zero.*** This is true for any constant term no matter what its value.

If the function we are considering has more than one term, for example, $y = x^3 + 3x^2 - 2$, then we simply ***apply the rule in sequence to each and every term***.

Example 4.61

Differentiate the following functions with respect to the variable:

(a) $y = 3x^3 - 6x^2 - 3x + 8$

(b) $y = \dfrac{3}{x} - x^3 + 6x^{-3}$ (c) $s = 3t^3 - \dfrac{16}{t^2} + 6t^{-1}$

(a) In this example, we can simply apply the rule $\dfrac{dy}{dx} = nx^{n-1}$ to each term in succession;

so $\dfrac{dy}{dx} = (3)(3)x^{3-1} - (2)(6)x^{2-1} - (1)(3)x^{1-1} + 0$. And remembering that any number raised to the power zero is one, that is, $x^0 = 1$, then:

$$\frac{dy}{dx} = 9x^{3-1} - 12x^{2-1} - 3x^{1-1} + 0$$

$$\frac{dy}{dx} = 9x^2 - 12x^1 - 3x^0$$

$$\frac{dy}{dx} = 9x^2 - 12x - 3$$

(b) In this example, we need to simplify before we use the rule. The simplification involves clearing fractions. Then remembering that $x = x^1$ and that ***from the laws of indices***, when

UNIT 4

you bring a number in index form over the fraction line, we **change its sign**:

$$y = \frac{3}{x} - x^3 + 6x^{-3} \text{ becomes:}$$

$$y = \frac{3}{x^1} - x^3 + 6x^{-3} \text{ or,}$$

$$y = 3x^{-1} - x^3 + 6x^{-3} \text{ and applying the rule,}$$

$$\frac{dy}{dx} = (-1)(3)x^{-1-1} - (3)x^{3-1} - (3)(6)x^{-3-1}$$

$$\frac{dy}{dx} = -3x^{-2} - 3x^2 - 18x^{-4}$$

Notice how we have dealt with **negative** indices. The rule can also be used when **fractional** indices are involved.

(c) The only change with this example is that it concerns different variables. In this case we are asked to differentiate the function s with respect to the variable t.

So, proceeding as before and **simplifying first**, we get:

$$s = 3t^3 - 16t^{-2} + 6t^{-1}.$$

Then differentiating we get,

$$\frac{ds}{dt} = (3)(3)t^{3-1} - (-2)(16)t^{-2-1} + (-1)(6)t^{-1-1}$$

$$\frac{ds}{dt} = 9t^2 + 32t^{-3} - 6t^{-2}$$

Note that you must take care with your signs!

> **KEY POINT**
>
> To find the first derivative of functions of the type $y = ax^n$, we use the rule $\dfrac{dy}{dx} = nax^{n-1}$

In the above example, we found in all cases **the first derivative**. If we wish to find the **second derivative of a function**, all we need to do is to differentiate again. So, in Example 4.61 question 1, for the function $y = 3x^3 - 6x^2 - 3x + 8$, then

$$\frac{dy}{dx} = 9x^2 - 12x - 3$$

and differentiating this function again, we get,

$$\frac{d^2 y}{dx^2} = (2)(9)x^{2-1} + (1)(-12)x^{1-1} = 18x - 12x^0 = 18x - 12$$

Notice the **Leibniz** terminology for the **second differential** in the above example.

Similarly, for the function $s = 3t^3 - 16t^{-2} + 6t^{-1}$,

$$\frac{ds}{dt} = 9t^2 + 32t^{-3} - 6t^{-2}$$

> **KEY POINT**
>
> $\dfrac{d^2 y}{dx^2}$ and $\ddot{x}$ are both ways of expressing the second derivative

and on differentiating this function again, we get,

$$\frac{d^2 s}{dt^2} = (2)(9)t^{2-1} + (-3)(32)t^{-3-1} + (-6)(-2)t^{-2-1}$$

$$= 18t - 96t^{-4} + 12t^{-3}$$

Notice once again, the care needed with signs in this example.

Rate of change

One application of the differential calculus is to find instantaneous rates of change. The example given at the beginning of this outcome concerned our ability to find how the speed of a motor vehicle changed at a particular point in time. ***In order to find the rate of change of any function, we simply differentiate that function (find its gradient) at the particular point concerned.***

So, for example, given that $y = 4x^2$ let us find its rate of change at the points $x = 2$ and $x = -4$. Then all we need do is differentiate the function and then substitute in the desired points.

Then, $\dfrac{dy}{dx} = (2)(4)x = 8x$ and when $x = 2$, $\dfrac{dy}{dx} = (8)(2) = 16$, thus the slope of the function at $x = 2$ is 16 and this tells us how the function is changing at this point.

Similarly, when $x = -4$, $\dfrac{dy}{dx} = 8x = (-4)(8) = -32$. In this case, the negative sign indicates a negative slope, so the function is changing in the opposite sense compared with what was happening when $x = 2$.

Example 4.62

The distance (s) covered by a missile is $4.905\,t^2 + 10\,t$. Determine its rate of change of distance with respect to time (its speed) after (a) 4 seconds and (b) 12 seconds have elapsed.

This is a simple rate of change problem, hidden in a rather wordy question!

To find *rate of change* of distance with respect to time we need to find the differential coefficient of the function then apply the rule:

$$\frac{ds}{dt} = (2)(4.905)t^{2-1} + 10t^{1-1} = 9.81t + 10.$$

Now, substitute for the desired times:

when $t = 4$, then $\dfrac{ds}{dt} = (9.81)(4) + 10 = \mathbf{49.24}$

and when $t = 12$, $\dfrac{ds}{dt} = (9.81)(12) + 10 = \mathbf{127.72}$.

Since $\dfrac{\boldsymbol{ds}}{\boldsymbol{dt}} = \boldsymbol{v}$ (speed), then what the above results tell us is that after 4 seconds the missile has reached a speed of $49.24\,\text{ms}^{-1}$ and after 12 seconds the missile reaches a speed of $127.72\,\text{ms}^{-1}$.

Thus, for very little effort, the differential calculus has enabled us to find instantaneous rates of change which are of practical use!

KEY POINT

The rate of change of distance with respect to time is speed

UNIT 4

TYK 4.14

1. Differentiate the following with respect to the appropriate variables
   (a) $y = x^3$ (b) $y = 6x^5$ (c) $s = 4t + t^2$ (d) $A = \pi r^2$

   (e) $y = 2\sqrt{x}$ (f) $p = q^2 - \dfrac{1}{q}$.

2. Differentiate the following with respect to the appropriate variables
   (a) $y = x^2 - 2x + 5$ (b) $s = 3t^5 - 3t^3 - 4t - 2$,

   (c) $p = q^{-3} + q^{\frac{1}{2}} - q^{-0.56}$ (d) $y = 20x^{1.5} - 2.5x^{-0.7}$

3. The distance s covered by a racing car is given by the equation $s = 2.7t^2 + 8t - 2$. Determine its rate of change of displacement with respect to time after (a) 3 seconds and (b) after 10 seconds.

Differentiation of elementary trigonometric and exponential functions

We have so far concentrated our attention on functions of the type $ax^n \pm ax^{n-1} \pm ax^{n-2} \pm \ldots \pm ax^3 \pm ax^2 \pm ax \pm a$. This general class of functions is known as **polynomials**.

There are, however, other mathematical functions that you have already met. These include **trigonometric functions** such as the sine and cosine. In addition, you have met the **exponential function** e^x and its mathematical inverse the **Napierian logarithm ln x**.

Finding the differential coefficient of these functions can be achieved by graphically differentiating them in a similar manner to the way we originally found the derived function for $y = x^2$. If we were to carry out this exercise, we would be able to establish patterns and subsequent rules, as we did for polynomial functions.

Table 4.8 Some standard derivatives

Rule number	y	$\dfrac{dy}{dx}$
1	x^n	nx^{n-1}
2	ax^n	nax^{n-1}
3	$\sin ax$	$a \cos ax$
4	$\cos ax$	$-a \sin ax$
5	e^{ax}	ae^{ax}
6	$\ln ax$	$\dfrac{\dfrac{dy}{dx}(ax)}{ax}$

Rather than going through this tedious process, you will be pleased to note that these rules have been listed in *Table 4.8* (without proof) for your convenience!

Now, one or two of the above rules may look a little complex, but in practice they are all fairly straightforward to use. The easiest way to illustrate their use is through the examples that follow.

Example 4.63

Differentiate the following with respect to the variable:

(a) $y = \sin 3x$ (b) $u = \cos 2\theta$ (c) $y = 5 \sin 2\theta - 3 \cos \theta$

(a) In this example, we may follow rule 3 in the table directly, noting that $a = 3$, then

$$\frac{dy}{dx} = a \cos ax = \boldsymbol{3 \cos 3x}$$

(b) The same approach is needed to solve this little problem, but noting that when we differentiate the cosine function it has a sign change. Also, we are differentiating the function u with respect to θ. Then the differential coefficient, using rule 4, is given as:

$$\frac{du}{d\theta} = \boldsymbol{-2 \sin 2\theta}$$

(c) With this final problem, we simply use rule 3 for differentiating sine, followed by rule 4 for differentiating cosine. Note that the numbers 5 and -3 are not the constant a given in the formulae in the table. We simply multiply these numbers by a when carrying out the differentiation process.

$$\text{So, } \frac{dy}{d\theta} = (2)(5)\cos 2\theta - 3(-1)\sin \theta = \boldsymbol{10 \cos 2\theta + 3 \sin \theta}$$

Note the effect of the sign change when differentiating the cosine function!

Example 4.64

(a) Find the differential coefficients of the following function, $y = e^{-2x}$.

(b) Find $\dfrac{d}{dx}(6\log_e 3x)$.

(c) Differentiate $v = \dfrac{e^{3\theta}}{2} - \pi \ln 4\theta$.

The above functions involve the use of rules 5 and 6.

(a) This is a direct application of rule 5 for the exponential function, where $a = -2$. Remember we are differentiating the function y with respect to the variable x. The base e is simply a constant (a number). As mentioned before the value of $e \simeq 2.71828$. It is a number like π: it has a limitless number of decimal places!

Then

$$\frac{dy}{dx} = (-2)e^{-2x} = \boldsymbol{-2e^{-2x}}$$

(b) This is yet another way of being asked to differentiate a function. What it is really saying is find $\dfrac{dy}{dx}$ of the function $y = 6\log_e 3x$. Remember that when dealing with the Napierian log function, $\log_e f(x) = \ln f(x)$, both methods of representing the Napierian log function are in common use; so all we need do is apply rule 6, where in this case the constant $a = 3$.

$$\frac{d}{dx}(6\log_e 3x) = (6)\frac{\dfrac{dy}{dx}(3x)}{3x} = \frac{(6)(3)}{3x} = \frac{18}{3x} = \frac{6}{x}.$$

Note when finding this differential we also had to apply rule 1 to the top part of the fraction. Providing you follow rule 6 exactly when laying out all your working, you should not make mistakes.

(c) For this example, we need to apply rule 5 to the exponential function and then rule 6 to the Napierian log function, noting that $-\pi$ is a constant and does not play any part in the differentiation; we simply multiply the differential by it at the end of the process. Therefore,

$$\frac{dv}{d\theta} = \frac{3e^{3\theta}}{2} + (-\pi)\frac{\dfrac{dy}{d\theta}(4\theta)}{4\theta} \quad \text{and} \quad \frac{dv}{d\theta} = 1.5e^{3\theta} + (-\pi)\frac{4}{4\theta}$$

$$\frac{dv}{d\theta} = 1.5e^{3\theta} - \frac{4\pi}{4\theta} \quad \text{and} \quad \frac{dv}{d\theta} = \boldsymbol{1.5e^{3\theta} - \frac{\pi}{\theta}}$$

This may look rather complicated but all we have done is followed rule 6, as before.

Now, being able to find the differential coefficient of the functions in the above examples is all very well, but what use is it all?

Well, as was the case with the general rule for differentiating polynomial functions, we can also apply these rules to solving simple rate of change problems. In our final example for the differential calculus, we apply rules 5 and 6 to the rate of change of current in an electrical circuit and rate of discharge from an electrical capacitor. This is not as difficult as it sounds!

Example 4.65

(a) A alternating voltage is given by the function $v = \sin 2\theta$, where θ is the angular distance travelled and v is the instantaneous voltage at that angular distance (in radians). Determine the way the voltage is changing with respect to distance at $\theta = 2$ and $\theta = 4$ rad.

(b) Suppose, the charge in a capacitor discharges according to the function $Q = \ln 3t$, where $Q =$ charge (in coulombs) and $t =$ time in seconds (s). Determine the *rate of discharge* at $t = 4$ milliseconds (ms).

(a) All we are being asked is to find the rate of change of the voltage after a particular angular distance has been covered by the alternating (sinusoidal) function. This means we need to find the differential coefficient (the rate of change function) and then simply substitute the appropriate values.

So, $\dfrac{dv}{d\theta} = 2\cos 2\theta$, which is the rate of change of voltage with respect to distance.

Then, at $\theta = 2\,\text{rad}$, remembering it is radian measure, we get,

$$\frac{dv}{d\theta} = 2\cos(2)(2) = 2\cos 4 = (2)(-0.653) = \boldsymbol{-1.3073}$$

and the voltage is changing negatively. This value is the slope of the graph of $v = \sin 2\theta$ at the point $\theta = 2\,\text{rad}$.

Similarly, at

$$\theta = 4\,\text{rad}, \frac{dv}{d\theta} = 2\cos(2)(4) = 2\cos 8 = (2)(-0.1455) = \boldsymbol{-0.291}.$$

This is again a negative slope, but with a shallower gradient.

(b) The rate of discharge in this case means the rate of change of charge with respect to time. So, it is a ***rate of change problem*** involving the differential coefficient of the function. Then following rule 6 and also using rule 1, we have:

$$\frac{dQ}{dt} = \frac{\dfrac{d}{dt}(3t)}{3t} = \frac{3}{3t} = \frac{1}{t}, \text{ then when } t = 4\text{ms or } 4\times 10^{-3} \text{ seconds,}$$

$$\frac{dQ}{dt} = \frac{1}{t} = \frac{1}{4\times 10^{-3}} = 250\,\text{C/s (coulombs per second).}$$

If you were to put in higher values of time you will find that the rate of discharge decreases.

KEY POINT

We always differentiate when finding rates of change

UNIT 4

TYK *4.15*

1. Differentiate the following with respect to the variable:
   (a) $y = \sin x$,
   (b) $y = \cos \theta$,
   (c) $p = 3\sin \beta$ and
   (d) $y = 4\sin 2\theta - 3\cos 3\theta$.

2. Differentiate the functions:
   (a) $y = e^{2\theta}$ (b) $y = 3e^{-3x}$ (c) $t = \dfrac{1}{3}e^{-3\theta}$

3. Differentiate the functions:
   (a) $y = 3\log_e 2x$ (b) $\rho = 0.24e^{2\theta} - 6\ln 3\theta$.

Integration

In this short section we are going to look at the *integral calculus*, which we mentioned earlier. It has something to do with finding areas and is also the inverse process of finding the derived function. The integral calculus is all about *summing things*, that is, finding the whole thing from its parts, as you will see shortly.

We start by considering *integration* (the arithmetic of the integral calculus) as the *inverse of differentiation*.

Integration as the inverse of differentiation

We know that for the function $y = x^2$ the derived function $\dfrac{dy}{dx} = 2x$.

So, reversing the process involves finding the function whose derived function is $2x$. One answer will be x^2, but is this the only possibility? The answer is *no*, because $2x$ is also the derived function of $y = 2x + 5$, $y = x^2 - 20.51$, $y = x^2 + 0.345$, etc. In fact, $2x$ is the derived function of $y = x^2 + c$, **where c is any constant**. So, when we are finding the inverse of the derived function, in other words, when we are integrating, we must always allow for the possibility of a constant being present by putting in this **arbitrary constant c**, which is known as the **constant of integration**. Then, in general terms, the inverse of the derived function $2x$ is $x^2 + c$.

Thus, whenever we wish to find the inverse of any derived function, that is, *whenever we integrate the derived function, we must include the constant of integration c*.

When carrying out the *anti-differentiation process or integration, we can only find a particular value for this constant c when we are given some additional information about the original function*. For example, if we are told that for $y = x^2 + c$, $y = 2$, when $x = 2$, then by substituting these values into the original function, we find that $2 = 2^2 + c$, from which we find that $c = -2$, so the particular function becomes $y = x^2 - 2$. This is now one of a *whole family* of functions $y = x^2 + c$, illustrated graphically in *Figure 4.56*.

Table 4.9 shows a few familiar *polynomial functions* on which has been carried out this inverse differentiation or *integration* process. When we integrate a derived function, the expression we obtain is often known as the *prime function (F)*. See if you can spot a pattern for the derivation of these prime functions.

Apart from the mandatory constant of integration, I hope you can see that the power or index of x increases by 1 over that of the derived function. Then we divide the prime function by this new power or index. Then, in general,

If $\dfrac{dy}{dx} = x^n$, then the prime function is $y = \dfrac{x^{n+1}}{n+1} + c$
This rule is valid for all values of n except $n = -1$.

If $n = -1$, then in finding the prime function we would be trying to divide by $n + 1 = -1 + 1 = 0$ and as you are well aware from your earlier study of the laws of arithmetic, division by 0 is not allowed! In this particular case we adopt a special rule, which is given below, without proof.

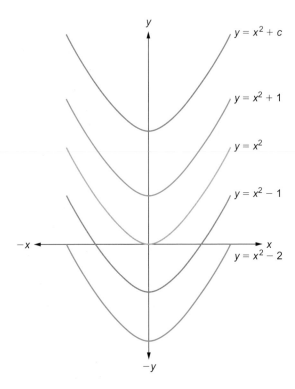

Figure 4.56 Family of curves $y = x^2 + c$

Table 4.9 Some polynomial derivatives with their prime functions

Derived function	Prime function (F)
$\dfrac{dy}{dx} = 1$	$y = x + c$
$\dfrac{dy}{dx} = x$	$y = \dfrac{x^2}{2} + c$
$\dfrac{dy}{dx} = x^2$	$y = \dfrac{x^3}{3} + c$
$\dfrac{dy}{dx} = x^3$	$y = \dfrac{x^4}{4} + c$

If $\dfrac{dy}{dx} = x^{-1} = \dfrac{1}{x}$, then the prime function is $y = \ln |x|$.

Notice that we have to take the ***modulus, or positive value of x***, when finding corresponding values of y. This is because *the **ln(log$_e$) function is not valid for all numbers ≤ 0 (less than or equal to zero)**.*

Notation for the integral

As with differentiation, when we carry out integration we need to use the appropriate mathematical notation in order to convey our desire to integrate.

If y is a function of x, then $\int y \, dx$ represents the integral of y with respect to the variable x. The integral sign $\int$ is the Greek letter S, and indicates that when carrying out the integration process, we are really carrying out a ***summing*** process.

Note that in the same way that the d cannot be separated from the y, in dy, neither can the $\int$ be separated from dx, if the integration is with respect to x.

So, for example, if we wish to find the prime function F, that is, if we wish to integrate the function x^2, then this is represented as $\int x^2 \, dx$ and using the general rule we see that $\int x^2 \, dx = \dfrac{x^{n+1}}{n+1} + c = \dfrac{x^{2+1}}{2+1} + c = \dfrac{x^3}{3} + c$, which is in agreement with the prime function or *the integral*, shown in the table.

The integration process

We have seen from above how to integrate elementary polynomial functions using the basic rule. In the example given next, we use the rule successively to integrate general polynomial expression with respect to the variable concerned.

Example 4.66

Integrate the following functions with respect to the variables given:

(a) $y = 3x^3 + 2x^2 - 6$

(b) $s = 5t^{-3} + t^4 - 2t^2$

(c) $p = r^{-1} + \dfrac{r^4}{4}$

(a) What we are being asked to do is to find the prime function $F(y)$. Using the conventional notation we have just learnt we must find:

$$F(y) = \int 3x^3 + 2x^2 - 6 \, dx$$

In this case, all we need to do is to successively apply the basic rule, that is,

$$\int 3x^3 + 2x^2 - 6 \, dx = (3)\frac{x^{3+1}}{3+1} + (2)\frac{x^{2+1}}{2+1} + (-6)x^{0+1} + c$$

$$= \frac{3x^4}{4} + \frac{2x^3}{3} - 6x + c$$

(b) In this question, we again apply the basic rule but in terms of the different variables, so

$$\int 5t^{-3} + t^4 - 2t^2 \, dt = (5)\frac{t^{-3+1}}{-3+1} + \frac{t^{4+1}}{4+1} + (-2)\frac{t^{2+1}}{2+1} + c = \frac{5t^{-2}}{-2} + \frac{t^5}{5} - \frac{2t^3}{3} + c$$

(c) With this question I hope you spotted immediately that for part of the function r^{-1}, we cannot apply the general rule but must apply the special case where $n = -1$. So, for the integration, we proceed as follows:

$$\int r^{-1} + \frac{r^4}{4} \, dr = \ln|r| + \left(\frac{1}{4}\right)\frac{r^{4+1}}{4+1} + c = \ln|r| + \frac{r^5}{20} + c$$

Notice also that dividing by 4 is the same as multiplying by a quarter and that we multiply tops by tops and bottoms by bottoms to obtain the final values.

KEY POINT

When finding indefinite integrals, we must always include the constant of integration

Some common integrals

We have now seen how to integrate polynomial expressions. We can also apply the inverse differentiation process to the sinusoidal, exponential and Napierian logarithm functions. *Table 4.10* shows the *prime functions* (the integral) for the basic functions we dealt with during our study of the differential calculus.

Table 4.10 Some standard integrals

Rule number	Function (y)	Prime function ($\int y\ dx$)
1	$x^n (n-1)$	$\dfrac{x^{n+1}}{n+1} + c$
2	$x^{-1} = \dfrac{1}{x}$	$\ln |x|$
3	$\sin ax$	$-\dfrac{1}{a}\cos ax$
4	$\cos ax$	$\dfrac{1}{a}\sin ax$
5	e^{ax}	$\dfrac{1}{a}e^{ax}$
6	$\ln x$	$x \ln x - x$

If you compare the integrals of the sine and cosine functions, you should be able to recognize that the integral is the inverse of the differential. This is also clearly apparent for the exponential function. The only 'strange' integral that seems to have little in common with its inverse is that for the Napierian logarithm function. The mathematical verification of this integral is beyond the level for this unit. However, you will learn the techniques of the calculus necessary for its proof if you study the further mathematics unit.

We will demonstrate the use of these *standard integrals* through the examples that follow.

Example 4.67

(a) Find $\int (\sin 3x + 3\cos 2x)\ dx$.

(b) Integrate the function $s = e^{4t} - 6e^{2t} + 2\ dt$.

(c) Find $\int 6\log_e t\ dt$.

(a) This integral involves using rules 3 and 4. Sequentially, the integral may be written as:

$$\int \sin 3x\ dx + \int 3\cos 2x\ dx = -\frac{1}{3}\cos 3x + (3)\frac{1}{2}\sin 2x + c = -\frac{1}{3}\cos 3x + \frac{3}{2}\sin 2x + c$$

Any integral involving expressions, separated by a plus or a minus sign, may be integrated separately. Note also that the constant multiplying the function, 3 in this case, does not play any part in the integration, it just becomes a multiple of the result.

(b) This is just a direct integral involving the successive use of rule 5 and the use of rule 1 for the last term.

Then $\int e^{4t} - 6e^{2t} + 2\, dt = \frac{1}{4}e^{4t} - (6)\left(\frac{1}{2}\right)e^{2t} + 2t + c = \frac{1}{4}e^{4t} - 3e^{2t} + 2t + c.$

(c) This integral demonstrates the direct use of rule 6, where the constant is taken behind the integral sign until the process is complete and then brought back in as the multiplier of the integral. Remembering also that $\log_e t = \ln t$, then

$$\int 6\log_e t\, dt = 6\int \log_e t\, dt = 6(t\log_e t - t) + c \text{ or } = \mathbf{6(t\ln t - t) + c}.$$

Simple applications of the integral

In the **differential calculus** we considered *rates of change*. One particular application involved determining *the rate of change of distance with respect to time*. In other words, differentiating the function involving distance to find the derived function, which gave the *velocity (speed in a given direction)*. Look back to Example 4.62, if you cannot remember this procedure. If we carry out the inverse operation, that is, we *integrate the velocity function*, we will get back to the *distance function*. Taking this idea one step further, if we differentiate the velocity function, we will find the rate of change of velocity with respect to time, in other words we will find the acceleration function (ms^{-2}). So again, if we integrate the acceleration function, we get back to the velocity function.

> **KEY POINT**
>
> If we integrate the acceleration function we obtain the velocity function and if we integrate the velocity function we obtain the distance function

Example 4.68

The acceleration of a missile moving vertically upwards is given by $a = 4t + 4$. Find the formulae for both the velocity and the distance of the missile, given that $s = 2$ and $v = 10$ when $t = 0$.

In this application it is important to recognize that *acceleration is rate of change of velocity*, or $\frac{dv}{dt} = 4t + 4$. This of course is a *derived function*; therefore, in order to find **v**, we need to carry out anti-differentiation, that is **integration**. When we do this we find the prime function **F(x)** by integrating both sides of the derived function, as follows:

$$\int \frac{dv}{dt} = \int 4t + 4\, dt \text{ and so } F(x) = v = \frac{4t^2}{2} + 4t + c = 2t^2 + 4t + c$$

We now have the original equation for the velocity

$$v = 2t^2 + 4t + c$$

We can now use the given information to find the **particular equation** for the velocity. We know that when the velocity $= 10$, the time $t = 0$. Therefore, substituting into our velocity equation gives $10 = (2)(0) + (4)(0) + c$ or $10 = c$. So, our particular equation for velocity is:

$$v = 2t^2 + 4t + 10$$

We are also asked to find the formula for the distance. Again, recognizing that velocity is the rate of change of distance with respect to time we may write the velocity equation in its derived form as:

$$\frac{ds}{dt} = 2t^2 + 4t + 10$$

then integrating as before to get back to distance, we get,

$$\int \frac{ds}{dt} = \int 2t^2 + 4t + 10 \; dt \text{ or } F(x) = s = \frac{2t^3}{3} + \frac{4t^2}{2} + 10t + c = \frac{2t^3}{3} + 2t^2 + 10t + c$$

We now have the original equation for distance:

$$s = \frac{2t^3}{3} + 2t^2 + 10t + c$$

Again, using the given information that **s** = 2 and **v** = 10, when **t** = 0, the particular equation for distance can be found. On substitution of time and distance into our distance equation we get that 2 = 0 + 0+0 + **c**, or **c** = 2. So, our particular equation for distance is:

$$s = \frac{2t^3}{3} + 2t^2 + 10t + 2$$

Area under a curve

Example 4.68 illustrates the power of the integral calculus in being able to find velocity from acceleration and distance from velocity. We know from:

$$\text{Velocity (speed in a given direction)} = \frac{\text{distance}}{\text{time}}$$
$$\text{therefore distance} = \text{velocity} \times \text{time}$$

So, if we set velocity against time on a velocity–time graph, the area under the graph (velocity × time) will be equal to the distance. Therefore, if we know the rule governing the motion, we could, in our case, find any distance covered within a particular time period by integrating the velocity–time curve over this period.

Consider *Figure 4.57*, which shows a velocity–time graph, where the motion is governed by the relationship:

$$v = -t^2 + 3t \quad \text{or} \quad \frac{ds}{dt} = -t^2 + 3t$$

Then to find the **distance** equation, for the motion, all we need to do is to **integrate the velocity equation**, as in Example 4.69.

The important point to note is that **when we integrate** and **find the distance equation**, this is the same as **finding the area under the graph**, because the area under the graph = velocity × time = distance.

> **KEY POINT**
>
> The area under a velocity–time curve is equal to the distance

UNIT 4

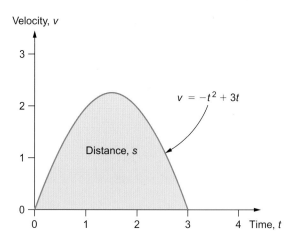

Figure 4.57 Velocity–time graph for the motion $v = -t^2 + 3t$

From the graph it can be seen that when time $t = 0$, the velocity $v = 0$, also that when $t = 3$, $v = 0$. So, the area of interest is contained between these two time *limits*.

Now, integrating our velocity equation in the normal manner gives:

$$\int v = \frac{ds}{dt} = \int -t^2 + 3t \, dt = s = \frac{-t^3}{3} + \frac{3t^2}{2} + c$$

This ***distance equation is equivalent to the area under the graph between time $t = 0$ and $t = 3$.***

At time $t = 0$ the distance travelled $s = 0$, from the graph. The ***constant of integration c*** can be found by substituting these values of time and distance into our distance equation:

$$s = \frac{-t^3}{3} + \frac{3t^2}{2} + c, \text{ so } 0 = 0 + 0c,$$

therefore $c = 0$ and our particular distance equation is

$$s = \frac{-t^3}{3} + \frac{3t^2}{2}$$

Now between our limits of time $t = 0$ to $t = 3$ seconds. The area under the graph indicates the distance travelled. So, at time $t = 0$, the distance travelled $s = 0$. At time $t = 3$, the area under the graph is found by substituting time $t = 3$ into our distance equation, then:

$$s = \frac{-t^3}{3} + \frac{3t^2}{2} = \frac{-(3)^3}{3} + \frac{(3)(3)^2}{2} = \frac{-27}{3} + \frac{27}{2} = -9 + 13.5 = 4.5$$

KEY POINT

When finding definite integrals the constant of integration is eliminated

Thus, in the above example, the area under the graph is 4.5 which is the distance travelled.

The definite integral

When we ***integrate between limits***, such as the time limits, we say that we are finding the ***definite integral***. All the integration that we have been doing up till now has involved the constant of integration and we refer to this type of integration as finding the ***indefinite integral*** that must contain an arbitrary constant ***c***.

The terminology for indefinite integration is that which we have used so far, for example:

$$\int -t^2 + 3t \, dt \text{ the } \textbf{\textit{indefinite integral}}.$$

When carrying out definite integration, we place limits on the integration sign, or summing sign, for example:

$$\int_0^3 -t^2 + 3t \, dt \text{ the } \textbf{\textit{definite integral}}.$$

To evaluate a definite integral, we first integrate the function, then we find the numerical value of the integral at its upper and lower limits and subtract the value of the lower limit from that of the upper limit to obtain the required result.

So, following this procedure for the definite integral shown above, which we used to find the distance s (area under a graph) from the velocity–time graph, we get:

$$s = \int_0^3 -t^2 + 3t \, dt = \left[\frac{-t^3}{3} + \frac{3t^2}{2} + c \right]_0^3$$

$$= \left(\frac{-27}{3} + \frac{27}{2} + c \right) - \left(\frac{0}{3} + \frac{0}{2} + c \right)$$

$$s = (-9 + 13.5 + c) - (0 + c) = 4.5 + c - c = 4.5$$

Thus $s = 4.5$. We have found the area under the graph using definite integration!

Note that when we subtract the upper limit value from the lower limit value, the constant of integration is eliminated. This will always be the case when evaluating definite integrals; therefore, it need not be shown from now on.

Example 4.69

(a) Evaluate $\displaystyle\int_{-1}^1 \frac{x^5 - 4x^3 + 6x}{x} dx$.

(b) Determine by integration the area enclosed by the curve, $y = 2x^2 + 2$, the x-axis and the ordinates $x = -2$ and $x = 2$.

(a) Before we integrate, it is essential to simplify the function as much as possible. So, in this case on division by x, we get

$$\int_{-1}^1 x^4 - 4x^2 + 6 \, dx = \left[\frac{x^5}{5} - \frac{4x^3}{3} + 6x \right]_{-1}^1 = \left(\frac{1}{5} - \frac{4}{3} + 6 \right) - \left(\frac{-1}{5} - \frac{-4}{3} - 6 \right) = 9\frac{11}{15}$$

Note that, in this case, it is easier to manipulate the upper and lower values as fractions!

(b) In order to get a picture of the area we are required to find, it is best to draw a sketch of the situation first. The area with the appropriate limits is shown in *Figure 4.58*.

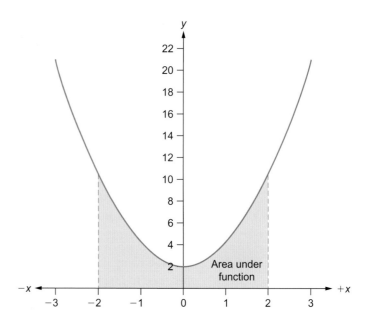

Figure 4.58 Graph of the function $y = 2x^2 + 2$

We are required to find the shaded area of the graph, between the limits $x = \pm 2$. Then,

$$\int_{-2}^{2} 2x^2 + 2\,dx = \left[\frac{2x^3}{3} + 2x\right]_{-2}^{2} = \left(\frac{2(2)^3}{3} + (2)(2)\right) - \left(\frac{2(-2)^3}{3} + (2)(-2)\right)$$

$$\left(\frac{16}{3} + 4\right) - \left(\frac{-16}{3} - 4\right) = \mathbf{18\frac{2}{3}} \text{ square units}$$

A final word of caution when finding areas under curves using integration: if the area you are trying to evaluate is part above and part below the x-axis, it is necessary to split the limits of integration for the areas concerned.

So, for the shaded area shown in *Figure 4.59*, we find the definite integral with the limits $(2, -2)$ and the definite integral with limits $(4,2)$. We then add the integral of the shaded area above the line to the integral of the shaded area below the line. The definite integral below the line (x-axis) always comes out as a negative value, where the minus sign is a consequence of the position of this area below the x-axis. ***Therefore, in order to find the total area we can either subtract a minus quantity $(- \times - = +)$ where from the laws of signs we are in effect adding together both values, or we can add the modulus (positive value) of the definite integral below the x-axis.***

Thus, the total shaded area in *Figure 4.59* $= \int_{-2}^{2} y\,dx - \int_{2}^{4} y\,dx$, as

shown in *Figure 4.59* or the total shaded area $\int_{-2}^{2} y\,dx + \left| \int_{2}^{4} y\,dx \right|$.

Notice that the higher value always sits at the top of the integral sign. The minus sign is always necessary before the integral of any area that sits *below the x-axis*.

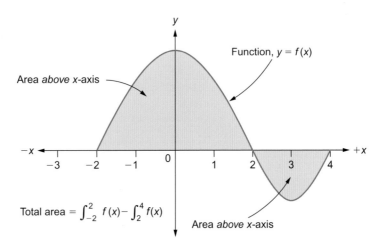

Figure 4.59 Function with areas above and below the *x*-axis

On this important point, we finish our study of the integral calculus and indeed your study of mathematics for the unit! You are left with a set of test your knowledge questions on this last topic and a set of end of chapter review questions that cover important aspects from all the outcomes within this chapter.

UNIT 4

TYK *4.16*

1. Find 　　(a) $\int x^3\,dx$ 　　(b) $3\int x^2\,dx$ 　　(c) $\int x^{-\frac{1}{2}}\,dx$

(d) $\int \frac{2}{\sqrt{x}}\,dx$.

2. Find 　　(a) $\int 3x^2 + 8x + 5\,dx$ 　　(b) $\int \frac{x^3}{4} - 3x^{-\frac{1}{2}} + \sqrt{x}\,dx$

(c) $\int (x + 2)^2\,dx$.

3. Find 　　(a) $\int -2.5\sin\theta\,d\theta$ 　　(b) $\int \frac{1}{3}\cos 4x\,dx$

(c) $\int 3\sin 2\theta - 4\cos 3\theta\,d\theta$.

4. Integrate the functions: 　　(a) $0.5e^{-0.5x}$ 　　(b) $6\ln x$ 　　(c) $\frac{3}{x}$.

Test your knowledge

TYK

5. Evaluate   (a) $\int_0^2 4x^2 + 2x^{-3} \, dx$ (b) $\int_0^1 \dfrac{3x^{\frac{1}{2}}}{6} - \sqrt{x} + x^{\frac{3}{2}} \, dx.$

6. The acceleration of a vehicle is given by the relationship $a = 3t + 4$. Find the formulae for the velocity and distance of the vehicle, given that $s = 0$ and $v = 8$ when $t = 0$. Also, find the distance travelled after a time of 10 seconds has elapsed.

7. Find the area under the curve $y = x + x^2$ between $x = 1$ and $x = 3$.

End of Unit Review

1. Multiply out the following brackets and simplify where possible:

(a) $(a - 4)(a + 4)$

(b) $(x - y)(x^2 + y^2)$

(c) $(x - 2)(x + 4)(x + 1)$.

2. Factorize the following expressions:

(a) $x^2 - 2x - 8$

(b) $8a^2bc + 2abcd^2 - 4a^2bcf + a^2b^2c^2$

(c) $6x^2 + 6x - 12$

(d) $w^2x^2y^2 + w^2x^2z^2 - w^2x^2p^2$.

3. Simplify the following expressions:

(a) $\dfrac{1}{3^2} \times 3^{-1} \times 3^{-\frac{3}{2}} \times 3^{\frac{5}{2}}$

(b) $\left[\left(\dfrac{w^3x^2y^4z^2}{w^2xy^5z^3} \right) \left(\dfrac{xyz^2}{wx^2z} \right) \right]^2$

(c) $(3a^2b^4c + 2a^2b^2c^2)(a^{-2}b^{-2}c^{-1})$.

4. Transpose the formula $\dfrac{1}{R_T} = \dfrac{1}{R_1} + \dfrac{1}{R_2}$ for R_2 and find the value when $R_T = 20$ ohms and $R_1 = 40$ ohms.

5. (a) Solve the following equations:

(i) $3x - 4 = 8$

(ii) $4(x + 4) - 2(x + 1) = 3x - 1$.

(b) Solve the following simultaneous equations analytically:

(i) $8x - 6y = 2$
$2x + 6y = 38$

(ii) $8x - 12y = -5$
$14x - 10y = -0.5$.

6. (a) Solve the following quadratic equation by factorization:

$$6x^2 - 22x = -12.$$

(b) Solve the following quadratic equation using the formulae:

$$3x^2 - 2x - 7 = 0.$$

7. The following data recorded from an experiment are believed to satisfy a law of the form $y = ax + b$:

x	0	1	2	3	4	5
y	5.8	8.7	12.5	15.0	18.4	20.9

(a) Plot a graph of the data and from your graph find an estimate for the values of the constants a and b.

(b) Write down the equation for the data and state whether or not the data satisfies the law or otherwise.

(c) From your graph determine an approximate value for y, when $x = 3.5$.

8. Solve graphically the simultaneous equations:
$x + 2y = 8$
$3x + 4y = 18$.

9. An equation connecting the final velocity of a cutting machine with the variables n and m is given by $v = 18^{n/m}$. Using the laws of logarithms transpose the formula for n and find its values when $v = 30$ and $m = 3$.

10. The pressure p and volume v of a gas at constant temperature are related by the law $p = av^{1.4}$, where a is a constant. Show that the experimental values given in the table follow this law and from the graph of the results, determine the value of the constant a.

Volume V (m^3)	2	2.5	3	3.5	4
Pressure P (bar)	5.278	7.213	9.311	11.554	13.929

11. (a) If the angle subtended at the centre of a circle by an arc length 8.5 cm is 78°, what is the radius of the circle?

(b) Find the angle in degrees of a sector of a circle of radius 25 cm and area 400 cm².

12. (a) Solve the triangle illustrated in the following figure:

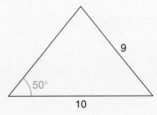

Figure 4.60 See End of unit review question 12

(b) Find all the angles of a triangle with sides $a = 10$ cm, $b = 16$ cm, $c = 14$ cm.

13. (a) Find the value of y for the following functions when $\theta = \dfrac{\pi}{4}, \dfrac{\pi}{3}, \dfrac{\pi}{2}$:

(b) $y = 3\sin \theta$ (ii) $y = 0.5\cos 2\theta$.

(c) Sketch the graphs of (i) $y = \sin 4\theta$ and (ii) $y = 4\cos \theta$ on the same set of axes and comment on their form.

14. A hexagonal bar is to be machined from a round bar. If the hexagonal cross-section just fits into the circular cross-section, as shown in the figure below and the radius of the bar is 1.5 cm, determine the cross-sectional area and volume of the material that needs to be removed, if the bar is 2 m in length.

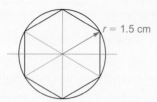

$r = 1.5$ cm

Figure 4.61 See End of unit review question 14

15. (a) A cylindrical paint can, whose height is equal to its diameter, has a capacity of 0.009 m³. Find the height of the can.

(b) Find the mass of a hemi-spherical copper bowl, with internal and external diameters of 18 cm and 25 cm, respectively. Take the density of copper as 8900 kg m⁻³.

16. *Figure 4.62* shows the cross-section of a gutter, find:

(a) the area of the cross-section.

(b) the volume of the water carried in a 12 m length of the gutter, when the water is 1 cm from the top of the gutter.

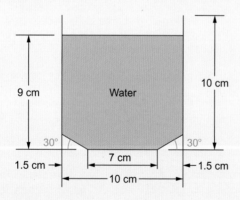

9 cm

Water

10 cm

30°   30°

7 cm

1.5 cm →   ← 1.5 cm

← 10 cm →

Figure 4.62 See End of unit review question 16

17. A sample of 100 tungsten filament light bulbs was time tested to failure, with the following results:

Results	Frequency
750–799	11
800–849	15
850–899	48
900–949	20
950–999	6

(a) Determine the arithmetic mean life to failure.

(b) Choosing suitable class intervals draw up a histogram for the data.

(c) From your histogram estimate the modal value.

(d) Produce a cumulative frequency curve (an ogive) and from it estimate the 50th- and 75th-percentile.

18. Differentiate the following expressions:

(a) $y = 2x^6 - 5x^2 - 2x$

(b) $y = 4x^{-\frac{5}{2}} - 2x^{\frac{1}{2}} + x^{-2}$

(c) $y = \sin 3\theta$

(d) $y = -2\cos 4\theta + 2\sin \theta$

(e) $y = 8\ln 3x$

(f) $y = 6e^{-3\theta}$.

19. (a) Determine the rate of change of the function

$$y = \frac{x^4}{2} - 3x^3 + x^2 - 3 \text{ at the point where } x = 2.$$

(b) Find the rate at which a capacitor is discharging at $t = 2.5$ ms and $t = 2.8$ ms, when the amount of charge on the capacitor is given by the function $Q = 2.2 \log_e t$.

(c) An alternating current is given by the function $i = \cos 2\theta$. Find the rate of change of current with θ when $\theta = 0.7$ rad.

20. (a) Integrate the following functions with respect to the variable given:

(i) $2x^2 - 3x + 4$

(ii) $t + \frac{1}{2}t^2$.

(b) Find:

(i) $\int -\sin 2x + \cos 3x \, dx$

(ii) $\int 4\log_e t \, dt$.

(c) Evaluate

(i) $\int_0^2 x^4 - \frac{3x^2}{4} + x \, dx$

(ii) $\int_0^{\frac{\pi}{2}} -3\sin 2\theta \, d\theta$.

(d) Determine by integration the area enclosed by the curve $y = 2x^2 + 4$ and the ordinates $x = -2, x = 2$.

The Trent 900 gas turbine engine is the largest engine ever built by Rolls Royce and has a fan diameter of 116 inches. It is the lead engine to power the Airbus A380 and has produced 90,000 lbs of thrust on the test bed. Using some of the fundamental principles explained in this unit, mechanical engineers are responsible for the design and build of this complex piece of turbo-machinery.

Mechanical Principles and Applications

The use and application of mechanical systems and machines is an essential part of modern day life. These systems and machines range from simple mechanical toys and implements to the ultra-sophisticated complex turbine engines used for the generation of electrical power and for the provision of thrust in large modern jet aircraft. The design, manufacture and maintenance of these systems and machines are the responsibilities of mechanical engineers and technicians. In this unit, you will study some of the fundamental underlying scientific principles and their applications needed by engineers to work with this technology. The work of great scientists such as Newton, Hooke, Young, Archimedes, Bernoulli, Boyle, Charles and Joule laid down the scientific principles that are still applied by engineers and technicians. Some of these principles and their engineering application form the focus of this unit and their study is considered essential for anybody wishing to practice as an engineer or technician.

Engineers need to select and use materials to produce a whole range of mechanical systems, machines and artefacts, and also need to know how these materials will stand up to the loads imposed on them when in use. You will be introduced to and use the principles laid down by Hooke and Young when you study the effects of loading static engineering systems in the first outcome of this unit. In the second outcome of this unit, Newton's laws of motion will underpin your study of dynamic systems. In the third outcome of the unit you will apply Archimedes' and Bernoulli's principles to your study of immersed bodies in fluids and the parameters associated with internal fluid flow in piped systems. Finally in the fourth outcome of this unit you will apply some of the principles laid down by Boyle, Charles and Joule when you study energy transfer in thermodynamic systems. These same thermodynamic principles were used by Sir Frank Whittle when he pioneered his Jet Engine – considered by some to be the greatest invention of the twentieth century.

No study of scientific principles would be complete without considering the units of measurement that underpin the subject. For this reason a brief introduction to the *Systeme International d'unites* (SI) has been included in Appendix 1.

The successful study of this unit will provide the necessary foundation for further study in areas such as mechanics, fluid dynamics, thermodynamics and other related applications of mechanical science.

BTEC National Engineering. DOI: 10.1016/B978-0-12-382202-4.00005-2

Loading in Static Engineering Systems

In this outcome we will first review the concept of force, the representation of forces as vector quantities and their vector addition. Those who have not met these topics before will need to study them thoroughly and ensure that they are able to successfully complete the set of TYK questions at the end of this section. We will then look at coplanar force systems and introduce the terminology and definitions associated with concurrent and non-concurrent coplanar forces. To end this section we consider the resolution of the forces (loads) that act on these systems and how they may be reduced to either one resultant force or to one resultant force and a turning moment. In the next section, we consider the simply supported beam as a loaded system and use the principle of moments and static equilibrium to determine system parameters. Finally, we look at the stresses and resulting strains imposed on static systems as a result of the loads imposed on them.

Force

In its simplest sense a force is a push or pull exerted by one object on another. In a member in a static structure, a push causes compression and a pull causes tension. Members subject to compressive and tensile forces have special names. A member of a structure that is in *compression* is known as a *strut* and a member in *tension* is called a *tie*.

Only rigid members of a structure have the capacity to act as both a strut and tie. Flexible members, such as ropes, wires or chains, can only act as ties.

KEY POINT

The action of a force always causes an opposite reaction

Force cannot exist without opposition, as you will see later when you study Newton's laws. An applied force is called an *action* and the opposing force it produces is called *reaction*.

The effects of any force depend on its three characteristics, which are illustrated in *Figure 5.1*.

In general, *Force (F) = mass (m) × acceleration (a) is used as the measure of force* and in symbols:

$$F = ma$$

The SI unit of force is the newton (N). The newton is thus defined as follows:

KEY POINT

Force = mass (*m*) × acceleration (*a*) and is a vector quantity

1 newton is the force that gives a mass of 1 kg an acceleration of $1\,ms^{-2}$.

Note that weight force is a special case where the acceleration acting on the mass is that due to gravity, so *weight force* may be defined as $F = mg$ and on the surface of the earth the gravitation acceleration is taken as $g = 9.81\,ms^{-2}$. This means, for example, that the weight of a body (with the same mass) will vary from that on earth if it is taken to the moon!

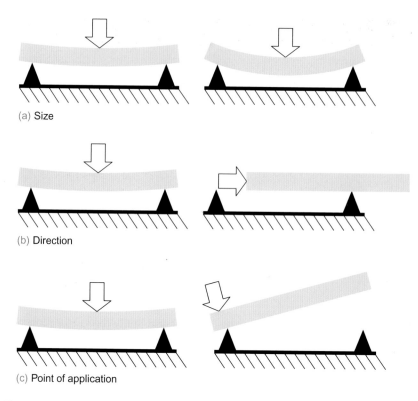

(a) Size

(b) Direction

(c) Point of application

Figure 5.1 Characteristics of force

KEY POINT

Weight = mass × the acceleration due to gravity and is a vector quantity

It can be seen from *Figure 5.1* that *a force has size (magnitude), direction and a point of application*. A force is thus a **vector quantity**, that is, it has magnitude and direction. A **scalar** quantity has only magnitude, for example mass and time. A force may, therefore, be represented graphically in two dimensions by drawing an arrow to scale with its length representing the magnitude of a force and the head of the arrow indicating the direction in relation to a set of previously defined axes. *Figure 5.2* illustrates the graphical representation of a force.

UNIT 5

TYK *5.1*

1. What are the three defining characteristics of any force?

2. Define (a) a scalar quantity, (b) a vector quantity and give an example of each.

3. Give the general definition of force and explain how weight force varies from this definition.

4. Complete the following statement: 'A strut is a member in _____ and a tie is a member in _____.'

5. A rocket launched into the earth's atmosphere is subject to acceleration due to gravity of $5.2\,ms^{-2}$. If the rocket has a mass of 120,000 kg, determine:
   (a) the weight of the rocket on earth
   (b) the weight of the rocket in orbit.

6. A body has a mass of 60 kg:
   (a) What force is required to give it an acceleration of $6\,ms^{-2}$?
   (b) What will be the reaction of the body when given this acceleration?

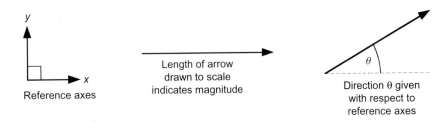

Figure 5.2 Graphical representation of a force

Vector representation and combination of forces

You have just been reminded of the concept of *force* and that the effect of a force was dependent on its magnitude (size), direction and point of application (*Figure 5.1*), and that a force may be represented on paper as a *vector* quantity (*Figure 5.2*).

We will now study the vector representation of a force or combination of forces, in more detail, noting that all vector quantities throughout this book will be identified using emboldened text.

In addition to possessing the properties of magnitude and direction from a given reference (*Figure 5.2*), vectors must obey the **parallelogram law** of combination. This law requires that two vectors v_1 and v_2 may be replaced by their equivalent vector v_T which is the diagonal of the parallelogram formed by v_1 and v_2 as shown in *Figure 5.3a*.

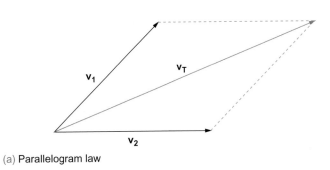

(a) Parallelogram law

Figure 5.3 Vector addition and subtraction

This vector sum is represented by the vector equation: $v_T = v_1 + v_2$.

Note that the plus sign in this equation refers to the addition of two vectors, and should not be confused with ordinary scalar addition, which is simply the sum of the magnitudes of these two vectors and is written as $v_T = v_1 + v_2$, in the normal way without emboldening.

Vectors may also be added head-to-tail using the **triangle law** as shown in *Figure 5.3b*. It can also be seen from *Figure 5.3c* that the order in which vectors are added does not affect their sum.

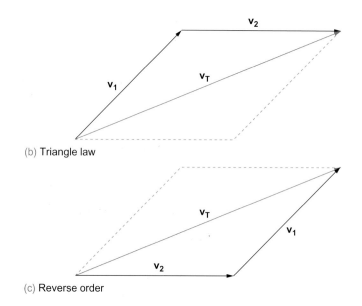

(b) Triangle law

(c) Reverse order

Figure 5.3 (Continued)

The vector difference $\mathbf{v_1} - \mathbf{v_2}$ is obtained by *adding* $-\mathbf{v_2}$ to $\mathbf{v_1}$. The effect of the minus sign is to reverse the direction of the vector $\mathbf{v_2}$ (*Figure 5.3d*).

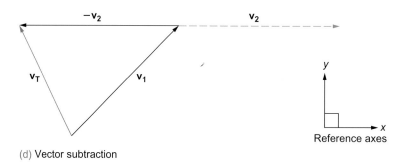

(d) Vector subtraction

Figure 5.3 (Continued)

The vectors $\mathbf{v_1}$ and $\mathbf{v_2}$ are known as the components of the vector $\mathbf{v_T}$.

In Examples 5.1 and 5.2 that follow, we are asked to find the resultant of the force systems. The **resultant is the single equivalent force that can replace all the forces in the system**.

Example 5.1

Two forces act at a point as shown in *Figure 5.4*. Find by vector addition their resultant.

UNIT 5

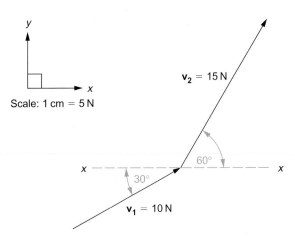

(a) Space diagram

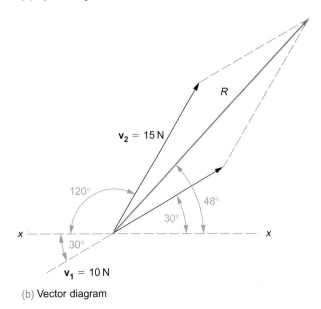

(b) Vector diagram

Figure 5.4 Vector addition using the parallelogram law

From the vector diagram the resultant vector **R** is 4.8 cm in magnitude that (from the scale) is equivalent to 24 N. So, the resultant vector **R** has a **magnitude of 24 N at an angle of 48°**.

KEY POINT

The resultant is the single equivalent force that replaces all the force in a force system

Note that a **space diagram** is first drawn to indicate the orientation of the forces with respect to the reference axes. These axes should always be shown. Also, note that the **line of action** of vector **v₁** passing through the point *0* is shown in the space diagram and can lie anywhere on this line, as indicated on the vector diagram.

KEY POINT

Vector addition of two or more forces acting at a point can be achieved using the head-to-tail rule

Example 5.2

Find the resultant of the system of forces shown in *Figure 5.5* using vector addition.

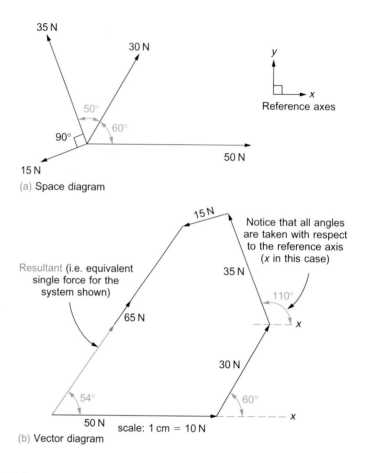

Figure 5.5 Vector addition, using polygon of forces method

From the diagram the resultant = 6.5 cm = 6.5 × 10 N = **65 N** acting at an angle of **54°** from the *x*-reference axis. This answer may be written mathematically as **resultant = 65 N∠54°**.

Note that for the force system in Example 5.2 vector addition has produced a **polygon**. Any number of forces may be added vectorially in any order, providing the *head-to-tail rule* is observed. In this example, if we were to add the vectors in reverse order, the same result would be achieved.

If a force, or system of forces, is acting on a body and is balanced by some other force, or system of forces, then the body is said to be in *equilibrium*; so, for example, a stationary body is in equilibrium.

The **equilibrant** *of a system of forces is that force which when added to a system produces equilibrium*. It has been shown in Examples 5.1 and 5.2 that the resultant is the single force that will replace an existing system of forces and produce the same effect. It therefore follows that if the equilibrant is to produce equilibrium it must be equal in magnitude and direction, but opposite in sense to the resultant *Figure 5.6* illustrates this point.

KEY POINT

The equilibrant of a system of forces is that force which when added to a system produces equilibrium

UNIT 5

KEY POINT

Bow's notation is an ideal way of identifying a number of forces acting at a point to avoid confusion

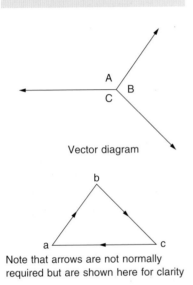

Vector diagram

Note that arrows are not normally required but are shown here for clarity

Figure 5.7 Bow's notation

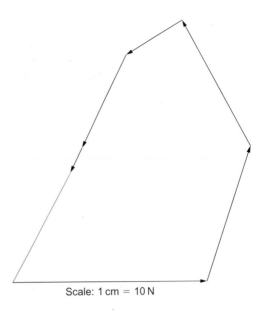

Scale: 1 cm = 10 N

Figure 5.6 Equilibrant for Example 5.6

Bow's notation is a convenient system of labelling the forces for ease of reference when there are three or more forces to be considered. Capital letters are placed in the space between forces in a clockwise direction, as shown in *Figure 5.7*.

Any force is then referred to by the letters that lay in the adjacent spaces either side of the vector arrow representing that force. The vectors representing the forces are then given the corresponding *lower case* letters. Thus, the forces AB, BC and CA are represented by the vectors **ab**, **bc** and **ca**, respectively. This method of labelling applies to any number of forces and their corresponding vectors. Arrowheads need not be used when this notation is adopted, but are shown in *Figure 5.7* for clarity.

TYK *5.2*

1. What is the essential difference in the use of the parallelogram rule and the triangle rule when combining two forces acting at a point?

2. What is the effect of the minus sign when finding vector differences?

3. With respect to force systems define (a) the equilibrant and (b) the resultant.

4. For the system of two forces shown in *Figure 5.8* determine, using the triangle rule, the resultant and equilibrant of the system.

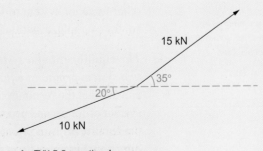

15 kN

35°

20°

10 kN

Figure 5.8 Figure for TYK 5.2 question 4

5. For the system of forces shown in *Figure 5.9*, determine using the polygon of forces method the magnitude and direction of the force that will put the system into equilibrium.

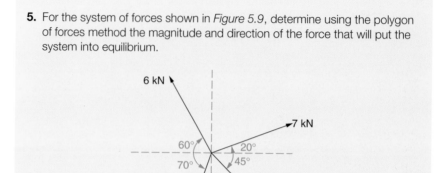

Figure 5.9 Figure for TYK 5.2 question 5

Coplanar force systems

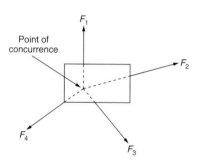

Figure 5.10 Concurrent coplanar force system

Forces that act within a two dimensional plane, such as the plane of this paper, are referred to as coplanar forces. If all the *lines of action* of these forces pass through the same point known as *the point of concurrence* (*Figure 5.10*) then we have a *concurrent coplanar force system*. All the force systems that we have considered so far, that have been solved using vectors, have involved only concurrent coplanar forces.

We know from our previous study of vector forces that a set of concurrent coplanar forces can be reduced to a single equivalent force: *the resultant* that represents the effect of the whole system. Also, as you will see when you study Newton's laws, that if a body is acted upon by a system of forces and is not constrained it will move in the direction of the line of action of the single resultant force. If we wish to stop this resulting movement we need to provide an equal and opposite force to that of the resultant in order to maintain the system in static equilibrium. This restoring force as you are already aware is known as *the equilibrant*.

We now need to consider another system of coplanar forces where the lines of action of the forces in the system do not pass through the same point. This system (*Figure 5.11a*) is known as a **non-concurrent coplanar force system**. In this system, there is not only a tendency for the force to move the body in a certain linear direction but also to make it rotate. This system may again be represented by a single resultant force but in addition, because the forces do not act through the same point, there must also be a *turning moment* created between the line of action of the resultant and the point about which the rotation of the body takes place. This idea is illustrated in *Figure 5.11b*.

The concept of turning moments and equilibrium is central to your understanding of non-coplanar force systems and their resolution. For this reason we focus next on the concept of turning moments, equilibrium and couples that up to now in your studies you may not have covered fully.

UNIT 5

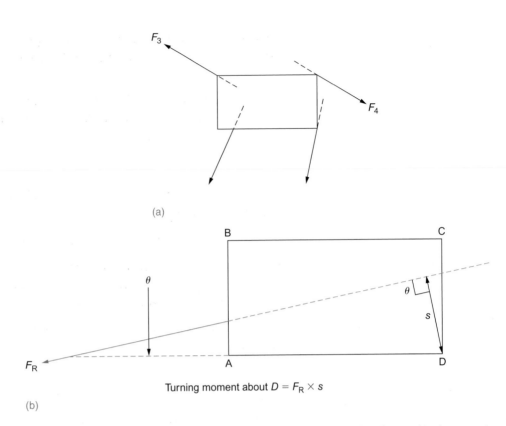

(a)

(b)

Turning moment about $D = F_R \times s$

Figure 5.11 (a) Non-concurrent coplanar force system and (b) a resultant force and turning moment

Turning moments and equilibrium

A moment is a turning force producing a turning effect. The magnitude of this turning force depends on the size of the ***force*** applied and the ***perpendicular distance*** from the pivot or axis to the line of action of the force (*Figure 5.12a*).

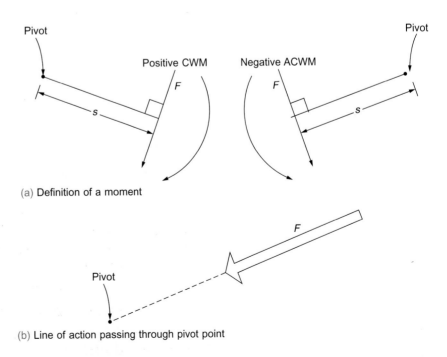

(a) Definition of a moment

(b) Line of action passing through pivot point

Figure 5.12 Moment of a force

Examples of a turning force are numerous; opening a door, using a spanner, turning the steering wheel of a motor vehicle are just three examples.

> The **moment of a force (*M*)** is defined as the product of the magnitude of force (*F*) and its perpendicular distance (*s*) from the pivot or axis to the line of action of the force.

This may be written mathematically as: ***M = Fs***. The SI units for a moment are ***newton-metres*** (Nm).

Moments are always concerned with perpendicular distances.

From *Figure 5.12a*, you should note that moments can be clockwise *CWM* or anticlockwise *ACWM*. Conventionally, we consider ***clockwise moments to be positive*** and ***anticlockwise moments to be negative***.

If the line of action of the force passes through the turning point it has no turning effect and so no moment. *Figure 5.12b* illustrates this point.

KEY POINT

If the line of action passes through the turning point, it has no effect because the effective distance of the moment is zero

Example 5.3

***Figure 5.13* shows a spanner being used to loosen a nut. Determine the turning effect on the nut.**

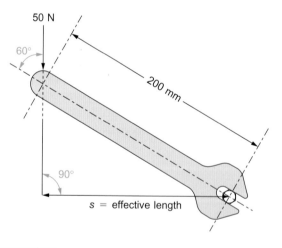

50 N

60°

200 mm

90°

s = effective length

Figure 5.13 Spanner and nut

The turning effect on the nut is equal to the moment of the 50 N force about the nut, that is,

$$M = Fs$$

Remembering that moments are always concerned with perpendicular distances, then the distance s is the perpendicular distance or effective length of the spanner. This length is found using trigonometric ratios:

$$s = 200 \sin 60°, \text{ therefore } s = (200)(0.866) = 173.2 \text{ mm}$$

$$\text{Then ACWM} = (50)(173.2) = 8660 \text{ N mm or } 8.66 \text{ Nm}$$

So, the *turning effect* of the 50 N force acting on a 200 mm spanner at 60° to the centre line of the spanner = ***8.66 Nm***.

UNIT 5

In engineering problems concerning forces acting on bodies and the resulting motions, you have already met the terms CWM and ACWM. Set out below are three more frequently used terms that you may encounter.

Fulcrum: The *fulcrum* is the point or axis about which rotation takes place. In Example 5.3, the geometrical centre of the nut is considered to be the fulcrum.

Moment arm: The perpendicular distance from the line of action of the force to the fulcrum is known as the *moment arm*.

Resulting moment: The *resulting moment* is the difference in magnitude between the total clockwise moment and the total anticlockwise moment. Note that if the body is in *static equilibrium* this *resultant will be zero*.

> **KEY POINT**
>
> For static equilibrium the algebraic sum of the moments is zero

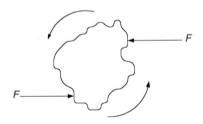

Figure 5.14 Non-equilibrium condition for equal and opposite non-coplanar forces acting on a body

When a body is in equilibrium there can be no *resultant force* acting on it. However, reference to *Figure 5.14* shows that a body subject to two equal and opposite non-coplanar forces is not necessarily in equilibrium even when there is no resultant force acting on it.

The resultant force on the body is zero, but two forces would cause the body to rotate, as indicated. Therefore, in the case illustrated a clockwise restoring moment would be necessary to bring the system into equilibrium. This leads us to a second condition that must be satisfied to ensure that a body is in static equilibrium. This is known as the *principle of moments*.

The principle of moments states that when a body is in static equilibrium under the action of a number of forces, the total CWM about any point is equal to the total ACWM about the same point.

This means that for static equilibrium the *algebraic sum of the moments must be zero*.

One other important fact needs to be remembered about bodies in static equilibrium. Consider the uniform beam (uniform here, means an equal cross-section along its total length) shown in *Figure 5.15*. We already know from the principle of moments that the sum of the CWM must equal the sum of ACWM. It is also true that the beam would sink into the ground or rise, if the upward forces did not equal the downward forces. So, a further necessary condition for static equilibrium is that: **upward forces = downward forces**. For the case shown in *Figure 5.15*, this means that $R_1 + R_2 = W_1 + W_2 + W_3 + W_4$, if we ignore the weight of the beam.

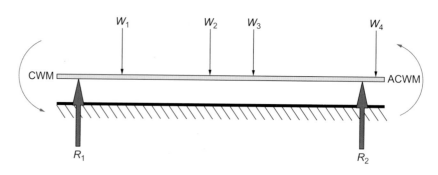

Figure 5.15 Conditions for static equilibrium

Couples

So far with respect to force systems we have restricted our problems on moments to the turning effect of forces taken one at a time.

A **couple** occurs when two equal forces acting in opposite directions have their lines of action parallel.

Example 5.4

Figure 5.16 shows the turning effect of a couple on a beam of regular cross-section. Determine the turning moment of the couple.

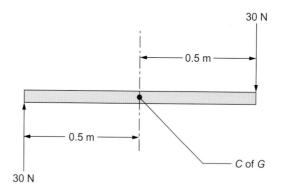

Figure 5.16 Turning effect of a couple with regular cross-section

Taking moments about the centre of gravity (**C** of **G**) that is the point at which all the weight of the beam is deemed to act, then we get:

$$(30 \times 0.5) + (30 \times 0.5) = \text{turning effect}$$

So, *turning effect (moment) of couple = 30 Nm*.

Example 5.5

Figure 5.17 shows the turning effect of a couple on a beam of irregular cross-section, which we will again try to revolve about its centre of gravity.

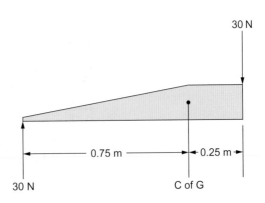

Figure 5.17 Turning effect of a couple with irregular cross-section beam

Again taking moments about the *C* of *G* gives:

$$(30 \times 0.75) + (30 \times 0.25) = \text{turning effect}$$

So, **moment of couple = 30 Nm**.

It can be seen from the above two examples that the moment is the same in both cases and is independent of the position of the fulcrum. Therefore, if the fulcrum is assumed to be located at the point of application of one of the forces: *the moment of a couple is equal to one of the forces multiplied by the perpendicular distance between them*. Thus, in both cases shown in Examples 5.4 and 5.5 the moment of the couple = 30 N $\times$ 1 m = 30 Nm, as before.

Another important application of the couple is its **turning moment or torque**.

Torque: The *torque* is the turning moment of a couple and is measured in newton-metres (Nm), that is, torque T = force $F \times$ radius r.

The turning moment of the couple given in Example 5.4 is $F \times r =$ $(30 \text{ N} \times 0.5 \text{ m}) = 15 \text{ Nm}$.

KEY POINT

The moment of a couple = force $\times$ distance between forces, the turning moment = force $\times$ radius

Example 5.6

A nut is to be torque loaded to a maximum of 100 Nm. What is the maximum force that may be applied perpendicular to the end of the spanner, if the spanner is of length 30 cm?

Since $T = F \times r$, then $F = T/r = 100/30$ therefore $F = 333.3 N$

Test your knowledge

?

TYK

TYK 5.3

1. Explain the difference between a concurrent and a non-concurrent force system.

2. 'If a non-coplanar force system is acted upon by equal and opposite forces, it must be in equilibrium.' State whether this statement is true or false giving a reason for your answer.

UNIT 5

3. Define the *moment of a force*.

4. If the line of action of a force passes through the turning point, explain why this force has no turning effect.

5. If a force acts other than at a perpendicular distance from the turning point, explain how its turning moment can be determined.

6. State the conditions for static equilibrium when a systems of forces act on a simply supported beam.

7. Define the terms (a) couple and (b) torque.

Resolution of forces for coplanar systems

Graphical solutions to problems involving vector forces are sufficiently accurate for many engineering problems and are invaluable for estimating approximate solutions to more complicated force problems. However, it is sometimes necessary to provide more accurate results in which case a mathematical method will be required. One such mathematical method is known as the **resolution of forces**. We look first at a number of examples of this method for **concurrent** coplanar force systems.

Consider a force F acting on a bolt (*Figure 5.18*). The force F may be replaced by two forces P and Q acting at right angles to each other, which together have the same effect on the bolt.

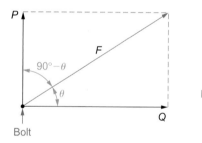

Figure 5.18 Resolving force F into its components

From your knowledge of the trigonometric ratios you will know that

$$\frac{Q}{F} = \cos\theta \text{ and so } Q = F\cos\theta$$

Also, $P/F = \cos(90 - \theta)$ and we know that $\cos(90 - \theta) = \sin\theta$

therefore, $P = F\sin\theta$.

So, from *Figure 5.18*, $P = F\sin\theta$ and $Q = F\cos\theta$.

So, the single force F has been resolved or split into two equivalent forces of magnitude $F\cos\theta$ and $F\sin\theta$ that act at right angles (they are said to be **orthogonal** to each other).

$F\cos\theta$ is known as the **horizontal component of F** and $F\sin\theta$ is known as the **vertical component of F**.

UNIT 5

Example 5.7

Three coplanar forces *A, B* and *C* are all applied to a pin joint (*Figure 5.19a*). Determine the magnitude and direction of the equilibrant for the system.

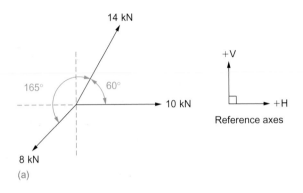

Figure 5.19a Space diagram for the force system

Each force needs to be resolved into its two orthogonal components, which act along the vertical and horizontal axes, respectively. Using the normal algebraic sign convention with our axes then, above the origin *V* is positive and below it is negative. Similarly, *H* is positive to the right of the origin and negative to the left. Using this convention we need only consider acute angles for the sine and cosine functions; these are tabulated below.

> **KEY POINT**
>
> Pythagoras theorem is used to find the single resultant force of two orthogonal forces

Magnitude of force	Horizontal component	Vertical component
10 kN	$+10\,\text{kN}\ (\rightarrow)$	0
14 kN	$+14\cos 60\,\text{kN}\ (\rightarrow)$	$+14\sin 60\,\text{kN}\ (\uparrow)$
8 kN	$-8\cos 45\,\text{kN}\ (\leftarrow)$	$-8\sin 45\,\text{kN}\ (\downarrow)$

Then total horizontal component $= 10 + 7 - 5.66\,\text{kN} = 11.34\,\text{kN}\ (\rightarrow)$

and total vertical component $= 0 + 12.22 - 5.66\,\text{kN} = 6.56\,\text{kN}\ (\uparrow)$.

Since both the horizontal and vertical components are positive the resultant force will act upwards to the right of the origin. The three original forces have now been reduced to two that act orthogonally. The magnitude of the resultant F_R or the equilibrant may now be obtained using Pythagoras theorem on the right angle triangle obtained from the orthogonal vectors, as shown in *Figure 5.19b*. From Pythagoras we get $R^2 = 6.56^2 + 11.34^2 = 171.629$ and so resultant $F_R = 13.10\,\text{kN}$, so the **magnitude** of the equilibrant also $= 13.10\,\text{kN}$.

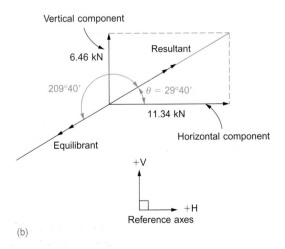

(b)

Figure 5.19b Resolution method

From the right angled triangle shown in *Figure 5.19b*, the angle θ that the resultant F_R makes with the given axes may be calculated using the trigonometric ratios. Then,

$$\tan\theta = \frac{6.56}{11.34} = 0.5244 \quad \text{and} \quad \theta = 30.05°$$

therefore the resultant $F_R = 13.10\,kN\angle30.05°$.

The **equilibrant** will act in the opposite sense and therefore $= 13.10\,kN\angle210.05°$.

To complete our study on the resolution of concurrent coplanar force systems, we consider one final example concerned with **equilibrium on a smooth plane**. Smooth in this case implies that the effects of friction may be ignored. A body is kept in equilibrium on a plane by the action of three forces as shown in *Figure 5.20*. These are:

1. the **weight W** of the body acting vertically down
2. the **reaction R** of the plane to the weight of the body. R is known as the **normal reaction**, normal in this sense means at right angles to the plane in this case
3. the **force P** acting in some suitable direction to prevent the body sliding down the plane.

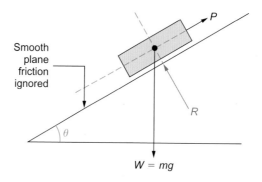

Figure 5.20 Equilibrium on a smooth plane

KEY POINT

Friction is considered small enough to ignore for bodies on a smooth inclined plane

Forces P and R are dependent on:

• the angle of inclination of the plane
• the magnitude of W, and
• the inclination of the force P to the plane.

It is, therefore, possible to express the magnitude of both P and R in terms of W and the trigonometric ratios connecting the angle θ.

In the example that follows we consider the case when the body remains in equilibrium as a result of the force P being applied parallel to the plane.

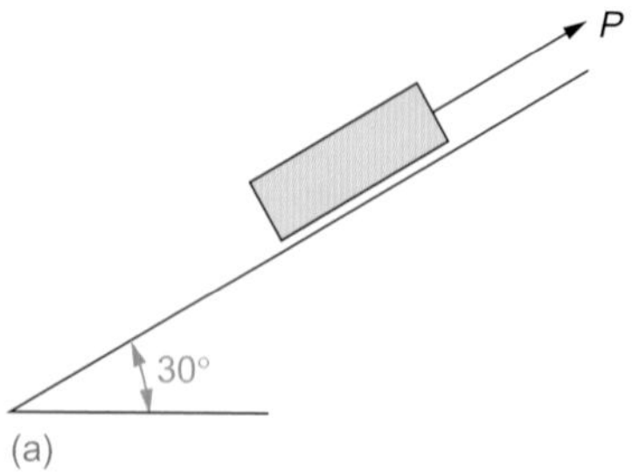

Figure 5.21a Crate acted on by force P

Example 5.8

A crate of mass 80 kg is held in equilibrium by a force *P* acting parallel to the plane as indicated in *Figure 5.21a*. Determine using the resolution method the magnitude of the force *P* and the normal reaction *R* ignoring the effects of friction.

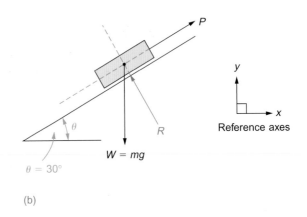

(b)

Figure 5.21b Space diagram

Figure 5.21b shows the space diagram for the problem clearly indicating the nature of the forces acting on the body.

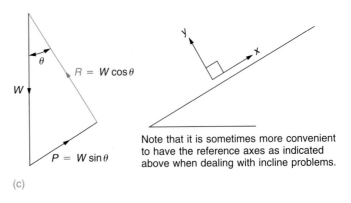

(c)

Figure 5.21c Vector components of forces acting on crate

W may, therefore, be resolved into the two forces P and R. The force component at right angles to the plane $= W \cos \theta$ and the force component parallel to the plane $= W \sin \theta$ (*Figure 5.21c*).

Equating forces gives $W \cos \theta = R$ and $W \sin \theta = P$

So, remembering the mass–weight relationship we have:

$$W = mg = (80)(9.81) = \textbf{784.8 N}$$

then,

$$R = 784.8 \cos 30° = \textbf{679.7 N}$$

and

$$P = 784.8 \sin 30° = \textbf{392.4 N}$$

KEY POINT

Non-concurrent coplanar force systems can be reduced to a resultant force and a turning moment

We now consider examples involving **non-concurrent** coplanar force systems and see the subtle differences in techniques that we must apply in order to solve such systems. In the first simple example on non-concurrent coplanar force systems we consider forces acting horizontally and vertically on a body.

Example 5.9

A flat metal plate is pivoted at its geometric centre and is acted upon by horizontal and vertical forces as shown in *Figure 5.22*.

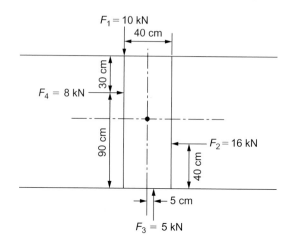

Figure 5.22 Pivoted metal plate

If the plate is free to rotate about the pivot determine the magnitude and direction of the resultant of the coplanar forces acting on the system and also find the perpendicular distance of its line of action from the pivot.

Now remembering that for non-coplanar systems there may not only be a resultant force but also a resulting turning moment that acts on the system, we must not only resolve forces but also find the resultant moment of the system by applying the principle of moments and so, in this case, find the perpendicular distance of the line of action of the resultant from the pivot. Thus a table of values of forces and moments can be set up and these values are calculated as shown below. Please note the use of the **sign convention**: positive forces are to the right and upwards, clockwise moments are positive and vice versa.

Force F (kN)	Horizontal component F_H	Vertical component F_V	Moment of F_H about pivot (kNm)	Moment of F_V about pivot (kNm)
10	0	−10	0	$-(-10 \times 0.2) = -2$
16	−16	0	$(-16 \times 0.2) = +3.2$	0
5	0	5	0	$-(5 \times 0.05) = -0.25$
8	8	0	$(8 \times 0.3) = 2.4$	0
Totals	−8	−5	+5.6 kNm	−2.25 kNm

Note that for the moment of each component in the table **we ignore the sign of the component as the moment itself is only dependent on whether it is clockwise (positive) or anticlockwise (negative)**, so in row one of the table the *ACWM* of the vertical component of the 10 kN force F_1 about the pivot is *negative*, as required.

The magnitude of the resultant of the force system is found using Pythagoras theorem and the angular direction of the resultant is found using the trigonometric ratios, in the same way as before.

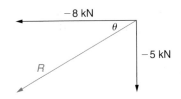

Figure 5.23

The sum of the horizontal and vertical forces are both negative; therefore, the line of action of the resultant force will act down towards the left, as shown.

Its magnitude will be: $R = \sqrt{(-8)^2 + (-5)^2} = \sqrt{89} = \textbf{9.43 kN}$ and the angle θ is found using the tangent ratio $\tan\theta = -5/-8 = 0.625$. Therefore, $\theta = \tan^{-1} 0.625 = \textbf{32°}$.

Now from the table the sum of the moments $= 5.6 - 2.25\,\text{kNm} = 3.35\,\text{kNm}$, then calling the perpendicular distance from the line of action of the resultant to the pivot point (d) we get the sum of the moments.

$$\sum M = R \times d \quad \text{or} \quad 3.35 = 9.43 \times d \quad \text{and} \quad d = \frac{3.35}{9.43} = \textbf{0.355 m.}$$

Now because the resulting turning moment is positive the line of action of the resultant must be below the pivot point, as illustrated in *Figure 5.24*.

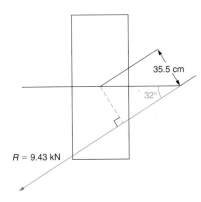

Figure 5.24 Line of action of the resultant with respect to the pivot

In the next example, we include two forces that are not horizontal or perpendicular, we resolve these forces using the techniques you have already learnt when dealing with concurrent coplanar forces.

Note: In Example 5.10, all angles are measured from the horizontal, so that all the horizontal components of the forces will then be given by $F_H = F \cos\theta$ and all vertical components are given by $F_V = F \sin\theta$. Do remember that this is only applicable if the components of the forces are measured in this way.

KEY POINT

The sign convention for coplanar force systems dictates that upward forces and clockwise turning moments are positive and vice versa

Example 5.10

Determine the magnitude and direction of the resultant force and turning moment about point A for the force system shown.

From *Figure 5.25* we tabulate the values as we did before, noting that we have resolved F_1 and F_3 into their horizontal and vertical components, giving each their appropriate sign.

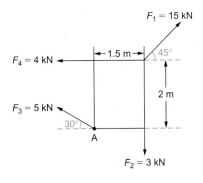

Figure 5.25 Non-concurrent force system

Force F (kN)	Horizontal component F_H	Vertical component F_V	Moment of F_H about point A (kNm)	Moment of F_V about point A (kNm)
15	$15\cos 45 = 10.6$	$15\sin 45 = 10.6$	$(10.6 \times 2) = 21.2$	$-(10.6 \times 1.5)$ $= -15.9$
3	0	-3	0	$(-3 \times 1.5) = 4.5$
5	$-5\cos 30 = -4.3$	$5\sin 30 = 2.5$	0	0
4	-4	0	$-(4 \times 2) = -8$	0
Totals	$+2.3$	$+10.1$	$+13.2\,\text{kNm}$	$-11.4\,\text{kNm}$

Note: When taking moments of the components about A take only the positive value (modulus) of the components. The sign of the moments is determined only by whether they are clockwise (positive) or anticlockwise (negative). This rule can be seen in operation for the vertical moment of the 3 kN force and the horizontal moment of the 4 kN force.

Then using Pythagoras theorem as before we have $R = \sqrt{(2.3)^2 + (10.1)^2} = \sqrt{107.3} = \textbf{10.36 kN}$, the direction is found using the tangent ratio. Then $\tan\theta = 10.1/2.3 = 4.391$ and $\theta = \tan^{-1} 4.391 = \textbf{77.17°}$.

The turning moment about A is the sum of the clockwise and anticlockwise turning moments given in the table. These are $\Sigma M = 13.2 - 11.4 = \boldsymbol{1.8\,kNm}$ (clockwise). The line of action is. Therefore, upwards towards the right.

We could go on to find the perpendicular distance from the line of action of the resultant to point A if we so wished in a similar manner to that given in Example 5.9.

In the next final example, you will meet a simple engineering application of the resolution of forces method for non-concurrent coplanar force systems. This example does not require turning moments to be found, this is left for you to complete in the TYK exercise that follows Example 5.11.

Example 5.11

A pulley holds a cable that laps it by an angle of 130 degrees as shown in *Figure 5.26*. The tensions in either side of the cable are 6 and 4 kN as indicated. What is the magnitude of the resultant force on the pulley?

All that is required to solve this problem is to resolve the tensile forces in the cable into their horizontal and vertical components. The diagram shows these components for the 4 kN force. I hope you can see from the geometry of the situation that the angle $A = 50°$. Then the horizontal and vertical components of F_2 are $F_{2H} = 4\cos 50 = 2.57\,kN$ and $F_{2v} = -4\sin 50 = -3.06\,kN$, respectively. Since the 6 kN force is horizontal then the total horizontal force components $= 2.57 + 6.0 = 8.57\,kN$ and the total vertical force components $= -3.06\,kN$. Therefore, by Pythagoras theorem the magnitude of the resultant force on the pulley is $= \sqrt{(8.57)^2 + (3.06)^2} = \boldsymbol{9.1\,kN}$.

Its direction although not asked for may be easily found using the tangent ratio as before. Its line of action will be down towards the right.

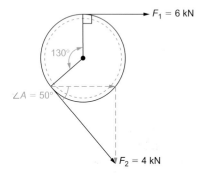

Figure 5.26 Pulley and cable assembly

TYK 5.4

1. With respect to the line of action of forces, what does the term orthogonal mean?

2. What is the sign convention for forces and for moments?

3. Sketch the free body diagram showing all the forces that act on a body on an inclined smooth plane, stating any assumptions made.

4. Find the magnitude and direction of the resultant and the equilibrant for the concurrent force system shown in *Figure 5.27*.

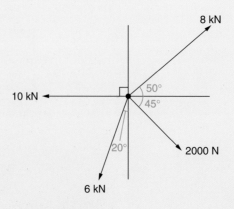

Figure 5.27 Concurrent force system

UNIT 5

5. A flat metal plate is acted upon by a system of forces and is free to pivot about A, as shown in *Figure 5.28*. Determine the magnitude and direction of the resultant force and its turning moment about A.

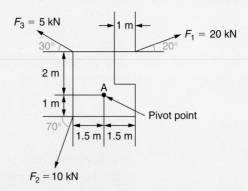

Figure 5.28 Plate force system

6. A drive belt is wrapped around a pulley assembly forming an angle of lap of 220° as shown in *Figure 5.29*. If the tensions in the belt are 10 and 8 kN, find the magnitude of the resultant force acting on the pulley.

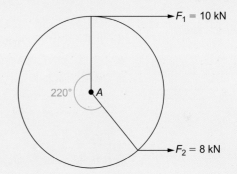

Figure 5.29 Drive belt system

7. A motor sports vehicle rounding a curve at speed is subject to two forces acting at its C of G, as shown in *Figure 5.30*. Find the resultant force on the vehicle and indicate where the line of action of the resultant cuts the vehicle axle X-X.

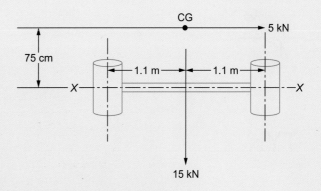

Figure 5.30 Sports vehicle rounding corner at speed

Simply supported beams

Beams are used throughout engineering as supports for loaded systems. For example, the rectangular section and I-section beams that are used in buildings to support the brickwork above windows and doors are classic examples of the use of beams. Rail tracks, conveyor system runners and bridge decking supports are yet more examples of the use of beams. We will be studying simply supported beams in this section, that is, beams with supports that are assumed to act at a point, such as wedges and encastered (built-in) supports at vertical walls.

In order to determine the nature and effect of the loads that act on these beams, we need to be familiar with the conditions for static equilibrium. In the previous section on coplanar forces you were introduced to these conditions and they are summarized below for your convenience.

Conditions for static equilibrium

1. Upward forces = downward forces
2. The sum of the CWM about any point = the sum of the ACWM at any point. Algebraically this may be written as $\Sigma CWM = \Sigma ACWM$ or that the algebraic sum of the moments is equal to zero, that is, $\Sigma CWM - \Sigma ACWM = 0$

KEY POINT

For static equilibrium, upward forces = downward forces and $\Sigma CWM - \Sigma ACWM = 0$

With all the information obtained from the previous sections you are in a very strong position to be able to follow the arguments presented in the following examples concerned with simply supported beams. In this first example we simply apply the principle of moments to the beam in order to find the unknown force. All forces or loads in this example are point loads acting vertically, they are referred to as **concentrated loads**, when applied to beams.

Example 5.12

A uniform horizontal beam is supported on a fulcrum (*Figure 5.31*). Calculate the force *F* necessary to ensure the beam remains in equilibrium.

We know that the sum of the CWM = the sum of the ACWM, therefore taking moments about the fulcrum we get

$$(F \times 1) + (50 \times 4) + (20 \times 2) = (200 \times 3) \text{ Nm}$$

then

$$(F \times 1) + 200 + 40 = 600 \text{ Nm}$$

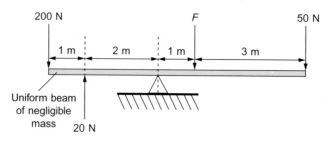

Figure 5.31 Uniform horizontal beam

UNIT 5

or

$$(F \times 1) = 600 - 200 - 40 \, \text{Nm}$$

so,

$$F = \frac{360 \, \text{Nm}}{1 \text{m}} = \mathbf{360 \, N}$$

Note (a) The 20 N force acting at a distance of 2 m from the fulcrum tends to turn the beam *clockwise* so is *added* to the sum of the CWM.

(b) The units of F are as required, that is they are in newton, because the RHS is in Nm and is divided by *1 m*.

(c) In this example, the weight of the beam has been ignored. If the beam is of uniform cross-section then its mass is deemed to act at its geometrical centre.

In the next example, we again consider not only concentrated loads acting on the beam but some that do not act at right angles (orthogonal) to the beam and so their vertical component needs to be found using the techniques you learnt earlier.

Example 5.13

For the beam system shown in *Figure 5.32*, ignoring the weight of the beam, determine the value of the load *F* to bring the beam into equilibrium.

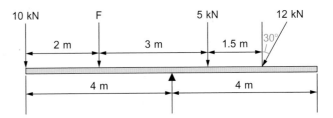

Figure 5.32 Loaded beam of uniform cross-section

We are told that the beam is of uniform cross-section and, therefore, its weight acts at its centre, where the fulcrum is, therefore the weight of the beam will produce no turning moment. We will take moments about the fulcrum in order to find the unknown force F, required for balance. However, before we do this we need to find the vertical component of the 12 kN load. This is $F_V = 12 \cos 30 = 10.4$ kN. Now, applying the principle of moments ΣCWM = ΣACWM and taking moments about the fulcrum we get

$$(5 \times 1) + (10.4 \times 2.5) = (10 \times 4) + (2 \times F)$$

or $5 + 26 = 40 + 2F$ and $F = -4.5$ kN. The minus sign is significant because it tells us that the load needs to be applied upwards, not downwards as shown in *Figure 5.32*. So, the balancing force F = **4.5 kN, acting upwards**.

In the next example, we provide an engineering application of beam theory and the resolution of forces when we consider a bell-crank lever as a beam.

Example 5.14

Figure 5.33 shows a motion control system crank lever ABC pivoted at B. AB is 20 cm and BC is 30 cm. Calculate the magnitude of the vertical rod force at C required to balance the horizontal control rod force of magnitude 10 kN applied at A.

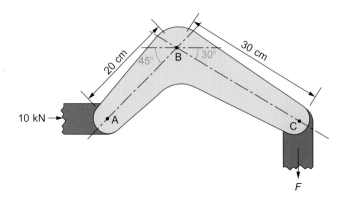

Figure 5.33 Bell-crank control lever

In order to achieve balance of the forces acting on the lever, the CWM about B must equal the ACWM about B. It can also be seen that the 10 kN force produces an ACWM about the fulcrum B. Therefore, the moment of 10 kN force about B = (10 × 0.2 sin 45°) kNm (note the manipulation of units) = (10)(0.2)(0.7071) kNm = 1.414 kNm.

If we now let the vertical force at C be of magnitude F, then F produces a CWM about fulcrum B. Therefore, the moment of force of magnitude F about B = F × (0.3 cos 30°) = 0.26F.

Then applying the principle of moments for equilibrium, we get 1.414 = 0.26F; therefore

$$F = \frac{1.414 \text{ kN}}{0.26 \text{ m}} = \textbf{5.44 kN}$$

Our final example on moments introduces the idea of the ***uniformly distributed load (UDL)***. In addition to being subject to point loads, beams can be subjected to loads that are distributed for all, or part, of the beam's length. For UDLs the whole mass of the load is assumed to act as a point load through the centre of the distribution.

Example 5.15

For the beam system shown in *Figure 5.34*, determine the reactions at the supports R_A and R_B, taking into consideration the weight of the beam.

So, from what has been said, the UDL acts as a point load of magnitude (1.5 kN × 5 = 7.5 kN) at the centre of the distribution, which is 5.5 m from R_A.

In problems involved with reactions, it is essential to eliminate one reaction from the calculations because only one equation is formed and only one unknown can be solved at any one time. This is achieved by taking moments about one of the reactions and then, since the distance from that reaction is zero, its moment is zero and it is eliminated.

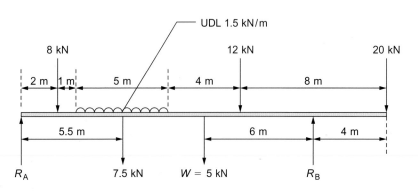

Figure 5.34 Beam system taking account of weight of beam

So, taking moments about R_A (thus eliminating R_A from the calculations), we get

$$(2 \times 8) + (5.5 \times 7.5) + (10 \times 5) + (12 \times 12) + (20 \times 20) = 16\,R_B$$

or $651.25 = 16\,R_B$ and the **reaction at B = 40.7 kN**

We could now take moments about B in order to find the reaction at A. However, at this stage, it is easier to use the fact that for static equilibrium:

upward forces $\quad$ = downward forces

so $\quad R_A + R_B \quad\quad\quad$ = $8 + 7.5 + 5 + 12 + 20$

$\quad\quad R_A + 40.7 \quad\quad$ = 52.5

and so the **reaction at A = 11.8 kN**

TYK 5.5

1. Ignoring the weight of the beam calculate the force *F* required to restore balance for the loaded uniform beam shown in *Figure 5.35*.

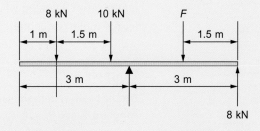

Figure 5.35 Loaded uniform beam

2. The cantilever beam (i.e, a beam rigidly supported at one end) shown in *Figure 5.36* is built into the wall and carries two loads as shown. Calculate the resulting moment of these forces about the point A in the beam at the face of the wall.

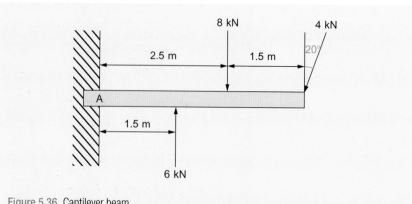

Figure 5.36 Cantilever beam

3. Taking into consideration the weight of the beam, determine the reactions at the supports of the loaded uniform cross-section beam shown in *Figure 5.37*.

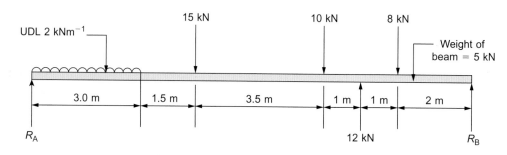

Figure 5.37 Loaded uniform cross-section beam

Loaded components

In this section, we consider engineering components and materials that are subject to loads. As you have already seen, when you were introduced to the concept of force, components that are subject to tensile loads in engineering structures are known as *ties*, while those subject to compressive loads are known as *struts*. In both cases the external loads cause internal extension (tensile loads) or compression (compressive loads) of the atomic bonds of the material from which the component is made. The intensity of the loads causing the atomic bonds to extend or compress is known as *stress* and any resulting deformation is known as *strain*. We will now look at the nature of this stress and strain in engineering components under load.

Stress

If a solid such as a metal bar is subjected to an external force (or load), a resisting force is set up within the bar and the material is said to be in a state of stress. There are three basic types of stress that are described next:

- *Tensile stress* – which is set up by forces tending to pull the material apart.
- *Compressive stress* – produced by forces tending to crush the material.
- *Shear stress* – produced by forces tending to cut through the material, that is tending to make one part of the material slide over the other.

Figure 5.38 illustrates these three types of stress.

KEY POINT

Structure components designed to carry tensile loads are known as *ties*, while components design to carry compressive loads are known as *struts*

UNIT 5

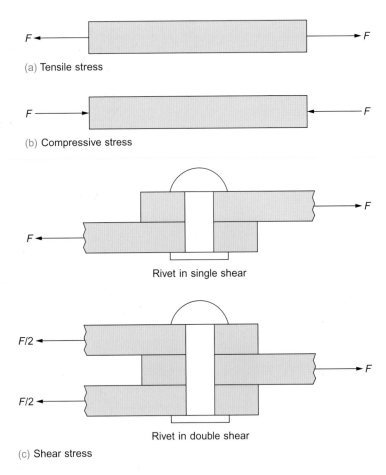

(a) Tensile stress

(b) Compressive stress

Rivet in single shear

Rivet in double shear

(c) Shear stress

Figure 5.38 Basic types of stress

KEY POINT

$1 \, \text{MNm}^{-2} = 1 \, \text{Nmm}^{-2}$

Stress is defined as force per unit area, that is, **stress** $\sigma = \dfrac{\text{force } F}{\text{area } A}$

The basic SI unit of stress is the Nm^{-2}. Other commonly used units include MNm^{-2}, Nmm^{-2} and the pascal (Pa).

Note: the Greek letter σ is pronounced sigma.

Strain

A material that is altered in shape due to the action of a force acting on it is said to be *strained*.

This may also mean that a body is strained internally even though there may be little measurable difference in its dimensions, just a stretching of the bonds at the atomic level. *Figure 5.39* illustrates three common types of strain resulting from the application of external forces (loads) that are described next.

- *Direct tensile strain* resulting from an axial tensile load being applied.
- *Compressive strain* resulting from an axial compressive load being applied.
- *Shear strain* resulting from equal and opposite cutting forces being applied.

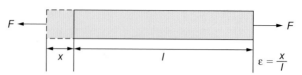

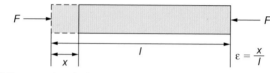

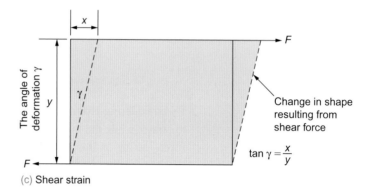

Figure 5.39 Types of strain

Direct strain may be defined as *the ratio of change in dimension (deformation) over the original dimension*, that is

$$\text{direct strain } \varepsilon = \frac{\text{deformation } x}{\text{original length } l} \text{ (both } x \text{ and } l \text{ are in metres)}.$$

The symbol ε is the Greek lower case letter **epsilon**. Note also that the deformation for tensile strain will be an extension and for compressive strain it will be a reduction.

Hooke's law and the elastic moduli

Hooke's law states that within the elastic limit of a material the change in shape is directly proportional to the applied force producing it.

A good example of the application of Hooke's law is the **spring**. A spring balance is used for measuring weight force, where an increase in weight will cause a corresponding extension (see *Figure 5.40*).

The **stiffness (k)** of a spring is the force required to cause a certain (unit deflection), that is

$$\text{stiffness } (k) = \frac{\text{force}}{\text{deflection}} \text{ SI units are } \mathbf{Nm^{-1}}$$

UNIT 5

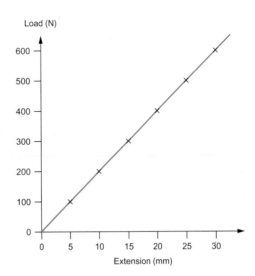

Figure 5.40 Load/extension graph for a spring

By considering Hooke's law, we can show that stress is directly proportional to strain, while the material remains *elastic*. That is, while the external forces acting on the material are only sufficient to stretch the atomic bonds without fracture, so that the material may return to its original shape after the external forces have been removed.

Hooke's law states that the extension is directly proportional to the applied load, providing the material remains within its elastic range. Since stress is load per unit area, then the load applied to any cross-sectional area of material can be converted into stress provided the cross-sectional area of the material is known. The corresponding extension of the stressed material divided by its original length will in fact provide a value of its strain. Therefore, we can quote Hooke's law in terms of stress and strain, rather than load and extension and say that ***stress is directly proportional to strain in the elastic range***, that is,

$$\text{Stress} \propto \text{strain} \quad \text{or} \quad \text{stress} = \text{strain} \times \text{a constant}$$

$$\text{so,} \quad \frac{\text{stress}}{\text{strain}} = \text{a constant}$$

This constant of proportionality will depend on the material and is given the *symbol E*.

Thus $\dfrac{\textbf{stress}}{\textbf{strain}} = \textbf{\textit{E}}$ **and \textit{E} is known as the modulus of elasticity**

KEY POINT

The elastic modulus of a material may be taken as a measure of the stiffness of that material

and because strain has no units the modulus of elasticity has the same units as stress. Modulus values tend to be very high, for this reason $\textbf{GNm}^{-2}$ or **GPa** are the preferred SI units.

The relationship between the shear stress (τ) and shear strain (γ) is known as the modulus of rigidity (G), that is

$$\textbf{Modulus of rigidity } G = \frac{\textbf{shear stress } \tau}{\textbf{shear strain } \gamma} \textbf{ (units GPa or GNm}^{-2}\textbf{)}$$

Note that the symbol τ is the lower case Greek letter ***tau*** and the symbol γ is the lower case Greek letter ***gamma***.

Example 5.16

A rectangular steel bar 10 mm × 16 mm × 200 mm long extends by 0.12 mm under a tensile force of 20 kN. Find:

(a) the stress
(b) the strain
(c) the elastic modulus of the bar material.

(a) Now the tensile stress $= \dfrac{\text{tensile force}}{\text{cross-sectional area}}$

Also tensile force $= 20\,\text{kN} = 20 \times 10^3\,\text{N}$ and cross-sectional area $= 10 \times 16 = 160\,\text{mm}^2$

Remember tensile loads act against the cross-sectional area of the material.

Then substituting in above formula we have,

$$\text{tensile stress } (\sigma) = \frac{20000\,\text{N}}{160\,\text{mm}^2} \quad \sigma = \mathbf{125\ Nmm^{-2}}$$

(b) Now, strain $\varepsilon = \dfrac{\text{deformation (extension)}}{\text{original length}}$

Also, extension $= 0.12\,\text{mm}$ and the original length $= 200\,\text{mm}$.

$$\text{Then substituting gives } \varepsilon = \frac{0.12\,\text{mm}}{200\,\text{mm}} = \mathbf{0.0006}$$

(c) $E = \dfrac{\text{stress}}{\text{strain}} = \dfrac{125\,\text{Nmm}^{-2}}{0.0006} = \mathbf{208{,}000\ Nmm^{-2}}$ or $\mathbf{208\ GNm^{-2}}$

Example 5.17

A 10-mm diameter rivet holds three sheets of metal together and is loaded as shown in *Figure 5.41*. Find the shear stress in the bar.

We know that the rivet is in double shear.

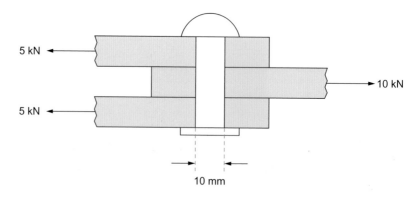

Figure 5.41 Rivet in double-shear

So, the area resisting shear $= 2 \times$ the cross-sectional area

$$= 2\pi r^2 = 2\pi 5^2 = \mathbf{157\ mm^2}$$

$$\text{Therefore, shear stress } \tau = \frac{10000}{157} = \mathbf{63.7\ N/mm^2} = \mathbf{63.7\ MNm^{-2}}$$

Note that when a rivet is in double shear, the area under shear is multiplied by 2. With respect to the load we know from Newton's laws that to every action there is an equal and opposite reaction, thus we only use the action or reaction of a force in our calculations, not both.

Factor of safety: engineering structures, components and materials need to be designed to cope with all normal working stresses. The *safety factor* is used in these materials to give a margin of safety and take account of a *certain factor of ignorance*. It may be defined as the ratio of the ultimate stress to that of the working stress, that is,

$$\text{factor of safety} = \frac{\text{ultimate stress}}{\text{working stress}}$$

Factors of safety vary in engineering design, dependent on the structural sensitivity of the member under consideration. They are often around 2.0, but can be considerably higher for joints, fittings, castings and primary load bearing static structures and somewhat lower for aircraft structures where weight saving is essential for efficient structural design.

Load–extension graphs

These show the results of mechanical tests used to determine certain properties of a material. For instance, as a check to see if heat treatment or processing has been successful, a sample from a batch would be used for such tests.

Load–extension graphs show certain phases when a material is tested to destruction. These include the elastic range, limit of proportionality, yield point, plastic stage and final fracture.

Figure 5.42 shows a typical load–extension curve for a specimen of mild steel, which is a ductile material.

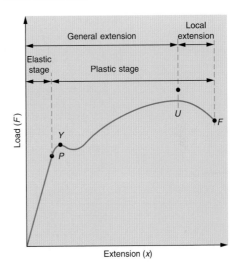

Figure 5.42 Load–extension curve for a mild steel test piece

The point P at the end of the straight line OP is called the *limit of proportionality*. Between the origin O and P the extension x is directly proportional to the applied force and in this range the material obeys

Hooke's law. The *elastic limit* is at or very near the limit of proportionality. When this limit has been passed the extension ceases to be proportional to the load, and at the *yield point Y* the extension suddenly increases and the material enters its plastic phase. At point *U*, the *ultimate tensile strength*, the load is greatest. The extension of the test piece has been general up to point *U* after which waisting or necking occurs and the subsequent extension is local (*Figure 5.43*).

Figure 5.43 Example of waisting, where extension is localized

Since the area at the waist is considerably reduced, then from *stress = force/area* the stress will increase resulting in a reduced load for a given stress and so fracture occurs at point *F* that is at a lower load value than at *U*.

Remember the elastic limit is at the end of the phase that obeys Hooke's law. After this Hooke's relationship is no longer valid, and full recovery of the material is not possible after removal of the load.

Figure 5.44 shows some typical load–extension curves for some common metals, where

HDB = hard drawn 70/30 brass
CI = cast iron
HDC = hard drawn copper
AA = aluminium alloy
AC = annealed copper.

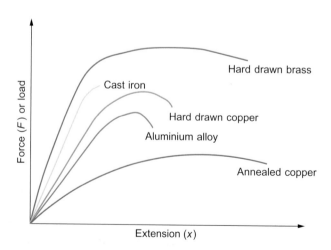

Figure 5.44 Some typical load–extension graphs

The above curves show that annealed copper is very ductile, while hard drawn copper is stronger but less ductile. Hard drawn 70/30 brass is both strong and ductile. Cast-iron can clearly be seen as brittle and it is for this reason that cast-iron is rarely used under tensile load. Aluminum alloy can be seen to be fairly strong yet ductile; it has excellent *structural efficiency* and it is for this reason that it is still used as one of the premier materials for aircraft construction.

TYK 5.6

1. Define: (a) tensile stress, (b) shear stress and (c) compressive stress.

2. State Hook's law and explain its relationship to the elastic modulus.

3. Define in detail the terms: (a) elastic modulus and (b) shear modulus.

4. Convert the following into N/m^2: (a) $240\,kN/m^2$, (b) $0.228\,GPa$, (c) $600\,N/mm^2$, (d) $0.0033\,N/mm^2$ and (e) $10\,kN/mm^2$.

5. What is the engineering purpose of the factor of safety?

6. If the ultimate tensile stress at which failure occurs in a material is $550\,MPa$ and a factor of safety of 4 is to apply, what will be the allowable working stress?

7. With respect to tensile testing and the resultant load–extension graph, define: (a) limit of proportionality (b) UTS, (c) yield point and (d) plastic range.

8. A metal tie rod $25\,mm$ in diameter and $200\,mm$ long extends $0.15\,mm$ under an axial load of $80\,kN$. Find: a) the value of E for the metal and b) the extension when the load is $100\,kN$.

Work, Power and Energy Transfer in Dynamic Systems

In this outcome we look first at the kinetic principles and parameters associated with linear motion. In particular we will study Newton's laws and the concepts of momentum and inertia. We then consider motion parameters (speed, velocity and acceleration) and the equations of linear motion and see how these are derived from Newton's laws. In the second part of this outcome we consider the dynamic principles and parameters associated with work, energy and power transfers in dynamic systems and machines. In the section on dynamic principles, we start by analysing the forces that act on dynamic systems and consider the use of free body diagrams in analysing such systems. We then study the conservation laws for momentum and energy as well as looking at the nature and use of *D'Alembert's principle*. The dynamic parameters we are to consider include mechanical work, kinetic and potential energy, instantaneous and average power dissipation and frictional resistance.

Newton's laws of motion

Kinetic parameters

KEY POINT

A body moving in a straight line with constant velocity is considered to be in equilibrium

Equilibrium: A body is said to be in equilibrium when its acceleration continues to be zero, that is, when it remains at rest or when it continues to move in a straight line with constant velocity.

Momentum: The quantity of motion of a body is the product of the mass of a body and its velocity (mv).

Any change in momentum requires a change in velocity, that is, acceleration. It may be said that for a fixed quantity of matter to be in equilibrium, it must have constant momentum. A more rigorous definition of momentum is discussed when we consider Newton's second law of motion.

All matter resists change. The inertia of a body depends on its mass; the greater the mass, the greater the inertia.

Inertia: The force resisting change in momentum (i.e., acceleration) is called inertia.

The inertia of a body is an innate force that only becomes effective when acceleration occurs.

An applied force acts against inertia so as to accelerate (or tend to accelerate) a body.

Before we consider each of Newton's laws we need to re-visit the concept of force. We already know that force cannot exist without opposition, that is, action and reaction. If we apply a 100 N pulling force to a rope, this force cannot exist without an opposing force being present within the material of the rope.

Force: Force is that which changes, or tends to change, the state of rest or uniform motion of a body.

Forces that act on a body may be external (applied from outside the body) such as weight, or internal (such as the internal resistance of a material subject to a compression) as illustrated in *Figure 5.45a*.

The difference between the forces tending to cause motion and those opposing motion is called the ***resultant*** or ***out-of-balance force***.

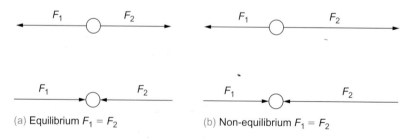

(a) Equilibrium $F_1 = F_2$ (b) Non-equilibrium $F_1 = F_2$

Figure 5.45 (a) Equilibrium forces and (b) non-equilibrium forces

A body that has no out-of-balance external force acting on it is in equilibrium and will not accelerate. A body that has such an out-of-balance force will accelerate at a rate dependent on the mass of the body and the

magnitude of the out-of-balance force. The necessary opposition that permits the existence of the out-of-balance force is provided by the force of inertia.

Newton's first law

Newton's first law of motion states that a body remains in a state of rest, or of uniform motion in a straight line, unless it is acted upon by some external resultant force.

Newton's second law

Newton's second law of motion states that the rate of change of momentum of a body is directly proportional to the force producing the change, and takes place in the direction in which the force acts.

KEY POINT

$F = ma$ is a consequence of Newton's second law of motion

We defined force earlier as force = mass × acceleration. We also know that *acceleration may be defined as change in velocity per unit time or rate of change in velocity*. If we assume that a body has an *initial velocity u* and a *final velocity v*, then the change in velocity is given by $(v - u)$ and so the rate of change of velocity or acceleration may be written as $(v - u)/t$, where t is unit time.

So since $F = ma$ then this may be written as $F = \dfrac{m(v - u)}{t}$

and multiplying out the brackets gives $F = \dfrac{mv - mu}{t}$

Now, we also know that momentum was defined earlier as *mass × velocity*. So, the product *mu* gives the initial momentum of the body prior to the application of the force, and *mv* gives the final momentum of the body. Thus the expression $(mv - mu)$ is the change in momentum and so $(mv - mu)/t$ is the rate of change of momentum and so Newton's second law may be expressed as

$$F = \frac{mv - mu}{t} \quad \text{or} \quad F = ma$$

Newton's third law

Newton's third law states that to every action there is an equal and opposite reaction.

So, for example, the compressive forces that result from the weight of a building, the *actions*, are held in equilibrium by the *reactions* that occur inside the materials of the building's foundation. Another example is that of propulsion. An aircraft jet engine produces a stream of high-velocity gases at its exhaust, the action, these gases act on the airframe of the aircraft causing a reaction, which enables the aircraft to accelerate and increase speed for flight.

TYK *5.7*

1. Define 'inertia' and quote its units.

2. Explain how the expression $F = ma$ is related to the rate of change of momentum with respect to Newton's second law.

3. Explain how a body that is moving can be in equilibrium.

4. Define momentum.

Linear equations of motion

Kinetic motion parameters

Linear **distance** with the symbol (s) in the SI system is measured in metres (m) and distance in a given direction is often referred to as **displacement**.

> **Speed** is *distance per unit time*. Speed takes no account of direction and is therefore a **scalar** quantity.

The standard derived SI unit of speed is the metre-per-second (ms^{-1}). Another commonly used unit of speed is the kilometre-per-hour (kph).

Example 5.18

Convert   (a) **150 kph into ms^{-1}**
             (b) **120 ms^{-1} into kph.**

We will derive the conversion factors for the above speeds by considering the **basic** units.

(a) We know that there are 10^3 metres in a kilometre; so, 150 kph = 150×10^3 metres per hour. Also, there are 3600 seconds in an hour. So, our required conversion factor is $10^3/3600 = 1/3.6$. So, 150 kph = $(150)(1/3.6) =$ **41.67 ms^{-1}**

Note that we did not change our conversion factor (1/3.6) into a decimal fraction, because this cannot be done exactly as $1/3.6 = 0.277777$ recurring.

(b) I hope you can see that all we need to do is to use the inverse of the conversion factor we have found above to convert ms^{-1} into kph; we are simply converting in reverse. Then the conversion factor becomes (3.6/1) = 3.6. So, 120 ms^{-1} = $(120)(3.6) =$ **432 kph**.

It will aid your understanding of unit conversion if you attempt to derive your own conversion factors from basic units.

KEY POINT

Speed is a scalar quantity and velocity is a vector quantity

> **Velocity** is defined as distance per unit time in a specified direction. Therefore, velocity is a **vector quantity**.

The SI units for the magnitude of velocity are the SI units for speed, that is, ms^{-1}.

The direction of a velocity is not always quoted but it should be understood that the velocity is in some defined direction, even though this direction may not be stated.

UNIT 5

Acceleration is defined as change in velocity per unit time or rate of change of velocity.

Acceleration is also a **vector quantity** and the SI unit of acceleration is ms^{-1}/s or ms^{-2}.

TYK

TYK 5.8

1. Convert the following units:
   (a) 600 km/h into ms^{-1}
   (b) 140 ms^{-1} into kph.

2. What is another name for distance travelled in a given time in a given direction?

3. State which of the following are vector quantities: speed, velocity, acceleration and time.

4. State the SI units of time, acceleration due to gravity and displacement.

Deriving the linear equations of motion graphically

You have already been introduced to the concept of a force, velocity, acceleration and Newton's laws. These concepts are needed in order to derive the linear equations of motion. Look back now and remind yourself of the relationship between mass, force, acceleration and Newton's laws.

The linear equations of motion rely for their derivation on one very important fact that is the **acceleration is assumed to be constant**. We will now consider the derivation of the four standard equations of motion using a graphical method.

Even simple linear motion, motion along a straight line, can be difficult to deal with mathematically. However, in the case where acceleration is constant, it is possible to solve problems of motion by use of a **velocity–time graph** without recourse to the calculus. The equations of motion use standard symbols to represent the variables, these are shown below:

$$s = \text{distance} \quad (m)$$
$$u = \text{initial velocity} \quad (ms^{-1})$$
$$v = \text{final velocity} \quad (ms^{-1})$$
$$a = \text{acceleration} \quad (ms^{-2})$$
$$t = \text{time} \quad (s)$$

The *velocity is plotted on the vertical axis and time on the horizontal axis*. Constant velocity is represented by a *horizontal straight line* and acceleration by a *sloping straight line*. **Deceleration** or **retardation** is also represented by a sloping straight line but with a *negative* slope.

KEY POINT

For the linear equations of motion to be valid, the acceleration is assumed to be constant

KEY POINT

Velocity is speed in a given direction and is a vector quantity

UNIT 5

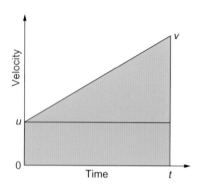

Figure 5.46 Velocity–time graph for uniform acceleration

By considering the velocity–time graph shown in *Figure 5.46*, we can establish the equation for distance.

The distance travelled in a given time is equal to the velocity ms^{-1} multiplied by the time s. This is found from the graph by the *area under the sloping line*. In *Figure 5.46*, a body is accelerating from a velocity u to a velocity v in time t seconds.

Now the distance travelled s = area under graph and

$$s = ut + \frac{(v-u)}{2} \times t$$

then

$$s = ut + \frac{(v-u)}{2} \times t \text{ and } s = ut + \frac{vt}{2} - \frac{ut}{2} \text{ or } s = \frac{(2u+v-u)t}{2}$$

giving $s = \dfrac{(u+v)t}{2}$.

In a similar manner to the above, one of the velocity equations can also be obtained from the velocity–time graph. Since the acceleration is the rate of change of velocity with respect to time, the value of the acceleration will be equal to the gradient of a velocity–time graph. Therefore, from *Figure 5.46*, we have

$$\text{gradient} = \frac{\text{acceleration}}{\text{time taken}} = \text{velocity}$$

therefore, acceleration is given by

$$a = \frac{(v-u)}{t}$$

and on transposing this formula in terms of the final velocity we get

$$v = u + at$$

The remaining equations of motion may be derived from the two equations found above. This is shown in Example 5.19.

Example 5.19

Use the equations $s = \dfrac{(u+v)t}{2}$ and $v = u + at$ to derive the equations:

(a) $s = ut + \dfrac{1}{2}at^2$ and (b) $v^2 = u^2 + 2as$.

(a) Let $s = \dfrac{(u+v)t}{2}$ (1)

$v = u + at$ (2)

Then substituting equation (2) into equation (1) gives $s = \dfrac{(u+u+at)t}{2}$ and on

simplification and multiplying out the bracket we get $s = \dfrac{2ut + at^2}{2}$ or $s = ut + \dfrac{1}{2}at^2$ as required

(b) Let $v = u + at$ (1)

and transposing $s = \dfrac{(u + v)t}{2}$ for t, let

$$t = \frac{2s}{u + v} \qquad (2)$$

Then substituting equation (2) into equation (1) gives

$$v = u + \frac{2as}{u + v}$$

and on simplification we get $v(u + v) = u(u + v) + 2as$. Then multiplying out the brackets gives $uv + v^2 = u^2 + uv + 2as$ and on final simplification we get $\boldsymbol{v^2 = u^2 + 2as}$ as required.

Example 5.20

A body starts from rest and accelerates with constant acceleration of $2.0\,\text{ms}^{-2}$ up to a speed of $9\,\text{ms}^{-1}$. It then travels at $9\,\text{ms}^{-1}$ for 15 seconds after which time it is retarded to a speed of $1\,\text{ms}^{-1}$. If the complete motion takes 24.5 seconds, find:

(a) the time taken to reach $9\,\text{ms}^{-1}$
(b) the retardation
(c) the total distance travelled.

The solution is made easier if we sketch a graph of the motion, as shown below.

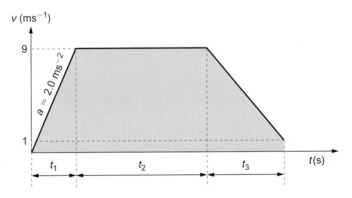

Figure 5.47 Velocity–time graph of the motion

(a) We first tabulate the known values: $u = 0\,\text{ms}^{-1}$ (we start from rest)

$$v = 9\,\text{ms}^{-1}$$
$$a = 2.0\,\text{ms}^{-2}$$
$$t = ?$$

All we need to do now is to select an equation which contains all the variables listed above, that is

$$v = u = at$$

and on transposing for t and substituting the variables, we get $t = \dfrac{9 - 0}{2}$

$$\boldsymbol{t = 4.5\,s}$$

(b) The retardation is found in a similar manner.

$$u = 9\,\text{ms}^{-1}, v = 2\,\text{ms}^{-1}, t = (24.5 - 4.5 - 15) = 5s, a = ?$$

We again select an equation that contains the variables, that is, $v = u + at$ and on transposing for a and substituting the variables we get $a = (1 - 9)/5$ so $a = -1.6\,\text{ms}^{-2}$ (the $-$ve sign indicates a retardation).

(c) The total distance travelled requires us to sum the component distances travelled for the times t_1, t_2 and t_3. Again we tabulate the variable for each stage:

$u_1 = 0\,\text{ms}^{-1}$ $u_2 = 9\,\text{ms}^{-1}$ $u_3 = 9\,\text{ms}^{-1}$

$v_1 = 9\,\text{ms}^{-1}$ $v_2 = 9\,\text{ms}^{-1}$ $v_3 = 1\,\text{ms}^{-1}$

$t_1 = 4.5\,s$ $t_2 = 15\,s$ $t_3 = 5\,s$

$s_1 = ?$ $s_2 = ?$ $s_3 = ?$

The appropriate equation is

$$s = \frac{(u + v)t}{2}$$

and in each case we get

$$s_1 = \frac{(0 + 9)4.5}{2} \quad s_2 = \frac{(9 + 9)15}{2} \quad s_3 = \frac{(9 + 1)5}{2}$$
$$s_1 = 20.25 \qquad s_2 = 135 \qquad s_3 = 25$$

Then total distance $S_T = 20.5 + 135 + 25 = \mathbf{180.25\,m}$.

TYK 5.9

With reference to the velocity–time graphs shown in *Figure 5.48*, answer questions 1 to 8.

1. The slope of the velocity–time graph measures _____.

2. The area under a velocity–time graph determines _____.

3. Average velocity may be determined by dividing the _____ _____ _____ by _____ _____.

4. Graph (a) is a graph of constant velocity, therefore, acceleration is _____ and the distance travelled is equal to _____.

5. Graph (b) shows uniformly accelerated motion, therefore, the distance travelled is equal to _____.

6. Graph (c) shows _____ _____ _____.

7. Graph (d) represents uniformly accelerated motion having initial velocity u, final velocity v and acceleration (a) so distance travelled is equal to _____.

8. Graph (e) represents _____ acceleration.

9. A vehicle accelerates at a constant rate from rest to $20\,\text{ms}^{-1}$ in a time of 4 seconds. Draw to scale a velocity–time graph and determine the acceleration and distance travelled.

Test your knowledge

TYK

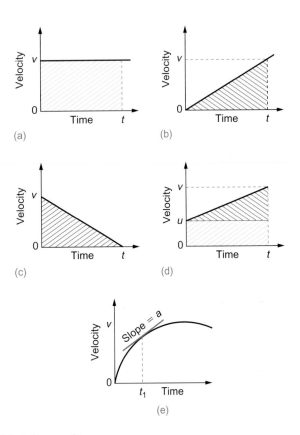

Figure 5.48 Velocity/time graphs

Using Newton's laws and the equations of motion

You saw earlier that **Newton's second law** can be defined as:

$$F = ma \ \text{ or } \ F = \frac{mv - mu}{t}$$

In words, we may say that force is equal to the rate of change of momentum of a body. Look back again and make sure you understand the relationship between force, mass and the momentum of a body. Remember also that the inertia force is such as to be equal and opposite to the accelerating force that produced it. We use these laws in the following example.

Example 5.21

A racing car of mass 1965 kg accelerates from 160 kph to 240 kph in 3.5 seconds. Find:

(a) the average acceleration

(b) the force required to produce the acceleration

(c) the inertia force on the car.

(a) We first need to convert the velocities to standard units:

$$u = 160\,\text{kph} = \frac{160 \times 1000}{60 \times 60} = 44.4\,\text{ms}^{-1}$$

$$v = 240\,\text{kph} = \frac{240 \times 1000}{60 \times 60} = 66.6\,\text{ms}^{-1}$$

Also $t = 3.5\,s$, and we are required to find the acceleration a.

Then using the equation $v = u + at$ and transposing for a, we get

$$a = \frac{v - u}{t}$$

and substituting values $a = \dfrac{66.6 - 44.4}{3.5}$ and so $\boldsymbol{a = 6.34\,ms^{-2}}$

(b) The accelerating force is readily found using Newton's second law, where

$$F = ma = 1965\,kg \times 6.34\,ms^{-2}$$

$$\text{Force} = \boldsymbol{12.46\,kN}$$

(c) From what has already been said you will be aware that the inertia force is equal and opposite to the accelerating force; therefore, the **inertia force $= -12.46\,kN$**. The minus sign for the inertia force indicates that it is acting in the opposite direction to the accelerating force.

Suppose in Example 5.21 we are told that the road resistance is 200 N per tonne. Then the total propulsive force required to cause the acceleration must be such that it overcomes both the inertia force and the resisting force of the road. This situation is illustrated in *Figure 5.49*.

From *Figure 5.49* the propulsive force $=$ inertia force $+$ road resistance.

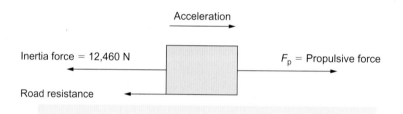

Figure 5.49 Acceleration of vehicle against inertia and road resistance

The road resistance $= 200\,N$ per tonne, so road resistance for this example is

$$\frac{200 \times 1965}{1000} = 393\,N.$$

So that the total propulsive force needed to cause the same acceleration is

$$F_p = 12460 + 393 = \boldsymbol{12.853\,kN.}$$

TYK 5.10

1. A thin cord can withstand a pulling force of 120 N. Calculate the maximum acceleration that can be given to a mass of 32 kg assuming that the force to accelerate the mass is transmitted through the cord.

2. A Cessna 172 aircraft and a Boeing 747 aircraft are each given an acceleration of $5\,\mathrm{ms^{-2}}$. To achieve this the thrust force produced by the Cessna's engines is $15\,\mathrm{kN}$ and the thrust force required by the Boeing 747 is $800\,\mathrm{kN}$. Find the mass of each aircraft.

3. A car starts from rest and accelerates uniformly at $2\,\mathrm{ms^{-2}}$ for 10 seconds. Calculate the distance travelled.

4. How long will it take a heavy vehicle to accelerate from 10 to $15\,\mathrm{ms^{-1}}$, if it covers a distance of $80\,\mathrm{m}$ in this time?

5. A train of mass 32 tonne accelerates uniformly from rest at $4\,\mathrm{ms^{-2}}$. If the average resistance to motion is $1\,\mathrm{kN}$ per $1000\,\mathrm{kg}$, determine: (a) the force required to produce the acceleration and (b) the inertia force on the train.

Dynamic parameters and principles

In this section we first look at energy, work and power, in particular mechanical work and energy. We then consider again momentum and see how the principle of the conservation of momentum can be used to solve problems involving elastic collisions. We then look at friction and frictional resistance and finally we look at D'Alembert's principle and its use in solving dynamic problems.

Introduction to mechanical work and energy

Energy and work are very closely related, where energy is considered to be the capacity to do work. Before we define mechanical work and energy, we should be aware of some of the other forms of energy that exist. Thus energy may exist in different forms, such as mechanical, electrical, nuclear, chemical, heat, light and sound. Energy cannot be created or destroyed, only changed from one form to another; this fact leads to the well-known principle of the conservation of energy.

> **Principle of the conservation of energy** states that energy may neither be created nor destroyed only changed from one form to another.

There are many engineering examples of devices that transform energy, these include:

- the loudspeaker, which transforms electrical to sound energy
- the petrol engine, which transforms heat to mechanical energy
- the microphone, which transforms sound to electrical energy
- the dynamo transforms mechanical to electrical energy
- the battery transforms chemical to electrical energy
- the filament bulb transforms electrical to light energy.

In our study of energy we start by looking at the various forms of mechanical energy and its conservation. Provided no mechanical energy is transferred to or from a body, the total amount of mechanical energy possessed by a body remains constant unless mechanical work is done. This concept is looked at next.

Mechanical work done

The energy possessed by a body is its capacity to do work. Mechanical work is done when a force overcomes a resistance and it moves through a distance.

> **Mechanical work done (J)** may be defined as force required to overcome the resistance (N) × distance moved against the resistance (s), in symbols $W = Fs$.

The SI unit of work is the newton-metre (Nm) or joule, where 1 Joule = 1 Nm.

Note:

(a) No work is done unless there is both resistance and movement.
(b) The resistance and the force needed to overcome it are equal.
(c) The distance moved must be measured in exactly the opposite direction to that of the resistance being overcome.

The more common resistances to be overcome include: friction, gravity (the weight of the body itself) and inertia (the resistance to the acceleration of the body), where

the work done (WD) against friction = friction force × distance moved

WD against gravity = weight × gain in height

WD against inertia = inertia force × distance moved.

Note: The inertia force is equal and opposite to the out-of-balance force causing the acceleration, that is

$$inertia\ force = -(mass \times acceleration)$$

In any problem involving calculation of work done, the first task should be to identify the type of resistance to overcome. If, and only if, there is motion between surfaces in contact, is work done against friction. Similarly, only where there is a gain in height is there work done against gravity and only if a body is accelerated is work done against inertia (look back at our definition of inertia).

Example 5.22

A body of mass 30 kg is raised from the ground at constant velocity through a vertical distance of 15 m. Calculate the work done (WD).

If we ignore air resistance, then the only WD is against gravity.

WD against gravity = weight × gain in height or WD = mgh (and assuming $g = 9.81\ ms^{-2}$)

then   WD = (30)(9.81)(15)

WD = **4414.5 J**   or   **4.415 kJ**

Work done may be represented graphically and, for linear motion, this is shown in *Figure 5.50a*, where the force needed to overcome the resistance is plotted against the distance moved. The *WD* is then given by the area under the graph.

Figure 5.50b shows the situation for a varying ***torque T*** in ***Nm*** when plotted against the angle turned through in ***radians***. Again for a torque, the area under the graph where the units are Nm × radian gives the *WD*. Then noting that the radian has no dimensions, the unit for work done remains as ***Nm or joules***.

Mechanical energy

Mechanical energy may be defined as the capacity to do work and is often subdivided into three different forms of energy. ***Potential energy, strain energy*** and ***kinetic energy***. In this book we confine our study to potential energy and kinetic energy.

> **Mechanical energy** may be defined as the capacity to do mechanical work.

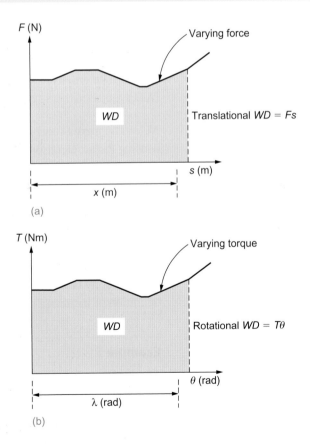

Figure 5.50 Work done

> **Potential energy (PE)** is energy possessed by a body by virtue of its position, relative to some datum. The change in PE is equal to its weight multiplied by the change in height.

KEY POINT

Potential energy is energy possessed by a body due to position, while kinetic energy is energy possessed by a body due to its motion

Since the weight of a body $= mg$, then the change in PE may be written as:

$$\textbf{change in PE} = \textit{mgh}$$

which of course is identical to the work done in overcoming gravity. So, the work done in raising a mass to a height is equal to the PE it possesses at that height, assuming no external losses.

Kinetic energy (KE) is energy possessed by a body by virtue of its motion.

Translational KE, that is, the KE of a body travelling in a linear direction (straight line) is

$$\text{Translational kinetic energy (J)} = \frac{\text{mass (kg)} \times (\text{velocity})^2 (\text{ms}^{-1})}{2}$$

$$\textbf{Translation KE} = \frac{1}{2}mv^2$$

Flywheels are heavy wheel-shaped masses fitted to shafts in order to minimize sudden variations in the rotational speed of the shaft, due to sudden changes in load. A flywheel is, therefore, a store of rotational KE. We **will not** be studying rotational kinetic energy in this unit but for completeness we define it here as

$$\textbf{Rotational KE of mass} = \frac{1}{2}I\omega^2$$

where **I = mass moment of inertia**.

Note: The moment of inertia of a rotating mass *I* can be defined in general terms by the expression **I = Mk²**, where **M** = the **total mass** of the rotating body and **k** = the **radius of gyration**, that is, the radius from the centre of rotation where all of the mass is deemed to act.

Example 5.23

Determine the translational kinetic energy of a car that has a mass of 800 kg and is travelling at 50 kph.

Then the linear KE $= \frac{1}{2}mv^2$, where $v = 50\,\text{kph} = (50 \times 1000/3600) = 13.89\,\text{m/s}$

$$= \left(\tfrac{1}{2}\right)(800)(13.89)^2 = \textbf{\textit{77.16 kJ}}$$

Conservation of mechanical energy

From the definition of the conservation of energy we can deduce that the total amount of energy within certain defined boundaries will remain the same. When dealing with mechanical systems, the potential energy

possessed by a body is frequently converted into kinetic energy and vice versa. If we ignore air frictional losses, then

Potential energy + kinetic energy = a constant

Thus, if a mass m falls freely from a height h above some datum, then at any height above that datum:

Total energy = potential energy + kinetic energy

This important relationship is illustrated in *Figure 5.51*, where at the highest level above the datum the potential energy is a maximum and is gradually converted into kinetic energy. As the mass falls towards the datum, immediately before impact when height $h = 0$, the potential energy is zero and the kinetic energy is equal to the initial potential energy.

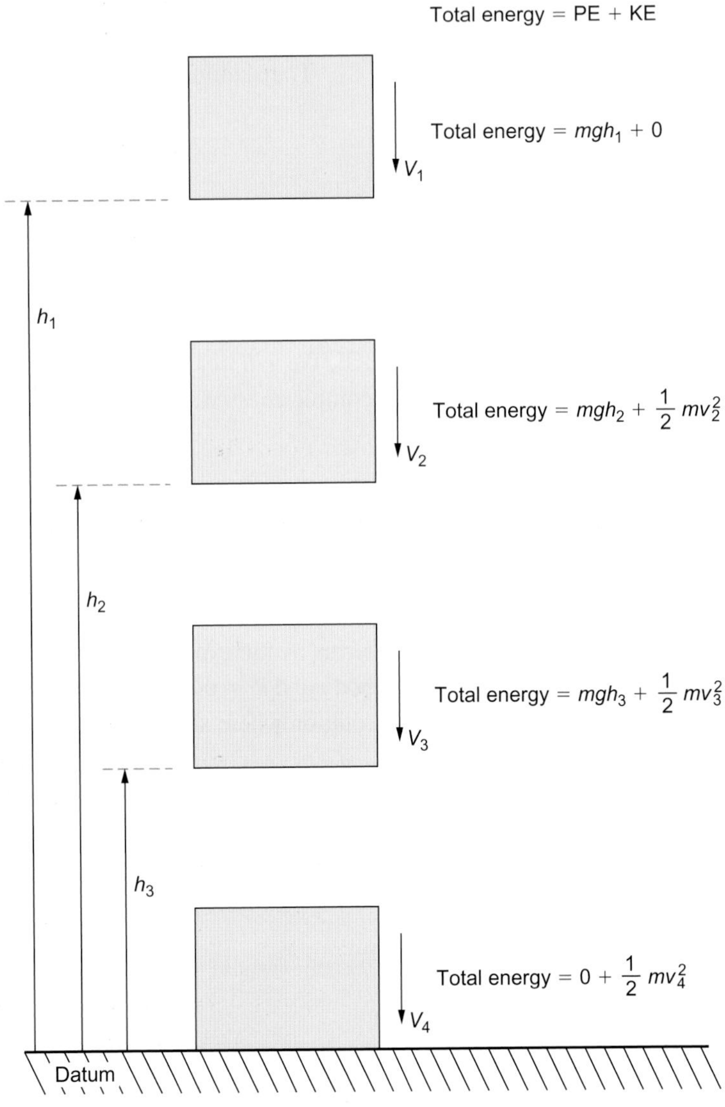

Total energy = PE + KE

Total energy = $mgh_1 + 0$ V_1

Total energy = $mgh_2 + \frac{1}{2}mv_2^2$ V_2

Total energy = $mgh_3 + \frac{1}{2}mv_3^2$ V_3

Total energy = $0 + \frac{1}{2}mv_4^2$ V_4

h_1, h_2, h_3, Datum

Figure 5.51 PE + KE equals a constant

Since the total energy is constant, then

$$mgh_1 = mgh_2 + \frac{1}{2}mv_2^2 = mgh_3 + \frac{1}{2}mv_3^2 = \frac{1}{2}mv_4^2$$

Immediately after impact with the datum surface, the mechanical kinetic energy is converted into other forms, such as heat, strain and sound.

If friction is present then work is done overcoming the resistance due to friction and this is dissipated as heat. Then

> ***Initial energy = final energy + work done in overcoming frictional resistance***

Note: Kinetic energy is not always conserved in collisions. Where kinetic energy is conserved in a collision we refer to the collision as ***elastic***, when kinetic energy is not conserved we refer to the collision as ***inelastic***.

Example 5.24

Cargo weighing 2500 kg breaks free from the top of the cargo ramp (*Figure 5.52*). Ignoring friction, determine the velocity of the cargo the instant it reaches the bottom of the ramp.

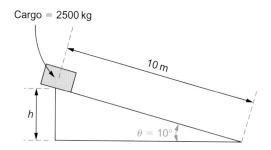

Figure 5.52 Cargo ramp

The vertical height *h* is found using the sine ratio, that is

$$10 \sin 10 = h \quad \text{so } h = 1.736\,\text{m}$$

and increase in potential energy $= mgh = (2500)(9.81)(1.736)\,\text{J} = 42,587.2\,\text{J}$

Now, using the relationship, PE + KE = total energy. Then immediately prior to the cargo breaking away KE = 0 and so PE = total energy also, immediately prior to the cargo striking the base of slope PE = 0 and KE = total energy (all other energy losses being ignored).

So, at base of slope: $42,587.2\,\text{J} = \text{KE}$

and $42,587.2 = \tfrac{1}{2}mv^2$

so $\dfrac{(2)(42,587.2)}{2500} = v^2$

and so velocity at bottom of ramp $= \boldsymbol{5.84\,ms^{-1}}$ (check this working for yourself!)

Power

> **Power** is a measure of the rate at which work is done or the rate of change of energy. *Power is therefore defined as the rate of doing work.*

KEY POINT

Power is the rate of doing work

The SI unit of power is the watt (W), that is

$$\textbf{Average power (W)} = \frac{\textbf{Work done (J)}}{\textbf{Time taken (s)}} = \frac{\textbf{Energy change (J)}}{\textbf{Time taken (s)}}$$

or, if the body moves with constant velocity,

$$\textbf{Instantaneous power (W)} = \textbf{Force used (N)} \times \textbf{Velocity (ms}^{-1}\textbf{)}$$

Note units are Nm/s = J/s = **Watt (W)**.

Example 5.25

A packing crate weighing 1000 N is loaded onto the back of a lorry by being dragged up an incline of 1 in 5 at a steady speed of 2 ms^{-1}. The frictional resistance to motion is 240 N. Calculate:

(a) the power needed to overcome friction
(b) the power needed to overcome gravity
(c) the total power needed.

(a) Power = friction force × velocity along surface
     = 240 × 2
     = **480 W**

(b) Power = weight × vertical component of velocity
     = 1000 × 2 × 1/5
     = **400 W**

(c) Since there is no acceleration and, therefore, no work done against inertia

     Total power = power for friction + power for gravity
     = 480 + 400
     = **880 W**

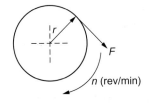

Figure 5.53 Power transmitted by a torque

Let us now consider power transmitted by a torque. You have already met the concept of torque, as being equivalent to the force multiplied by the radius of a rotating body. *Figure 5.53* shows a force F (N) applied at radius r (m) from the centre of a shaft that rotates at n rev/min.

Since the work done is equal to the force multiplied by the distance, then the WD in one revolution is given by

$$\text{WD in one revolution} = F \times 2\pi r \text{ J}$$

but force × radius (**Fr**) is the torque **T** applied to the shaft; therefore, the work done in one revolution is given by

$$\text{WD in one revolution} = 2\pi T \text{ J}$$

In 1 minute the work done = work done per revolution multiplied by the number of rev/min (n)

$$= 2\pi n T$$

and WD in 1 second = $2\pi nT/60$ and since WD per second is equal to power ($1\,\text{Js}^{-1} = 1\,\text{W}$)

then **power (W) transmitted by a torque** = $\mathbf{2\pi nT/60}$

TYK

TYK 5.11

1. Define work done.
2. Write down the equation for work done against gravity, stating SI units.
3. State the principle of the conservation of energy.
4. Detail the forms of energy input and output for the following devices: (a) generator, (b) a reciprocating piston engine, (c) battery and (d) loudspeaker.
5. Write down the formulae for translational kinetic energy and rotational kinetic energy and explain the meaning of each of the symbols within these formulae.
6. Define instantaneous power and average power.
7. Machine A delivers 45,000 J of energy in 30 seconds, machine B produces 48 kN m of work in 31 seconds, which machine is more powerful and why?
8. Find the kinetic energy of a mass of 2000 kg moving with a velocity of 40 km/h.
9. A crane raises a load of 1640 N to a height of 10 m in 8 seconds. Calculate the average power developed.
10. A motor vehicle starting from rest free wheels down a slope whose gradient is 1 in 8. Neglecting all resistances to motion, find its velocity after travelling a distance of 200 m down the slope.

Conservation of linear momentum

Momentum was defined earlier, when you studied Newton's laws as mass multiplied by velocity, with units kg m/s or newton second (Ns). Now, you will remember that Newton's first law states that 'a body remains in a state of rest or of uniform motion in a straight line, unless acted upon by an external force'. Therefore, from this law, the momentum of a body will remain constant unless there is a change in velocity resulting from the action of an external force. This is also true when two or more bodies forming a system collide. When no external forces act on a system it is said to be closed and the total momentum in a given direction prior to a collision is equal to the total momentum in that direction after the collision.

> **The principle of the conservation of momentum** may be stated as the total linear momentum of a closed system in any given direction is a constant or momentum before impact = momentum after impact.

Thus, for the following system (*Figure 5.54*) we have two masses (M_1, M_2), travelling with two initial velocities (u_1, u_2) prior to impact and two final velocities (v_1, v_2) after impact.

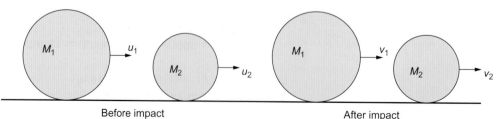

Before impact   After impact

Figure 5.54 Conservation of momentum before and after impact

Then from the conservation of momentum law:

momentum before impact = momentum after impact

that is,

$$M_1 u_1 + M_2 u_2 = M_1 v_1 + M_2 v_2$$

Note: Bodies moving towards the right are considered to have a positive momentum and moving towards the left negative.

Example 5.26

A body of mass 20 tonne is moving with a velocity of $10\,\text{ms}^{-1}$ to the right and collides with a body of 5 tonne moving at $12\,\text{ms}^{-1}$ in the opposite direction. If the bodies move off with a common velocity after impact, calculate this velocity.

Then remembering that there are 1000 kg in a tonne and the total momentum before impact = the total momentum after impact and letting the velocity after impact = **v**, we have

$$(20 \times 1000 \times 10) + (5 \times 1000 \times -12) = (25 \times 1000)v$$

so

$$200{,}000 - 60{,}000 = 25{,}000v$$

and the common velocity after impact

$$v = \frac{200{,}000 - 60{,}000}{25{,}000} = \textbf{5.6}\,\textbf{ms}^{-1}$$

KEY POINT

In elastic collisions both energy and momentum are conserved

Impulse

When a force is applied to a body suddenly as in a collision the change in time is very small and difficult to measure. In such cases the effect of the force is measured by *the change in momentum* it produces. Remembering Newton's second law where 'the rate of change of momentum is directly proportional to the applied force producing the change', the forces that act for very short periods of time are known as *impulse forces*.

Impulse of a force is a product of the impulsive force and the time during which it acts and is equal to the change in momentum produced by the impulsive force.

Impulse = applied force × time = change in linear momentum

or **Impulse** = $Ft = m(v - u)$. The units are the same as those for momentum, that is $\text{kg}\,\text{ms}^{-1}$ or Ns.

In practice, when impulsive forces occur, some energy is lost through change into heat and sound and momentum is not entirely conserved. This is

accounted for by *the coefficient of restitution (e)* where

$$e = -\frac{\text{relative velocity after impact}}{\text{relative velocity before impact}}$$

The minus sign takes account of the fact that the relative velocities before and after impact are of the opposite sense. The value of *e* is then positive and for perfectly elastic collisions $e = 1$ and for completely inelastic collision (where the two bodies stick together) it takes the value zero. You will not be required to use this coefficient in this unit but you should be able to define *e* and know its purpose.

Example 5.27

The hammer of a pile driver of mass 1500 kg falls a distance of 2 m onto a pile. The impact takes place in 0.02 seconds and the hammer does not rebound. Determine the average applied force exerted on the pile by the hammer.

In order to solve this problem, we first need to determine the velocity of impact using the equations of motion, then we can use the relationship that the impulse equals the change in linear momentum, which can be found because we will know the mass, the initial velocity and final velocity.

Using $v^2 = u^2 + 2as$ then $v^2 = 0^2 + 2(9.81)(2) = 39.24$ and $v = 6.26 \, \text{ms}^{-1}$

Now, since the impact is inelastic (no rebound) then the momentum lost by the hammer = change in momentum = $m(v - u) = (1500)(6.26) = 9.39 \, \text{kNs}$.

Then since impulse $I = Ft$ the average impulsive force = $\dfrac{9.39}{0.02} = \boldsymbol{469.2 \, kN}$.

TYK 5.12

1. Define 'impulse'.

2. Define 'inelastic collisions' and explain how compensation is made for such collisions.

3. A bullet of mass 10 g is fired into a block of wood of mass 390 g, lying on a smooth frictionless surface. The block of wood and the bullet move off together with a velocity of 10 ms^{-1}. What was the velocity of the bullet?

4. The hammer of a pile driver of mass 2 tonne falls a distance of 1.5 m onto a pile. The impact takes place in 0.025 seconds and the hammer does not rebound. Determine the average applied force exerted on the pile by the hammer.

Test your knowledge

TYK

Friction

We have already met *friction*, in terms of the *frictional force that tends to oppose relative motion*, but up till now we have not fully defined the nature of friction.

When a surface is moved over another surface with which it is in contact, a resistance is set up opposing this motion. The value of the resistance will depend on the materials involved, the condition of the two surfaces, and

UNIT 5

KEY POINT

Static friction is always less than sliding or dynamic friction

the force holding the surfaces in contact, but the opposition to motion will always be present. This resistance to movement is said to be the result of *friction* between the surfaces.

We require a slightly greater force to start moving the surfaces (*static friction*) than we do to keep them moving (*sliding friction*). As a result of numerous experiments involving different surfaces in contact under different forces, a set of rules or laws has been established which, for all general purposes, materials in contact under the action of forces seem to obey. These rules are detailed below, together with one or two limitations for their use.

Laws of friction

1. The frictional force always opposes the direction of motion or the direction in which a body is tending to move.

2. The sliding friction resisting force F_R opposing motion, once motion has started, is proportional to the normal force N that is pressing the two surfaces together, that is, $F_R \propto N$.

3. The sliding frictional resisting force is independent of the area of the surfaces in contact. Thus, two pairs of surfaces in contact made of the same materials and in the same condition, with the same forces between them, but having different areas, will experience the same frictional forces opposing motion.

4. The frictional resistance is independent of the relative speed of the surfaces. This is not true for very low speeds or in some cases for fairly high speeds.

5. The frictional resistance at the start of sliding, static friction, is slightly greater than that encountered as motion continues, sliding friction.

KEY POINT

Friction always opposes the motion that produces it

6. The frictional resistance is dependent on the nature of the surfaces in contact, for example, the type of material, surface geometry, surface chemistry, etc.

Solving problems involving friction

From the above laws, we have established that the sliding frictional force (resistance) F_R is proportional to the normal force N pressing the two surfaces together, that is $F_R \propto N$. You will remember from your mathematical study of proportion that in order to equate these forces we need to insert a constant, the constant of proportionality, that is $F_R = \mu N$. This constant μ is known as *the coefficient of friction* and in theory it

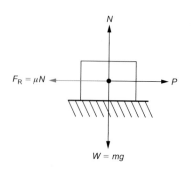

Figure 5.55 Forces on two horizontal surfaces in contact

has a maximum value of 1. *Figure 5.55* shows the space diagram for the arrangement of forces on two horizontal surfaces in contact.

You may find the solution of problems involving friction rather difficult. This is because it is often difficult to visualize the nature and direction of all the forces that act on two bodies in contact, as well as resolving these forces into their component parts. Problems involving friction may be solved by calculation or by drawing. The following generalized example involving the simple case of a block in contact with a horizontal surface should help you understand both methods of solution.

Example 5.28

(a) Solution by calculation

Consider again the arrangement of forces shown in *Figure 5.55*. If the block is in equilibrium, that is, just on the point of moving, or moving with constant velocity then we can equate the horizontal and vertical forces as follows:

resolving horizontally gives	$P = F_R$	(1)
resolving vertically	$N = mg$	(2)
but from the laws of dry friction	$F_R = \mu N$	(3)
substituting (2) into (3) gives	$F_R = \mu mg$	(4)
substituting (4) into (1) gives	$P = \mu mg$	(5)

(b) Solution by vector drawing

You know from your previous work on resolution of coplanar forces that a single resultant force in a vector diagram can replace two forces. The space diagram for our horizontal block is shown, where F_R and N can be replaced by a resultant R at an angle ϕ to the normal force N.

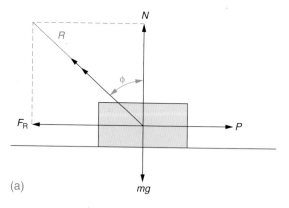

(a)

Figure 5.56a Space diagram

From the above diagram it can be seen that:

$$\frac{F_R}{R} = \sin\phi$$
$$F_R = R\sin\phi$$
$$\text{and} \quad \frac{N}{R} = \cos\phi$$
$$N = R\cos\phi$$
$$\frac{F_R}{N} = \frac{R\sin\phi}{R\cos\phi} = \tan\phi$$

However $F_R = \mu N$

therefore $\mu = \tan \phi$ and ϕ is known as the **angle of friction**.

Once F and N have been replaced by R the problem becomes one of three coplanar forces mg, P and R and can, therefore, be solved using the triangle of forces you met earlier.

Then choosing a suitable scale the vector diagram is constructed as shown in *Figure 5.56b*.

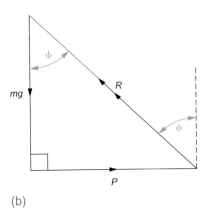

(b)

Figure 5.56b Vector diagram

Example 5.29

For the situation illustrated in *Figure 5.57a*, find the value of the force P to maintain equilibrium.

(a) Solution by calculation

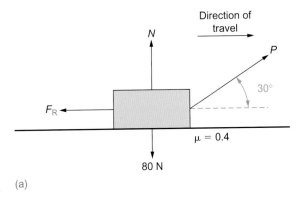

(a)

Figure 5.57a General situation

We can solve this problem by calculation resolving the forces into their horizontal and vertical components or we can solve by drawing. Both methods of solution are detailed below.

Resolving forces horizontally	$F_R = P \cos 30$
resolving forces vertically	$N + P \sin 30 = 80$
but	$F_R = \mu N$
and substituting for N from above gives	$F_R = \mu(80 - P \sin 30)$

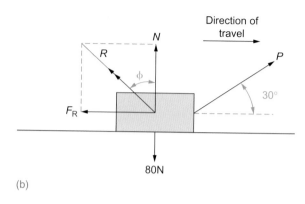

(b)

Figure 5.57b Magnitude and direction of all known forces

We are told that $\mu = 0.4$ and replacing F_R in the above equation by $P \cos 30$, in a similar manner to the general example, gives

$$P \cos 30 = 0.4 (80 - P \sin 30)$$

and by multiplying out the brackets and rearrangement we get

$$P \cos 30 + 0.4 P \sin 30 = 0.4 \times 80$$

so $$P (\cos 30 + 0.4 \sin 30) = 32$$

and   **$P = 30.02 \, N$**

Make sure you can follow the above trigonometric and algebraic argument.

(b) Solution by drawing

The magnitude and direction of all known forces for our block is shown in *Figure 5.57b* that was used for the calculation method.

Remembering that $\mu = \tan \phi$, then $\tan \phi = \mu = 0.4$, so, $\phi = \tan^{-1} 0.4$ (the angle whose tangent is) and the angle $\phi = 21.8°$.

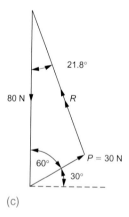

(c)

Figure 5.57c Vector diagram

Now after drawing the resulting vector diagram we find that **$P = 30 \, N$**.

KEY POINT

The coefficient of friction is given by the tangent of the angle of friction

TYK

TYK *5.13*

1. On what variables does the value of frictional resistance depend?

2. The frictional resistance is independent of the relative speed of the surfaces, under all circumstances'. Is this statement true or false? You should give reasons for your decision.

UNIT 5

3. Explain how the angle of friction and the coefficient of friction are related.

4. Sketch a space diagram that shows all the forces that act on a body moving with uniform velocity along a horizontal surface.

5. A wooden packing case of mass 80 kg rests on a concrete floor. The coefficient of friction is 0.3. Calculate the least horizontal force required to move the case.

6. A drilling machine of mass 300 kg is pulled along a horizontal floor by means of a rope inclined at 20° to the horizontal. If a tension in the rope of 600 N is just sufficient to move the machine, determine the coefficient of friction between the machine and the floor.

D'Alembert's principle

You will recall that *inertia force* is the resistance of the body *that opposes the out-of-balance force* causing the acceleration of the body, and that from Newton's second law we obtained the relationship that $F = ma$, where the force F is that required to cause the acceleration. The inherent inertia force (resistance) of the mass *opposes* this acceleration.

> **D'Alembert's principle states:** If the inertia resistance to motion is treated as an external force, a body can be treated as if it is in static equilibrium

Thus, by treating the inherent inertia resistance of a body to acceleration as an *external force* we are able to apply static principles, rather than the more difficult dynamic principles, to many engineering problems that involve the acceleration of a moving load or body.

In fact, the use of this principle is a lot easier than it might first appear. Consider the situation shown in *Figure 5.58*, where by applying D'Alembert's principle, the free body diagram for a load being lifted vertically over a pulley and at the same time being subject to acceleration is shown.

In *Figure 5.58*, it can be seen that the inertia resistance is shown as an external load. Therefore, applying D'Alembert's principle, that is, using the conditions for static equilibrium and equating upward forces and downward forces, then we get

applied force = gravitational force + inertia resistance + frictional resistance

or in symbols $F = mg + ma + F_R$

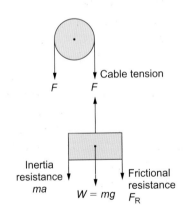

Figure 5.58 Free body diagram showing the forces acting on load

Cable tension
F F

Inertia resistance
ma $W = mg$ Frictional resistance F_R

Example 5.30

A load of mass 60 kg is hoisted vertically by means of a light rope running over a pulley and is accelerated from 3 to 6 ms^{-1} while travelling through a distance of 6 m. If the resistance to motion is 50 N determine:

(a) the pulling force in the rope

(b) work done

(c) average power required by the operator to accelerate this load.

(a) For this problem there is a gravitational force, inertia resistance and frictional resistance to be overcome in order for the load to be accelerated over the given distance. This is the same situation as that illustrated in *Figure 5.58*.

Using $F = mg + ma + F_R$, the only unknown that is required to determine the pulling force is the acceleration of the load. This can be found using the equation of motion $v^2 = u^2 + 2as$, where after transposition

$$a = \frac{v^2 - u^2}{2s} = \frac{6^2 - 3^2}{(2)(6)}$$

and $a = 2.25 \text{ ms}^{-2}$.

Then pulling force $F = (60)(9.81) + (60)(2.25) + (50) = \textbf{773.6 N}$.

(b) The work done during the period of the acceleration is given by the pulling force multiplied by the linear distance moved, that is, WD $= Fd = (773.6)(6) = \textbf{4641.6 Nm}$.

(c) The average power required by the operator, assuming no other losses, is equal to the work done divided by the time taken. To find the time taken we again use an equation of motion so that from $v = u + at$, then

$$t = \frac{v - u}{a} = \frac{6 - 3}{2.25} = 1.33 \text{ s}$$

and

$$P = \frac{\text{WD}}{t} = \frac{4641.6}{1.33} = \textbf{3489.2 W}$$

Let us now consider another example that can be tackled either by using D'Alembert's principle or by applying an energy method. First the theory. Consider the car shown in *Figure 5.59* that is being accelerated up the slope by a ***tractive effort*** force that is required to overcome the gravitational force, inertia resistance and the frictional resisting force. In fact the tractive effort is the force being applied where the tyre meets the road. A ***torque*** is created by the drive shaft and this acts at the extremity of the wheel, that is, at the tread of the tyre. The tractive effort force acts to move the vehicle linearly against the rotation of the wheel that has been created by the torque from the drive shaft. This force will be a minimum on a wet slippery road with worn tyres, and a maximum on a dry road with ample tread on the tyres.

KEY POINT

The tractive effort force acts between the drive wheels and the surface in contact

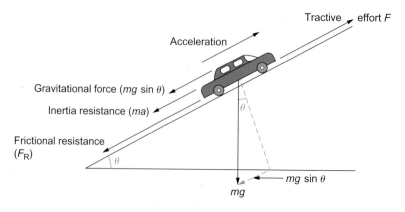

Figure 5.59 Free body diagram for a car being accelerated up an incline

UNIT 5

We will not be considering the application of D'Alembert's principle to the torques in this unit, only to the forces and acceleration shown in *Figure 5.59*. Any example involving vehicles moving up or down a slope and being accelerated or retarded can be solved using D'Alembert's principle. Other applications of D'Alembert's principle include vehicles turning and banking when subject to centripetal accelerations and forces. The solution of these latter applications involves use of angular motion theory, which is beyond the remit of this unit and will therefore not be covered.

We know from the application of D'Alembert's principle that for the situation illustrated in *Figure 5.59* the:

Tractive effort = gravitational force + inertia resistance + frictional resistance

and when the forces have been resolved parallel to the slope, this is equivalent to:

$$F = mg \sin \theta + ma + F_R.$$

The significance of the term $\sin \theta$ when dealing with slopes is easily seen if we consider any slope with, say, an incline of 1 in 5. This means that we go up 1 vertically and go 5 along the slope. Therefore, as the incline ratio involves the opposite and the hypotenuse, then finding the angle of slope or the vertical height involves the sine ratio and in this case we have $\sin 0.2$ and its inverse give the angle $\theta = 11.53°$.

Example 5.31

A car of mass 1500 kg accelerates up a road with a slope of 1 in 8, increasing its speed from 4.0 ms^{-1} to 16 ms^{-1} while travelling up the road a distance of 80 m against a frictional resistance of 1.0 kN. Calculate:
(a) the tractive effort between the driving wheels and the road surface
(b) the work done during the period of the acceleration
(c) the average power developed.
Figure 5.60 illustrates the situation for this problem.

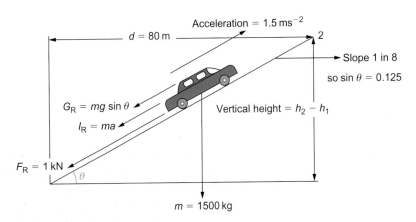

Figure 5.60 Car being accelerated up an incline

(a) We use the formula $F = mg \sin \theta + ma + F_R$ in order to determine the tractive effort, where the only unknown is the acceleration of the car. This is easily found by again using the equation $v^2 = u^2 + 2as$ and after transposition

$$a = \frac{v^2 - u^2}{2s} = \frac{16^2 - 4^2}{(2)(80)} = 1.5 \, ms^{-2}$$

and also from the slope $\sin \theta = 0.125$.

Then $F = (1500)(9.81)(0.125) + (1500)(1.5) + (1000) = \mathbf{5.089\,kN}$

(b) The WD $= (5.089)(80) = \mathbf{407.12\,kJ}$.

(c) The average power developed is given by the work done divided by the time taken, where time taken can be found from $v = u + at$ and so

$$t = \frac{v - u}{a} = \frac{16 - 4}{1.5} = 8.0 \, s$$

and

$$P = \frac{WD}{t} = \frac{407.12}{8} = \mathbf{50.89\,kW}$$

If we do work (W) on a mechanical system, say, on our car in accelerating it up the hill, then from ***the principle of the conservation of energy***, this work input can cause changes in potential energy (PE), kinetic energy (KE) and the work needed to overcome friction ***that equal*** the work input. Since our car is being accelerated up the slope and applying the conservation of energy principle, these changes will all be increases that equal the work input to the car (our system). This work input takes place from when the tractive effort was first applied to cause the car to accelerate. Thus

> **Work input = change in PE + change in KE**
> **+ work done to overcome friction**

where from our previous work on energy, we may write this equation in symbols as

$$W = mg(h_2 - h_1) + \tfrac{1}{2}m(v_2^2 - v_1^2) + (F_R d)$$

Let us now consider Example 5.31 again where we use the energy method to solve the problem rather than D'Alembert's principle.

Example 5.32

A car of mass 1500 kg accelerates up a road with a slope of 1 in 8, increasing its speed from 4.0 to 16 ms^{-1} while travelling up the road a distance of 80 m against a frictional resistance of 1.0 kN. Using an energy method calculate:

(a) the work done during the period of the acceleration

(b) the tractive effort between the driving wheels and the road surface

(c) the average power developed.

UNIT 5

(a) To find the work done, we use the formula

$$W = mg(h_2 - h_1) + \tfrac{1}{2}m(v_2^2 - v_1^2) + (F_R d),$$

where in this case the only unknown is the change in height of the car during it travel up the slope.

Since the slope is 1 in 8, then the change in vertical height

$$h_2 - h_1 = \frac{80}{8} = 10\,\text{m}.$$

If this method of finding the vertical height is not immediately clear to you then use the sine ratio, where vertical height

$$h_2 - h_1 = d \sin\theta = (80)(0.125) = 10\,\text{m},$$

where d is the distance up the slope (the hypotenuse).

Then

$$W = (1500)(9.81)(10) + \tfrac{1}{2}(1500)(16^2 - 4^2) + (1000)(80)$$
$$= \textbf{407,150 J} \quad \text{or} \quad \textbf{407.15 kJ}.$$

(b) The tractive effort is found by dividing the work done by the distance travelled and we find that the tractive effort

$$\frac{407.15}{80} = \textbf{5,089 kN.}$$

Note that using this energy method we found the work done first and then the tractive effort force.

(c) The average power may be found from the relationship that power is equal to the tractive effort force multiplied by the average velocity, where the average velocity

$$\frac{16 + 4}{2} = 10\,\text{ms}^{-1}$$

and therefore the average power

$$P = (F)(\overline{v}) = (50.89)(10) = \textbf{50.89 kW.}$$

Note that all solutions using this method are very much in accord with those found in Example 5.31 where we applied D'Alembert's principle.

The above examples conclude our short study on D'Alembert's principle.

Note: The free body diagrams and/or force diagrams shown for the above examples do not include the torques that act on these systems. The total torque applied to any system will need to overcome the rotational torque due to the load or the tractive torque and the inertia torque and friction torque, among others. All or some of these torques would appear in a complete free body diagram for any system involving rotation. As mentioned earlier, solving these torques is beyond the remit of this unit.

TYK 5.14

1. When a load is being hoisted using a pulley system, what forces need to be overcome to accelerate the load vertically upwards?

2. Draw a free body diagram showing all the forces that act linearly on a vertical pulley system when a moving load is being accelerated vertically.

3. A load of mass 80 kg is pulled vertical by means of a cable running over a pulley and is accelerated from 2 to 6 ms^{-1} while travelling upwards through a distance of 3 m. If the resistance to motion is 60 N determine using an energy method:
   (a) the work done
   (b) the average power required by the operator to accelerate the load.

4. A vehicle of mass 4000 kg accelerates up a road that has a slope of 1 in 10, increasing its speed from 2 to 10 ms^{-1} while travelling up the road a distance of 200 m against a frictional resistance of 5 kN. Determine using D'Alembert's principle:
   (a) the tractive effort between the driving wheel and the road surface
   (b) the work done during the period of the acceleration and
   (c) the average power developed.

Parameters and Principles of Fluid Systems

In this outcome we will look at a number of fluid system parameters, in particular for hydrostatic systems that are concerned with liquids such as water and oil, rather than gases, leaving our study of gases until the next outcome. Thus density, relative density, hydrostatic pressure, fluid pressure factors, centre of pressure and thrust will be considered, together with the parameters concerned with fluid flow that include: flow velocity, volume flow rate and mass flow rate.

The fluid principles that are to be covered include Archimedes, thrust on immersed surfaces, buoyancy and the continuity of mass and volume for incompressible flow and their application to closed pipe systems.

Basic parameters of fluid systems

Density and relative density

> The **density rho (ρ) of a body** is defined as its mass per unit volume.

KEY POINT

The density of pure water at 4°C is taken as 1000 kg m^{-3}

Thus density is a standard measure of the mass of a substance that is contained in a volume of one metre cubed (1 m^3). Combining the SI units for mass and volume gives the unit of density as **kg m^{-3}**. Using symbols the formula for density is given as:

$$\rho = \frac{m}{V}$$

where again the mass is in kilograms (kg) and the volume in m^3.

UNIT 5

The amount of a fluid contained in a cubic metre will vary with temperature, although the expansion of liquids with rising temperature is insignificant when compared to gases. The standard densities quoted for liquids will be valid for all temperatures unless stated otherwise.

Density may be expressed in relative terms, that is, in comparison to some datum value. The datum that forms the basis of relative density for solid and liquid substances is water.

> The **relative density of a solid or liquid body** is the ratio of the density of the body with that of the density of pure water measured at 4°C. That is,
>
> $$\text{relative density} = \frac{\text{density of substance}}{\text{density of water}}$$

KEY POINT

To find the relative density of any liquid or solid substance, divide its density by $1000\,\text{kg}\,\text{m}^{-3}$

The density of water under these conditions is $1000\,\text{kg}\,\text{m}^{-3}$. Since relative density is a ratio it has **no units**. The old name for relative density was **specific gravity (SG)** and this is something you need to be aware of in case you meet this terminology in the future. Thus, to find the relative density of any liquid or solid material, divide its density by $1000\,\text{kg}\,\text{m}^{-3}$.

Note: The relative density that forms the basis for gases is air, so that the relative density of both water and air is 1.00. The standard sea-level value for the density of air is taken as $1.2256\,\text{kg}\,\text{m}^{-3}$.

The density and relative density of some of the more common engineering substances are laid out in *Table 5.1*.

Table 5.1 Density and relative density of some common engineering substances

Substance	Density ($\text{kg}\,\text{m}^{-3}$)	Relative density
Gases		
Hydrogen	0.085	0.0695
Helium	0.169	0.138
Nitrogen	1.185	0.967
Air	1.2256	1.0
Oxygen	1.3543	1.105
Liquids		
Petroleum	720	0.72
Aviation fuel	780	0.78
Alcohol	790	0.79
Kerosene	820	0.82
Hydraulic oil	850	0.85
Water	1000	1.00
Mercury	13,600	13.6

(Continued)

Table 5.1 *(Continued)*

Substance	Density (kg m^{-3})	Relative density
Solids		
Ice	920	0.92
Polyethylene	1300–1500	1.3–1.5
Rubber	860–2000	0.86–2.0
Magnesium	1740	1.74
Concrete	2400	2.4
Glass	2400–2800	2.4–2.8
Aluminium	2700	2.7
Titanium	4507	4.507
Tin	7300	7.3
Steel	7800–7900	7.8–7.9
Iron	7870	7.87
Copper	8960	8.96
Lead	11,340	11.34
Gold	19,320	19.32

Example 5.33

A hydraulic cylindrical shock absorber has an internal diameter of 10 cm and is 1.2 m in length. It is half filled with hydraulic oil and half filled with nitrogen gas, at standard atmospheric pressure. Determine (taking appropriate values from the table) the weight of the nitrogen and the weight of the hydraulic oil within the shock absorber.

All that is required is that we first find the internal volume of the shock absorber:

$$V = \left(\frac{\pi d^2}{4}\right)l$$

then

$$V = \left(\frac{\pi(0.1)^2}{4}\right)1.2 \quad \text{and} \quad V = 0.009425\,\text{m}^3.$$

Therefore, the volume of the nitrogen and the volume of the oil are both $= 0.004713\,\text{m}^3$. Now the mass of each may be found using their densities. From

$$\rho = \frac{m}{V}$$

then

$$m = \rho V$$

and for nitrogen $m = (1.185)(0.004713) = 0.005585\,\text{kg}$. For the hydraulic oil the mass

$$m = (850)(0.004713) = 4.006\,\text{kg}.$$

Then the weight of the nitrogen = *0.0548 N* and the weight of the hydraulic oil = *39.299 N*.

It can be seen from Example 5.33 that the weight of the nitrogen is virtually insignificant when compared with that of the oil. In a real shock absorber

the nitrogen would be highly pressurized increasing it density and mass many fold. Pressure–volume relationships are an important consideration when dealing with gases and these will be covered in some detail when you study the final outcome on thermodynamics.

TYK *5.15*

1. What is the SI unit of density?

2. What is the essential difference between finding the relative density of solids and liquids compared with gases?

3. What is likely to happen to the density of pure water as its temperature increases?

4. Why does relative density have no units?

5. A rectangular container has dimensions of 1.6 m × 1.0 m × 0.75 m. A certain liquid that fills the container has a mass of 864 kg. What is the density and relative density of this liquid?

Pressure and pressure laws

You may already have met the concept of pressure before when considering pressure acting on solids, where it was defined as force (N) per unit area (m²). There are in fact several types of pressures that you may not have met. These include ***hydrostatic pressure***, that is the pressure created by stationary or slow moving bulk liquid, ***atmospheric pressure*** due to weight of the air above the earth acting on its surface, and ***dynamic pressure*** due to fluid movement, as well as the pressure applied to solids mentioned above.

There are several common units for pressure and the more common SI units are shown in *Table 5.2*.

Table 5.2 Some common SI units for pressure

Use	SI units
Pressure applied to solids	Nm^{-2}, MNm^{-2}
Fluid dynamic and hydrostatic pressure	Pascal (Pa) = 1 N/m², kPa, MPa
Atmospheric pressure	Bar = 10^5 Pa = 10^5 N/m², mbar Millimetres of mercury (mmHg)

There are four basic factors or laws that govern the pressure within fluids.

With reference to *Figure 5.61* these four laws may be defined as follows.

1. Pressure at a given depth in a fluid is equal in all directions.
2. Pressure at a given depth in a fluid is independent of the shape of the containing vessel in which it is held. In *Figure 5.61b* the pressure at point *X*, *Y* and *Z* is the same.
3. Pressure acts at right angles to the surfaces of the containing vessel.
4. When a pressure is applied to a fluid, this pressure is transmitted equally in all directions.

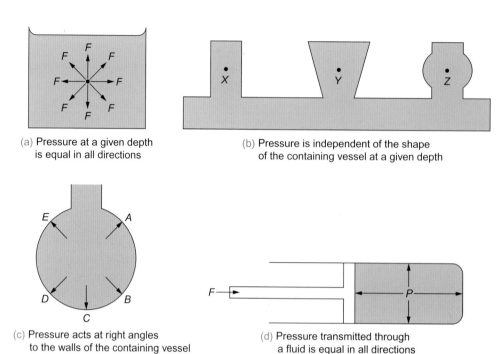

(a) Pressure at a given depth is equal in all directions

(b) Pressure is independent of the shape of the containing vessel at a given depth

(c) Pressure acts at right angles to the walls of the containing vessel

(d) Pressure transmitted through a fluid is equal in all directions

Figure 5.61 Illustration of fluid pressure laws

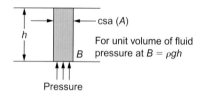

Figure 5.62 Pressure at a point in a liquid

KEY POINT

ρgh = gauge pressure

The consequences of these laws or properties will become clear as you continue your study of fluids and in particular hydrostatic systems.

Hydrostatic and atmospheric pressure

Pressure at a point (depth) in a liquid can be determined by considering the weight force of a fluid above the point. Consider *Figure 5.62*.

If the density of the liquid is known, then we may express the *mass* of the liquid in terms of its density and volume where the volume $V = Ah$. Thus, the mass of liquid (m) = density $\times$ cross-sectional area (A) $\times$ height (h) or in symbols:

$$m = \rho Ah$$

Now, since the weight is equal to the mass multiplied by the acceleration due to gravity then the weight is given by $W = \rho Agh$ and it follows that the pressure due to the weight of the liquid (hydrostatic pressure) is equal to the weight divided by area A, that is

hydrostatic pressure due to weight of liquid $= \rho gh$

If standard SI units are used for density (kg m^{-3}), acceleration due to gravity (9.81 ms^{-2}) and the height (m), then the pressure may be expressed in Nm^{-2} or pascal (Pa).

Note that the **atmospheric pressure** above the liquid was ignored. The above formula refers to **gauge pressure**. This should always be remembered when using this formula.

UNIT 5

Example 5.34

Find the head *h* of mercury corresponding to a pressure of 101,325 kNm^{-2}. Take the density of mercury as 13,600 kgm^{-3}.

Then since pressure $p = \rho g h$ and so $h = \dfrac{p}{\rho g}$ then using standard SI units

$$h = \frac{101{,}325}{(13{,}600)(9.81)} = \textbf{\textit{0.76 m}} \text{ or } \textbf{\textit{760 mm of Hg}}$$

Therefore, this is the height of mercury needed to balance standard atmospheric pressure.

The air surrounding the earth has mass and is acted upon by the earth's gravity, thus it exerts a force over the earth's surface. This force per unit area is known as **atmospheric pressure**. At the earth's surface at sea level, this pressure is found by measurement to be *101,325 N/m^2*.

Outer space is a vacuum and is completely devoid of matter, consequently there is no pressure in a vacuum. Therefore, pressure measurement relative to a vacuum is *absolute*. For most practical purposes it is only necessary to know how pressure varies from the earth's atmospheric pressure. A pressure gauge is designed to read zero when subject to atmospheric pressure; therefore, if a gauge is connected to a pressure vessel it will only read *gauge* pressure. So, to convert gauge pressure to absolute pressure, atmospheric pressure must be added to it, that is

Absolute pressure = gauge pressure + atmospheric pressure

Example 5.35

Taking atmospheric pressure as 101,325 N/m^2, convert the following gauge pressures into absolute pressures giving your answers in kPa.

(a) 400 kNm^{-2} (b) 20 MNm^{-2} (c) 5000 Pa

We know from above that absolute pressure is equal to gauge pressure plus atmospheric pressure; therefore, the only real problem here is to ensure the correct conversion of units:

$$\text{atmospheric pressure} = 101.325 \, \text{kNm}^{-2}$$

(a) $400 + 101.325 = \textbf{\textit{501.325 kNm}}^{-2}$

(b) $20{,}000 + 101.325 = \textbf{\textit{20,101.325 kNm}}^{-2}$

(c) $5 + 101.325 = \textbf{\textit{106.325 kNm}}^{-2}$ (remember that $1 \, \text{Pa} = 1 \, \text{Nm}^{-2}$)

Parameters and principles of hydrostatic systems

In this section, we look at parameters and principles associated with hydraulic transmission systems, pressure measurement systems and systems associated with buoyancy and immersed surfaces.

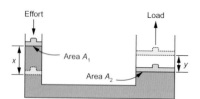

Figure 5.63 The Bramah press

Hydraulic press

An application of the use of fluid pressure to a hydrostatic system can be found in the **hydraulic press**, often referred to as the **Bramah press**. This hydraulic machine may be used for dead weight testing, hydraulic actuators, lifting loads or for compression and shear testing. *Figure 5.63* shows the general arrangement for such a machine. Since the fluid involved in this machine is liquid hydraulic oil, and therefore virtually incompressible, the fluid displaced by the effort piston must be equal to the amount of fluid displaced at the load piston.

In other words, the volumes A_1x and A_2y must be the same. Therefore, the **velocity ratio VR** is given by:

$$VR = \frac{x}{y} = \frac{A_2}{A_1}$$

or in words

$$\frac{\text{distance moved by effort piston } x}{\text{distance moved by load piston } y} = \frac{\text{area of load piston } A_2}{\text{area of effort piston } A_1}$$

Example 5.36

(a) A force of 500 N is applied to the small cylinder of a hydraulic press. The smaller cylinder has a cross-sectional area (csa) of 10 cm²; The large cylinder has a csa of 180 cm². What load can be lifted by the larger piston if the pistons are at the same level?

(b) What load can be lifted by the larger piston if the larger piston is 0.75 m below the smaller?

Taking the density of the oil in the press as 850 kg m⁻³.

The situation for both cases is shown in *Figure 5.64*.

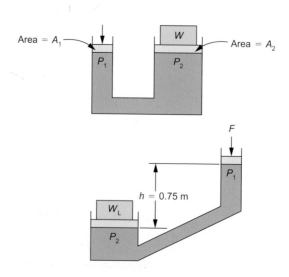

Figure 5.64

(a) We know that $P_1 = P_2$, since pressure is applied equally in all directions. Therefore

$$\frac{F}{A_1} = \frac{W}{A_2} \text{ or } F = WA_1/A_2 \text{ then } W = FA_2/A_1 \text{ and substituting values gives}$$

$$\text{load } W = \frac{(500)(180 \times 10^{-4})}{1 \times 10^{-3}} = \boldsymbol{9000\,N}$$

(b) If the larger piston is 0.75 m below the smaller piston, then pressure P_2 will be greater than P_1 due to the head of liquid. Then

$$P_2 = P_1 + \rho g h$$

$$P_1 = \frac{F}{A_1} = \frac{500}{1 \times 10^{-3}} = 50.0 \times 10^4\,N/m^2$$

then $\quad P_2 = (50 \times 10^4) + (850 \times 9.81 \times 0.75)$

$$P_2 = 50.6254 \times 10^4\,N/m^2$$

and $\quad W_L = P_2 A_2 = (50.6254 \times 10^4)(180 \times 10^{-4}) = \boldsymbol{9112.57\,N}$

TYK

TYK 5.16

1. Define: (a) hydrostatic, (b) atmospheric and (c) absolute pressure.

2. Convert the following quantities into MPa (a) $2034\,kNm^{-2}$, (b) $9 \times 10^7\,Pa$ and (c) 26 bar.

3. Determine the gauge pressure that is equivalent to 0.8 m of mercury. Take the density of mercury from *Table 5.1*.

4. Convert the following gauge pressures into absolute pressures, giving your answers in kNm^{-2}: (a) 847,677 Pa, (b) 0.346 MPa (c) 70 bar and (d) 1010 mbar.

5. If a hydraulic press has a VR = 180 and the load piston is raised 10 cm, determine the distance travelled by the effort piston in metres.

Measurement of pressure

Devices used to measure pressure will depend on the magnitude (size) of the pressure, the accuracy of the desired readings and whether the pressure is static or dynamic. Here we are concerned with barometers to measure atmospheric pressure and the manometer to measure low pressure changes, such as might be encountered in a laboratory or from variations in flow through a wind tunnel.

The two most common types of barometer used to measure atmospheric pressure are the mercury and aneroid types. The simplest type of *mercury barometer* is illustrated in *Figure 5.65*. It consists of a mercury filled tube, which is inverted and immersed in a reservoir of mercury.

The atmospheric pressure acting on the mercury reservoir is balanced by the pressure *ρgh* created by the mercury column. Thus, the atmospheric

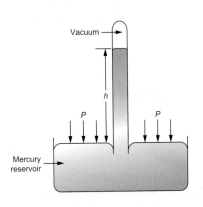

Figure 5.65 Simple mercury barometer

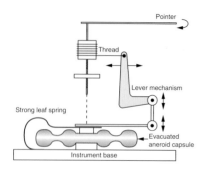

Figure 5.66 Aneroid barometer mechanism

pressure can be calculated from the height of the column of mercury it can support.

The mechanism of an ***aneroid barometer*** is shown in *Figure 5.66*. It consists of an evacuated aneroid capsule, which is prevented from collapsing by a strong spring.

Variations in pressure are felt on the capsule that causes it to act on the spring. These spring movements are transmitted through gearing and amplified causing a pointer to move over a calibrated scale.

A common laboratory device used for measuring low pressures is the U-tube manometer, shown in *Figure 5.67*. A fluid is placed in the tube to a certain level. When both ends of the tube are open to atmosphere the level in the fluid of the two arms is equal. If one of the arms is connected to the source of pressure to be measured it causes the fluid in the manometer to vary in height. This height variation is proportional to the pressure being measured.

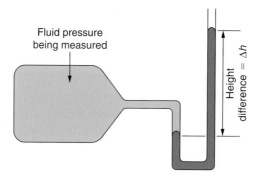

Figure 5.67 The U-tube manometer

The magnitude of the pressure being measured is the product of the difference in height between the two arms Δh, the density of the liquid in the manometer and the acceleration due to gravity, that is pressure being measured ***gauge pressure*** $= \rho g \Delta h$.

Example 5.37

A mercury manometer is used to measure the pressure above atmospheric of a water pipe, the water being in contact with the mercury in the left-hand arm of the manometer. If the right-hand arm of the manometer is 0.4 m above the left-hand arm, determine the gauge pressure of the water. Take the density of mercury as 13,600 kg/m³.

We know that gauge pressure $= \rho g \Delta h$
$$= (13,600)(9.81)(0.4)$$
$$= \mathbf{53,366\,Nm^{-2}}$$

Buoyancy and Archimedes principle

It is well known that a piece of metal placed in water will sink and that a piece of cork placed below the surface of the water will rise. Also, that a steel ship having a large volume of empty space in the hull will float. The

UNIT 5

study of floating, sinking or rising bodies immersed in a fluid is known as **buoyancy**. We know from our study of fluid pressure that there will be an increase in pressure of a fluid with depth, irrespective of the nature of the fluid. This means that eventually there will be a greater pressure pushing up on the body from underneath than there is pushing down on it from above. So, dependent on the relative densities of the fluids and bodies involved, equivalence will be reached when the upthrust force due to the fluid equals the weight force exerted by the body immersed in it.

> **Archimedes** expresses this relationship very succinctly in his principle that states that when a body is immersed in a fluid it experiences an upthrust, or apparent loss of weight, equal to the weight of the fluid displaced by the body.

This equality relationship is illustrated in *Figure 5.68*, where it can be seen that the body immersed in the fluid floats when the upthrust force (equal to the weight of the fluid displaced) equals the weight of the body. If the weight of the body is less that the weight of the fluid it displaces then it will rise (upthrust).

This principle and the concept of buoyancy enable us to determine why and when airships, balloons, ships and submarines will float. As an example, consider the buoyancy of a helium balloon. The density of the atmosphere reduces with altitude, so when the upthrust force (per unit area) created by the atmospheric air is equal to the weight of the helium and balloon, then the balloon will float at a specified altitude. This assumes of course that the balloon does not burst first!

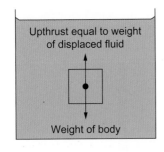

Figure 5.68 Illustration of Archimedes principle

Thrust on partially and fully immersed surfaces

It can be shown that *the total thrust force that acts on an immersed surface is the product of the pressure at the centroid (centre of mass) and the total area of surface in contact*. The general proof of this statement is beyond the remit of this unit but the meaning of this statement is easily explained.

Consider a vertical plate that is immersed in a fluid, such as water, with its top edge in line with the free surface of the water and one surface of area A in complete contact with the water *Figure 5.69*.

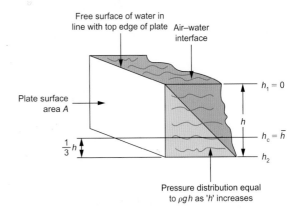

Figure 5.69 Plate with one surface immersed vertically in water

Then from *Figure 5.69* it can be seen that the **gauge pressure** of the water at any point is found by using the expression (**ρgh**), where the height of the water varies from zero at the air–water interface (h_1) to the depth of the plate (h_2). The pressure distribution is triangular and as such the centroid of this pressure distribution ($\bar{h}$) can be shown to act one-third of the way up from its base and for **vertical surfaces** this is also coincident with the centre of pressure (h_c) of the fluid (water in this case).

Therefore, since *force-pressure × area* the **thrust force** acting on the immersed surface is $F = \rho A g \bar{h}$ and because for vertical surfaces the centre of pressure acts at the centroid, that is

$$\bar{h} = h_c$$

then in this case

$$\boldsymbol{F = \rho A g h_c}.$$

Knowledge of the second moment of area is required to show that the centroid of a triangular section acts one-third from the base, so this is given without proof.

The above theory may be usefully applied to hydrostatic systems, such as lock gates, dams and storage tank separator plates, where the horizontal forces acting on partially or totally immersed surfaces can be calculated as well as determining the line of action of these forces and the likely out-of-balance moments that must be accounted for when designing these systems.

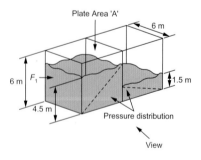

Figure 5.70 Fuel storage tank with separator plate

Example 5.38

A vertical plate is used to separate two areas of a large industrial rectangular fuel storage tank as shown in *Figure 5.70*. Taking the density of the fuel as 780 kg m^{-3} for the system shown, find the net horizontal thrust on the separator plate and the position of its line of action from the base of the tank.

From *Figure 5.70* it can be seen that the area of fuel acting on the left-hand side (LHS) of the separator plate = 4.5 × 6 = 27 m².

Area of fuel acting on the right-hand side (RHS) of separator plate = 1.5 × 6 = 9 m².

Then,

$$\text{LHS thrust} = F_1 = \rho g A_1 h_1 = (780)(9.81)(27)(1.5) = 309.898 \text{ kPa,}$$

where $h_1 = (1/3)(4.5) = 1.5$ m.

$$\text{RHS thrust} = F_2 = (780)(9.81)(9)(0.5) = 34.43 \text{ kPa,}$$

where $h_2 = (1/3)(1.5) = 0.5$ m.

Therefore, net horizontal thrust force acting on separator plate = 309.898 − 34.43 = **275.468 kPa.**

UNIT 5

Now, in order to establish the line of action of the net thrust force we need to take the moments of all the thrust forces about the base and equate them using **the principle of moments**, so that,

$$\text{LHS thrust moment} = (309.898)(1.5) = 464.847 \, \text{kNm (clockwise)}$$

$$\text{RHS thrust moment} = (34.43)(0.5) = 17.215 \, \text{kNm (anticlockwise)}$$

Net thrust moment acting at a distance z from the base of the tank = $275.468z$.

Now, equating moments $275.468z = 464.847 - 17.215 \, \text{kNm}$ and the line of action of net thrust from the base

$$z = \frac{464.847 - 17.215}{275.468} = \textbf{1.625 m}$$

and line of action acts as shown in *Figure 5.71*.

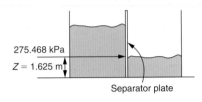

275.468 kPa

$Z = 1.625 \, \text{m}$

Separator plate

Figure 5.71 Net resultant thrust force on separator plate and its line of action

Test your knowledge

? TYK

TYK 5.17

1. Describe the operation of a mercury barometer.

2. If the difference in height of the mercury between the two arms of a U-tube manometer is 30 cm determine: (a) the gauge pressure and (b) the absolute pressure being measured.

3. State Archimedes principle and explain how this principle relates to buoyancy.

4. A rectangular block of mild steel (density 7800 kg m^{-3}), having dimensions 0.25 m × 1.5 m × 3 m, is suspended completely immersed in water by a wire with its longest side vertical. Calculate (a) the weight of block and (b) the tension in the rope.

5. A hollow rectangular box having dimensions 10 cm × 20 cm × 40 cm floats upright in water with three-quarters of its longest side immersed in the water. Find (a) the upthrust on the box and (b) the mass of the box.

6. With respect to immersed or partially immersed surfaces, define (a) centroid and (b) total thrust force.

7. A lock gate is 8 m wide and has water at a depth of 4 m acting on one side and 8 m acting on the other as shown in *Figure 5.72*. Determine the net thrust force acting on the sides of the lock gate and the line of action of this force from the base of the gate, taking the normal value for the density of water.

Width = 8 m

8 m

4 m

Side view

Figure 5.72 Water acting on a lock gate

Internal fluid flow in piped systems

Introduction

In order to study fluid systems such as hydraulic, water works, pneumatic and fuel systems, a basic understanding of fluids in motion is necessary. The study of fluid in motion is known as *fluid dynamics* and in this very short section we will look only at the continuity of flow through piped systems. This is important for hydraulic engineers when, for example, estimating

flow rates in order to size the pipelines through which the oil or water is to flow.

Flow through piped systems may be laminar or turbulent. **Laminar flow** is flow in which the fluid particles move in an orderly manner and retain the same relative positions in successive cross-sections. An example of laminar flow is when the water from a hosepipe exits the pipe in a single streamlined jet. We represent this type of flow as a series of parallel lines (*Figure 5.73a*).

Turbulent flow is a flow in which the fluid particles may move perpendicular as well as parallel to the surface of the body and undergo eddying or unsteady motions (*Figure 5.73b*). This may cause problems in piped systems where severe turbulence not only creates heavy energy losses but may also cause a phenomena known as **water hammer**, where severe vibration is set up that in long pipe runs can cause physical damage to the system.

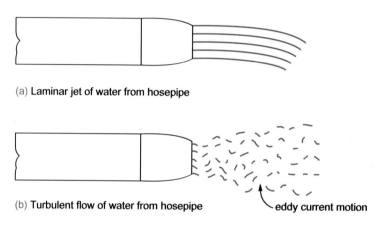

(a) Laminar jet of water from hosepipe

(b) Turbulent flow of water from hosepipe — eddy current motion

Figure 5.73 Laminar and turbulent flow

Incompressible flow is a flow in which the density does not change from point to point. We will base the remainder of our work on fluids on the assumption that they are incompressible. For most liquids this is a justified assumption, as compressibility effects in liquids are very small. This is clearly not the case for air, where compressibility effects would need to be taken into consideration when the air travels at higher velocities.

A **stream tube** or **tube of flow** (*Figure 5.74*) is considered to be an imaginary boundary defined by streamlines drawn so as to enclose a tubular region of fluid.

No fluid crosses the boundary of such a tube. This type of representation is particularly useful for external flows. For internal flows the system is bounded by the walls of the pipe work and components.

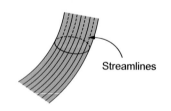

Streamlines

Figure 5.74 Stream tube

Equation of continuity

The **continuity equation** is based on the conservation of mass and simply states that **fluid mass flow rate is constant**; that is, fluids where the density at successive cross-sections through the system is constant.

Figure 5.75 shows an incompressible fluid flowing through a tapered pipe, in the same manner as through a theoretical stream tube, where the density

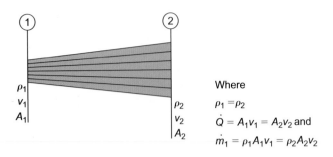

Where

$p_1 = p_2$

$\dot{Q} = A_1 v_1 = A_2 v_2$ and

$\dot{m}_1 = \rho_1 A_1 v_1 = \rho_2 A_2 v_2$

Figure 5.75 Flow through a tapered pipe at two different cross-sections

at the inlet *1* is constant and equal to the density at the outlet *2*. v_1 and A_1 are the velocity and cross-sectional area (csa) at cross-section *1*, and v_2 and A_2 are the velocity and csa at cross-section *2*.

It is also a fact that the volume of the fluid entering the stream tube per second must be equal to the volume of the fluid leaving the stream tube per second. This follows from the conservation of mass and our stipulation that flow is incompressible. Then from what has just been said:

at inlet, volume entering per second, Vs^{-1} = area × velocity = $A_1 v_1$

at outlet, volume leaving per second Vs^{-1} = area × velocity = $A_2 v_2$

therefore, $\dot{Q} = A_1 v_1 = A_2 v_2$

where $\dot{Q}$ = volumetric flow rate ($m^3 s^{-1}$). This equation is known as the ***continuity equation for volume flow rate***.

You should ensure that you understand why the units are the same on both sides of this equation. We can also measure mass flow rate as well as volume flow rate by remembering that density is equal to mass divided by volume $\rho = m/V$, so density × volume = ρV.

Therefore, to obtain mass flow rate, all we need to do is to multiply the volume flow rate by the density. Then,

$$\dot{m} = \rho_1 A_1 v_1 = \rho_2 A_2 v_2$$

where $\dot{m}$ is the mass flow rate ($kg\ s^{-1}$). This equation is known as the ***continuity equation for mass flow rate***.

Make sure that you do not mix up the symbols for velocity and volume! For velocity, we use lower case (v) and for volume we use upper case (V).

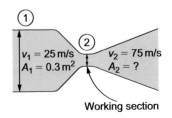

Figure 5.76 Wind tunnel

Example 5.39

In the wind tunnel shown in *Figure 5.76*, the air passes through a converging duct just prior to the working section. The air velocity entering the converging duct is 25 ms^{-1} and the duct has a csa of 0.3 m^{-2}. If the speed of flow in the working section is to be 75 ms^{-1}, calculate the csa of the working section. Assume air density is constant at 1.225 kgm^{-3}.

So, we may use our equation for incompressible fluid flow, since $\rho_1 = \rho_2$, therefore,

$$A_1v_1 = A_2v_2 \quad \text{so } A_2 = A_1v_1/v_2$$

and

$$A_2 = (0.3)(25)/75 = \textbf{0.1}\,\boldsymbol{m^2}\,\textbf{(area)}$$

You will note that the continuity equation is far easier to use than to verify!

Example 5.40

In a water supply system, the mains supply pipe is connected to a tributary pipe via a tapered connector. The inlet diameter to the connector is 0.5 m and the velocity of flow of the water at the inlet is 1.5 ms^{-1} (*Figure 5.77*).

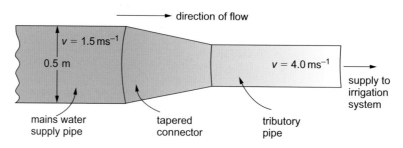

Figure 5.77 A water supply system

The supply from the tributary pipe is used to supply water to an irrigation system, where the flow velocity from the tributary pipe must be maintained at 4 ms^{-1} in order for the irrigation system to operate efficiently. Assuming that the flow is incompressible and taking the density of the water as 1000 kgm^{-3}, determine:
(a) the volume flow rate for the system
(b) the mass flow rate for the system
(c) the diameter of the tributary pipe to the irrigation system.

(a) If the flow is incompressible then the density of the water is the same on the inlet and outlet sides of the connector, that is, $\rho_1 = \rho_2 = 1000\,\text{kgm}^{-3}$. Also, since the mass of the water flowing must be the same throughout the system (since continuity of mass), then the volume flow rate either side of the connector must also be the same.

Volume flow rate at inlet to connector $\dot{Q} = A_1v_1$ and the area of the mains supply pipe

$$A_1 = \frac{\pi D^2}{4} = \frac{\pi (0.5)^2}{4} = 0.196\,\text{m}^2$$

so that

$$\dot{Q} = (0.196)(1.5) = \boldsymbol{0.294\,m^3s^{-1}}$$

(b) The mass flow rate is given by

$$\dot{m} = \rho_1 A_1 v_1 = (1000)(0.294) = \boldsymbol{294\,kg\,s^{-1}}$$

(c) Since the volume flow rate

$$\dot{Q} = A_2v_2 \text{ then } A_2 = \frac{Q}{v_2} = \frac{0.294}{4} = 0.0735\text{m}^2;$$

therefore,

$$0.0735 = \frac{\pi d^2}{4} \quad \text{and} \quad d = \sqrt{\frac{(4)(0.0735)}{\pi}} = \textbf{0.306 m.}$$

TYK 5.18

1. Define: (a) laminar flow, (b) turbulent flow and (c) stream tube.

2. Explain using the equation of continuity how volume flow rate and mass flow rate vary and the effect on these relationships when fluid flow is incompressible.

3. A tapered diverging (getting larger in the direction of flow) duct has an inlet diameter of 10 cm and an outlet diameter of 12 cm. If oil of density 850 kgm^{-3} enters the duct at a velocity of 6 ms^{-1}, then assuming incompressible flow, determine:
   (a) the volume flow rate
   (b) the mass flow rate
   (c) the velocity of the oil at the duct outlet.

Energy Transfer in Thermodynamic Systems

Heat transfer

In this section, we first consider the idea of temperature, its measurement and thermal expansion, then we look at heat energy and the mechanisms by which heat may be transferred. We next consider the properties and laws that define the thermodynamic systems in which heat transfer takes place. Then the heat energy transfer between specific materials and their inherent specific heat energies is considered. Finally, we look at the change of state (or phase) of the material that results from heat energy flow.

Temperature, its measurement and thermal expansion

You would already have met the idea of temperature in your previous studies, but you may not have formally defined it. A formal definition of temperature is as follows:

> **Temperature** is a measure of the quantity of energy possessed by a body or substance. It measures the vibration of the molecules that form the substance.

These molecular vibrations only cease when the temperature of the substance reaches *absolute zero*, that is, *−273.15°C. Thermodynamic temperature* (in the SI system) *is always measured in Kelvin and is known as absolute temperature*.

You would also have met the Celsius temperature scale and the way in which we convert degrees centigrade into kelvin and vice versa, but to remind you see Example 5.41.

KEY POINT

Temperature measures the energy possessed by the vibration of the molecules that go to make up a substance

UNIT 5

Example 5.41

Convert 60°C into Kelvin and 650 K into degrees centigrade.

You need to remember that 1°C = 1 K and that to convert degrees Celsius into kelvin, we simple add 273 then 60°C + 273 = **333 K**.

Also to convert kelvin into degrees centigrade we subtract 273, then 650 − 273 = **377°C**.

Note that to be strictly accurate we should add 273.15, but for all practical purposes the approximate value of 273 is adequate.

The method used to measure temperature depends on the degree of hotness of the body or substance being measured. Measurement apparatus include liquid-in-glass thermometers, resistance thermometers, thermistors and thermocouples.

All **thermometers** are based on some property of a material that changes when the material becomes colder or hotter. Liquid-in-glass thermometers use the fact that most liquids expand slightly when they are heated. Two common types of liquid-in-glass thermometer are mercury thermometers and alcohol thermometers, both have relative advantages and disadvantages.

Alcohol thermometers are suitable for measuring temperatures down to −115°C and have a higher expansion rate than mercury, so a larger containing tube may be used. They have the disadvantage of requiring the addition of a colouring in order to be seen easily. Also, the alcohol tends to cling to the side of the glass tube and may separate.

Mercury thermometers conduct heat well and respond quickly to temperature change. They do not wet the sides of the tube and so flow well in addition to being easily seen. Mercury has the disadvantage of freezing at −39°C and so is not suitable for measuring low temperatures. Mercury is also poisonous and special procedures must be followed in the event of spillage.

Resistance thermometers are based on the principle that current flow becomes increasing more difficult with increase in temperature. They are used where a large temperature range is being measured: approximately −200°C to 1200°C. **Thermistors** work along similar lines, except in this case they offer less and less resistance to the flow of electric current as temperature increases.

Thermocouple thermometers are based on the principle that when two different metal wires are joined at two junctions and each junction is subjected to a different temperature, a small current will flow. This current is amplified and used to power an analogue or digital temperature display. Thermocouple temperature sensors are often used to measure the temperatures inside engines. They can operate over a temperature range from about −200°C to 1600°C.

We have mentioned in our discussion on thermometers that certain liquids expand with increase in temperature. This is also the case with solids. **Thermal expansion** is dependent on the nature of the material and the magnitude of the temperature increase. We normally measure the linear expansion of solids,

UNIT 5

such as the increase in length of a bar of the material. With gases (as you have already seen) we measure volumetric or cubic expansion.

Every solid has a *linear expansivity value that is the amount the material will expand in metres per kelvin or per degree celsius*. This expansivity value is often referred to as the **coefficient of linear expansion (α)**. Some typical values of α are given in *Table 5.3*.

Table 5.3 Some linear expansion coefficients for engineering materials

Material	Linear expansion coefficient $\alpha/°C$
Glass	9×10^{-6}
Cast iron	10×10^{-6}
Concrete	1.1×10^{-6}
Steel	12×10^{-6}
Copper	17×10^{-6}
Brass	19×10^{-6}
Aluminium	24×10^{-6}

Given the length of a material (l), its linear expansion coefficient (α) and the temperature rise (Δt), the *increase in its length* can be calculated using

$$\text{increase in length} = \alpha l(t_2 - t_1)$$

Note that we are using lower case t to indicate temperature because when we find a temperature difference (Δt) we do not need to convert to kelvin. However, if you are unsure about when to use celsius, then for all thermodynamic problems always convert to **Kelvin** (absolute thermodynamic temperature).

For solids, an *estimate* of the cubic or volumetric expansion may be found using

$$\text{change in volume} = 3\alpha V(t_2 - t_1)$$

where V is the original volume.

A similar relationship exists for surface expansion, where a body experiences a change in area. In this case, the linear expansion coefficient is multiplied by 2, therefore

$$\text{change in area} = 2\alpha A(t_2 - t_1)$$

where A is the original area.

Example 5.42

An aluminium rod is 0.9 m long at a temperature of 10°C. What will be the length of the rod at 120°C?

The increase in length is $\alpha/(t_2 - t_1) = (24 \times 10^{-6})(0.9)(110) = 2.376 \times 10^{-3}\,\text{m}$

So length of rod at 120°C = **0.90238 m**.

Example 5.43

A steel bar has a length of 4.0 m at 10°C. (a) What will be the length of the bar when it is heated to 350°C? (b) If a sphere of diameter 15 cm is made from the same material what will be the percentage increase in surface area, if the sphere is subject to the same initial and final temperatures?

(a) Using $\alpha = 12 \times 10^{-6}$ from *Table 5.3*, the increase in length of the bar is given by

$$x = \alpha l(t_2 - t_1) = (12 \times 10^{-6})(4.0)(350 - 10) = 0.0163\,\text{m}$$

This can now be added to the original length, so final length $= 4.0 + 0.0163 = \textbf{4.0163 m}$.

(b) Increase in surface area of the sphere is $= 2\alpha A(t_2 - t_1)$

We first need to find the original surface area, which is given by

$$A = 4\pi r^2 = 4\pi \times (0.075)^2 = 0.0707\,\text{m}^2$$

and from above the increase in surface area $= (2)(12 \times 10^{-6})(0.0707)(340) = 5.769 \times 10^{-4}\,\text{m}^2$.

$$\text{Therefore, percentage increase in area} = \left(\frac{\text{increase in area}}{\text{original area}}\right) \times 100$$

$$= \frac{5.769 \times 10^{-4}}{0.0707} \times 100\% = \textbf{0.82\%}$$

Test your knowledge

TYK

TYK 5.19

1. Define temperature.

2. Convert (a) $-70°C$ into Kelvin and (b) convert $288\,\text{K}$ into °C.

3. You wish to record temperatures between $-80°C$ and $20°C$. Which type of thermometer would you choose and why?

4. The coefficient of linear expansion for copper is 17×10^{-6}. What is the approximation for the superficial expansion coefficient for copper?

5. A metal rod is 30 cm long at a temperature of 18°C and 30.04 cm long when the temperature is increased to 130°C. What is the value of the coefficient of linear expansion of the metal?

KEY POINT

Heat and work is energy in transit and cannot be stored by matter

Heat

The study of *heat energy* is a necessary foundation for an understanding of the area of science known as *thermodynamics*. Engineering thermodynamics is concerned with the relationship between heat, work and the properties of systems. As engineers we are concerned with the machines (engines) that convert heat energy from fuels into useful mechanical work. It is therefore appropriate to start our study of heat energy by considering the concept of *heat* itself.

Energy is the most important and fundamental physical property of the universe.

Energy may be defined as the capacity to do work; more accurately it may be defined as the capacity to produce an effect. These effects are apparent during the process of energy transfer.

A modern idea of heat is that it is energy in transition and cannot be stored by matter.

> **KEY POINT**
>
> Heat energy can only travel from a hot body to a cold body

Heat (Q) may be defined as transient energy brought about by the interaction of bodies by virtue of their temperature difference when they communicate.

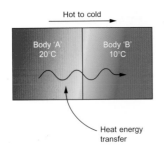

Figure 5.78

Matter possesses stored energy but not transient (moving) energy such as heat or work. Heat energy can only travel or transfer from a hot body to a cold body, it cannot travel up hill. *Figure 5.78* illustrates this fact.

When the temperature between the two bodies illustrated in *Figure 5.78* are equal and no change of state is taking place (e.g. water to ice), then we say that they are in *thermal equilibrium*.

> **KEY POINT**
>
> Heat flow when measured in joules per second is in fact thermal power in watts

Zeroth law of thermodynamics states that if two bodies are each in thermal equilibrium with a third body, they must also be in thermal equilibrium with each other.

The unit of heat energy is the *joule*, which is identical to that of the *unit of work* the **Nm** as you know from your study of mechanical energy. When heat flows then a certain quantity of heat flows into or out of a material in a certain time.

Therefore heat flow $\dot{Q}$ = energy/time or $\dot{Q}$ = J/s = power in watts.

Thus, when heat flows into or out of a thermal system, power is either being absorbed or given out. *Heat energy flowing into a system is considered to be positive and heat flow from a system is negative.*

Heat energy transfer mechanisms

Literature on heat transfer generally recognizes three distinct modes of heat transmission, the names of which will be familiar to you, that is, *conduction*, *convection* and *radiation*. Technically only conduction and radiation are true heat transfer processes, because both of these depend totally and utterly on a temperature difference being present. Convection also depends on the transportation of a mechanical mass. Nevertheless, since convection also accomplishes transmission of energy from high to low temperature regions, it is conventionally regarded as a heat transfer mechanism.

Thermal conduction in solids and liquids seems to involve two processes, the first is concerned with atoms and molecules (*Figure 5.79*), and the second with free electrons.

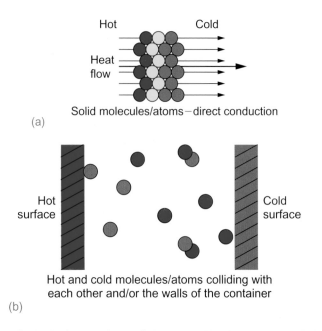

Figure 5.79 (a) Conduction by molecular transfer in solids and liquids and (b) conduction in gases

Atoms at high temperatures vibrate more vigorously about their equilibrium positions than their cooler neighbours. Since atoms and molecules are bonded to one another, they pass on some of their vibrational energy. This energy transfer occurs from atoms of high vibrational energy to those of low vibrational energy, without appreciable displacement. This energy transfer has a knock-on effect, since high vibrational energy atoms increase the energy in adjacent low vibrational energy atoms, which in turn causes them to vibrate more energetically, causing thermal conduction to occur. In solids and liquids (*Figure 5.79a*), the energy transfer is by direct contact between one molecule and another. In gases (*Figure 5.79b*), the conduction process occurs as a result of collisions between hot and cold molecules and the surface of the containing vessel.

The second process involves material with a ready supply of free electrons. Since electrons are considerably lighter than atoms, then any gain in energy by electrons results in an increase in the electron's velocity and it is able to pass this energy on quickly to cooler parts of the material. This phenomenon is one of the reasons why electrical conductors that have many free electrons are also good thermal conductors. Do remember that metals are not the only good thermal conductors; the first mechanism described above which does not rely on free electrons is a very effective method of thermal conduction, especially at low temperatures.

Heat transfer by **convection** consists of two mechanisms. In addition to energy transfer by random molecular motion (diffusion), there is also energy being transferred by the bulk motion of the fluid.

So, in the presence of a temperature difference large numbers of molecules are moving together in bulk (*Figure 5.80*), at the same time as the individual motion of the molecules takes place. The cumulative effect of both of these energy transfer methods is referred to as heat transfer by convection.

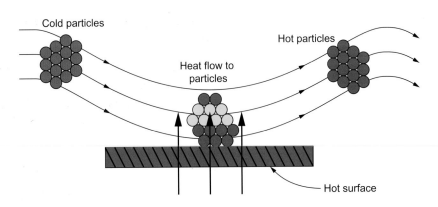

Figure 5.80 Heat transfer by convection

KEY POINT

Radiated heat does not require a medium through which to travel and can travel through a vacuum

Radiation may be defined as the transfer of energy that does not require a medium through which to pass, thus radiation can be transferred through empty space.

Thermal radiation is attributed to the electron energy changes within atoms or molecules. As electron energy levels change energy is released, which travels in the form of electromagnetic waves of varying wavelength. You will meet electromagnetic waves again if you study radio communications or light. When striking a body the emitted radiation is absorbed by, reflected by, or transmitted through the body.

TYK

TYK 5.20

1. State the units and give the sign convention for heat flowing into and out of a system.

2. Write down the symbol for a quantity of heat energy.

3. Explain the nature of the two mechanisms of heat transfer by conduction.

4. What is the essential difference between heat transfer by convection and heat transfer by radiation?

5. With respect to engineering, why is the study of thermodynamics important?

KEY POINT

A thermodynamic system is essentially a thermodynamic substance surrounded by an identifiable boundary

Thermodynamic systems

Thermodynamic systems may be defined as particular amounts of a thermodynamic substance, normally compressible fluids such as vapours and gases, which are surrounded by an identifiable boundary.

We are particularly interested in thermodynamic systems that involve *working fluids* (rather than solids) because these fluids enable the system to do work or have work done upon it. Only transient energies in the form of heat (Q) and work (W) can cross the system boundaries, and as a result there

will be a change in the stored energy of the contained substance (working fluid).

The **property** of a working fluid is an observable quantity, such as pressure, temperature, etc. The state of a working fluid may be defined by any two unique and independent properties. *Boyle's law* defines the state of the fluid by specifying the independent thermodynamic properties of volume and pressure.

A thermodynamic system must contain the following elements:

- a working fluid or substance, that is the matter, which may or may not cross the system boundaries, such as water, steam, air, etc.
- a heat source
- a cold body to promote heat flow and enable heat energy transfer
- system boundaries, which may or may not be fixed.

Some of the more common properties (with units) that are used to define the state of the working matter within a thermodynamic system are shown in *Table 5.4*.

Table 5.4 Thermodynamic properties

	Properties	Units
1	Absolute pressure	Pascal (Pa)
2	Absolute temperature	Kelvin (K)
3	Volume	Cubic metres (m^3)
4	Specific volume	$m^3 kg^{-1}$
5	Specific internal energy	$J kg^{-1}$
6	Specific enthalpy	$J kg^{-1}$

KEY POINT

Extensive or specific properties of a thermodynamic system are dependent on the mass of the working substance

The first three properties given in *Table 5.4* are known as **intensive** properties because they are independent of the mass of the working substance. The last three are **extensive or specific** properties because they are dependent on the mass of the working substance, as can be seen by their units.

When a system working fluid is subject to a **process**, then the fluid will have started with one set of properties and ended with another, irrespective of how the process took place or what happened between the start and end states. For example, if a fluid within a system has an initial pressure (p_1) and temperature (T_1) and is then compressed producing an increase in pressure (p_2) and temperature (T_2), then we say that the fluid has undergone a **process** from state 1 to state 2. A system that undergoes a set of processes whereby the state of the working fluid return to its original state is said to have undergone **a cycle** of processes.

A **closed system** may be defined as a system in which no mass transfer of the working fluid takes place.

The energy exchange that takes place in a closed system with fixed system boundaries is illustrated in *Figure 5.81*.

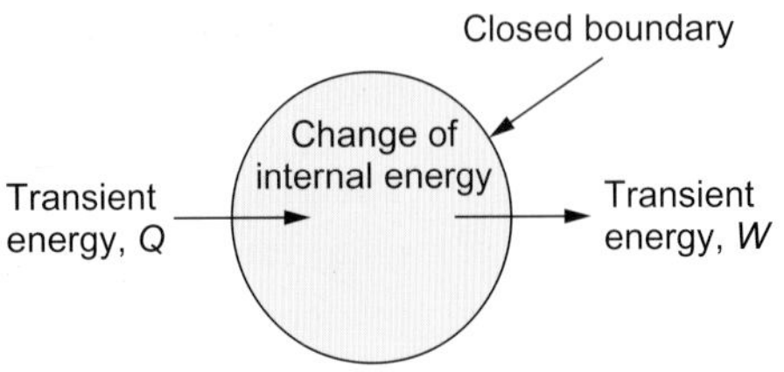

Closed boundary

Change of
internal energy

Transient
energy, Q

Transient
energy, W

Figure 5.81 Closed system energy exchange

The boundary of a closed system is not necessarily rigid. What makes the system closed is the fact that no mass transfer of the system fluid takes place, while an interchange of heat and work may take place.

A classic example of a **closed system** with **moving boundaries** that undergoes a cycle of processes is **the internal reciprocating piston combustion engine**. We say that work is transferred in a thermodynamic system **if there is movement of the system boundaries**; this would be the case in an internal combustion engine when the piston within the cylinder moves, as a result of the combustion process or when acted upon by the crankshaft. When work **is done by the system** we refer to this as **positive work**, when **work is done on the system** this is **negative work**.

Consider the cylinder and piston assembly of an internal combustion engine in *Figure 5.82*.

The crown of the piston, the cylinder walls and the cylinder head form the closed boundary with the valves closed. The **transient energy** is in the form of combustible fuel, which creates a sudden pressure wave that forces the piston down. Therefore, as the piston moves the boundaries of the system move. This movement causes the system to do positive **work** (force × distance) on its surroundings. In this case the piston connecting rod drives a crank to provide motive power.

Notice that in a closed system, it requires movement of the system boundary for work to be done by the system or on the system, thus as we have said previously work (like heat) is a transient energy; it is not contained within the system.

The first law of thermodynamics applies the principle of the conservation of energy to open and closed thermodynamic systems. Formally, it may be stated as follows: **when a system undergoes a thermodynamic cycle then the net heat energy transferred to the system from its surroundings is equal to the net heat energy transferred from the system to its surroundings.** You have already been introduced to the idea of a thermodynamic *cycle* where the working fluid of the system undergoes a series of processes and finally returns to its initial state. In this unit, we will only consider the application of the first law to closed systems.

The principle of the conservation of energy, that is, the first law of thermodynamics, applied to a closed system states that: given a total amount of energy in a system and its surroundings this total remains the same irrespective of the changes of form that may occur.

KEY POINT

In a closed system there is no mass transfer of the system fluid

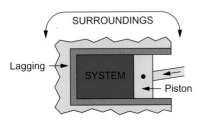

SURROUNDINGS

Lagging

SYSTEM

Piston

Figure 5.82 Cylinder and piston engine of the internal combustion engine

In other words, ***the total energy entering a system must be equal to the total energy leaving the system***. This is represented diagrammatically in *Figure 5.83*, where the initial internal energy is U_1, and the final internal energy is U_2, so the change in internal energy is shown as $U_2 - U_1$ or ΔU. So in symbol form: $U_1 + Q = U_2 + W$ (i.e. total energy in = total energy out).

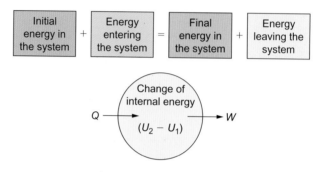

Figure 5.83 First law of thermodynamics applied to a closed system

In its more normal form: $\boldsymbol{Q - W = \Delta U}$

Thus, the above equation represents the concept of the first law of thermodynamics applied to a closed system. This equation is also known as the **Non-flow energy equation (NFEE)**.

You should also note from *Figure 5.83* that heat supplied *to* the system and work done *by* the system are both ***positive***.

Example 5.44

During a non-flow thermodynamic process the internal energy possessed by the working fluid within the system was increased from 10 to 30 kJ, while 40 kJ of work was done by the system. What is the magnitude and direction of the heat energy transfer across the system during the process?

Using $Q - W = U_2 - U_1$, where $U_1 = 10\,\text{kJ}$, $U_2 = 30\,\text{kJ}$ and $W = 40\,\text{kJ}$ (positive work), then $Q - 40 = 30 - 10$ and $\boldsymbol{Q = 60\,kJ}$.

Since Q is *positive*, it must be heat supplied to the system, which may be represented by an arrow pointing into the system, as shown in *Figure 5.83*.

According to our previous definition for the first law, when a system undergoes a complete cycle, then the net heat energy supplied is equal to the net work done, and this definition was based on the principle of the conservation of energy, which is a universal law determined from the observation of natural events. The **second law of thermodynamics** extends this idea. It tells us that although the net heat supplied is equal to the net work done, the *total or **gross heat supplied must be greater than the net work done***. This is because some heat must be ***rejected*** (lost) by the system during the cycle. Thus, in a ***heat engine*** (*Figure 5.84*), such as the internal combustion engine, the heat energy supplied by the fuel must be greater than the work done by the crankshaft. During the cycle, heat energy is rejected or lost to the surroundings of the system through friction, bearing drag, component wear, etc.

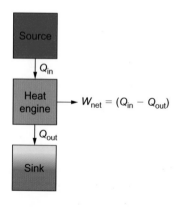

Figure 5.84 The heat engine

A heat engine is a system operating in a complete cycle and developing net work from a supply of heat. The second law implies that there is a need for a heat source and a means of rejection or absorption of heat from the system. The heat rejected within the system is often referred to as the heat **sink**. We know from the second law that for a complete cycle, the net heat supplied is equal to the net work done, then from *Figure 5.84* using the symbols:

$$Q_{in} - Q_{out} = W_{net}$$

We also know from the second law that the total heat supplied (heat in) has to be greater than the net work done, that is, $Q_{in} > W$.

Now the ***thermal efficiency (η)*** of a heat engine is given by:

$$\text{Thermal efficiency } (\eta) = \frac{\text{net work done } (W_{net})}{\text{total heat supplied } (Q_{in})}$$

or

$$\textbf{Thermal efficiency } (\eta) = \frac{Q_{in} - Q_{out}}{Q_{in}}$$

So, for example, from the above formula if the heat supplied to a thermal system is 50 MJ and the net work output of the system in this case is 30 MJ, then the efficiency of the system $\eta = \frac{30}{50} = 60\%$.

Note: The 'net work' is $W_{net} = Q_{in} - Q_{out} = 30$ MJ in this case.

There are many examples of the heat engine, designed to minimize thermal losses, predicted by the second law. These include among others: the steam turbine, refrigerators and air-conditioning units. The internal combustion engine is not strictly a heat engine because the heat source is mixed directly with the working fluid, but because it is used as the basis for so many thermodynamic engineering systems it has been mentioned here.

TYK 5.21

1. Define heat and explain the difference between heat energy and heat flow.
2. Define a thermodynamic system using your own words.
3. What elements are necessary for a thermodynamic system?
4. Explain the difference between intrinsic and extensive properties of the working fluid.
5. 'A closed system is one in which no mass transfer takes place and one that must have fixed system boundaries.' State whether this statement is true or false and give a reason to qualify your answer.

6. What is the essential difference between a closed system and an open system?

7. Write down the non-flow energy equation (NFEE) and explain the meaning of each term in this equation.

8. If 35 kJ of heat energy is supplied to a non-flow system, while the internal energy in the system is increased from 5 to 45 kJ, determine the magnitude and direction of the work done.

9. What does the second law of thermodynamics tell us about the efficiency of a heat engine?

10. A heat engine is supplied with 150 MJ of heat energy. If during this time the net work done by the heat engine is 65,500 kJ, determine its thermal efficiency.

Specific heats and energy transfer

From what has been said so far about heat transfer and the expansion of heated materials, it will be apparent that different materials have different capacities for absorbing and transferring thermal energy. The thermal energy needed to produce a temperature rise depends on the mass of the material, the type of material and the temperature rise to which the material is subjected. Therefore, the inherent ability of a material to absorb heat for a given mass and temperature rise is dependent on the material itself. This property of the material is known as its **specific heat capacity**.

In the **SI system**, the specific heat capacity of a material is the same as the thermal energy required to produce a 1 K rise in temperature in a mass of 1 kg.

UNIT 5

From the definition it can be seen that the SI units of specific heat capacity are joules per kilogramme Kelvin or in symbols $J\,kg^{-1}K^{-1}$. Some specific heat capacities for *engineering solids and liquids* are shown in *Table 5.5*. It should be noted that the value of specific heat capacities may vary slightly dependent on the temperature at which the heat transfer takes place, therefore the table only gives average values.

Another way of defining the specific heat capacity of any substance is the amount of heat energy required to raise the temperature of unit mass of the substance through one degree, under specific conditions.

In thermodynamics, two specified conditions are used, those of *constant volume* and *constant pressure*. With *gases* the two specific heats do not have the same value and it is essential that we distinguish between them.

If 1 kg of a gas is supplied with an amount of heat energy sufficient to raise the temperature by 1 degree centigrade or Kelvin while the volume of the gas remains constant, then the amount of heat energy supplied is known as the *specific heat capacity at constant volume* and is denoted by c_v.

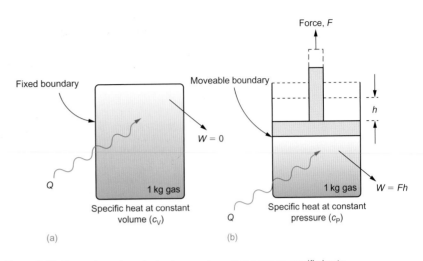

Figure 5.85 Comparison of constant volume and constant pressure specific heats

Note that under these circumstances (*Figure 5.85a*) no work is done, but the gas has received an increase in internal energy (*U*). The specific heat at constant volume for air is $c_v = 718\,J\,kg^{-1}K^{-1}$. This constant is well worth memorizing!

If 1 kg of a gas is supplied with a quantity of heat energy sufficient to raise the temperature of the gas by 1 degree centigrade or Kelvin while the pressure is held constant, then the amount of heat energy supplied is known as the specific heat capacity at constant pressure and is denoted by c_p.

This implies that when the gas has been heated it will expand a distance *h* (*Figure 5.85b*), so work has been done. Thus, for the same amount of heat energy there has been an increase in internal energy (*U*) plus work. The value of c_p is, therefore, greater than the corresponding value of c_v.

The specific heat capacity at constant pressure for air is $c_p = 1005\,J/kg\,K$. Again this is a constant worth remembering.

Some common specific heat capacities for engineering solids, liquids and gases are shown in *Tables 5.5* and *5.6*. Note that in the table for gases both the specific heats at constant volume and constant pressure are given.

KEY POINT

For a gas the specific heat at constant pressure will be greater than the specific heat at constant volume because work is done

Table 5.5 Specific heat capacities for solids and liquids

Substance	Specific heat capacity ($J\,kg^{-1}K^{-1}$)
Aluminium	920
Cast iron	540
Steel	460
Copper	390
Alcohol	230
Ice	700
Mercury	140
Water	4200

UNIT 5

Table 5.6 Specific heat capacities for gases

Gas	Specific heat capacity (J kg^{-1}K^{-1})	
	c_v	c_p
Air	718	1005
Argon	520	710
Carbon dioxide	640	860
Nitrogen	743	1040
Oxygen	660	920
Propane	1490	1680

Therefore, knowing the mass of any substance and its specific heat capacity, we are now in a position to be able to calculate the thermal energy required to produce any given temperature rise, from the equation:

$$\textbf{Thermal energy } Q = mc\Delta T$$

where c is the specific heat capacity of the material, with units J/kg K, and ΔT (delta T) is the temperature difference. Note that in many books you will see ΔT written as $T_2 - T_1$ indicating the temperature difference, so that the above formula for heat quantity is given as $Q = mc(T_2 - T_1)$. Also, you will remember from our earlier discussion on heat that heat flow

$$\dot{Q} = \frac{Q}{t} = \frac{\text{heat transfer (joules)}}{\text{time taken (seconds)}}$$

or, thermal power. So that we may write

$$\textbf{Thermal power } \dot{Q} = \frac{\textbf{mc}\Delta\textbf{T}}{\textbf{t}} \textbf{ Watts}$$

Example 5.45

How much thermal energy is required to raise the temperature of 5 kg of aluminium from 20°C to 40°C? Take the specific heat capacity for aluminium as 900 /kg K.

All that is required is to substitute the appropriate values directly into the equation:

$$Q = mc\Delta T = (5)(900)(40 - 20) = 90,000 \text{ J} = \textbf{\textit{90 kJ}}$$

Latent heat

The addition or rejection of thermal energy does not always give rise to a measurable temperature change, *instead a change in state or phase change may occur*, where the three states of matter are solid, liquid and gas. So, for example, when an electric kettle boils water, steam is given off. Water boils

at 100°C (373 K) and it will remain at this temperature even though heat energy is being added to the water by the element of the kettle, until all the water has turned to steam. Since heat has been added and no measurable temperature rise has taken place we refer to this heat energy as **latent heat** or hidden heat. When the addition of heat causes a measurable temperature rise in the substance this is referred to as the addition of *sensible heat* or measurable heat. *Figure 5.86* shows the effect of these state changes for water, when it is heated from ice to steam. The same graph may be applied to all matter that can exist as a solid, liquid and gas or vapour.

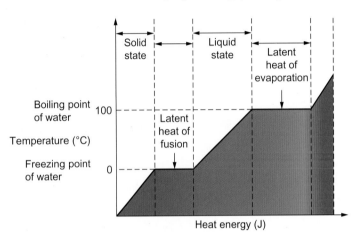

Figure 5.86 Change in temperature and state for water with continuous heat addition

We refer to the thermal energy required to change a solid material into a liquid as the latent heat of fusion. For water, 334 kJ of thermal energy is required to change 1 kg of ice at 0° C into water at the same temperature. Thus, the specific latent heat of fusion for water is 334 kJ. In the case of latent heat, *specific*, refers to unit mass of the material, that is per kilogramme. So, the **specific latent heat of fusion** of a substance **is the thermal energy required to turn 1 kg of a substance from a liquid into a solid without change in temperature**.

Note: The specific latent heat of fusion is also known as the specific *enthalpy (h)* of fusion.

If you study thermodynamics further you will see that enthalpy is used as a measure of the heat energy in a fluid when it is flowing and it is the sum of the internal energy *U* of the fluid and its pressure–volume energy (*pv*). In symbols, $h = u + pv$; we will not be using this relationship in this unit.

If we wish to find the thermal energy required to change any amount of a substance from a solid into a liquid, then we use the relationship $Q = mL$, where **L** is the *specific latent heat* of the substance.

In a similar manner to the above argument: *the thermal energy required to change 1 kg of a substance from a liquid into a gas without change in temperature is known as* **the specific latent heat of vaporization**. Again, if we wish to find the thermal energy required to change any amount of a substance from a liquid into a gas we use the relationship $Q = mL$, but in this case L = the specific latent heat of vaporisation. The specific latent heat of vaporization for water is 2.26 MJ $kg^{-1} K^{-1}$.

Example 5.46

(a) How much heat energy is required to change 3 kg of ice at 0°C into water at 30°C?

(b) What thermal energy is required to condense 0.2 kg of steam into water at 100°C?

(a) The thermal energy required to convert ice at 0°C into water at 0°C is calculated using the equation: $Q = mL$ and substituting values we get $Q = (3)(334 \times 10^3) = 1.002$ MJ.

The 3 kg of water formed has to be heated from 0°C to 30°C. The thermal energy required for this is calculated using the equation $Q = mc\Delta T$. You have already met this equation when we studied specific heat earlier.

$$\text{So, in this case } Q = (3)(4200)(30) = 378,000 \text{ J}$$

$$\text{Then total thermal energy required} = 1.002 + 0.378 = \textbf{\textit{1.38 MJ}}.$$

(b) In this case we simply use $Q = mL$ since we are converting steam to water at 100°C, which is the vaporization temperature for turning water into steam.

$$\text{Then } Q = (0.2)(2.26 \times 10^6) = \textbf{\textit{452 kJ}}$$

Note the large amounts of thermal energy required to change the state of a substance.

A liquid does not have to boil in order for it to change state, the nearer the temperature is to the boiling point of the liquid, then the quicker the liquid will turn into a gas. At much lower temperatures the change may take place by a process of *evaporation*. The steam rising from a puddle, when the sun comes out after a rainstorm, is an example of evaporation where water vapour forms as steam, well below the boiling point of the water.

TYK 5.22

1. Define 'specific heat' for a solid and explain the difference between the specific heat capacities for solids and those for gases.

2. Why for a gas is the specific heat at constant pressure greater than that at constant volume?

3. Write down the formula for thermal energy and state the meaning of each term.

4. Using the tables of specific heat capacities, determine the thermal energy required to raise the temperature of a copper container of mass 2 kg from 288 to 348 K. If the time taken to achieve the temperature rise in the copper was 2 minutes what is the amount of thermal power being absorbed by the copper during this time?

5. If 3 kg of aluminium requires 54 kJ of energy to raise the temperature from 10°C to 30°C, find the specific heat capacity of the aluminium.

6. How much heat energy is required to change 2 kg of ice at 273 K into water at 313 K? Take the specific latent heat of fusion of water as 334 kJ $kg^{-1}K^{-1}$ and the specific heat of water as 4200 J $kg^{-1}K^{-1}$.

Expansion and compression of gases

In this short section we consider the expansion and compression of gases, rather than solids or liquids. Gas behaviour may be modelled using *the*

ideal gas laws, where it is assumed that these gas processes (expansion, compression, etc.) take place without any external energy changes outside of the thermodynamic system being considered. We say that *a gas is a perfect gas if it obeys the ideal gas laws*.

The gas laws

In the study of gases we have to consider the interactions between temperature, pressure and volume (remembering that density is mass per unit volume). A change in one of these characteristics always produces a corresponding change in at least one of the other two.

Unlike liquids and solids, gases have the characteristics of being easily compressible and of expanding or contracting readily in response to changes in temperature. Although the characteristics themselves vary in degree for different gases, certain basic laws can be applied to what we call a perfect gas. **A *perfect or ideal gas* is simply one which has been shown, through experiment, to follow or adhere very closely to these gas laws. In these experiments one factor, for example volume, is kept constant while the relationship between the other two is investigated. In this way it can be shown that:

> 1. **The pressure of a fixed mass of gas is directly proportional to its absolute temperature, providing the volume of the gas is kept constant.**
>
>   In symbols: $P \propto T$ or, for a fixed mass of gas $\dfrac{P}{T} = $ **constant**
>
>   (providing V remains constant)
>   The above relationship is known as the **pressure law**.

Gas molecules are in a state of perpetual motion, constantly bombarding the sides of the gas-containing vessel. Each molecule produces a minute force as it strikes the walls of the container. Since many billion molecules hit the container every second, this produces a steady outward pressure.

Figure 5.87 shows how the pressure of the gas varies with temperature.

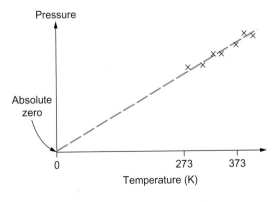

Figure 5.87 Pressure/temperature relationship of a gas

> **KEY POINT**
>
> When dealing with the gas equations or any thermodynamic relationship we always use absolute temperature (T) measured in kelvin

If the graph is 'extrapolated' downwards, in theory we will reach a temperature where the pressure is zero. This temperature is known as

absolute zero and is approximately equal to −273 K. You have already met this value when you studied temperature earlier. The relationship between the kelvin scale and the celsius scale is shown in *Figure 5.88*.

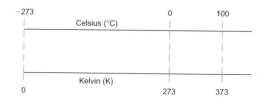

Figure 5.88 Kelvin-Celsius scales

Returning to the gas laws, it can also be shown experimentally that

2. **The volume of a fixed mass of gas is directly proportional to its absolute temperature providing the pressure of the gas remains constant.**

   So, for a fixed mass of gas:

   $$\frac{V}{T} = \text{constant}$$

   (providing *m* is fixed and *p* remains constant)
   This relationship is known as **Charles' law**.

A further relationship occurs if we keep the temperature of the gas constant. This states that

3. **The volume of a fixed mass of gas is inversely proportional to its pressure providing the temperature of the gas is kept constant.**
   In symbols:

   $$p \propto \frac{1}{V} \text{ or, for a fixed mass of gas:}$$

   $$pV = \text{constant}$$

   This relationship is better known as **Boyle's law**. It is illustrated in *Figure 5.89*.

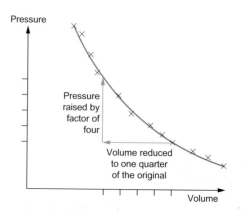

KEY POINT

A perfect gas is one that is assumed to obey the ideal gas laws

Figure 5.89 Boyle's pressure/volume relationship

In dealing with problems associated with the gas laws, remember that we assume that all gases are *ideal* or *perfect*. In reality no gas is ideal but at low and medium pressures and temperatures, most gases behave in an ideal way.

The pressure law, Charles' law and Boyle's law can all be expressed in terms of one single equation known as the *combined gas equation*, this is for a fixed mass of gas:

$$\frac{pV}{T} = \text{constant}$$

If we consider a fixed mass of gas before and after changes have taken place, then from the combined gas equation it follows that

$$\frac{p_1 V_1}{T_1} = \frac{p_2 V_2}{T_2}$$

where the subscript 1 is used for the initial state and subscript 2 for the final state of the gas.

The above relationship is very useful when solving problems concerned with the gas laws.

Example 5.47

A quantity of gas occupies a volume of 0.5 m³. The pressure of the gas is 300 kPa when its temperature is 30°C. What will be the pressure of the gas if it is compressed to half its volume and heated to a temperature of 140°C?

When solving problems involving several variables, always tabulate the information given in appropriate units.

$$p_1 = 300 \, \text{kPa} \quad p_2 = ?$$
$$V_1 = 0.5 \, \text{m}^3 \quad V_2 = 0.25 \, \text{m}^3$$
$$T_1 = 303 \, \text{K} \quad T_2 = 413 \, \text{K}$$

Remember to convert temperature to Kelvin, by adding 273°C.

Using the combined gas equation and after rearrangement:

$$p_2 = \frac{p_1 V_1 T_2}{T_1 V_2} = \frac{(300)(0.5)(413)}{(303)(0.25)} = \boldsymbol{818 \, kPa}$$

The characteristic gas equation

The combined gas law, which you met earlier, stated that for a perfect gas with unit mass:

$$\frac{pV}{T} = \text{a constant}$$

This relationship is of course true for any fixed mass of gas and so we can write that

$$\frac{pV}{T} = \text{mass} \times \text{a constant} \; (R)$$

Now, for any perfect gas that obeys the ideal gas laws, this constant R is specific to that particular gas, that is, **R is the characteristic gas constant** or specific gas constant for the individual gas concerned.

The **characteristic gas equation** may be written as

$$\frac{pV}{T} = mR \quad \text{or} \quad pV = mRT$$

The units for the characteristic gas constant is the joule per kilogramme kelvin ($J\,kg^{-1}K^{-1}$).

Note that when the above equation is used both *absolute pressure and absolute temperature must be used.*

The characteristic gas constant for a number of gases is given in *Table 5.7.*

Table 5.7 Some characteristic gas constants

Gas	Characteristic gas constant ($J\,kg^{-1}K^{-1}$)
Hydrogen	4124
Helium	2077
Nitrogen	297
Air	287
Oxygen	260
Argon	208
Carbon dioxide	189

The characteristic gas constant for air, from the above table, is **$R = 287\,J/kg\,K$**. This is related to the *specific heat capacities* (that you met earlier) for air in the following way, that is, **$R = c_p - c_v$**. You should check this relationship by noting the above values of R, c_p and c_v for air. This relationship **($R = c_p - c_v$)** is not only valid for air, but it is also valid for any perfect gas that follows the ideal laws.

Example 5.48

0.22 kg of gas at a temperature of 20°C and pressure of 103 kNm^{-2} occupies a volume of 0.18 m^3.
If the c_v for the gas $= 720\,J\,kg^{-1}K^{-1}$, find:
(a) the characteristic gas constant
(b) the specific heat capacity at constant pressure.

(a) Using $pV = mRT$

then on rearrangement,

$$R = pV/mT = \frac{(103 \times 10^3)(0.18)}{(0.22)(293)} = \textbf{288}\,\textbf{J\,kg}^{-1}\,\textbf{K}^{-1}$$

(b) From $R = c_p - c_v$,

then $c_p = R + c_v = 288 + 720 = \textbf{1008}\,\textbf{J\,kg}^{-1}\textbf{K}^{-1}$

TYK 5.23

1. Define 'perfect gas'.

2. When using the 'pressure law' what parameter is assumed to remain constant?

3. According to Boyle's law what happens to the volume of gas if the pressure is tripled?

4. Write down the SI units for the characteristic gas constant and explain their meaning in words.

5. The volume of a fixed mass of gas is $60\,cm^3$ at 27°C. If the pressure is kept constant, what is the volume at 127°C?

6. A sealed oxygen cylinder has a pressure of 10 bar at 0°C. Calculate:
   (a) the new pressure when the temperature is 90°C
   (b) the temperature when the pressure increases by 50%.

7. A quantity of gas occupies a volume of $0.5\,m^3$. The pressure of the gas is $300\,kPa$, when the temperature is 300 K and the pressure is $900\,kPa$ when the temperature is raised to 400 K. What is the new volume of the gas?

8. 0.25 kg of a gas at a temperature of 300 K and pressure of 111.4 kPa occupies a volume of $0.2\,m^3$. If the c_v for the gas = 743 J kg^{-1}K^{-1}, find:
   (a) the characteristic gas constant
   (b) the specific heat capacity at constant pressure.

End of Unit Review

1. (a) What is the sign convention for:

(i) forces?

(ii) moments?

(b) Look at the diagram below of a non-concurrent coplanar force system and determine:

(i) The magnitude and direction of the resultant force

(ii) The turning moment of the resultant force.

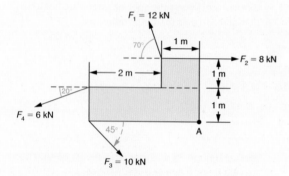

Figure 5.90 Non-concurrent coplanar force system for End of unit review question 1

2. (a) State the conditions for static equilibrium when a system of forces act on a simply supported beam

(b) Determine the reactions at the supports for the loaded uniform cross-section beam shown below.

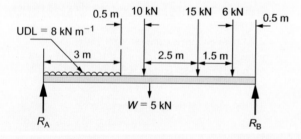

Figure 5.91 Simply supported beam for End of unit review question 2

3. A circular steel tie rod 25 cm long has a diameter of 14 mm and extends 0.15 mm under a load of 22 kN, determine:

(a) the stress

(b) the strain

(c) the modulus of elasticity, **E**, of the steel material from which the tie rod is made.

4. A workshop guillotine is use to shear a sheet of metal 1.5 m across and 1.6 mm thick. If a shear force of 840 kN is required to shear the metal, determine:

(a) The ultimate shear stress of the metal

(b) The working stress in shear, if the metal is used for a component that has a safety factor of 2.

5. The maximum load in a tensile test on a mild steel specimen of diameter 12.5 mm is 95 kN. Calculate the ultimate tensile stress. Also , determine the working stress and greatest allowable load on a rod of the same material 25 mm in diameter, given that the factor of safety is to be 3.5.

6. A body starts from rest and is subject to a constant acceleration of 4 ms^{-2} up to a speed of 20 ms^{-1}. It then travels at 20 ms^{-1} for 30 seconds after which time it is retarded to a speed of 4 ms^{-1}. If the complete motion takes 50 seconds, find:

(a) the time taken to first reach 20 ms^{-1}

(b) the retardation

(c) the total distance travelled.

7. A motor vehicle of mass 2 tonne accelerates from 40 to 110 km/h in 15 seconds. Determine the change in linear kinetic energy of the vehicle and the power required for the acceleration, assuming no external losses.

8. (a) Define:

(i) linear momentum

(ii) impulse

(b) A railway engine of mass 30 tonne and velocity 5 ms^{-1} collides with a railway wagon of mass 15 tonne moving in the same direction with a velocity of 2 ms^{-1}, during a shunting operation. When coupled, the railway engine and wagon move off with a common velocity. Calculate this velocity, ignoring friction and all other external effects.

9. A hammer of a pile driver of mass 2.5 tonne falls 2 m onto a pile. The impact takes 0.025 seconds and the hammer does not rebound. Determine the impulse and the average applied force exerted on the pile by the hammer.

10. (a) A load of mass 70 kg is hoisted vertically by means of a light rope running over a pulley. It is accelerated from 2 to 4 ms^{-1} while travelling through a distance of 5 m. If the resistance to motion is 55 N, determine using *both* an energy method and D'Alembert's principle:

(i) the pulling force in the rope

(ii) work done

(iii) average power required to accelerate the load.

(b) Compare and contrast the appropriateness of using an energy method or D'Alembert's principle to solve the problem in part (a).

11. (a) State and explain the significance of Archimedes' principle

(b) An air balloon has a total volume of 5500 m^3 and its weight and content total 80,000 N. If standard sea-level density of air is 1.2256 kg m^{-3}, calculate

(i) the upthrust

(ii) the mass that needs to be removed from the balloon in order that the balloon rises.

UNIT 5

12. A vertical rectangular sluice plate is used to prevent river water flowing onto a water mill wheel. If on a particular day the sluice plate is fully closed and the water level is in contact with the whole of one side of a sluice plate that is 8 m in height and 3 m wide, determine the thrust force exerted on the plate by the water and the subsequent overturning moment from the base of the plate. Take the density of the water as 1000 kg m^{-3}.

13. A tapered converging fuel pipe in the diagram below has an inlet diameter of 20 cm and fuel enters the inlet to the pipe at a velocity of 2 ms^{-1}. The duct has been designed to deliver fuel to the outlet, at a velocity between 4.0 and 5.0 ms^{-1}.

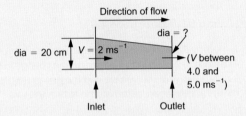

Figure 5.92 A converging duct for End of unit review question 13

Assuming the fuel flow is incompressible and the relative density of the fuel is 0.78, determine:

(a) volume flow rate for the system

(b) the mass flow rate for the system

(c) the outlet duct diameter range to allow fuel to flow between minimum and maximum design velocities.

14. A cast iron surface table has a length of 1.2 m at 15°C.

(a) What will be the length of the surface table at 365°C, given that the linear expansion coefficient for cast iron is 10×10^{-6}/°C?

(b) If the width of the same cast iron surface table is 0.4 m at 15°C, estimate the area of the surface table after being subject to the same temperature rise.

15. A rigid steel strut 400 mm long is fixed rigidly between two solid unyielding supports in the figure below

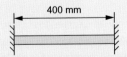

Figure 5.93 Rigidly constrained steel strut for End of unit review question 15

The strut is subject to a temperature rise of 80°C. Given that the coefficient of linear expansion of the steel is $\alpha = 12 \times 10^{-6}$/°C and its modulus of elasticity is 208 GPa determine:

(a) the change in length due to the temperature rise that would occur if the strut was unconstrained

(b) the strain

(c) the stress set up in the strut due to the temperature rise.

16. (a) Define the specific latent heat of fusion *and* of vaporization.

(b) How much thermal energy is required to turn 5 kg of ice at 0°C into water at 50°C. Take the specific latent heat of ice as 334 kJ kg^{-1}K^{-1} and the specific heat of water as 4200 J kg^{-1}K^{-1}.

(c) A 200 W electric heater is embedded in a 1 kg block of ice. Taking the specific latent heat of ice as that given in part (b), determine:

(i) the mass of ice that has melted after 10 minutes

(ii) the length of time it takes for all the ice to melt.

17. (a) Define:

(i) temperature

(ii) heat

(iii) heat flow.

(b) Explain how heat energy transfer by conduction may be treated as a two stage process.

(c) You wish to measure, with reasonable accuracy, the temperature of the exhaust gases as they leave the cylinder of an internal reciprocating piston combustion engine. Suggest a suitable temperature measuring system, giving reasons for your choice.

18. (a) Explain what the first law of thermodynamics states about energy transfer when applied to a closed system and how the second law modifies and refines the first law.

(b) A heat engine is supplied with 180 MJ of heat energy in 40 seconds. If during this time the work output by the heat engine is 95,000 kJ, determine:

(i) the power output from the heat engine

(ii) the engines thermal efficiency.

19. (a) Define:

(i) a perfect gas

(ii) Boyle's law

(iii) combined gas law.

(b) The air pressure of a motor vehicle tyre is 2 bar at 290 K. The temperature rises to 320 K after the vehicle has been travelling. Assuming there is no change in the volume of the tyre, determine the new tyre pressure in pascal.

20. (a) Define:

(i) c_v

(ii) c_p

(iii) R

(b) 0.06 m^3 of air is contained in a sealed cylinder at a pressure of 1.39 bar and temperature of 17°C. A quantity of heat energy is supplied to increase the temperature of the air to 97°C. Given that the

characteristic gas constant for air is 287 J kg^{-1} K^{-1} and $c_v = 718$ J kg^{-1} K^{-1} determine:

 (i) the mass of the air

 (ii) the density of the air at the initial conditions

(iii) the heat energy supplied during the process

(iv) the final pressure of the air.

(c) Suppose that the cylinder described in part (b) was not sealed but free to expand during the time that the heat energy was being supplied, explain how this change would effect the air and the output from the system.

This electric rapid transit vehicle uses electrical energy to provide power for its motors and electronics to control its movement. Even more exciting is the prospect of using microcomputer-based systems to make vehicles like this fully automatic so that they do not even need a driver. Engineers need to know how electricity and electronics can be used to provide control of motion as well as countless other applications in the home, industry and telecommunications.

Unit 6

Electrical and Electronic Principles

In today's world, electricity is something that we all take for granted. So, before we get started, it is worth thinking about what electricity means to you and, more importantly, how it affects *your* life. Think, for a moment, about where and how electricity is used in your home, car, workplace or college. You will quickly conclude that electricity is a means of providing heat, light, motion and sound. You should also conclude that electricity is invisible – we only know that it is there by looking at what it does!

Closely related to electricity is the world of electronics in which devices such as transistors and integrated circuit 'microchips' operate using tiny electric charges and electrons in motion. Electronics underpins the operation of a wide variety of consumer electronic equipment such as CD and DVD players, TV receivers, computers and mobile phones. In addition, cars, motorbikes, aircraft, trains and ships all rely on electronic equipment for their safe operation. For example, a modern aircraft could not even get off the ground without the aid of the electronic systems that control its engines, operate its flight control systems, provide communications with the ground and with other aircraft, and operate the radar and other navigation aids.

When you have completed this unit you will have an understanding of fundamental electrical and electronic principles and you will be able to apply this to a variety of simple electronic circuits using components such as resistors, capacitors, inductors, transformers and diodes. You will be able to apply circuit theory to determine the voltage, current and resistance and an electric circuit and you will understand how components work together in simple timing and rectifier circuits.

Many electrical and electronic circuits use alternating currents (AC). This unit provides you with an introduction to the use of this type of electric current as well as the direct current (DC) that comes from batteries and low-voltage power supplies. This unit will also provide you with plenty of opportunity to put your new found knowledge and understanding into practice when investigating and making measurements on real electronic circuits.

BTEC National Engineering. DOI: 10.1016/B978-0-12-382202-4.00006-4
Copyright © 2010 Mike Tooley and Lloyd Dingle. Published by Elsevier Ltd. All rights reserved.

Direct Current Circuits

Electricity

To understand what electricity is we need to take a look inside the atoms that make up all forms of matter. Since we cannot actually do this with a real atom we will have to use a model. Fortunately, understanding how this model works is not too difficult – just remember that what we are talking about is very, very small!

As you already know, all matter is made up of atoms or groups of atoms (*molecules*) bonded together in a particular way. In order to understand something about the nature of electrical charge we need to consider a simple model of the atom. This model known as the Bohr model (see *Figure 6.1*) shows a single atom consisting of a central nucleus with orbiting electrons.

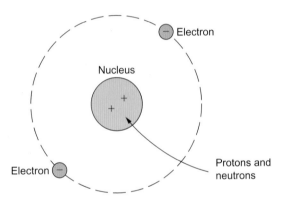

Figure 6.1 The Bohr model of the atom

Within the nucleus there are **protons** which are positively charged and **neutrons** which as their name implies are electrical neutral and have no charge. Orbiting the nucleus are electrons that have a negative charge, equal in magnitude (size) to the charge on the proton. These electrons are approximately 2,000 times lighter than the protons and neutrons in the nucleus.

In a stable atom the number of protons and electrons are equal, so that overall, the atom is neutral and has no charge. However, if we rub two particular materials together, electrons may be transferred from one to another. This alters the stability of the atom, leaving it with a net positive or negative charge. When an atom within a material loses electrons it becomes positively charged and is known as a **positive ion**, when an atom gains an electron it has a surplus negative charge and so is known as a **negative ion**. These differences in charge can cause **electrostatic** effects. For example, combing your hair with a nylon comb may result in a difference in charge between your hair and the rest of your body, resulting in your hair standing on end when your hand or some other differently charged body is brought close to it.

The number of electrons occupying a given orbit within an atom is predictable and is based on the position of the element within the periodic table. The electrons in all atoms sit in a particular position (shell) dependent on their energy level. Each of these shells within the atom is filled by electrons from the nucleus outwards, as shown in *Figure 6.2*. The first, inner most, of these shells can have up to two electrons, the second shell can have up to eight and the third up to 18.

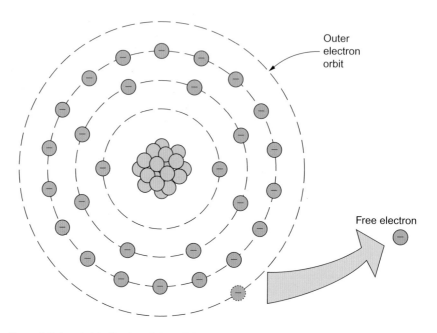

Figure 6.2 A material with a loosely bound electron in its outer shell

Conductors and insulators

A material which has many free electrons available to act as charge carriers and thus allows current to flow freely is known as a **conductor**. Examples of good conductors include aluminium, copper, gold and iron. *Figure 6.2* shows a material with one outer electron that can become easily detached from the parent atom. It requires a small amount of external energy to overcome the attraction of the nucleus. Sources of such energy may include heat, light or electrostatic fields. The atom once detached from the atom is able to move freely around the structure of the material and is called a **free electron**. It is these free electrons that become the **charge carriers** within a material. Materials that have large numbers of free electrons make good **conductors** of electrical energy and heat.

In a material containing free electrons their direction of motion is random, as shown in *Figure 6.3a*, but if an external force is applied that causes the free electrons to move in a uniform manner (*Figure 6.3b*) an electric **current** is said to flow.

Metals are the best conductors, since they have a very large number of free electrons available to act as charge carriers. Materials that do not conduct charge are called *insulators*, their electrons are tightly bound to the nuclei

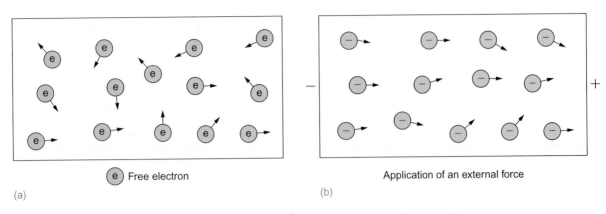

(a)   (b)

Figure 6.3 Free electrons and the application of an external force

KEY POINT

Metals like copper, aluminium and silver are good conductors of electricity and they readily support the flow of current. Plastics, rubber and ceramic materials are insulators and do not support the flow of current

KEY POINT

Semiconductors are pure insulating materials with a small amount of an impurity element present. Typical examples are silicon and germanium

of their atoms. Examples of insulators include plastics, glass, rubber and ceramic materials.

The effects of electric current flow can be detected by the presence of one or more of the following effects: light, heat, magnetism, chemical, pressure and friction. Thus, for example, if a piezoelectric crystal is subject to an electrical current it can change its shape and exert pressure. Heat is another, more obvious effect, from electric heating elements.

Semiconductors

Some materials combine some of the electrical properties of conductors with those of insulators. They are known as **semiconductors**. In these materials there may be a number of free electrons sufficient to allow a small current to flow. It is possible to add foreign atoms (called **impurity atoms**) to the semiconductor material that modify the properties of the semiconductor. Varying combinations of these additional atoms are used to produce various electrical devices such as diodes and transistors. Common types of semiconductor material are silicon, germanium, selenium and gallium.

Temperature effects

As stated earlier, all materials offer some resistance to current flow. In conductors the free electrons, rather than passing unobstructed through the material, collide with the relatively large and solid nuclei of the atoms. As the temperature increases, the nuclei vibrate more energetically further obstructing the path of the free electrons, causing more frequent collisions. The result is that the resistance of conductors increases with temperature.

Due to the nature of the bonding in insulators, there are no free electrons, except that when thermal energy increases as a result of a temperature increase, a few outer electrons manage to break free from their fixed positions and act as charge carriers. The result is that the resistance of insulators decreases as temperature increases.

Semiconductors behave in a similar manner to insulators. At absolute zero ($-273°C$) both types of material act as perfect insulators. However, unlike the insulator, as temperature increases in a semiconductor large numbers

of electrons break free to act as charge carriers. Therefore, as temperature increases, the resistance of a semiconductor decreases rapidly.

By producing special alloys, such as eureka and manganin that combine the effects of insulators and conductors, it is possible to produce a material where the resistance remains constant with increase in temperature. *Figure 6.4* shows how the resistance of insulators, semiconductors and conductors change with temperature.

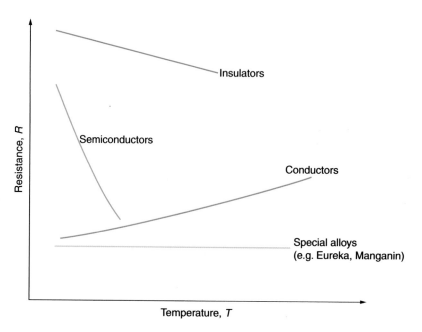

Figure 6.4 Variation of resistance with temperature for various materials

TYK

TYK 6.1

1. Name two materials that are classed as electrical conductors.

2. Name two materials that are classed as electrical insulators.

3. Name two semiconductor materials.

UNIT 6

Voltage and resistance

The ability of an energy source (e.g. a battery) to produce a current within a conductor may be expressed in terms of **electromotive force** (e.m.f.). Whenever an e.m.f. is applied to a circuit a **potential difference** (p.d.) exists. Both e.m.f. and p.d. are measured in volts (V). In many practical circuits there is only one e.m.f. present, the **power source** (i.e. a battery or power supply). When the power source is connected, a p.d. will be developed across each component present in the circuit.

The **conventional flow** of current in a circuit is from the point of more positive potential to the point of greatest negative potential (note that electrons move in the *opposite* direction!). **Direct current** results from the application of a direct e.m.f. (derived from batteries or a DC power supply). An essential characteristic of these supplies is that the applied e.m.f. does not change its polarity (even though its value might be subject to some fluctuation).

For any conductor, the current flowing is directly proportional to the e.m.f. applied. The current flowing will also be dependent on the physical dimensions (length and cross-sectional area) and material of which the conductor is composed.

The amount of current that will flow in a conductor when a given e.m.f. is applied is inversely proportional to its **resistance**. Resistance, therefore, may be thought of as an opposition to current flow; the higher the resistance the lower the current that will flow (assuming that the applied e.m.f. remains constant).

Figure 6.5 shows a simple electric circuit in which a battery provides a source of e.m.f. The battery is connected to resistor in a closed circuit in which current will flow. The e.m.f. supplied by the battery (E) will be equal to the potential difference (V) developed across the resistor (R).

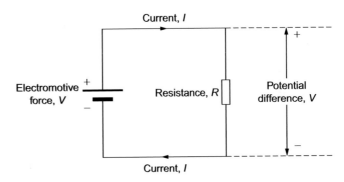

Figure 6.5 Simple circuit to illustrate the relationship between voltage (V), current (I) and resistance (R). Note that the direction of conventional current flow is from positive to negative

Ohm's Law

Provided that temperature does not vary, the ratio of p.d. across the ends of a conductor to the current flowing in the conductor is a constant. This relationship is known as *Ohm's Law* and it may be expressed as follows:

$$\frac{V}{I} = \text{a constant} = R$$

where V is the potential difference (or voltage drop) in volts (V), I is the current in amperes (A) and R is the resistance in ohms (Ω) (see *Figure 6.5*).

> **Ohm's law states that,** provided temperature remains constant, the potential difference between the ends of a conductor is directly proportional to the current flowing through it.

The formula may be arranged to make V, I or R the subject, as follows:

$$V = I \times R, \quad I = \frac{V}{R} \quad \text{and} \quad R = \frac{V}{I}$$

The triangle shown in *Figure 6.6* should help you remember these three important relationships. However, it is worth noting that, when performing calculations of currents, voltages and resistances in practical circuits it is seldom necessary to work with an accuracy of better than $\pm 1\%$ simply because component tolerances are usually greater than this. Furthermore, in calculations involving Ohm's Law, it can sometimes be convenient to work in units of kΩ and mA (or MΩ and μA) in which case potential differences will be expressed directly in V.

KEY POINT

The flow of conventional current is from positive to negative. Electron flow is in the opposite direction, from negative to positive

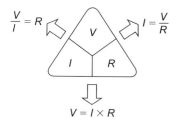

$$V = I \times R$$

Figure 6.6 Triangle showing the relationship between V, I and R

Example 6.1

A 12 Ω resistor is connected to a 6V battery. What current will flow in the resistor?

Here we must use $I = \dfrac{V}{R}$ (where $V = 6$V and $R = 12\,\Omega$):

$$I = \frac{V}{R} = \frac{6\,V}{12\,\Omega} = 0.5\,A \text{ (or 500mA)}$$

Hence a current of 500 mA will flow in the resistor.

Example 6.2

A current of 100 mA flows in a 56 Ω resistor. What voltage drop (potential difference) will be developed across the resistor?

Here we must use $V = I \times R$ and ensure that we work in units of V, A and Ω.

$$V = I \times R = 0.1A \times 56\,\Omega = 5.6\,V$$

Note that 100 mA is the same as 0.1 A.

This calculation shows that a potential difference of 5.6V will be developed across the resistor.

Example 6.3

A voltage drop of 15V appears across a resistor in which a current of 1 mA flows. What is the value of the resistance?

$$R = \frac{V}{I} = \frac{15\,V}{0.001A} = 15{,}000\,\Omega = 15k\Omega$$

Note that it is often more convenient to work in units of mA and V which will produce an answer directly in kΩ, i.e.

$$R = \frac{V}{I} = \frac{15\,V}{1mA} = 15k\Omega$$

UNIT 6

KEY POINT

Good electrical conductors have very low values of resistivity. Examples of such materials include copper, silver and aluminium

Resistance and resistivity

The resistance of a metallic conductor is directly proportional to its length and inversely proportional to its area. The resistance is also directly proportional to its **resistivity** (or **specific resistance**). Resistivity is defined as the resistance measured between the opposite faces of a cube having sides of 1 cm.

The resistance, R, of a conductor is thus given by the formula:

$$R = \frac{\rho l}{A}$$

where R is the resistance (Ω), ρ is the resistivity (Ωm), l is the length (m) and A is the area (m^2).

Table 6.1 shows the electrical properties of some common metals.

Table 6.1 Properties of some common metals

Metal	Resistivity (at 20°C) (Ωm)	Relative conductivity (copper = 1)	Temperature coefficient of resistance (per °C)
Silver	1.626×10^{-8}	1.06	0.0041
Copper (annealed)	1.724×10^{-8}	1.00	0.0039
Copper (hard drawn)	1.777×10^{-8}	0.97	0.0039
Aluminium	2.803×10^{-8}	0.61	0.0040
Mild steel	1.38×10^{-7}	0.12	0.0045
Lead	2.14×10^{-7}	0.08	0.0040
Nickel	8.0×10^{-8}	0.22	0.0062

Example 6.4

A coil consists of an 8 m length of annealed copper wire having a cross-sectional area of 1 mm^2. Determine the resistance of the coil.

We will use the formula $R = \dfrac{\rho l}{A}$

The value of ρ for annealed copper given in *Table 6.1* is $1.724 \times 10^{-8}\,\Omega$m. The length of the wire is 4 m while the area is 1 mm^2 or 1×10^{-6} m^2 (note that it is important to be consistent in using units of metres for length and square metres for area).

Hence the resistance of the coil will be given by:

$$R = \frac{1.724 \times 10^{-8} \times 8}{1 \times 10^{-6}} = 13.792 \times 10^{(-8+6)}$$

Thus, $R = 13.792 \times 10^{-2}$ or $0.13792\,\Omega$

Example 6.5

A wire having a resistivity of $1.6 \times 10^{-8}\,\Omega$m, length 20 m and cross-sectional area 1 mm^2 carries a current 5 A. Determine the voltage drop between the ends of the wire.

First we must find the resistance of the wire (as in Example 6.4):

$$R = \frac{\rho l}{A} = \frac{1.6 \times 10^{-8} \times 20}{1 \times 10^{-6}} = 32 \times 10^2 = 0.32\,\Omega$$

The voltage drop can now be calculated using Ohm's Law:

$$V = I \times R = 5\text{A} \times 0.32\,\Omega = 1.6\,\text{V}$$

This calculation shows that a potential of 1.6 V will be dropped between the ends of the wire.

TYK 6.2

Which one of the following metals is the best conductor of electricity: aluminium, copper, silver or mild steel? Why?

TYK

TYK 6.3

A wire wound resistor is made from a 4-m length of aluminium wire ($\rho = 2.8 \times 10^{-8}\,\Omega$m). Determine the resistance of the wire if it has a cross-sectional area of 0.2 mm^2.

TYK

TYK 6.4

1. A resistor of 270 Ω is connected across a 9 V DC supply. What current will flow?

2. A current of 56 µA flows in a 120 kΩ resistor. What voltage drop will appear across the resistor?

3. A voltage drop of 13.2 V appears across a resistor when a current of 4 mA flows in it. What is the value of the resistor?

TYK

Energy and power

At first you may be a little confused about the difference between energy and power. Put simply, energy is the ability to do work while power is the rate at which work is done. In electrical circuits, energy is supplied by batteries or generators. It may also be stored in components such

as capacitors and inductors. Electrical energy is converted into various other forms of energy by components such as resistors (producing heat), loudspeakers (producing sound energy) and light emitting diodes (producing light).

Energy is the ability to do work while power is the rate at which work is done. Energy is measured in Joules (J) and power is measured in Watts (W).

The unit of energy is the Joule (J). Power is the rate of use of energy and it is measured in watts (W). A power of 1 W results from energy being used at the rate of 1 J/s. Thus,

$$P = \frac{W}{t}$$

where P is the power in watts (W), W is the energy in Joules (J), and t is the time in seconds (s).

The power in a circuit is equivalent to the product of voltage and current. Hence,

$$P = I \times V$$

where P is the power in W, I is the current in A and V is the voltage in V.

The formula may be arranged to make P, I or V the subject as follows:

$$P = I \times V, \quad I = \frac{P}{V} \quad \text{and} \quad V = \frac{P}{I}$$

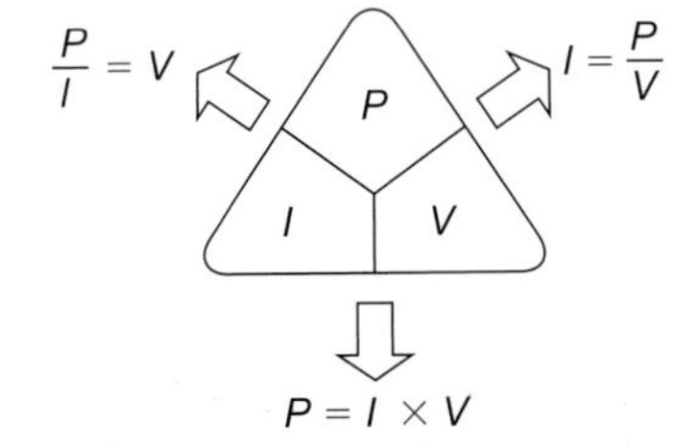

The triangle shown in *Figure 6.7* should help you remember these relationships.

The relationship, $P = I \times V$, may be combined with that which results from Ohm's Law ($V = I \times R$) to produce two further relationships. First, substituting for V gives:

$$P = I \times (I \times R) = I^2 R$$

Secondly, substituting for I gives:

$$P = \left(\frac{V}{R}\right) \times V = \frac{V^2}{R}$$

$\frac{P}{I} = V$ $I = \frac{P}{V}$

$P = I \times V$

Figure 6.7 Triangle showing the relationship between P, I and V

Example 6.6

A current of 1.5 A is drawn from a 3 V battery. What power is supplied?

Here we must use $P = I \times V$ (where $I = 1.5$ A and $V = 3$ V).

$$P = I \times V = 1.5\text{A} \times 3\text{V} = 4.5\text{W}$$

Example 6.7

A voltage drop of 4 V appears across a resistor of 100 Ω. What power is dissipated in the resistor?

Here we use $P = \dfrac{V^2}{R}$ (where $V = 4\,V$ and $R = 100\,\Omega$).

$$P = \frac{V^2}{R} = \frac{(4\,V \times 4\,V)}{100\,\Omega} = 0.16\,W \text{ (or 160 mW).}$$

Example 6.8

A current of 20 mA flows in a 1 kΩ resistor. What power is dissipated in the resistor?

Here we use $P = I^2 \times R$ but, to make life a little easier, we will work in mA and kΩ (in which case the answer will be in mW).

$$P = I^2 \times R = (20\,mA \times 20\,mA) \times 1k\Omega = 400\,mW$$

TYK 6.5

1. A power supply is rated at 15 V, 1 A. What value of load resistor would be required to test the power supply at its full rated output?

2. A current of 25 mA flows in a 47 Ω resistor. What power is dissipated in the resistor?

3. A 9 V battery supplies a circuit with a current of 75 mA. What power is consumed by the circuit?

4. A resistor of 150 Ω is rated at 0.5 W. What is the maximum current that can be applied to the resistor without exceeding its rating?

TYK 6.6

A heating element is rated at 24 V, 12 A. Determine the power dissipated by the heating element and the energy consumed if the element is used for a total period of 1 hour and 40 minutes

Resistors

Earlier we introduced the notion of resistance as opposition to current flow. Conventional forms of resistor obey a straight line law when voltage is plotted against current (see *Figure 6.8*) and this allows us to use resistors as a means of converting current into a corresponding voltage drop, and vice versa (note that doubling the applied current will produce double the voltage drop, and so on). Therefore, resistors provide us with a means of controlling the currents and voltages present in electronic circuits. They can also act as **loads** to simulate the presence of a circuit during testing (e.g. a suitably

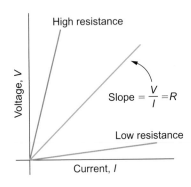

Figure 6.8 Voltage plotted against current for three different values of resistor

rated resistor can be used to replace a loudspeaker when an audio amplifier is being tested).

The specifications for a resistor usually include the value of resistance expressed in ohms (Ω), kilohms (kΩ) or megohms (MΩ), the accuracy or tolerance (quoted as the maximum permissible percentage deviation from the marked value), and the power rating (which must be equal to, or greater than, the maximum expected power dissipation).

Other practical considerations when selecting resistors for use in a particular application include temperature coefficient, stability and ambient temperature range. *Table 6.2* summarizes the properties of six of the most common types of resistor. *Figure 6.9* shows a typical selection of fixed resistors with values from 15 Ω to 4.7 kΩ.

Preferred values

The value marked on the body of a resistor is not its *exact* resistance. Some minor variation in resistance value is inevitable due to production tolerance. For example, a resistor marked 100 Ω and produced within a tolerance of $\pm10\%$ will have a value which falls within the range 90–110 Ω. A similar component with a tolerance of $\pm1\%$ would have a value that falls within the range 99–101 Ω. Thus, where accuracy is important it is essential to use close tolerance components.

Table 6.2 Properties of common types of resistor

Property	Resistor type					
	Carbon film	Metal film	Metal oxide	Ceramic wirewound	Vitreous wirewound	Metal clad
Resistance range (Ω)	10 to 10M	1 to 1M	10 to 10M	0.47–22K	0.1–22K	0.05–10K
Typical tolerance (%)	±5	±1	±2	±5	±5	±5
Power rating (W)	0.25–2	0.125–0.5	0.25–0.5	4–17	2–4	10–300
Temperature coefficient (ppm/°C)	−250	+50 to +100	+250	+250	+75	+50
Stability	Fair	Excellent	Excellent	Good	Good	Good
Ambient temperature range (°C)	−45 to +125	−45 to +125	−45 to +125	−45 to +125	−45 to +125	−55 to +200
Typical applications	General purpose	Amplifiers, test equipment, etc., requiring low-noise high-tolerance components		Power supplies, loads, medium and high power applications		Very high power applications

Resistors are available in several series of fixed decade values, the number of values provided with each series being governed by the tolerance involved. In order to cover the full range of resistance values using resistors having a $\pm20\%$ tolerance it will be necessary to provide six basic values,

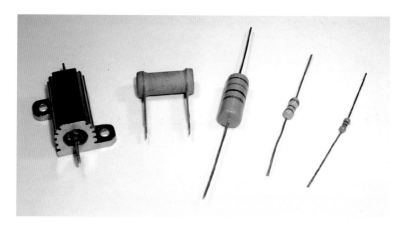

Figure 6.9 A selection of resistors, including high-power metal clad, ceramic wirewound, carbon and metal film types with values ranging from 15 Ω to 4.7 kΩ

KEY POINT

Resistors are available 'off the shelf' in one of several series of preferred values according to the tolerance. The E6 series is available in multiples of six basic values whilst the E12 and E24 series comprise 12 and 24 basic values, respectively

also known as the **E6 series**. More values will be required in the series which offers a tolerance of ±10% and consequently the **E12 series** provides 12 basic values. The **E24 series** for resistors of ±5% tolerance provides no fewer than 24 basic values and, as with the E6 and E12 series, decade multiples (i.e. ×1, ×10, ×100, ×1 k, ×10 k, ×100 k and ×1 M) of the basic series. *Figure 6.10* shows the relationship between the E6, E12 and E24 series.

Power ratings

Resistor power ratings are related to operating temperatures and resistors should be derated at high temperatures. Where reliability is important resistors should be operated at well below their nominal maximum power dissipation.

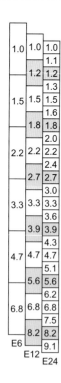

Figure 6.10 Relationship between the E6, E12 and E24 series

Example 6.9

A resistor has a marked value of 220 Ω. Determine the tolerance of the resistor if it has a measured value of 207 Ω.

The difference between the marked and measured values of resistance (the *error*) is $(220\,\Omega - 207\,\Omega) = 13\,\Omega$. The tolerance is given by:

$$\text{Tolerance} = \frac{\text{error}}{\text{marked value}} \times 100\%$$

The tolerance is thus $(13/220) \times 100 = 5.9\%$.

Example 6.10

A 9 V power supply is to be tested with a 39 Ω load resistor. If the resistor has a tolerance of 10% find:

(a) the nominal current taken from the supply;

(b) the maximum and minimum values of supply current at either end of the tolerance range for the resistor.

(a) If a resistor of *exactly* 39 Ω is used the current will be:

$$I = \frac{V}{R} = \frac{9\,V}{39\,\Omega} = 231 \text{mA}$$

The lowest value of resistance would be $(39\,\Omega - 3.9\,\Omega) = 35.1\,\Omega$.

UNIT 6

In which case the current would be:

$$I = \frac{V}{R} = \frac{9\,V}{35.1\,\Omega} = 256.4\,mA$$

At the other extreme, the highest value would be $(39\,\Omega + 3.9\,\Omega) = 42.9\,\Omega$.

In this case the current would be:

$$I = \frac{V}{R} = \frac{9\,V}{42.9\,\Omega} = 209.8\,mA$$

(b) The maximum and minimum values of supply current will thus be 256.4 mA and 209.8 mA, respectively.

Example 6.11

A current of 100 mA (±20%) is to be drawn from a 28 V DC supply. What value and type of resistor should be used in this application?

The value of resistance required must first be calculated using Ohm's Law:

$$R = \frac{V}{I} = \frac{28\,V}{100\,mA} = 280\,\Omega$$

The nearest preferred value from the E12 series is 270 Ω (which will actually produce a current of 103.7 mA (i.e. within ±4% of the desired value). If a resistor of ±10% tolerance is used, current will be within the range 94 to 115 mA (well within the ±20% accuracy specified).

The power dissipated in the resistor (calculated using $P = I \times V$) will be 2.9 W and thus a component rated at 3 W (or more) will be required. This would normally be a vitreous enamel coated wire wound resistor (see Table 6.2).

Resistor markings

Carbon and metal oxide resistors are normally marked with colour codes which indicate their value and tolerance. Two methods of colour coding are in common use; one involves four coloured bands (see *Figure 6.11*) while the other uses five colour bands (see *Figure 6.12*).

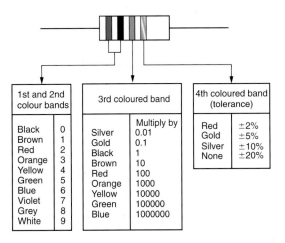

Figure 6.11 Four band resistor colour code

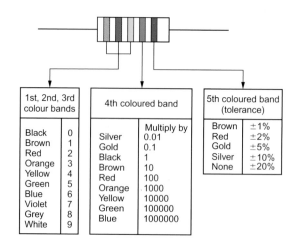

Figure 6.12 Five band resistor colour code

Example 6.12

A resistor is marked with the following coloured stripes: brown, black, red, silver. What is its value and tolerance?

See *Figure 6.13.*

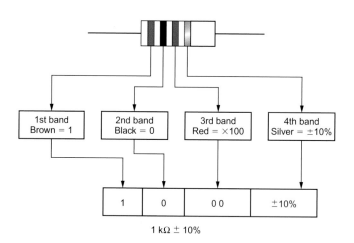

Figure 6.13 See Example 6.12

Example 6.13

A resistor is marked with the following coloured stripes: red, violet, orange, gold. What is its value and tolerance?

See *Figure 6.14.*

Example 6.14

A resistor is marked with the following coloured stripes: green, blue, black, gold. What is its value and tolerance?

See *Figure 6.15.*

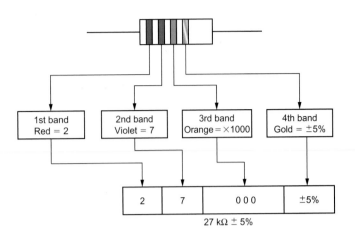

Figure 6.14 See Example 6.13

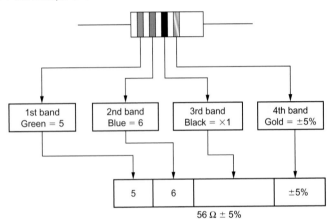

Figure 6.15 See Example 6.14

Example 6.15

A resistor is marked with the following coloured stripes: red, green, black, black, brown. What is its value and tolerance?

See *Figure 6.16*.

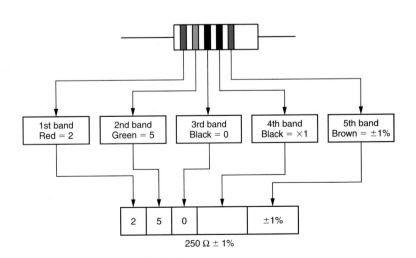

Figure 6.16 See Example 6.15

UNIT 6

TYK 6.7

A 2.2 kΩ resistor of ±2% tolerance is required. What four band colour code should this resistor have?

BS 1852 resistor coding

Some types of resistor have markings based on a system of coding defined in BS 1852. This system involves marking the position of the decimal point with a letter to indicate the multiplier concerned as shown in *Table 6.3*. A further letter is then appended to indicate the tolerance as shown in *Table 6.4*.

Table 6.3 BS 1852 resistor multiplier markings

Letter	Multiplier
R	1
K	1000
M	1000000

Table 6.4 BS 1852 resistor tolerance markings

Letter	Multiplier
F	±1%
G	±2%
J	±5%
K	±10%
M	±20%

UNIT 6

Example 6.16

A resistor is marked coded with the legend 4R7K. What is its value and tolerance?

4R7 indicates that the value is 4.7 Ω whilst K shows that the tolerance is ±10%. Thus the resistor has a value of 4.7 Ω ± 10%.

Example 6.17

A resistor is marked coded with the legend 330RG. What is its value and tolerance?

330 R indicates that the value is 330 Ω whilst G shows that the tolerance is ±2%. Thus, the resistor has a value of 330 Ω ± 2%.

Example 6.18

A resistor is marked coded with the legend R22M. What is its value and tolerance?

R22 indicates that the value is 0.22 Ω whilst M shows that the tolerance is ±20%. Thus, the resistor has a value of 0.22 Ω ± 20%.

Test your knowledge

TYK

TYK 6.8

Determine the value and tolerance of resistors marked with the following coloured bands:
(a) red, violet, yellow, gold
(b) brown, black, black, silver
(c) blue, grey, green, gold
(d) orange, white, silver, gold
(e) red, red, black, brown, red.

Test your knowledge

TYK

TYK 6.9

Resistors in a batch are all marked yellow, violet, black, gold. If a resistor is selected from this batch, within what range would you expect its value to be?

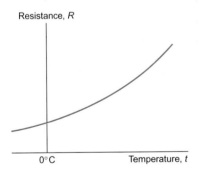

Figure 6.17 Variation of resistance with temperature for a metal conductor

KEY POINT

The resistance of a metal conductor increases with temperature. Materials that increase in resistance with temperature are said to have a positive temperature coefficient. Other materials (such as carbon) have a resistance that decreases with temperature and are said to have a negative temperature coefficient

Resistance and temperature

Figure 6.17 shows how the resistance of a metal conductor (e.g. copper) varies with temperature. Since the resistance of the material increases with temperature, this characteristic is said to exhibit a **positive temperature coefficient (PTC)**. Not all materials have a PTC characteristic. The resistance of a carbon conductor falls with temperature and it is therefore said to exhibit a **negative temperature coefficient (NTC)**.

The resistance of a conductor at a temperature, t, is given by the equation:

$$R_t = R_0(1 + \alpha t + \beta t^2 + \gamma t^3 \ldots)$$

where α, β, γ, etc. are constants and R_0 is the resistance at 0°C.

The coefficients β, γ, etc. are quite small and since we are normally only dealing with a relatively restricted temperature range (e.g. 0–100°C) we can usually approximate the characteristic shown in *Figure 6.17* to the straight line law shown in *Figure 6.18*. In this case, the equation simplifies to:

$$R_t = R_0(1 + \alpha t)$$

where α is known as the **temperature coefficient** of resistance. *Table 6.5* shows some typical values for α (note that α is expressed in Ω/Ω/°C or just /°C).

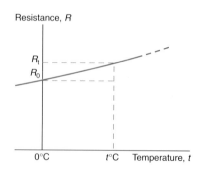

Resistance, R

R_t
R_0

$0°C$ $t°C$ Temperature, t

Figure 6.18 Straight line approximation of Figure 6.16

Table 6.5 Temperature coefficient of resistance

Material	Temperature coefficient of resistance, α (/°C)
Platinum	+0.0034
Silver	+0.0038
Copper	+0.0043
Iron	+0.0065
Carbon	−0.0005

Example 6.19

A resistor has a temperature coefficient of 0.001/°C. If the resistor has a resistance of 1.5 kΩ at 0°C, determine its resistance at 80°C.

Now $R_t = R_0(1 + \alpha t)$

thus

$$R_t = 1.5\,\text{k}\Omega \times [1 + (0.001 \times 80)]$$

Hence

$$R_t = 1.5 \times 1.08 = 1.62\,\text{k}\Omega$$

Example 6.20

A resistor has a temperature coefficient of 0.0005/°C. If the resistor has a resistance of 680 Ω at 20°C, what will its resistance be at 90°C?

First we must find the resistance at 0°C. Rearranging the formula for R_t gives:

$$R_0 = \frac{R_t}{1+\alpha t} = \frac{680}{1 + (1 + 0.0005 \times 20)} = \frac{680}{1 + 0.01} = 673.3\,\Omega$$

Now

$$R_t = R_0(1 + \alpha t)$$

thus

$$R_{90} = 673.3 \times [1 + (0.0005 \times 90)]$$

Hence

$$R_{90} = 673.3 \times 1.045 = 704\,\Omega$$

Example 6.21

A resistor has a resistance of 40 Ω at 0°C and 44 Ω at 100°C. Determine the resistor's temperature coefficient.

First we need to make α the subject of the formula:

$$R_t = R_0(1 + \alpha t)$$

UNIT 6

Now

$$\alpha = \frac{1}{t}\left(\frac{R_t}{R_0} - 1\right) = \frac{1}{100}\left(\frac{44}{40} - 1\right)$$

from which

$$\alpha = \frac{1}{100}(1.1 - 1) = \frac{1}{100} \times 0.1 = 0.001/°C$$

TYK 6.10

A resistor has a temperature coefficient of 0.0008/°C. If the resistor has a resistance of 390 Ω at 0°C, determine its resistance at 55°C.

TYK 6.11

A resistor has a temperature coefficient of 0.004/°C. If the resistor has a resistance of 82 kΩ at 20°C, what will its resistance be at 75°C?

Thermistors

With conventional resistors we would normally require resistance to remain the same over a wide range of temperatures (i.e. α should be zero). On the other hand, there are applications in which we could use the effect of varying resistance to detect a temperature change. Components that allow us to do this are known as **thermistors**. The resistance of a thermistor changes markedly with temperature and these components are widely used in temperature sensing and temperature compensating applications. Two basic types of thermistor are available, NTC and PTC (see *Figure 6.19*). A typical glass encapsulated thermistor is shown in *Figure 6.21*.

Typical NTC thermistors have resistances that vary from a few hundred (or thousand) ohms at 25°C to a few tens (or hundreds) of ohms at 100°C. On the other hand, PTC thermistors usually have a resistance-temperature characteristic which remains substantially flat (typically at around 100 Ω) over the range 0°C to around 75°C. Above this and at a critical temperature (usually in the range 80–120°C) their resistance rises very rapidly to values of up to, and beyond, 10 kΩ.

A typical application of PTC thermistors is over-current protection. Provided the current passing through the thermistor remains below the threshold current, the effects of self-heating will remain negligible and the resistance of the thermistor will remain low (i.e. approximately the same as the resistance quoted at 25°C). Under fault conditions, the current exceeds

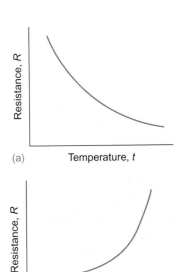

Figure 6.19 Characteristics of (a) NTC and (b) PTC thermistors

the threshold value by a considerable margin and the thermistor starts to self-heat. The resistance then increases rapidly and, as a consequence, the current falls to the rest value. Typical values of threshold and rest currents are 200 and 8 mA, respectively, for a device which exhibits a nominal resistance of 25 Ω at 25°C.

Light-dependent resistors

Light-dependent resistors (LDR) use a semiconductor material (i.e. a material that is neither a conductor nor an insulator) whose electrical characteristics vary according to the amount of incident light. The two semiconductor materials used for the manufacture of LDRs are cadmium sulphide (CdS) and cadmium selenide (CdSe). These materials are most sensitive to light in the visible spectrum, peaking at about 0.6 μm for CdS and 0.75 μm for CdSe. A typical CdS LDR exhibits a resistance of around 1 MΩ in complete darkness and less than 1 kΩ when placed under a bright light source (see *Figure 6.20*). A typical LDR is shown in *Figure 6.21*.

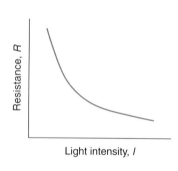

Figure 6.20 Typical characteristic of LDR

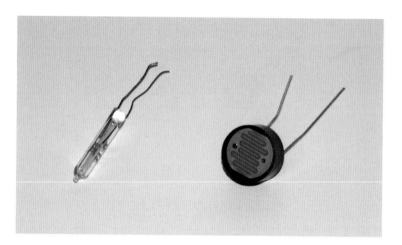

Figure 6.21 A typical thermistor (left) and LDR (right)

Variable resistors

Variable resistors are available in several forms including those that use carbon tracks and those that use a wire resistance element (known as 'wirewound' types). In either case, a moving slider makes contact with the resistance element. Most variable resistors have three (rather than two) terminals and as such are more correctly known as **potentiometers**. Carbon potentiometers are available with linear or semilogarithmic law tracks (see *Figure 6.22*) and in rotary or slider formats. Ganged controls, in which several potentiometers are linked together by a common control shaft, are also available. *Figure 6.23* shows a selection of variable resistors.

You will also encounter various forms of preset resistors that are used to make occasional adjustments (e.g. for calibration). Various forms of preset resistor are commonly used including open carbon track skeleton presets and fully encapsulated carbon and multi-turn ceramic metal ('Cermet') types, as shown in *Figure 6.24*.

UNIT 6

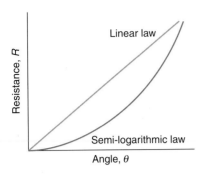

Figure 6.22 Characteristics of linear and semi-logarithmic law variable resistors

Figure 6.23 A selection of common types of carbon and wirewound variable resistors/potentiometers

Figure 6.24 A selection of common types of standard and miniature preset variable resistors/potentiometers

UNIT 6

Kirchhoff's Laws

In many cases, Ohm's Law alone is insufficient to determine the magnitude of the voltages and currents present in a circuit. In this section we shall introduce you to two of the most useful laws of electronics that will allow you to solve more complex circuits in which several voltages and currents are present.

Kirchhoff's Laws relate to the algebraic sum of currents at a junction (or **node**) or voltages in a network (or **mesh**). The term 'algebraic' simply indicates that the polarity of each current or voltage drop must be taken into account by giving it an appropriate sign, either positive (+) or negative (−).

Kirchhoff's Current Law

Kirchhoff's Current Law states that the algebraic sum of the currents present at a junction (node) in a circuit is zero (see *Figure 6.25*).

KEY POINT

Kirchhoff's Laws can be used in conjunction with Ohm's Law to find the currents and voltages in complex circuits

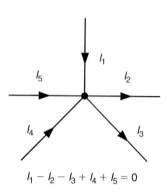

$$I_1 - I_2 - I_3 + I_4 + I_5 = 0$$

Current flowing towards the junction is positive (+)
Current flowing away from the junction is negative (−)

Figure 6.25 Kirchhoff's Current Law

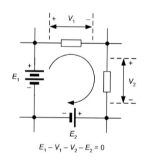

$$E_1 - V_1 - V_2 - E_2 = 0$$

Move clockwise around the circuit starting with the positive terminal of the largest e.m.f.
Voltages acting in the same sense are positive (+)
Voltages acting in the opposite sense are negative (−)

Figure 6.27 Kirchhoff's Voltage Law

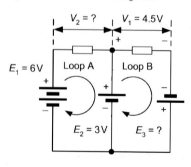

Figure 6.28 See Example 6.23

Example 6.22

In *Figure 6.26*, use Kirchhoff's Current Law to determine:

(a) the value of current flowing between A and B
(b) the value of I_3.

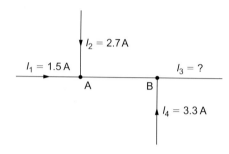

Figure 6.26 See Example 6.22

(a) I_1 and I_2 both flow towards Node A so, applying our polarity convention, they must both be positive. Now, assuming that a current I_5 flows between A and B and that this current flows away from the junction (obvious because I_1 and I_2 both flow towards the junction) we arrive at the following Kirchhoff's Current Law equation:

$$+I_1 + I_2 - I_5 = 0$$

From which:

$$I_5 = I_1 + I_2 = 1.5 + 2.7 = 4.2 \text{ A}$$

(b) Moving to Node B, let us assume that I_3 flows outwards so we can say that:

$$+I_4 + I_5 - I_3 = 0$$

From which:

$$I_3 = I_4 + I_5 = 3.3 + 4.2 = 7.5 \text{ A}$$

Kirchhoff's Voltage Law

Kirchhoff's Voltage Law states that the algebraic sum of the potential drops in a closed network (or 'mesh') is zero (see *Figure 6.27*).

Example 6.23

In *Figure 6.28*, use Kirchhoff's Voltage Law to determine:
(a) the value of V_2
(b) the value of E_3.

(a) In Loop A, and using the conventions shown in *Figure 6.27*, we can write down the Kirchhoff's Voltage Law equations:

$$E_1 - V_2 - E_2 = 0$$

UNIT 6

From which:

$$V_2 = E_1 - E_2 = 6 - 3 = 3\,\text{V}$$

(b) Similarly, in Loop B, we can say that:

$$E_2 - V_2 + E_3 = 0$$

From which:

$$E_3 = V_2 - E_2 = 4.5 - 3 = 1.5\,\text{V}$$

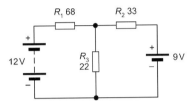

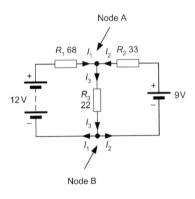

Figure 6.29 See Example 6.24

Figure 6.30 See Example 6.24

Example 6.24

Determine the currents and voltages in the circuit of *Figure 6.29*.

In order to solve the circuit shown in *Figure 6.29*, it is first necessary to mark the currents and voltages on the circuit, as shown in *Figures 6.30* and *6.31*.

By applying Kirchhoff's Current Law at Node A that we have identified in *Figure 6.27*:

$$+I_1 + I_2 - I_3 = 0$$

Therefore:

$$I_1 = I_3 - I_2 = 0 \tag{i}$$

By applying Kirchhoff's Voltage Law in Loop A we obtain:

$$12 - V_1 - V_3 = 0$$

From which:

$$V_1 = 12 - V_3 \tag{ii}$$

By applying Kirchhoff's Voltage Law in Loop B we obtain:

$$9 - V_2 - V_3 = 0$$

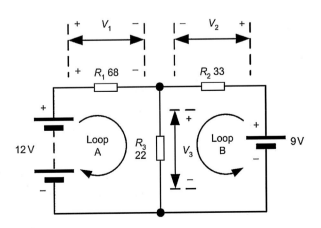

Figure 6.31 See Example 6.24

From which:

$$V_2 = 9 - V_3 \qquad \text{(iii)}$$

Next we can generate three further relationships by applying Ohm's Law:

$$V_1 = I_1 R_1 \text{ from which } I_1 = \frac{V_1}{R_1}$$

$$V_2 = I_2 R_2 \text{ from which } I_2 = \frac{V_2}{R_2}$$

and

$$V_3 = I_3 R_3 \text{ from which } I_3 = \frac{V_3}{R_3}$$

Combining these three relationships with the Current Law equation (i) gives:

$$\frac{V_1}{R_1} = \frac{V_3}{R_3} - \frac{V_2}{R_2}$$

from which:

$$\frac{V_1}{110} = \frac{V_3}{22} - \frac{V_2}{33} \qquad \text{(iv)}$$

Combining (ii) and (iii) with (iv) gives:

$$\frac{(12 - V_3)}{110} = \frac{V_3}{22} - \frac{(9 - V_3)}{33}$$

Multiplying both sides of the expression by 330 gives:

$$\frac{330(12 - V_3)}{110} = \frac{330 V_3}{22} - \frac{330(9 - V_3)}{33}$$

from which

$$3(12 - V_3) = 15V_3 - 10(9 - V_3)$$

$$36 - 3V_3 = 15V_3 - 90 + 10V_3$$

$$36 + 90 = 15V_3 + 10V_3 + 3V_3$$

thus

$$126 = 28V_3 \quad \text{so} \quad V_3 = 126/28 = 4.5\,\text{V}$$

from (ii)

$$V_1 = 12 - V_3 \quad \text{so} \quad V_1 = 12 - 4.5 = 7.5\,\text{V}$$

from (iii)

$$V_2 = 9 - V_3 \quad \text{so} \quad V_2 = 9 - 4.5 = 4.5\,\text{V}$$

Using the Ohm's Law equations that we met earlier gives:

$$I_1 = \frac{V_1}{R_1} = \frac{7.5}{110} = 0.068\,\text{A} = 68\,\text{mA}$$

UNIT 6

$$I_2 = \frac{V_2}{R_2} = \frac{4.5}{33} = 0.136\,A = 136\,mA$$

$$I_3 = \frac{V_3}{R_3} = \frac{4.5}{22} = 0.204\,A = 204\,mA$$

Finally, it is worth checking these results with the Current Law equation (i):

$$+I_1 + I_2 - I_3 = 0$$

Inserting our values for I_1, I_2 and I_3 gives:

$$+0.068 + 0.136 - 0.204 = 0$$

Since the left and right hand sides of the equation are equal we can be reasonably confident that our results are correct.

TYK 6.12

Determine the unknown currents and voltages in *Figure 6.32*.

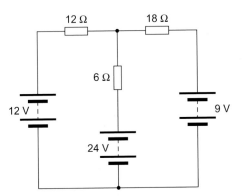

Figure 6.32 See TYK 6.12

Series and parallel combinations of resistors

We can use Ohm's Law and Kirchhoff's Laws to determine the equivalent resistance of more complicated circuits where resistors are connected in **series** and in **parallel**, as shown in *Figure 6.33*.

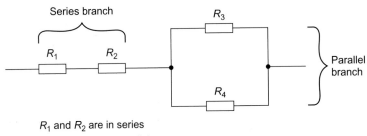

R_1 and R_2 are in series
R_3 and R_4 are in parallel

Figure 6.33 A circuit with series and parallel branches

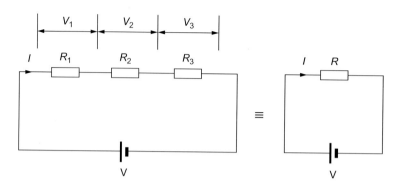

Figure 6.34 Equivalent resistance of three resistors connected in series

In the case of the series arrangement of three resistors shown in *Figure 6.34*, the same current will flow through each resistor and, from Kirchhoff's Voltage Law, the supply voltage, *V*, will be equal to the sum of the individual voltage drops. Hence,

$$V = IR_1 + IR_2 + IR_3$$

If *R* is the equivalent resistance that could replace R_1, R_2 and R_3 in series, we arrive at:

$$V = IR$$

Hence,

$$IR = IR_1 + IR_2 + IR_3$$

Dividing both sides by *I* gives:

$$R = R_1 + R_2 + R_3$$

In the case of the parallel arrangement of three resistors shown in *Figure 6.35*, the same voltage will appear across each resistor and, from

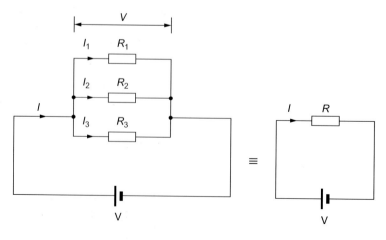

Figure 6.35 Equivalent resistance of three resistors connected in parallel

UNIT 6

Kirchhoff's Current Law, the supply current, I, will be equal to the sum of the currents flowing in the three resistors. Hence:

$$I = I_1 + I_2 + I_3$$

If R is the equivalent resistance that could replace R_1, R_2 and R_3 connected in parallel, we arrive at:

$$I = \frac{V}{R}$$

Hence:

$$\frac{V}{R} = \frac{V}{R_1} + \frac{V}{R_2} + \frac{V}{R_3}$$

Dividing both sides by V gives:

$$\frac{1}{R} = \frac{1}{R_1} + \frac{1}{R_2} + \frac{1}{R_3}$$

In order to obtain a particular value of resistance, fixed resistors may be arranged in either series or parallel as shown in *Figures 6.36 and 6.37*. The effective resistance of each of the series circuits shown in *Figure 6.36* is simply equal to the sum of the individual resistances. So, for the circuit shown in *Figure 6.36a*:

$$R = R_1 + R_2$$

while for *Figure 6.36b*

$$R = R_1 + R_2 + R_3$$

Turning to the parallel resistors shown in *Figure 6.37*, the reciprocal of the effective resistance of each circuit is equal to the sum of the reciprocals of the individual resistances. Hence, for *Figure 6.37a*:

$$\frac{1}{R} = \frac{1}{R_1} + \frac{1}{R_2}$$

while for *Figure 6.37b*

$$\frac{1}{R} = \frac{1}{R_1} + \frac{1}{R_2} + \frac{1}{R_3}$$

In the former case (and *only* the former case), the formula can be more conveniently re-arranged as follows:

$$R = \frac{R_1 \times R_2}{R_1 + R_2}$$

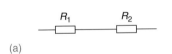

(a)

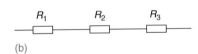

(b)

Figure 6.36 Resistors in series

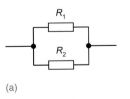

(a)

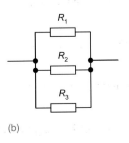

(b)

Figure 6.37 Resistors in parallel

UNIT 6

You can remember this as the *product* of the two resistance values *divided by* the *sum* of the two resistance values.

Example 6.25

Resistors of 22, 47 and 33 Ω are connected (a) in series and (b) in parallel. Determine the effective resistance in each case.

(a) In the series circuit $R = R_1 + R_2 + R_3$, thus $R = 22\,\Omega + 47\,\Omega + 33\,\Omega = 102\,\Omega$

(b) In the parallel circuit:

$$\frac{1}{R} = \frac{1}{R_1} + \frac{1}{R_2} + \frac{1}{R_3}$$

thus

$$\frac{1}{R} = \frac{1}{22} + \frac{1}{47} + \frac{1}{33}$$

or

$$\frac{1}{R} = 0.045 + 0.021 + 0.03 = 0.096$$

From which R = 10.31 Ω.

Example 6.26

Determine the effective resistance of the circuit shown in *Figure 6.38*.

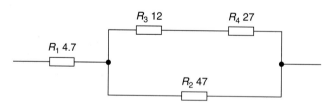

Figure 6.38 See Example 6.26

The circuit can be progressively simplified as shown in *Figure 6.39*. The stages in this simplification are:

(a) R_3 and R_4 are in series and they can be replaced by a single resistance (R_A) of $(12\,\Omega + 27\,\Omega) = 39\,\Omega$.

(b) R_A appears in parallel with R_2. These two resistors can be replaced by a single resistance (R_B) of $(39\,\Omega \times 47\,\Omega)/(39\,\Omega + 47\,\Omega) = 21.3\,\Omega$.

(c) R_B appears in series with R_1. These two resistors can be replaced by a single resistance (R) of $(21.3\,\Omega + 4.7\,\Omega) = 26\,\Omega$.

(a)

(b)

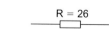

(c)

Figure 6.39 See Example 6.26

Example 6.27

A resistance of 50 Ω rated at 2 W is required. What parallel combination of preferred value resistors will satisfy this requirement? What power rating should each resistor have?

Two 100 Ω resistors may be wired in parallel to provide a resistance of 50 Ω as shown below:

$$R = \frac{R_1 \times R_2}{R_1 + R_2} = \frac{100 \times 100}{100 + 100} = \frac{10,000}{200} = 50 \, \Omega$$

Note, from this, that when two resistors of the same value are connected in parallel the resulting resistance will be half that of a single resistor.

Having shown that two 100 Ω resistors connected in parallel will provide us with a resistance of 50 Ω we now need to consider the power rating. Since the resistors are identical, the applied power will be shared equally between them. Hence each resistor should have a power rating of 1 W.

Activity 6.1

Resistors of 27, 33, 56 and 68 Ω are available. Show, with the aid of labelled circuit sketches, how can two or more of these be arranged to produce the following resistance values?

(a) **60 Ω**
(b) **14.9 Ω**
(c) **124 Ω**
(d) **11.7 Ω**
(e) **128 Ω.**

TYK

TYK 6.13

Three 100 Ω resistors are connected as shown in *Figure 6.40*. Determine the effective resistance of the circuit.

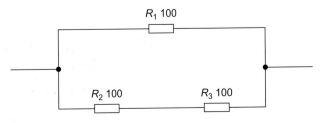

Figure 6.40 See TYK 6.13

TYK 6.14

Determine the effective resistance of the circuit shown in *Figure 6.41*.

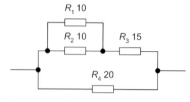

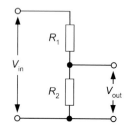

Figure 6.41　See TYK 6.14

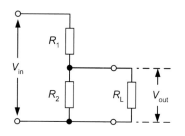

Figure 6.42　Potential divider circuit

The potential divider

The potential divider circuit (see *Figure 6.42*) is commonly used to reduce voltages in a circuit. The output voltage produced by the circuit is given by:

$$V_{out} = V_{in} \frac{R_2}{R_1 + R_2}$$

It is, however, important to note that the output voltage (V_{out}) will fall when current is drawn from the arrangement.

Figure 6.43 shows the effect of **loading** the potential divider circuit. In the loaded potential divider (*Figure 6.43*) the output voltage is less than that of the unloaded circuit (some current will be drawn away from the output into the **load resistor**, R_L). The output voltage of the loaded potential divider will be given by:

$$V_{out} = V_{in} \frac{R_P}{R_1 + R_P}$$

where

$$R_P = \frac{R_2 \times R_L}{R_2 + R_L}$$

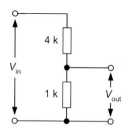

Figure 6.43　Loaded potential divider circuit

Example 6.28

The potential divider shown in *Figure 6.44* is used as a simple **voltage calibrator** to be used with an input voltage of 5 V. Determine the output voltage produced by the circuit (a) when the output terminals are left open circuit (i.e. when no load is connected); and (b) when the output is loaded by a resistance of 10 kΩ.

(a) In the first case we can simply apply the formula $V_{out} = V_{in} \frac{R_2}{R_1 + R_2}$, where $V_{in} = 5\,V, R_1 = 4\,k\Omega$ and $R_2 = 1\,k\Omega$.

Hence

$$V_{out} = 5 \times \frac{1}{4 + 1} = 1\,V$$

(b) In the second case we need to take into account the effect of the 10 kΩ resistor connected to the output terminals of the potential divider.

First we need to find the equivalent resistance of the parallel combination of R_2 and R_L:

$$R_P = \frac{R_2 \times R_L}{R_2 + R_L} = \frac{1 \times 10}{1 + 10} = \frac{10}{11} = 0.909\,k\Omega$$

Figure 6.44　See Example 6.28

UNIT 6

Then we can determine the output voltage from:

$$V_{out} = V_{in} \frac{R_p}{R_1 + R_p} = 5 \times \frac{0.909}{4 + 0.909} = 0.925\,V$$

The current divider

The current divider circuit (see *Figure 6.45*) is used to divert a known proportion of the current flowing in a circuit. The output current produced by the circuit is given by:

$$I_{out} = I_{in} \frac{R_1}{R_1 + R_2}$$

It is, however, important to note that the output current (I_{out}) will fall when the load connected to the output terminals has any appreciable resistance.

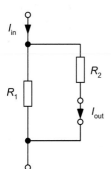

Figure 6.45 Current divider circuit

Figure 6.46 See Example 6.29

Example 6.29

A moving coil meter requires a current of 1 mA to provide full-scale deflection. If the meter coil has a resistance of 100 Ω and is to be used as a milliammeter reading 5 mA full scale, determine the value of parallel shunt resistor required. This problem may sound a little complicated so it is worth taking a look at the equivalent circuit of the meter (*Figure 6.46*) and comparing it with the current divider shown in *Figure 6.45*.

We can apply the current divider formula, replacing I_{out} with I_m (the meter full-scale deflection current) and R_2 with R_m (the meter resistance). R_1 is the required value of shunt resistor, R_s. Hence:

$$I_{out} = I_{in} \frac{R_s}{R_s + R_m}$$

Re-arranging the formula gives:

$$I_m \times (R_s + R_m) = I_{in} \times R_s$$

thus

$$I_m R_s + I_m R_m = I_{in} R_s$$

or

$$I_{in} R_s - I_m R_s = I_m R_m$$

From which

$$R_s(I_{in} - I_m) = I_m R_m$$

so

$$R_s = \frac{I_m R_m}{(I_{in} - I_m)}$$

Now $I_{in} = 1\,mA$, $R_m = 100\,Ω$ and $I_{in} = 5\,mA$, thus

$$R_s = \frac{1 \times 100}{5 - 1} = \frac{100}{4} = 25\,Ω$$

Activity 6.2

A battery manufacturer has asked you to design a battery tester for 1.5 V cells. The tester is to apply nominal loads of 1 A, 500 mA and 250 mA for D, C and AA cells respectively and is to incorporate a moving coil indicator having a full-scale deflection of 1 mA and an internal resistance of 100 Ω. The indicator is to display the on-load voltage provided by the cell under test and is to read 1.6 V full scale. Design a circuit for the battery tester and draw a fully labelled circuit diagram. Specify the value of all components required (including power ratings).

Semiconductors

Earlier we described the simplified structure of an atom and showed that it contains both negative charge carriers (electrons) and positive charge carriers (protons). Electrons each carry a single unit of negative electric charge while protons each exhibit a single unit of positive charge. Since atoms normally contain an equal number of electrons and protons, the net charge present will be zero. For example, if an atom has 11 electrons, it will also contain 11 protons. The end result is that the negative charge of the electrons will be exactly balanced by the positive charge of the protons.

Electrons are in constant motion as they orbit around the nucleus of the atom. Electron orbits are organized into shells. The maximum number of electrons present in the first shell is 2, in the second shell 8, and in the third, fourth and fifth shells it is 18, 32 and 50, respectively. In electronics, only the electron shell furthermost from the nucleus of an atom is important. It is important to note that the movement of electrons only involves those present in the outer **valence shell**.

If the valence shell contains the maximum number of electrons possible the electrons are rigidly bonded together and the material has the properties of an insulator. If, however, the valence shell does not have its full complement of electrons, the electrons can be easily loosened from their orbital bonds, and the material has the properties associated with an electrical conductor.

An isolated silicon atom contains four electrons in its valence shell. When silicon atoms combine to form a solid crystal, each atom positions itself between four other silicon atoms in such a way that the valence shells overlap from one atom to another. This causes each individual valence electron to be shared by two atoms, as shown in *Figure 6.47*. By sharing the electrons between four adjacent atoms each individual silicon atom *appears* to have eight electrons in its valence shell. This sharing of valence electrons is called **covalent bonding**.

In its pure state, silicon is an insulator because the covalent bonding rigidly holds all of the electrons leaving no free (easily loosened) electrons to conduct current. If, however, an atom of a different element (i.e. an **impurity**) is introduced that has five electrons in its valence shell, a surplus electron will be present, as shown in *Figure 6.48*. These **free electrons** become available for use as **charge carriers** and they can be made to move through the lattice by applying an external potential difference to the material.

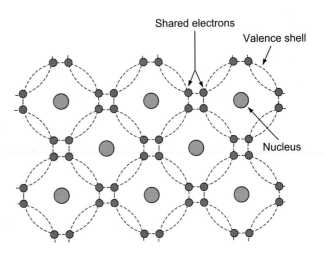

Figure 6.47 Lattice showing covalent bonding

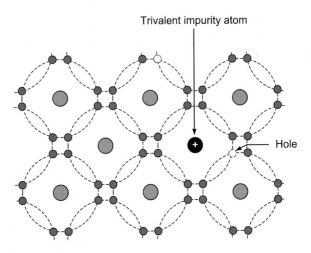

Figure 6.48 Free negative charge carriers (electrons) produced by introducing a pentavalent impurity

Similarly, if the impurity element introduced into the pure silicon lattice has three electrons in its valence shell, the absence of the fourth electron needed for proper covalent bonding will produce a number of gaps into which electrons can fit, as shown in *Figure 6.49*. These gaps are referred to as **holes**. Once again, current will flow when an external potential difference is applied to the material.

Regardless of whether the impurity element produces surplus electrons or holes, the material will no longer behave as an insulator, neither will it have the properties that we normally associate with a metallic conductor. Instead, we call the material a **semiconductor** – the term simply indicates that the substance is no longer a good insulator or a good conductor but is somewhere in between!

The process of introducing an atom of another (impurity) element into the lattice of an otherwise pure material is called **doping**. When the pure material is doped with an impurity with five electrons in its valence

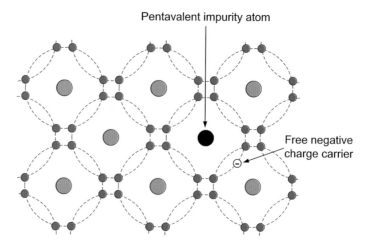

Figure 6.49 Holes produced by introducing a trivalent impurity

KEY POINT

Small amounts of impurity elements are added to pure semiconductor materials in order to give them N-type or P-type properties. This process is called doping

KEY POINT

A diode can be forward or reverse biased according to the polarity of the voltage applied to it. When forward biased, a diode will conduct heavily (i.e. it will pass current). When reverse biased, a diode will not conduct (i.e. it will block current)

shell (i.e. a **pentavalent impurity**) it will become an **N-type** material. If, however, the pure material is doped with an impurity having three electrons in its valence shell (i.e. a **trivalent impurity**) it will become **P-type** material. N-type semiconductor material contains an excess of negative charge carriers, and P-type material contains an excess of positive charge carriers.

Diodes

When a junction is formed between N-type and P-type semiconductor materials, the resulting device is called a diode. This component offers an extremely low resistance to current flow in one direction and an extremely high resistance to current flow in the other. This characteristic allows the diode to be used in applications that require a circuit to behave differently according to the direction of current flowing in it.

An ideal diode would pass an infinite current in one direction and no current at all in the other direction. In addition, the diode would start to conduct current when the smallest of voltages was present. In practice, a small voltage must be applied before conduction takes place. Furthermore, a small **leakage current** will flow in the **reverse direction**. This leakage current is usually a very small fraction of the current that flows in the **forward direction**.

If the P-type semiconductor material is made positive relative to the N-type material by an amount greater than its **forward threshold voltage** (about 0.6 V if the material is silicon and 0.2 V if the material is germanium), the diode will freely pass current. If, on the other hand, the P-type material is made negative relative to the N-type material, virtually no current will flow unless the applied voltage exceeds the maximum (breakdown) voltage that the device can withstand. Note that a normal diode will be destroyed if its **reverse breakdown voltage** is exceeded.

A semiconductor junction diode is shown in *Figure 6.50*. The connection to the P-type material is referred to as the **anode** while that to the

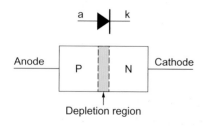

Figure 6.50 A P–N junction diode

N-type material is called the **cathode**. With no externally applied potential, electrons from the N-type material will cross into the P-type region and fill some of the vacant holes. This action will result in the production of a region either side of the junction in which there are no free charge carriers. This zone is known as the **depletion region**.

Figure 6.51a shows a junction diode in which the anode is made positive with respect to the cathode. In this **forward-biased** condition, the diode freely passes current. *Figure 6.51b* shows a diode with the cathode made positive with respect to the cathode. In this **reverse-biased** condition, the diode passes a negligible amount of current. In the freely conducting forward-biased state, the diode acts rather like a closed switch. In the reverse-biased state, the diode acts like an open switch.

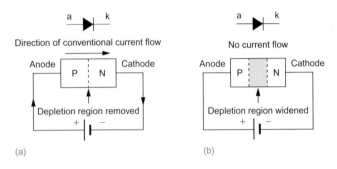

Figure 6.51   (a) Forward-biased P–N junction and (b) reverse-biased P–N junction

If a positive voltage is applied to the P-type material, the free positive charge carriers will be repelled and they will move away from the positive potential towards the junction. Likewise, the negative potential applied to the N-type material will cause the free negative charge carriers to move away from the negative potential towards the junction.

When the positive and negative charge carriers arrive at the junction, they will attract one another and combine (recall that unlike charges attract). As each negative and positive charge carrier combine at the junction, a new negative and positive charge carrier will be introduced to the semiconductor material from the voltage source. As these new charge carriers enter the semiconductor material, they will move toward the junction and combine. Thus, current flow is established and it will continue for as long as the voltage is applied.

As stated earlier, the **forward threshold voltage** must be exceeded before the diode will conduct. The forward threshold voltage must be high enough to completely remove the depletion layer and force charge carriers to move across the junction. With silicon diodes, this forward threshold voltage is approximately 0.6–0.7 V. With germanium diodes, the forward threshold voltage is approximately 0.2–0.3 V.

Figure 6.52 shows typical characteristics for small germanium and silicon diodes. It is worth noting that diodes are limited by the amount of forward current and reverse voltage they can withstand. This limit is based on the physical size and construction of the diode.

UNIT 6

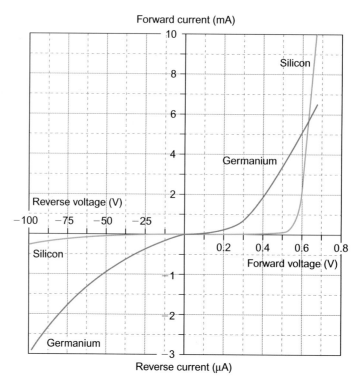

Figure 6.52 Typical diode characteristics

In the case of a reverse-biased diode, the P-type material is negatively biased relative to the N-type material. In this case, the negative potential applied to the P-type material attracts the positive charge carriers, drawing them away from the junction. Likewise, the positive potential applied to the N-type material attracts the negative charge carriers away from the junction. This leaves the junction area depleted, and virtually no charge carriers exist. Therefore, the junction area becomes an insulator, and current flow is inhibited.

The reverse bias potential may be increased to the reverse breakdown voltage for which the particular diode is rated. As in the case of the maximum forward current rating, the reverse breakdown voltage is specified by the manufacturer. The reverse breakdown voltage is usually very much higher than the forward threshold voltage. A typical general-purpose diode may be specified as having a forward threshold voltage of 0.6 V and a reverse breakdown voltage of 200 V. If the latter is exceeded, the diode may suffer irreversible damage. It is also worth noting that, where diodes are designed for use as rectifiers, manufacturers often quote **peak inverse voltage (PIV)** or **maximum reverse repetitive voltage (V_{RRM})** rather than maximum reverse breakdown voltage.

Figure 6.53 shows a test circuit for obtaining diode characteristics (note that the diode must be reverse connected in order to obtain the reverse characteristic).

UNIT 6

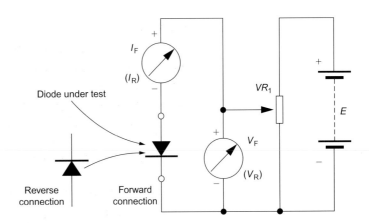

Figure 6.53 Diode test circuit

Example 6.30

The characteristic shown in *Figure 6.54* refers to a germanium diode. Determine the resistance of the diode when (a) the forward current is 2.5 mA and (b) when the forward voltage is 0.65 V.

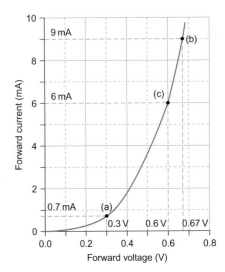

Figure 6.54 See Example 6.30

(a) When $I_F = 2.5$ mA the corresponding value of V_F can be read from the graph. This shows that $V_F = 0.43$ V. Using Ohm's Law, the resistance of the diode at this point on the characteristic will be given by:

$$R = \frac{V_F}{I_F} = \frac{0.43 \text{ V}}{2.5 \text{ mA}} = 172\,\Omega$$

(b) When $V_F = 0.65$ V the corresponding value of I_F can be read from the graph. This shows that $I_F = 7.4$ mA. Using Ohm's Law, the resistance of the diode at this point on the characteristic will be given by:

$$R = \frac{V_F}{I_F} = \frac{0.65 \text{ V}}{7.4 \text{ mA}} = 88\,\Omega$$

This example shows how the resistance of a diode does not remain constant but instead changes according to the point on the characteristic at which it is operating.

Diode types

Diodes are often divided into **signal** or **rectifier** types according to their principal field of application. Signal diodes require consistent forward characteristics with low forward voltage drop. Rectifier diodes need to be able to cope with high values of reverse voltage and large values of forward current, consistency of characteristics is of secondary importance in such applications. *Table 6.6* summarizes the properties of some common semiconductor diodes whilst a selection of diodes is shown in *Figure 6.55*.

Table 6.6 Properties of some common semiconductor diodes

Device	Material	PIV	I_F max.	I_R max.	Application
1N4148	Silicon	100V	76mA	25nA	General purpose
1N914	Silicon	100V	75mA	25nA	General purpose
AA113	Germanium	60V	10mA	200μA	RF detector
OA47	Germanium	25V	110mA	100μA	Signal detector
OA91	Germanium	115V	50mA	275μA	General purpose
1N4001	Silicon	50V	1A	10μA	Low-voltage rectifier
1N5404	Silicon	400V	3A	10μA	High-voltage rectifier
BY127	Silicon	1250V	1A	10μA	High-voltage rectifier

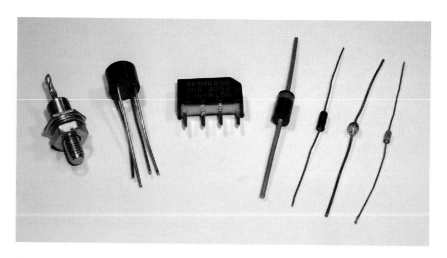

Figure 6.55 A selection of diodes including power diodes, bridge rectifiers and signal diodes

Later in this unit we will show how diodes are used in simple power supply circuits. Before we do this, however, we will be taking a look at some important components that rely on using electrostatics and electromagnetism for their operation.

TYK *6.15*

Sketch the symbol for a semiconductor diode and label the anode and cathode connections.

UNIT 6

TYK 6.16

Explain each of the following terms in relation to a semiconductor diode:

1. Doping
2. Depletion region
3. Forward bias
4. Reverse bias
5. Forward threshold voltage.

TYK

TYK 6.17

Sketch the forward and reverse characteristics for a typical semiconductor diode. Labels the axes and include typical values of current and voltage.

Activity 6.3

The following data refers to a signal diode:

V_F (V)	I_F (mA)
0.0	0.0
0.1	0.2
0.2	0.5
0.3	1.4
0.4	2.6
0.5	5.8
0.6	9.8
0.7	13.8

Plot the characteristic and use it to determine:

(a) **the forward current when $V_F = 0.35\,$V**
(b) **the forward voltage when $I_F = 15\,$mA**
(c) **the resistance of the diode when $V_F = 0.65\,$V.**

Capacitance

Electrostatics

If a conductor has a deficit of electrons, it will exhibit a net positive charge. On the other hand, if it has a surplus of electrons, it will exhibit a net negative charge. An imbalance in charge can be produced by friction (removing or depositing electrons using materials such as silk and fur,

respectively) or induction (by attracting or repelling electrons using a second body which is, respectively, positively or negatively charged).

Force between charges

Coulomb's Law states that, if charged bodies exist at two points, the force of attraction (if the charges are of opposite polarity) or repulsion (if the charges have the same polarity) will be proportional to the product of the magnitude of the charges divided by the square of their distance apart. Thus,

$$F = \frac{kQ_1Q_2}{r^2}$$

where Q_1 and Q_2 are the charges present at the two points (in Coulombs), r is the distance separating the two points (in m), F is the force (in Newtons) and k is a constant depending upon the medium in which the charges exist.

In vacuum or 'free space',

$$k = \frac{1}{4\pi\varepsilon_0}$$

where ε_0 is the **permittivity of free space** ($8.854 \times 10^{-12} \, C/Nm^2$).

Combining the two previous equations gives:

$$F = \frac{kQ_1Q_2}{4\pi \times 8.854 \times 10^{-12} \times r^2}$$

Electric fields

The force exerted on a charged particle is a manifestation of the existence of an electric field. The electric field defines the direction and magnitude of a force on a charged object. The field is invisible to the human eye but can be drawn by constructing lines which indicate the motion of a free positive charge within the field; the number of field lines in a particular region being used to indicate the relative strength of the field at the point in question.

Figures 6.56 and *6.57* show the electric fields between charges of the same and opposite polarity while *Figure 6.58* shows the field which exists between two charged parallel plates. You will see more of this particular arrangement when we introduce capacitors a little later in this section.

Electric field strength

The strength of an electric field (E) is proportional to the applied potential difference and inversely proportional to the distance between the two conductors. The electric field strength is given by:

$$E = \frac{V}{d}$$

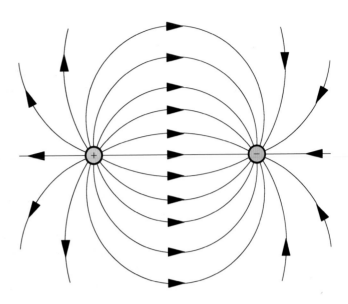

Figure 6.56 Electric field between two unlike electric charges

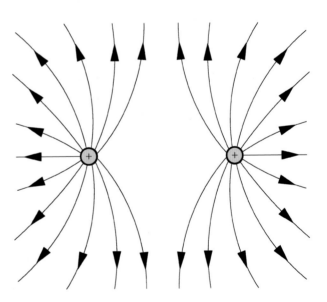

Figure 6.57 Electric field between two like electric charges (in this case both positive)

where E is the electric field strength (V/m), V is the applied potential difference (V) and d is the distance (m).

Example 6.31

Two parallel conductors are separated by a distance of 25 mm. Determine the electric field strength if they are fed from a 600 V DC supply.

The electric field strength will be given by:

$$E = \frac{V}{d} = \frac{600}{25 \times 10^{-3}} = 24 \times 10^3 = 24\,\text{kV/m}$$

TYK

TYK 6.18

Determine the electric field strength that appears in the space between two parallel plates separated by an air gap of 4 mm if a potential of 2.5 kV exists between them.

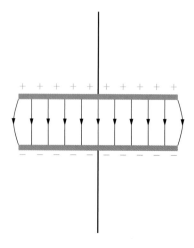

Figure 6.58 Electric field between two parallel plates

Permittivity

The amount of charge produced on the two plates shown in *Figure 6.58* for a given applied voltage will depend not only on the physical dimensions but also on the insulating dielectric material that appears between the plates. Dielectric materials need to have a very high value of resistivity (they must not conduct charge and allow it to drain away) coupled with an ability to withstand high voltages without breaking down.

A more practical arrangement is shown in *Figure 6.59*. In this arrangement the ratio of charge, Q, to potential difference, V, is given by the relationship:

$$\frac{Q}{V} = \frac{\varepsilon A}{d}$$

where A is the surface area of the plates (in m), d is the separation (in m), and ε is a constant for the dielectric material known as the **absolute permittivity** of the material (sometimes also referred to as the **dielectric constant**).

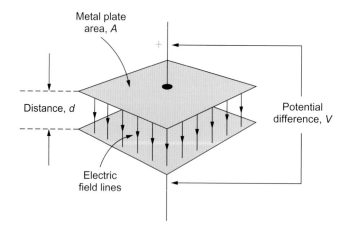

Figure 6.59 Parallel plates with an insulating dielectric material

The absolute permittivity of a dielectric material is the product of the permittivity of free space (ε_0) and the **relative permittivity** (ε_r) of the material. Thus,

$$\varepsilon = \varepsilon_0 \times \varepsilon_r \quad \text{and} \quad \frac{Q}{V} = \frac{\varepsilon_0 \varepsilon_r A}{d}$$

The **dielectric strength** of an insulating dielectric is the maximum electric field strength that can safely be applied to it before breakdown (conduction)

UNIT 6

Table 6.7 Properties of some common insulating dielectric materials

Dielectric material	Relative permittivity (free space = 1)	Dielectric strength (kV/mm)
Vacuum, or free space	1	∞
Air	1	3
Polythene	2.3	50
Paper	2.5–3.5	14
Polystyrene	2.5	25
Mica	4–7	160
Pyrex glass	4.5	13
Glass ceramic	5.9	40
Polyester	3.0–3.4	18
Porcelain	6.5	4
Titanium dioxide	100	6
Ceramics	5–1000	2–10

occurs. *Table 6.7* shows values of relative permittivity and dielectric strength for some common dielectric materials.

Capacitors

A capacitor is a device for storing electric charge. In effect, it is a reservoir into which charge can be deposited and then later extracted. Typical applications include reservoir and smoothing capacitors for use in power supplies, coupling AC signals between the stages of amplifiers, and decoupling supply rails (i.e. effectively grounding the supply rails as far as AC signals are concerned).

A capacitor can consist of nothing more than two parallel metal plates as shown earlier in *Figure 6.59*. To understand what happens when a capacitor is being charged and discharged take a look at *Figure 6.60*. If the switch is left open (position A), no charge will appear on the plates and in this condition there will be no electric field in the space between the plates nor will there be any charge stored in the capacitor.

When the switch is moved to position B, electrons will be attracted from the positive plate to the positive terminal of the battery. At the same time, a similar number of electrons will move from the negative terminal of the battery to the negative plate. This sudden movement of electrons will manifest itself in a momentary surge of current (conventional current will flow from the positive terminal of the battery towards the positive terminal of the capacitor).

Eventually, enough electrons will have moved to make the e.m.f. between the plates the same as that of the battery. In this state, the capacitor is said to be **fully charged** and an electric field will be present in the space between the two plates.

If, at some later time the switch is moved back to position A, the positive plate will be left with a deficiency of electrons whilst the negative plate will

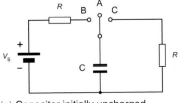

(a) Capacitor initially uncharged

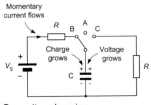

(b) Capacitor charging

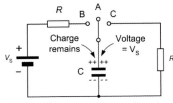

(c) Capacitor remains charged

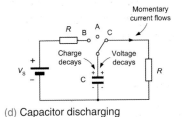

(d) Capacitor discharging

Figure 6.60 Capacitor charging and discharging

be left with a surplus of electrons. Furthermore, since there is no path for current to flow between the two plates the capacitor will remain charged and a potential difference will be maintained between the plates.

Now assume that the switch is moved to position C. The excess electrons on the negative plate will flow through the resistor to the positive plate until a neutral state once again exists (i.e. until there is no excess charge on either plate). In this state the capacitor is said to be **fully discharged** and the electric field between the plates will rapidly collapse. The movement of electrons during the discharging of the capacitor will again result in a momentary surge of current (current will flow from the positive terminal of the capacitor and into the resistor).

Figure 6.61 shows the direction of current flow in the circuit of *Figure 6.60* during charging (switch in position B) and discharging (switch in position C). It should be noted that current flows momentarily in both circuits even though you may think that the circuit is broken by the gap between the capacitor plates!

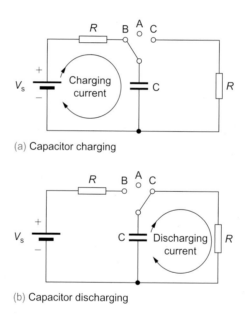

(a) Capacitor charging

(b) Capacitor discharging

Figure 6.61 Current flow during charging and discharging

Capacitance

The unit of capacitance is the Farad (F). A capacitor is said to have a capacitance of 1 F if a current of 1 A flows in it when a voltage charging at the rate of 1 V/s is applied to it. The current flowing in a capacitor will thus be proportional to the product of the capacitance, C, and the rate of charge of applied voltage. Hence:

$$i = C \times \text{rate of charge of voltage}$$

Note that we have used a small i to represent the current flowing in the capacitor. We have done this because the current is charging and does not remain constant.

The rate of charge of voltage is often represented by the expression dv/dt, where dv represents a very small charge in voltage and dt represents the corresponding small charge in time. Expressing this mathematically gives:

$$i = C\frac{\mathrm{d}v}{\mathrm{d}t}$$

We will return to the mathematics of charge and discharge a little later in this section.

Example 6.32

A voltage is charging at a uniform rate from 10 to 50 V in a period of 0.1 second. If this voltage is applied to a capacitor of 22 μF, determine the current that will flow in the capacitor.

Now the current flowing will be given by:

$$i = C \times \text{rate of charge of voltage}$$

Thus

$$i = 22 \times 10^{-6} \times \frac{50 - 10}{0.1} = 22 \times 400 \times 10^{-6}\,\text{mA}$$

From which

$$i = 8.8 \times 10^{-3}\,\text{A} = 8.8\,\text{mA}$$

KEY POINT

Charge is the quantity of electricity that can be stored in a capacitor. The charge in a capacitor is directly proportional to the product of the capacitance and the applied potential difference

Charge, capacitance and voltage

The charge or **quantity of electricity** that can be stored in the electric field between the capacitor plates is proportional to the applied voltage and the capacitance of the capacitor (see *Figure 6.62*). Thus,

$$Q = CV$$

where Q is the charge (coulombs), C is the capacitance (F) and V is the potential difference (V).

Example 6.33

A 10 μF capacitor is charged to a potential of 250 V. Determine the charge stored.

The charge stored will be given by:

$$Q = CV = 10 \times 10^{-6} \times 250 = 2.5\,\text{mC}$$

TYK

TYK 6.19

Determine the charge in a capacitor of 470 μF when a potential difference of 22 V appears across its plates.

UNIT 6

TYK 6.20

A capacitor of 150 µF is required to store a charge of 400 µC. What voltage should be applied to the capacitor?

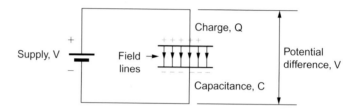

Figure 6.62 Capacitance, charge and voltage

Energy storage

When charge, Q, is plotted against voltage, V, for a particular value of capacitance, C, it follows the linear law shown in *Figure 6.63a*. The slope of the line (Q/V) indicates the capacitance whilst the area below the line (shown as shaded portion in *Figure 6.63b*) is a measure of the energy stored in the capacitor. The larger this area is the more energy is stored.

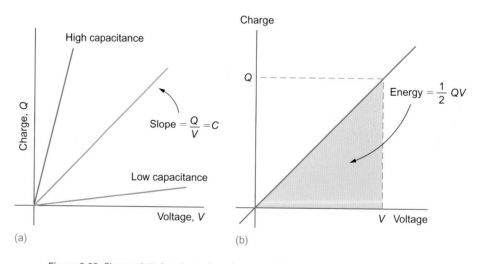

Figure 6.63 Charge plotted against voltage for a capacitor

KEY POINT

The energy stored in a capacitor is proportional to the product of the capacitance and the square of the potential difference

In *Figure 6.63*, the shaded area can be found by considering the area to be a triangle in which the area is the product of half the base and the height. Hence,

Energy stored, $$W = \tfrac{1}{2}QV$$

Combining this relationship with $Q = CV$ gives:

$$\text{Energy stored} = \tfrac{1}{2}(CV) \times V = \tfrac{1}{2}CV^2$$

where W is the energy (J), C is the capacitance (F) and V is the potential difference (V).

UNIT 6

Example 6.34

A capacitor of 47 μF is required to store 4 J of energy. Determine the potential difference that must be applied to the capacitor.

The foregoing formula can be re-arranged to make *V* the subject as follows:

$$V = \sqrt{\frac{2W}{C}} = \sqrt{\frac{2 \times 4}{47 \times 10^{-6}}}$$

from which

$$V = \sqrt{0.170 \times 10^{6}} = 0.412 \times 10^{3} = 412\,V$$

Test your knowledge
TYK

TYK 6.21

Determine the energy stored in a 68 μF when it is charged to a potential of 200 V.

Capacitance and physical dimensions

The capacitance of a capacitor depends upon the physical dimensions of the capacitor (i.e. the size of the plates and the separation between them) and the dielectric material between the plates (see *Figure 6.64*). By combining the relationships that we met earlier in this section:

$Q = CV$ (from which) $C = Q/V$ and $Q/V = \varepsilon_0\varepsilon_r A/d$ we can obtain the following expression for the capacitance of a parallel plate capacitor:

$$C = \frac{\varepsilon_0\varepsilon_r A}{d}$$

where *C* is the capacitance (F), ε_0 is the permittivity of free space, ε_r is the **relative permittivity** of the dielectric medium between the plates and *d* is the separation between the plates (m).

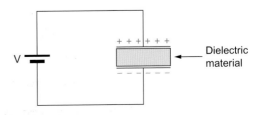

Figure 6.64 Use of a dielectric material between the plates of a capacitor

Example 6.35

A capacitor of 1 nF is required. If a dielectric material of thickness 0.1 mm and relative permittivity 5.4 is available, determine the required plate area.

Re-arranging the formula

$$C = \frac{\varepsilon_0 \varepsilon_r A}{d}$$

to make A the subject gives:

$$A = \frac{Cd}{\varepsilon_0 \varepsilon_r} = \frac{1 \times 10^{-9} \times 0.1 \times 10^{-3}}{8.854 \times 10^{-12} \times 5.4}$$

From which

$$A = \frac{0.1 \times 10^{-12}}{47.8116 \times 10^{-12}}$$

thus

$$A = 0.00209 \, \text{m}^2 \text{ or } 20.9 \, \text{cm}^2$$

In order to increase the capacitance of a capacitor, many practical components employ multiple plates, as shown in *Figure 6.65a*. The capacitance is then given by:

$$C = \frac{\varepsilon_0 \varepsilon_r (n-1)A}{d}$$

where C is the capacitance (F), ε_0 is the permittivity of free space, ε_r is the relative permittivity of the dielectric medium between the plates, d is the separation between the plates (m) and n is the total number of plates. The construction of a practical tubular capacitor is shown in *Figure 6.65b*.

(a) Basic interleaved plate arrangement

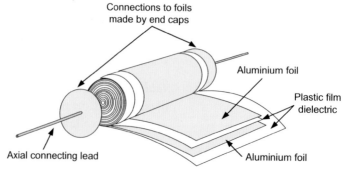

Connections to foils
made by end caps

Aluminium foil

Plastic film
dielectric

Axial connecting lead

Aluminium foil

(b) Typical construction of fubular capacitor

Figure 6.65 Capacitor construction

Example 6.36

A capacitor consists of six plates each of area 20 cm² separated by a dielectric of relative permittivity 4.5 and thickness 0.2 mm. Determine the value of capacitance.

Using

$$C = \frac{\varepsilon_0 \varepsilon_r (n-1) A}{d}$$

gives

$$C = \frac{8.854 \times 10^{-12} \times 4.5 \times (6-1) \times 20 \times 10^{-4}}{0.2 \times 10^{-3}}$$

from which

$$C = \frac{3984.3 \times 10^{-16}}{0.2 \times 10^{-3}} = 1.992 \times 10^{-9}$$

thus

$$C = 1.992 \times 10^{-9}\, \text{F} \ \text{or} \ 1.992\, \text{nF}$$

Test your knowledge

TYK

TYK 6.22

A capacitor consists of five plates each of area 12 cm² separated by a polythene dielectric of thickness 0.16 mm. Determine the capacitance and maximum working voltage for the capacitor. (*Hint*: Refer to Table 6.7 on page 488 for details of the dielectric material.)

Capacitor specifications

The specifications for a capacitor usually include the value of capacitance (expressed in microfarads, nanofarads or picofarads), the voltage rating (i.e. the maximum voltage which can be continuously applied to the capacitor under a given set of conditions), and the accuracy or tolerance (quoted as the maximum permissible percentage deviation from the marked value).

Other practical considerations when selecting capacitors for use in a particular application include temperature coefficient, leakage current, stability and ambient temperature range.

Table 6.8 summarizes the properties of five of the most common types of capacitor. Note that **electrolytic** capacitors require the application of a **polarizing voltage** in order to start the chemical action on which they depend for their operation. The polarizing voltages used for electrolytic capacitors can range from as little as 1 V to several hundred volts depending upon the working voltage rating for the component in question.

Figure 6.66 shows some typical non-electrolytic capacitors (including polyester, polystyrene, ceramic and mica types) whilst *Figure 6.67* shows

KEY POINT

Electrolytic capacitors use a chemical dielectric and they must be connected with the correct polarity in order for them to work correctly

UNIT 6

Table 6.8 Properties of some common types of capacitor

Property	Capacitor type				
	Ceramic	Electrolytic	Polyester	Mica	Polystyrene
Capacitance range (F)	2.2 p to 100n	100n to 10m	10n to 2.2μ	0.47 p to 10n	10 p to 22n
Typical tolerance (%)	±10 and ±20	−10 to +50	±10	±1	±5
Typical voltage rating (W)	50 to 200V	6.3 to 400V	100 to 400V	350V	100V
Temperature coefficient (ppm/°C)	+100 to −4700	+1000 typical	+100 to +200	+50	+250
Stability	Fair	Poor	Good	Excellent	Good
Ambient temperature range (°C)	−85 to +85	−40 to +80	−40 to +100	−40 to +125	−40 to +100
Typical applications	High frequency and low cost	Smoothing and decoupling	General purpose	Tuned circuits and oscillators	General purpose

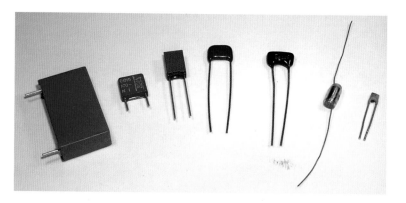

Figure 6.66 A typical selection of non-polarized capacitors (including polyester, polystyrene, ceramic and mica types) with values ranging from 10 pF to 470 nF and working voltages from 50 to 250V

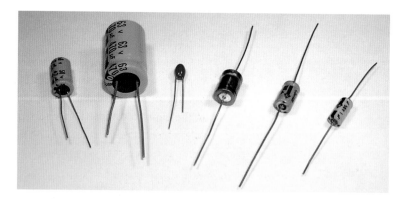

Figure 6.67 A typical selection of electrolytic (polarized) capacitors with values ranging from 1 μF to 470 μF and working voltages from 10 to 63V

a selection of electrolytic (polarized) capacitors. An air-spaced variable capacitor is shown in *Figure 6.66*.

Variable capacitors

By moving one set of plates relative to the other, a capacitor can be made variable. The dielectric material used in a variable capacitor can be either air (see *Figure 6.68*) or plastic (the latter tend to be more compact). Typical

Figure 6.68 An air-spaced variable capacitor. This component (used for tuning an AM radio) has two separate variable capacitors (each of 500 pF maximum) operated from a common control shaft

values for variable capacitors tend to range from about 25 to 500 pF. These components are commonly used for tuning radio receivers.

Capacitor markings

The vast majority of capacitors employ written markings which indicate their values, working voltages and tolerance. The most usual method of marking resin dipped polyester (and other) types of capacitor involves quoting the value (μF, nF or pF), the tolerance (often either 10% or 20%), and the working voltage (often using – and ~ to indicate DC and AC, respectively). Several manufacturers use two separate lines for their capacitor markings and these have the following meanings:

First line: capacitance (pF or μF) and tolerance ($K = 10\%$, $M = 20\%$)
Second line: rated DC voltage and code for the dielectric material

A three-digit code is commonly used to mark monolithic ceramic capacitors (*Figure 6.69*). The first two digits of this code correspond to the first two digits of the value while the third digit is a multiplier which gives the number of zeros to be added to give the value in picofarads. Other capacitors may use a colour code similar to that used for marking resistor values (see *Figure 6.70*).

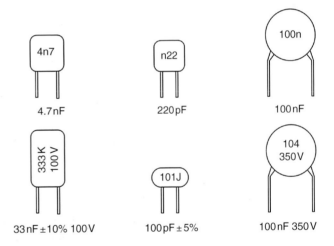

Figure 6.69 Examples of capacitor markings

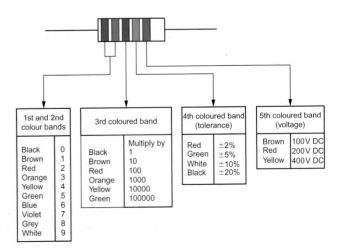

Figure 6.70 Capacitor colour code

Example 6.37

A monolithic ceramic capacitor is marked with the legend '103 K'. What is its value?

The value (pF) will be given by the first two digits (10) followed by the number of zeros indicated by the third digit (3). The value of the capacitor is thus 10,000 pF or 10 nF. The final letter (K) indicates that the capacitor has a tolerance of 10%.

Example 6.38

A tubular capacitor is marked with the following coloured stripes: brown, green, brown, red and brown. What is its value, tolerance and working voltage?

See *Figure 6.71*.

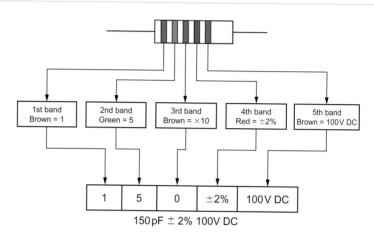

Figure 6.71 See Example 6.38

TYK 6.23

List three different types of capacitor. For each type specify a typical range of available values of capacitance, tolerance, and working voltage.

Series and parallel combination of capacitors

We can use $Q = CV$ and Kirchhoff's Voltage Laws to determine the equivalent capacitance of more complicated circuits where capacitors are connected in **series** and in **parallel**.

In the case of the series arrangement of three resistors shown in *Figure 6.72*, the same charge will appear in each capacitor. Hence,

$$Q = Q_1 = Q_2 = Q_3$$

From Kirchhoff's Voltage Law, the supply voltage, V, will be equal to the sum of the individual voltage drops. Hence,

$$V = V_1 + V_2 + V_3$$

UNIT 6

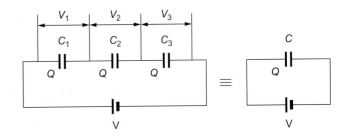

Figure 6.72 Equivalent capacitance of three capacitors connected in series

$$\text{Now } V_1 = \frac{Q_1}{C_1} = \frac{Q}{C_1}, \; V_2 = \frac{Q_2}{C_2} = \frac{Q}{C_2}, \text{ and } V_3 = \frac{Q_3}{C_3} = \frac{Q}{C_3}$$

Combining these relationships with the previous equation gives:

$$V = \frac{Q}{C_1} + \frac{Q}{C_2} + \frac{Q}{C_3}$$

If C is the equivalent capacitance that could replace C_1, C_2 and C_3 connected in series, we arrive at:

$$C = \frac{Q}{V}$$

from which

$$V = \frac{Q}{C}$$

Hence,

$$\frac{Q}{C} = \frac{Q}{C_1} + \frac{Q}{C_2} + \frac{Q}{C_3}$$

Finally, dividing both sides of the equation by Q gives:

$$\frac{1}{C} = \frac{1}{C_1} + \frac{1}{C_2} + \frac{1}{C_3}$$

In the case of the parallel arrangement of three capacitors shown in *Figure 6.73*, the same voltage will appear across each capacitor.

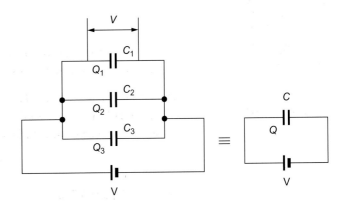

Figure 6.73 Equivalent resistance of three capacitors connected in parallel

Resistor and, from Kirchhoff's Current Law, the supply current, I, will be equal to the sum of the currents flowing in the three resistors. Hence,

$$Q_1 = C_1V, Q_2 = C_2V, \text{ and } Q_3 = C_3V$$

The total charge in the parallel circuit will be divided between the three capacitors. Hence,

$$Q = Q_1 + Q_2 + Q_3$$

If C is the equivalent capacitance that could replace C_1, C_2 and C_3 connected in parallel we arrive at:

$$Q = CV$$

Hence,

$$CV = C_1V + C_2V + C_3V$$

Dividing both sides by V gives:

$$C = C_1 + C_2 + C_3$$

In order to obtain a particular value of capacitance, fixed capacitors may be arranged in either series or parallel as shown in *Figures 6.74 and* 6.75. In the case of the series connected capacitors shown in *Figure 6.74*, the reciprocal of the effective capacitance of each circuit is equal to the sum of the reciprocals of the individual capacitances. Hence, for *Figure 6.74a*:

$$\frac{1}{C} = \frac{1}{C_1} + \frac{1}{C_2}$$

while for *Figure 6.74b*

$$\frac{1}{C} = \frac{1}{C_1} + \frac{1}{C_2} + \frac{1}{C_3}$$

In the former case (and *only* the former case), the formula can be more conveniently re-arranged as follows:

$$C = \frac{C_1 \times C_2}{C_1 + C_2}$$

You can remember this as the *product* of the two capacitor values *divided by* the *sum* of the two values – just as you did for two resistors in parallel.

Turning to the parallel connected capacitors shown in *Figure 6.75*, the effective capacitance of each of the parallel circuits shown in *Figure 6.75* is simply equal to the sum of the individual capacitances. So, for the circuit shown in *Figure 6.75a*:

$$C = C_1 + C_2$$

while for *Figure 6.75b*

$$C = C_1 + C_2 + C_3$$

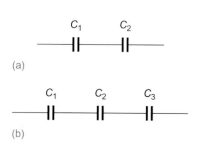

Figure 6.74 Capacitors in series

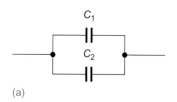

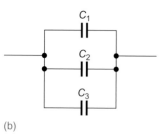

Figure 6.75 Capacitors in parallel

UNIT 6

Example 6.39

Determine the effective capacitance of the circuit shown in *Figure 6.76*.

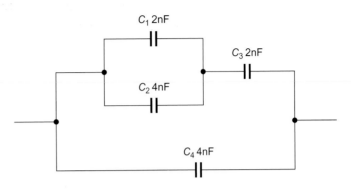

Figure 6.76 See Example 6.39

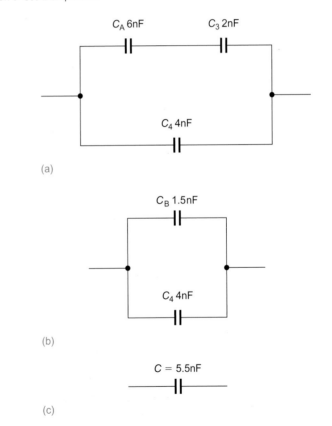

Figure 6.77 See Example 6.39

The circuit of *Figure 6.76* can be progressively simplified as shown in *Figure 6.77*. The stages in this simplification are:

(a) C_1 and C_2 are in parallel and they can be replaced by a single capacitor (we have called this C_A) of $2 + 4 = 6\,\text{nF}$.

(b) C_A appears in series with C_3. These two resistors can be replaced by a single capacitor (we have labelled this C_B) of $\dfrac{6 \times 2}{6 + 2} = \dfrac{12}{8} = 1.5\,\text{nF}$

(c) Finally, C_B appears in parallel with C_4. These two capacitors can be replaced by a single capacitance, C, of $1.5 + 4 = 5.5\,\text{nF}$.

Example 6.40

A capacitance of 50 μF (rated at 100 V) is required. What series combination of preferred value capacitors will satisfy this requirement? What voltage rating should each capacitor have?

Two 100 μF capacitors wired in series will provide a capacitance of 50 μF, as follows:

$$\frac{1}{100} = \frac{1}{C} + \frac{1}{C} = \frac{2}{C}$$

Thus

$$C = \frac{100}{2} = 50\,\mu F$$

Since the capacitors are of equal value, the applied DC potential will be shared equally between them. Thus, each capacitor should be rated at 50 V. Note that, in a practical circuit, we could take steps to ensure that the DC voltage was shared equally between the two capacitors by wiring equal, high-value (e.g. 1 MΩ) resistors across each capacitor.

TYK 6.24

Three 15 μF capacitors are connected (a) in series and (b) in parallel. Determine the effective capacitance of each combination and the charge that will appear in each capacitor when connected to a 30 V supply.

TYK 6.25

Capacitors of 3 and 6 μF are connected in series across a 150 V supply. Determine (a) the voltage that will appear across each capacitor and (b) the energy stored in each capacitor.

Activity 6.4

Capacitors of 1, 3.3, 4.7 and 10 μF are available. How can two or more of these capacitors be arranged to realize the following capacitance values?
(a) 8 μF
(b) 11 μF
(c) 19 μF
(d) 0.91 μF
(e) 1.94 μF.

C–R circuits

Networks of capacitors and resistors (known as *C–R* circuits) form the basis of many timing and pulse shaping circuits and are thus often found in practical electric and electronic circuits.

Charging

A simple C–R circuit is shown in *Figure 6.78*. In this circuit C is charged through R from the constant voltage source, V_s. The voltage, v_c, across the (initially uncharged) capacitor voltage will rise exponentially as shown in *Figure 6.79*. At the same time, the current in the circuit, i, will fall, as shown in *Figure 6.80*.

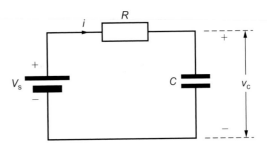

Figure 6.78 A C–R circuit in which C is charged through R

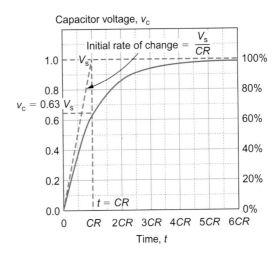

Figure 6.79 Exponential growth of capacitor voltage, v_c, in Figure 6.78

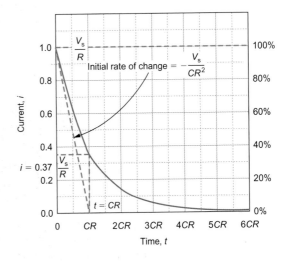

Figure 6.80 Exponential decay of current, i, in Figure 6.78

The rate of growth of voltage with time (and decay of current with time) will be dependent upon the product of capacitance and resistance. This value is known as the **time constant** of the circuit. Hence:

Time constant, $\qquad t = C \times R$

where C is the value of capacitance (F), R is the resistance (Ω) and t is the time constant (s).

The voltage developed across the charging capacitor, v_c, varies with time, t, according to the relationship:

$$v_c = V_s\left(1 - e^{\frac{-t}{CR}}\right)$$

where v_c is the capacitor voltage, V_s is the DC supply voltage, t is the time and CR is the time constant of the circuit (equal to the product of capacitance, C, and resistance, R).

The capacitor voltage will rise to approximately 63% of the supply voltage, V_s, in a time interval equal to the time constant.

At the end of the next interval of time equal to the time constant (i.e. after an elapsed time equal to $2CR$) the voltage will have risen by 63% of the remainder, and so on. In theory, the capacitor will **never** become fully charged. However, after a period of time equal to $5CR$, the capacitor voltage will to all intents and purposes be equal to the supply voltage. At this point the capacitor voltage will have risen to 99.3% of its final value and we can consider it to be fully charged. *Figure 6.81* shows a typical application for C–R circuits with different time constants selected by a rotary switch.

Figure 6.81 *C–R* circuits are widely used in electronics. In this oscilloscope, for example, a rotary switch is used to select different *C–R* combinations in order to provide the various timebase ranges (adjustable from 500 ms/cm to 1 µs/cm). Each *C–R* time constant corresponds to a different timebase range

UNIT 6

During charging, the current in the capacitor, i, varies with time, t, according to the relationship:

$$i = \frac{V_s}{R} e^{\frac{-t}{CR}}$$

where V_s is the DC supply voltage, t is the time, R is the series resistance and C is the value of capacitance.

The current will fall to approximately 37% of the initial current in a time equal to the time constant. At the end of the next interval of time equal to the time constant (i.e. after a total time of $2CR$ has elapsed) the current will have fallen by a further 37% of the remainder and so on.

Example 6.41

An initially uncharged 1 µF capacitor is charged from a 9 V DC supply via a 3.3 MΩ resistor. Determine the capacitor voltage 1 s after connecting the supply.

The formula for exponential growth of voltage in the capacitor is:

$$v_C = V_s \left(1 - e^{\frac{-t}{CR}}\right)$$

Here we need to find the capacitor voltage, v_C, when $V_s = 9$ V, $t = 1$ s, $C = 1$ µF and $R = 3.3$ MΩ. The time constant, CR, will be given by:

$$CR = 1 \times 10^{-6} \times 3.3 \times 10^{6} = 3.3 \text{ s}$$

Thus,

$$v_C = 9 \left(1 - e^{\frac{-t}{3.3}}\right)$$

and

$$v_C = 9(1 - 0.738) = 9 \times 0.262 = 2.358 \text{ V}$$

Example 6.42

A 100 µF capacitor is charged from a 350 V DC supply through a series resistance of 1 kΩ. Determine the initial charging current and the current that will flow 50 and 100 ms after connecting the supply. After what time is the capacitor considered to be fully charged?

At $t = 0$ the capacitor will be uncharged ($v_C = 0$) and all of the supply voltage will appear across the series resistance. Thus, at $t = 0$:

$$i = \frac{V_s}{R} = \frac{350}{1000} = 0.35 \text{ A}$$

When t = 50 ms, the current will be given by:

$$i = \frac{V_s}{R} e^{\frac{-t}{CR}}$$

where $V_s = 350\,V$, $t = 50\,ms$, $C = 100\,\mu F$, $R = 1\,k\Omega$. Hence:

$$i = \frac{350}{1000}e^{\frac{-0.05}{0.1}} = 0.35e^{-0.5} = 0.35 \times 0.607 = 0.21\,A$$

When $t = 100\,ms$ (using the same equation but with $t = 0.1\,s$) the current is given by:

$$i = \frac{350}{1000}e^{\frac{-0.1}{0.1}} = 0.35e^{-1} = 0.35 \times 0.368 = 0.129\,A$$

The capacitor can be considered to be fully charged when $t = 5CR = 5 \times 100 \times 10^{-6} \times 1 \times 10^3 = 0.5\,s$. Note that, at this point the capacitor voltage will have reached 99% of its final value.

Discharge

Having considered the situation when a capacitor is being charged, let us consider what happens when an already charged capacitor is discharged.

When the fully charged capacitor from *Figure 6.78* is connected as shown in *Figure 6.82*, the capacitor will discharge through the resistor, and the capacitor voltage, V_c, will fall exponentially with time, as shown in *Figure 6.83*.

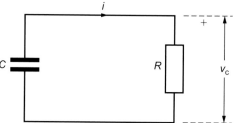

Figure 6.82 A *C–R* circuit in which *C* is initially charged and then discharges through *R*

KEY POINT

When an initially uncharged capacitor is being charged the voltage across its plates will rise exponentially. At the same time, the current supplied will fall exponentially. When an initially charged capacitor is being discharged the voltage across its plates will fall exponentially. At the same time, the current supplied delivered by the capacitor will fall exponentially

KEY POINT

The time constant of a *C–R* circuit is a measure of how quickly the capacitor will charge or discharge. The time constant is simply the product of the values of capacitance (expressed in F) and the resistance (expressed in Ω). In practice, a capacitor is usually considered to be fully charged or discharged after a period of five times the time constant (5*CR*)

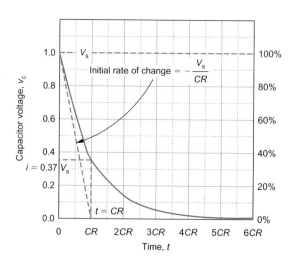

Figure 6.83 Exponential decay of capacitor voltage, v_c, in Figure 6.82

The current in the circuit, i, will also fall, as shown in *Figure 6.84*. The rate of discharge (i.e. the rate of decay of voltage with time) will once again be governed by the time constant of the circuit, $C \times R$.

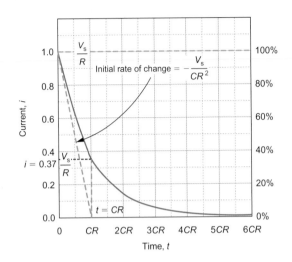

Figure 6.84 Exponential decay of current, i, in Figure 6.82

The voltage developed across the discharging capacitor, v_c, varies with time, t, according to the relationship:

$$v_c = V_s e^{\frac{-t}{CR}}$$

where V_s is the supply voltage, t is the time, C is the capacitance and R is the resistance.

The capacitor voltage will fall to approximately 37% of the initial voltage in a time equal to the time constant. At the end of the next interval of time equal to the time constant (i.e. after an elapsed time equal to $2CR$) the voltage will have fallen by 37% of the remainder and so on.

In theory, the capacitor will **never** become fully discharged. However, after a period of time equal to $5CR$, however, the capacitor voltage will to all intents and purposes be zero.

At this point the capacitor voltage will have fallen below 1% of its initial value. At this point we can consider it to be fully discharged.

As with charging, the current in the capacitor, i, varies with time, t, according to the relationship:

$$v_c = \frac{V_s}{R} e^{\frac{-t}{CR}}$$

where V_s is the supply voltage, t is the time, C is the capacitance and R is the resistance.

The current will fall to approximately 37% of the initial value of current, V_s/R, in a time equal to the time constant.

At the end of the next interval of time equal to the time constant (i.e. after a total time of $2CR$ has elapsed) the voltage will have fallen by a further 37% of the remainder and so on.

Example 6.43

A 10 μF capacitor is charged to a potential of 20 V and then discharged through a 47 kΩ resistor. Determine the time taken for the capacitor voltage to fall below 10 V.

The formula for exponential decay of voltage in the capacitor is:

$$v_c = V_s e^{\frac{-t}{CR}}$$

where $V_s = 20\,$V and $CR = 10\,\mu F \times 47\,k\Omega = 0.47\,$s.

We need to find t when $v_c = 10\,$V. Rearranging the formula to make t the subject gives:

$$t = -CR \times \ln\left(\frac{v_c}{V_s}\right)$$

thus

$$t = -0.47 \times \ln\left(\frac{10}{20}\right) = -0.47 \times \ln(0.5)$$

or

$$t = -0.47 \times (-0.693) = 0.325\,\text{s}$$

In order to simplify the mathematics of exponential growth and decay, Table 6.9 provides an alternative tabular method that may be used to determine the voltage and current in a C–R circuit.

Example 6.44

A 150 μF capacitor is charged to a potential of 150 V. The capacitor is then removed from the charging source and connected to a 2 MΩ resistor. Determine the capacitor voltage 1 minute later.

We will solve this problem using Table 6.9 rather than the exponential formula.

First we need to find the time constant, t:

$$t = C \times R = 150\,\mu F \times 2\,M\Omega = 300\,\text{s}$$

Next we find the ratio of t to CR:

After 1 minute, $t = 60\,$s therefore the ratio of t to CR is 60/300 or 0.2. Table 6.9 shows that when $t/CR = 0.2$, the ratio of instantaneous value to final value (k in Table 6.9) is 0.8187.

Thus

$$v_c/V_s = 0.8187$$

or

$$v_c = 0.8187 \times V_s = 0.8187 \times 150\,V = 122.8\,V$$

Table 6.9 Exponential growth and decay

t/CR	k (growth)	k (decay)
0.0	0.0000	1.0000
0.1	0.0951	0.9048
0.2	0.1812	0.8187 (1)
0.3	0.2591	0.7408
0.4	0.3296	0.6703
0.5	0.3935	0.6065
0.6	0.4511	0.5488
0.7	0.5034	0.4965
0.8	0.5506	0.4493
0.9	0.5934	0.4065
1.0	0.6321	0.3679
1.5	0.7769	0.2231
2.0	0.8647 (2)	0.1353
2.5	0.9179	0.0821
3.0	0.9502	0.0498
3.5	0.9698	0.0302
4.0	0.9817	0.0183
4.5	0.9889	0.0111
5.0	0.9933	0.0067

Notes: (1) See Example 6.44
(2) k is the ratio of the value at time, t, to the final value (e.g. v_c/V_s)

TYK 6.26

An initially uncharged 220 μF capacitor is charged from a 20 V supply through a resistor of 120 k.
(a) After what period of time would you consider the capacitor to be fully charged?
(b) Determine the time taken for the capacitor voltage to reach 10 V.

TYK 6.27

An unmarked capacitor is charged to 350 V and then discharged through a 1 MΩ. If it takes 20 seconds for the capacitor voltage to fall to 210 V, determine the value of capacitance.

Electromagnetism

When a current flows through a conductor a magnetic field is produced in the vicinity of the conductor. The magnetic field is invisible but its presence can be detected using a compass needle (which will deflect from its normal North–South position). If two current-carrying conductors are placed in the vicinity of one another, the fields will interact with one another and the conductors will experience a force of attraction or repulsion (depending upon the relative direction of the two currents).

Force between two current-carrying conductors

The mutual force which exists between two parallel current-carrying conductors will be proportional to the product of the currents in the two conductors and the length of the conductors but inversely proportional to their separation. Thus,

$$F = \frac{k I_1 I_2 l}{d}$$

where I_1 and I_2 are the currents in the two conductors (A), l is the parallel length of the conductors (m), d is the distance separating the two conductors (m), F is the force (Newtons) and k is a constant depending upon the medium in which the charges exist.

In vacuum or 'free space', $k = \dfrac{\mu_0}{2\pi}$

where μ_0 is a constant known as the **permeability of free space** ($4\pi \times 10^{-7}$ or 12.57×10^{-7} H/m).

Combining the two previous equations gives:

$$F = \frac{\mu_0 I_1 I_2 l}{2\pi d}$$

or

$$F = \frac{4\pi \times 10^{-7} I_1 I_2 l}{2\pi d}$$

or

$$F = \frac{2 \times 10^{-7} I_1 I_2 l}{d}$$

Magnetic fields

The field surrounding a straight current-carrying conductor is shown in *Figure 6.85*. The magnetic field defines the direction of motion of a free North pole within the field. In the case of *Figure 6.85*, the lines of flux are concentric and the direction of the field determined by the direction of current flow) is given by the right-hand rule.

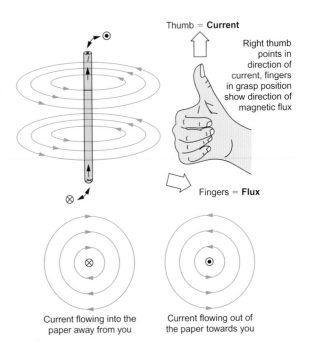

Figure 6.85 Magnetic field surrounding a straight conductor

Magnetic field strength

The strength of a magnetic field is a measure of the density of the flux at any particular point. In the case of *Figure 6.85*, the field strength will be proportional to the applied current and inversely proportional to the perpendicular distance from the conductor. Thus,

$$B = \frac{kI}{d}$$

where B is the magnetic flux density (T), I is the current (A), d is the distance from the conductor (m) and k is a constant.

Assuming that the medium is vacuum or 'free space', the density of the magnetic flux will be given by:

$$B = \frac{\mu_0 I}{2\pi d}$$

where B is the **flux density** (in T), μ_0 is the permeability of 'free space' ($4\pi \times 10^{-7}$ or 12.57×10^{-7}), I is the current (A) and d is the distance from the centre of the conductor (m).

The flux density is also equal to the total flux divided by the area of the field. Thus,

$$B = \frac{\Phi}{A}$$

where Φ is the flux (in Webers (Wb)) and A is the area of the field (m²).

In order to increase the strength of the field, a conductor may be shaped into a loop (*Figure 6.86*) or coiled to form a solenoid (*Figure 6.87*). Note, in the latter case, how the field pattern is exactly the same as that which surrounds a bar magnet.

KEY POINT

Whenever a current flows in a conductor a magnetic field appears in the space around it. The magnetic field can be concentrated by forming the conductor into a coil around a material having a high magnetic permeability, such as iron, steel or ferrite

UNIT 6

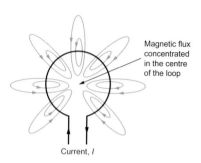

Figure 6.86 Forming a conductor into a loop increases the strength of the magnetic field in the centre of the loop

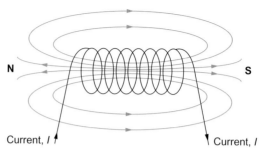

(a) Magnetic field around a solenoid

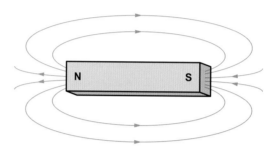

(b) Magnetic field around a permanent magnet

Figure 6.87 The magnetic field surrounding a solenoid coil resembles that of a permanent magnet

Example 6.45

Determine the flux density produced at a distance of 50 mm from a straight wire carrying a current of 20 A.

Applying the formula $B = \dfrac{\mu_0 I}{2\pi d}$ gives:

$$B = \frac{12.57 \times 10^{-7} \times 20}{2 \times 3.142 \times 50 \times 10^{-3}} = \frac{251.4 \times 10^{-7}}{314.2 \times 10^{-3}}$$

from which:

$$B = 0.8 \times 10^{-4} \text{ T}$$

Thus, $B = 80 \times 10^{-6}\,\text{T}$ or $B = 80\,\mu\text{T}.$

Example 6.46

**A flux density of 2.5 mT is developed in free space over an area of 20 cm².
Determine the total flux.**

Re-arranging the formula $B = \Phi/A$ to make Φ the subject gives $\Phi = B \times A$ thus:

$\Phi = (2.5 \times 10^{-3}) \times (20 \times 10^{-4}) = 50 \times 10^{-7}\,\text{Wb}$ from which $B = 5\,\mu\text{Wb}$

Test your knowledge

TYK

TYK 6.28

Determine the current that must be applied to a straight wire conductor in order to produce a flux density of 200 μT at a distance of 12 mm in free space.

UNIT 6

TYK 6.29

A flux density of 1.2 mT is developed in free space over an area of 50 cm². Determine the total flux present.

Magnetic circuits

Materials such as iron and steel possess considerably enhanced magnetic properties. Hence, they are employed in applications where it is necessary to increase the flux density produced by an electric current. In effect, magnetic materials allow us to channel the electric flux into a 'magnetic circuit', as shown in *Figure 6.88*.

In the circuit of *Figure 6.88b* the **reluctance** of the magnetic core is analogous to the resistance present in the electric circuit shown in *Figure 6.88a*. We can make the following comparisons between the two types of circuit (see *Table 6.10*).

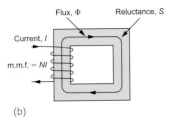

(a)

(b)

Figure 6.88 Comparison of (a) electric and (b) magnetic circuits

Table 6.10 Comparison of electric and magnetic circuits

Electric circuit (Figure 6.88a)	Magnetic circuit (Figure 6.88b)
Electromotive force, e.m.f. = V	Magnetomotive force, m.m.f. = $N \times I$
Resistance = R	Reluctance = S
Current = I	Flux = Φ
e.m.f. = current $\times$ resistance	m.m.f. = flux $\times$ reluctance
$V = I \times R$	$NI = S\Phi$

In practice, not all of the magnetic flux produced in a magnetic circuit will be concentrated within the core and some **leakage flux** will appear in the surrounding free space (as shown in *Figure 6.89*). Similarly, if a gap appears within the magnetic circuit, the flux will tend to spread out as shown in *Figure 6.90*. This effect is known as **fringing**.

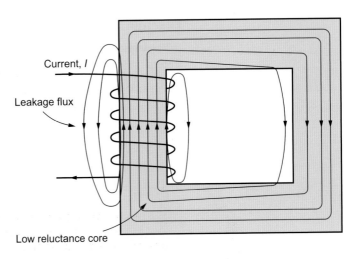

Figure 6.89 Leakage flux in a magnetic circuit

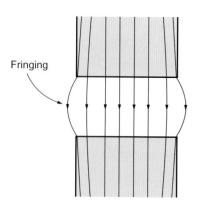

Fringing

Figure 6.90 Fringing of the magnetic flux at an air gap in a magnetic circuit

Reluctance and permeability

The reluctance of a magnetic path is directly proportional to its length and inversely proportional to its area. The reluctance is also inversely proportional to the **absolute permeability** of the magnetic material. Thus,

$$S = \frac{l}{\mu A}$$

where S is the reluctance of the magnetic path, l is the length of the path (in m), A is the cross-sectional area of the path (in m^2) and μ is the absolute permeability of the magnetic material.

The absolute permeability of a magnetic material is the product of the permeability of free space (μ_0) and the **relative permeability** of the magnetic medium (μ_r). Thus,

$$\mu = \mu_0 \mu_r \quad \text{and} \quad S = \frac{l}{\mu_0 \mu_r A}$$

The permeability of a magnetic medium is a measure of its ability to support magnetic flux and it is equal to the ratio of flux density (B) to **magnetizing force** (H). Thus,

$$\mu = \frac{B}{H}$$

where B is the flux density (in T) and H is the magnetizing force (in A/m). The magnetizing force (H) is proportional to the product of the number of turns and current but inversely proportional to the length of the magnetic path:

$$H = \frac{NI}{l}$$

where H is the magnetizing force (in A/m), N is the number of turns, I is the current (in A) and l is the length of the magnetic path (in m).

UNIT 6

Test your knowledge
TYK

TYK 6.30

A ferrite rod has a length of 250 mm and a diameter of 10 mm. Determine the reluctance of the rod if the material that is made from has a relative permeability of 2500.

Test your knowledge
TYK

TYK 6.31

A coil of 400 turns is wound on a closed mild steel core (see *Figure 6.91*) having a length 400 mm and cross-sectional area 480 mm^2. Determine the current required to establish a flux of 0.6 mWb in the core.

B–H curves

Figure 6.91 shows four typical *B–H* (flux density plotted against permeability) curves for some common magnetic materials. If you look carefully at these curves you will notice that they flatten off due to magnetic **saturation** and that the slope of the curve (indicating the value of μ corresponding to a particular value of H) falls as the magnetizing force increases. This is important since it dictates the acceptable working range for a particular magnetic material when used in a magnetic circuit.

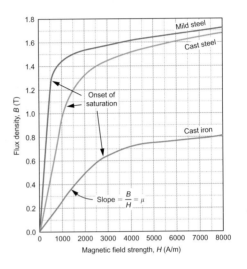

Figure 6.91   B–H curves for three ferromagnetic materials

Example 6.47

Estimate the relative permeability of cast steel (see *Figure 6.92*) at (a) a flux density of 0.6 T and (b) a flux density of 1.6 T.

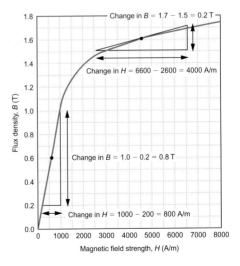

Figure 6.92   *B–H* curve for a sample of cast steel

From *Figure 6.92*, the slope of the graph at any point gives the value of μ at that point. We can easily find the slope by constructing a tangent at the point in question and then finding the ratio of vertical change to horizontal change.

(a) The slope of the graph at 0.6 T is $\dfrac{0.6}{800} = 0.75 \times 10^{-3}$

since $\mu = \mu_0 \times \mu_r$, at 0.6 T, $\mu_r = \dfrac{\mu}{\mu_0} = \dfrac{0.75 \times 10^{-3}}{12.57 \times 10^{-7}} = 597$

(b) The slope of the graph at 1.6 T is $\dfrac{0.2}{4000} = 0.05 \times 10^{-3}$

since $\mu = \mu_0 \times \mu_r$, at 1.6 T, $\mu_r = \dfrac{\mu}{\mu_0} = \dfrac{0.05 \times 10^{-3}}{12.57 \times 10^{-7}} = 39.8$

Note how this example clearly shows the effect of saturation on the permeability of a magnetic material.

Example 6.48

A coil of 800 turns is wound on a closed mild steel core having a length 600 mm and cross-sectional area 500 mm². Determine the current required to establish a flux of 0.8 mWb in the core.

Now $B = \dfrac{\Phi}{A} = \dfrac{0.8 \times 10^{-3}}{500 \times 10^{-6}} = 1.6\ \text{T}$

From *Figure 6.91*, a flux density of 1.6 T will occur in mild steel when $H = 3500\ \text{A/m}$.

The current can now be determined by re-arranging $H = \dfrac{NI}{l}$ as follows:

$$H = \frac{NI}{l}$$

$$I = \frac{Hl}{N} = \frac{(3500)(0.6)}{800} = 2.63\ \text{A}$$

Inductors

Figure 6.93 A small toroidal inductor

Inductors provide us with a means of storing electrical energy in the form of a magnetic field. Typical applications include chokes, filters and (in conjunction with one or more capacitors) frequency selective circuits. The electrical properties of an inductor are determined by a number of factors including the material of the core (if any), the number of turns and the physical dimensions of the coil. *Figure 6.93* shows the construction of a typical toroidal inductor wound on a ferrite (high permeability) core.

Note that with all practical inductors their coil has some resistance due to the resistivity of the material used for its windings. Although this resistance is distributed throughout the component it is much more convenient to consider it to be grouped together into a single resistor, as shown in *Figure 6.94*. Furthermore, since this resistance is something that we do not usually want, we often refer to this as **loss resistance**.

Inductance, *L*   Series loss
                                   resistance, *R*

Figure 6.94 A practical coil contains inductance, *L*, and a small amount of series loss resistance, *R*

UNIT 6

To understand what happens when a changing current flows through an inductor, take a look at the circuit shown in *Figure 6.95a*. If the switch is left open, no current will flow and no magnetic flux will be produced by the inductor. If the switch is closed, as shown in *Figure 6.95b*, current will begin to flow as energy is taken from the supply in order to establish the magnetic field. However, the change in magnetic flux resulting from the appearance of current creates a voltage (an **induced e.m.f.**) across the coil which opposes the applied e.m.f. from the battery.

The induced e.m.f. results from the changing flux and it effectively prevents an instantaneous rise in current in the circuit. Instead, the current increases slowly to a maximum at a rate which depends upon the ratio of inductance (*L*) to resistance (*R*) present in the circuit.

After a while, a steady state condition will be reached in which the voltage across the inductor will have decayed to zero and the current will have reached a maximum value determined by the ratio of *V* to *R* (i.e. Ohm's Law). This is shown in *Figure 6.95c*.

If, after this steady state condition has been achieved, the switch is opened, as shown in *Figure 6.95d*, the magnetic field will suddenly collapse and the energy will be returned to the circuit in the form of an induced **back e.m.f.** which will appear across the coil as the field collapses. For large values of magnetic flux and inductance this back e.m.f. can be extremely large!

Inductance

Inductance is the property of a coil which gives rise to the opposition to a change in the value of current flowing in it. As a result of inductance, any change in the current applied to a coil/inductor will result in an **induced voltage** or **induced e.m.f.** appearing across it. The unit of inductance is the henry (H) and a coil is said to have an inductance of 1 H if a voltage of 1 V is induced across it when a current changing at the rate of 1 A/s is flowing in it.

The voltage induced across the terminals of an inductor will thus be proportional to the product of the inductance (*L*) and the rate of change of applied current. Hence,

$$e = -L \times \text{rate of change of current}$$

Note that the minus sign indicates the polarity of the voltage, i.e. opposition to the change.

The rate of change of current is often represented by the expression d*i*/d*t*, where d*i* represents a very small change in current and d*t* represents the corresponding small change in time. Using mathematical notation to write this we arrive at:

$$e = -L\frac{\mathrm{d}i}{\mathrm{d}t}$$

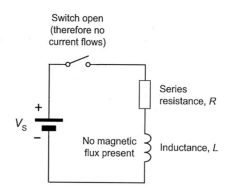

(a) No current in the inductor

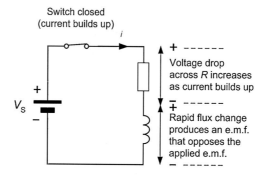

(b) Current builds up

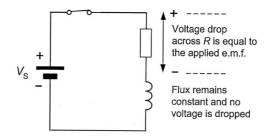

(c) Current remains constant

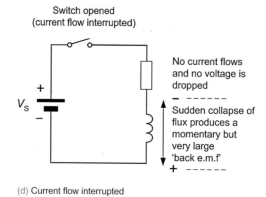

(d) Current flow interrupted

Figure 6.95 Flux and e.m.f. generated when a changing current is applied to an inductor

Example 6.49

A current increases at a uniform rate from 2 to 6 A in a period of 250 ms. If this current is applied to an inductor of 600 mH, determine the voltage induced.

Now the induced voltage will be given by:

$$e = -L \times (\text{rate of change of current})$$

Thus,

$$e = -L\left(\frac{\text{change in current}}{\text{change in time}}\right) = -60 \times 10^{-3}\left(\frac{6-2}{250 \times 10^{-3}}\right)$$

From which

$$e = -60 \times 10^{-3}\left(\frac{4}{0.25}\right) = -60 \times 10^{-3} \times 16$$

so

$$e = -9.6 \, \text{V}$$

TYK

TYK 6.32

The current in a 2.5 H inductor increases uniformly from 0 to 50 mA in 400 ms. Determine the e.m.f. induced.

KEY POINT

The energy stored in an inductor is proportional to the product of the inductance and the square of the current flowing in it

Energy storage

The energy stored in an inductor is proportional to the product of the inductance and the square of the current flowing in it. Thus,

$$W = \tfrac{1}{2}LI^2$$

where W is the energy (J), L is the capacitance (H) and I is the current flowing in the inductor (A).

Example 6.50

An inductor of 20 mH is required to store 2.5 J of energy. Determine the current that must be applied.

The foregoing formula can be re-arranged as follows:

$$I = \sqrt{\frac{2W}{L}} = \sqrt{\frac{2 \times 2.5}{20 \times 10^{-3}}}$$

From which

$$I = \sqrt{\frac{5}{20 \times 10^{-3}}} = \sqrt{0.25 \times 10^{-3}} = \sqrt{250} = 15.81 \text{A}$$

TYK 6.33

A current of 4 A flows in a 60 mH inductor. Determine the energy stored.

Inductance and physical dimensions

The inductance of an inductor depends upon the physical dimensions of the inductor (e.g. the length and diameter of the winding), the number of turns, and the permeability of the material of the core. The inductance of an inductor is given by:

$$L = \frac{\mu_0 \mu_r n^2 A}{l}$$

where L is the inductance (H), μ_0 is the permeability of free space, μ_r is the relative permeability of the magnetic core, l is the mean length of the core (m) and A is the cross-sectional area of the core (m^2).

Example 6.51

An inductor of 100 mH is required. If a closed magnetic core of length 20 cm, cross-sectional area 15 cm^2 and relative permeability 500 is available, determine the number of turns required.

First we must re-arrange the formula $L = \frac{\mu_0 \mu_r n^2 A}{l}$ in order to make n the subject:

$$n = \sqrt{\frac{L \times l}{\mu_0 \mu_r A}}$$

From which

$$n = \sqrt{\frac{100 \times 10^{-3} \times 20 \times 10^{-2}}{12.57 \times 10^{-7} \times 500 \times 15 \times 10^{-4}}} = \sqrt{\frac{2 \times 10^{-2}}{94275 \times 10^{-11}}} = \sqrt{21215} = 146$$

Hence the inductor requires 146 turns of wire.

TYK 6.34

An inductor has 200 turns of wire wound on a closed magnetic core of mean length 24 cm, cross-sectional area 10 cm^2 and relative permeability 650. Determine the inductance of the inductor.

Inductor specifications

Inductor specifications normally include the value of inductance (expressed in henries, millihenries or microhenries), the current rating (i.e. the maximum current which can be continuously applied to the inductor under a given set of conditions) and the accuracy or tolerance (quoted as the

UNIT 6

maximum permissible percentage deviation from the marked value). Other considerations may include the temperature coefficient of the inductance (usually expressed in parts per million, p.p.m., per unit temperature change), the stability of the inductor, the DC resistance of the coil windings (ideally zero), the *Q*-factor (quality factor) of the coil, and the recommended working frequency range. *Table 6.11* summarizes the properties of four common types of inductor. Some typical small inductors are shown in *Figure 6.96*. These have values of inductance ranging from 15 μH to 1 mH.

Table 6.11 Properties of common types of inductor

Property	Inductor type			
	Air cored	**Ferrite cored**	**Ferrite pot cored**	**Iron cored**
Core material	Air	Ferrite rod	Ferrite pot	Laminated steel
Inductance range (H)	50 n to 100 μ	10 μ to 1 m	1 m to 100 m	20 m to 20
Typical DC resistance (Ω)	0.05 to 5	0.1 to 10	5 to 100	10 to 200
Typical tolerance (%)	±5	±10	±10	±20
Typical Q-factor	60	80	40	20
Typical frequency range (Hz)	1 M to 500 M	100 k to 100 M	1 k to 10 M	50 to 10 k
Typical applications	Tuned circuits and filters	Filters and HF transformers	LF and MF filters and transformers	Smoothing chokes and filters

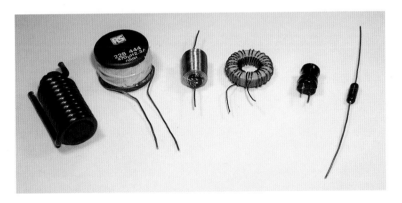

Figure 6.96 A selection of small inductors with values ranging from 15 μH to 1 mH

Inductor markings

As with capacitors, the vast majority of inductors use written markings to indicate values, working current and tolerance. Some small inductors are marked with coloured stripes to indicate their value and tolerance (in which case the standard colour values are used and inductance is normally expressed in microhenries).

Alternating Current Circuits

This section is designed to provide you with an introduction to basic alternating current theory. We discuss the terminology used to describe alternating waveforms and the behaviour of resistors, capacitors, inductors and transformers when an alternating current is applied to them.

Alternating current generation

When a conductor is moved through a magnetic field, an e.m.f. will be induced across its ends. An induced e.m.f. will also be generated if the conductor remains stationary whilst the field moves. In either case, cutting at right angles through the lines of magnetic flux (see *Figure 6.97*) results in a generated e.m.f., the magnitude of which will be given by:

$$E = Blv$$

where B is the magnetic flux density (in T), l is the length of the conductor (in m) and v is the velocity the field (in m/s).

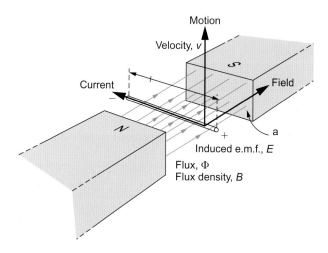

Figure 6.97 Generating an e.m.f. by moving a conductor through a magnetic field

If the field is cut at an angle θ (rather than at right angles), the generated e.m.f. will be given by:

$$E = Blv \sin \theta$$

where θ is the angle between the direction of motion of the conductor and the field lines.

Example 6.52

A conductor of length 20 cm moves at 0.5 m/s through a uniform perpendicular field of 0.6 T. Determine the e.m.f. generated.

Since the field is perpendicular to the conductor, the angle is 90° ('perpendicular' means the same as 'at right angles') so we can use the basic equation:

$$E = Blv$$

where $B = 0.6\,T$, $l = 20\,cm = 0.2\,m$ and $v = 0.5\,m/s$.

Thus, $E = Blv = 0.6 \times 0.2 \times 0.5 = 0.06\,V = 60\,mV$

TYK *6.35*

A wire of length 1.2 m moves at a constant velocity of 2.5 m/s inside a perpendicular magnetic field. Determine the magnetic flux density if an e.m.f. of 200 mV appears across the ends of the wire.

KEY POINT

In a simple AC generator a loop of wire rotates inside the magnetic field produced by two opposite magnetic poles. Contact is made to the loop as it rotates by means of slip rings and brushes

A simple AC generator

Being able to generate a voltage by moving a conductor through a magnetic field is extremely useful as it provides us with an easy way of generating electricity. Unfortunately, moving a wire at a constant linear velocity through a uniform magnetic field presents us with a practical problem simply because the mechanical power that can be derived from an aircraft engine is available in rotary (rather than linear) form!

The solution to this problem is that of using the rotary power available from the engine (via a suitable gearbox and transmission) to rotate a conductor shaped into the form of loop as shown in *Figure 6.98*. The loop is made to rotate inside a permanent magnetic field with opposite poles (N and S) on either side of the loop.

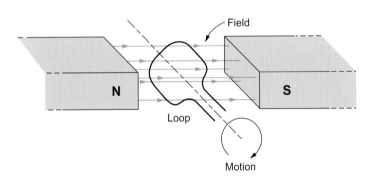

Figure 6.98 A loop rotating within a magnetic field

There now remains the problem of making contact with the loop as it rotates inside the magnetic field but this can be overcome by means of a pair of carbon brushes and copper slip rings (*Figure 6.99*). The brushes are spring loaded and held against the rotating slip rings so that, at any time, there is a path for current to flow from the loop to the load to which it is connected.

The opposite sides of the loop consist of conductors that move through the field. At 0° (with the loop vertical as shown in *Figure 6.100*) the opposite sides of the loop will be moving in the same direction as the lines of flux. At that instant, the angle θ, at which the field is cut is 0° and since the sine of 0° is 0 the generated voltage (from $E = Blv \sin \theta$) will consequently also be zero.

If the loop has rotated to a position which is 90° from that shown at an angle of 0° in *Figure 6.100*, the two conductors will effectively be moving at right angles to the field. At that instant, the generated e.m.f. will take a maximum value (since the sine of 90° is 1).

At 180° from the starting position the generated e.m.f. will have fallen back to zero since, once again, the conductors are moving along the flux lines (but in the direction opposite to that at 0°).

At 270° the conductors will once again be moving in a direction which is perpendicular to the flux lines (but in the direction opposite to that at 90°). At this point, a maximum generated e.m.f. will once again be produced. It is, however, important to note that the e.m.f. generated at this instant will be of opposite polarity to that which was generated at 90°. The reason for this is simply that the relative direction of motion (between the conductors and flux lines) has effectively been reversed.

Since $E = Blv \sin \theta$, the e.m.f. generated by the arrangement shown in *Figure 6.100* will take a sinusoidal form, as shown in *Figure 6.101*. Note that the maximum values of e.m.f. occur at 90° and 270° and that the generated voltage is zero at 0°, 180° and 360°.

In practice, the single loop shown in *Figure 6.99* would comprise a coil of wire wound on a suitable non-magnetic former. This coil of wire effectively increases the length of the conductor within the magnetic field and the generated e.m.f. will then be directly proportional to the number of turns on the coil.

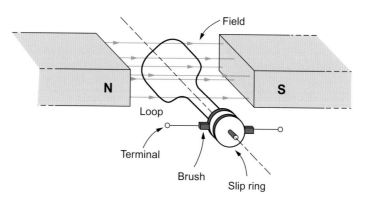

Figure 6.99 Brush arrangement

Alternating versus direct current

Direct currents are currents which, even though their magnitude may vary, essentially flow only in one direction. In other words, direct currents are **unidirectional**. Alternating currents, on the other hand, are **bidirectional** and continuously reverse their direction of flow. The polarity of the e.m.f. which produces an alternating current must consequently also be changing from positive to negative, and vice versa.

Periodic time and frequency

The periodic time (or **period**) of a waveform is the time taken for one complete cycle of the wave (see *Figure 6.102*). The frequency of a repetitive waveform is the number of cycles of the waveform which occur in unit time. Frequency is expressed in Hertz (Hz) and a frequency of 1 Hz is equivalent to one cycle per second. Hence, if a voltage has a frequency of 400 Hz, 400 cycles of it will occur in every second.

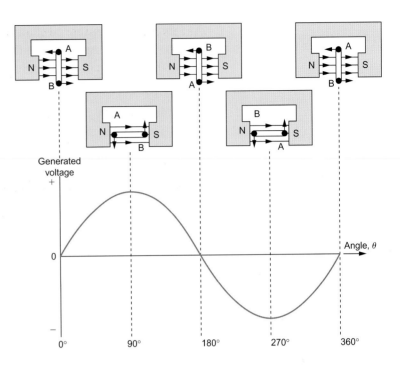

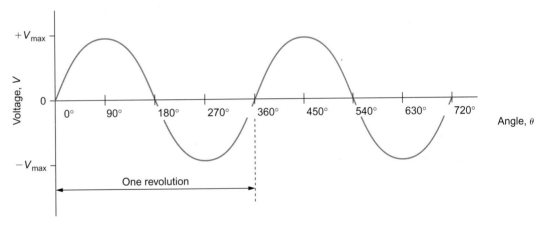

Figure 6.100 e.m.f. generated at various angles

Figure 6.101 Sinusoidal voltage produced by the rotating loop

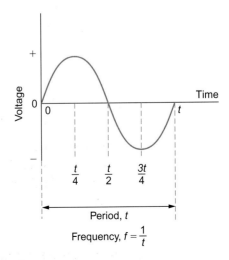

$$\text{Frequency}, f = \frac{1}{t}$$

Figure 6.102 One cycle of a sine wave voltage

The relationship between **periodic time** and frequency is thus:

$$t = \frac{1}{f} \quad \text{or} \quad f = \frac{1}{t}$$

where t is the periodic time (in s) and f is the frequency (in Hz).

Example 6.53

A waveform has a frequency of 400 Hz. What is the periodic time of the waveform?

$$t = \frac{1}{f} = \frac{1}{400} = 0.0025\,\text{s (or 2.5\,ms)}$$

Example 6.54

A waveform has a periodic time of 40 ms. What is its frequency?

$$f = \frac{1}{t} = \frac{1}{40 \times 10^{-3}} = 25\,\text{Hz}$$

Test your knowledge

TYK

TYK 6.36

(a) An AC generator produces a supply voltage with a period of 22 ms. What is the frequency of the supply?

(b) A radio signal has a frequency of 200 kHz. What is the periodic time of the signal?

Equation for a sinusoidal voltage

The equation for the voltage shown in *Figure 6.101* at an angle, θ, time, t, is:

$$v = V_{max} \sin \theta$$

where v is the instantaneous voltage, V_{max} is the maximum (or peak) voltage of the sine wave and θ is the angle at which the voltage is measured.

It is often more convenient to express a sinusoidal voltage in terms of time rather than angle. Since we know that one cycle of the wave corresponds to an angle of 360° (or 2π radians) we can replace θ in the foregoing formula with:

$$\theta = \omega t$$

where ω is the angular velocity (i.e. the rate at which the coil is turning expressed in degrees per second).

UNIT 6

Now, when ω is expressed in terms of radians per second (more convenient units) we arrive at:

$$\omega = 2\pi f$$

and so:

$$v = V_{max}\sin(\omega t) = V_{max}\sin(2\pi f t)$$

where v is the instantaneous voltage, V_{max} is the maximum (or peak) voltage of the sine wave and f is the frequency of the sine wave.

Example 6.55

A sine wave voltage has a maximum value of 20 V and a frequency of 50 Hz. Determine the instantaneous voltage present (a) 2.5 ms and (b) 15 ms from the start of the cycle.

We can find the voltage at any instant of time using:

$$v = V_{max}\sin(2\pi f t)$$

where $V_{max} = 20$ V and $f = 50$ Hz.

In (a), $t = 2.5$ ms, hence:

$$v = 20\sin(2\pi \times 50 \times 0.0025) = 20\sin(0.785)$$
$$= 20 \times 0.707 = 14.14\,\text{V}$$

In (b), $t = 15$ ms, hence:

$$v = 20\sin(2\pi \times 50 \times 0.0015) = 20\sin(4.71)$$
$$= 20 \times -1 = -20\,\text{V}$$

Test your knowledge

TYK

TYK 6.37

A sine wave voltage has a maximum value of 160 V and a frequency of 60 Hz. Write down an equation for the instantaneous voltage and use it to determine the instantaneous voltage when (a) $t = 2$ ms and (b) $t = 15$ ms.

Average, peak, peak–peak, and r.m.s. values

The **average value** of an alternating current which swings symmetrically above and below zero will be zero when measured over a long period of time. Hence average values of currents and voltages are invariably taken over one complete half-cycle (either positive or negative) rather than over one complete full-cycle (which would result in an average value of zero).

The **amplitude** (or **peak value**) of a waveform is a measure of the extent of its voltage or current excursion from the resting value (usually zero). In other words, the amplitude is the same as the **maximum value** that we met earlier.

The **peak-to-peak value** for a wave which is symmetrical about its resting value is twice its peak value (see *Figure 6.103*).

The **r.m.s.** (or **effective**) **value** of an alternating voltage or current is the value that would produce the same heat energy in a resistor as a direct voltage or current of the same magnitude. Since the r.m.s. value of a waveform is very much dependent upon its shape, values are only meaningful when dealing with a waveform of known shape. Where the shape of a waveform is not specified, r.m.s. values are normally assumed to refer to sinusoidal conditions.

For a given waveform, a set of fixed relationships exist between average, peak, peak–peak and r.m.s. values. The required multiplying factors are summarized for sinusoidal voltages and currents in *Table 6.12*.

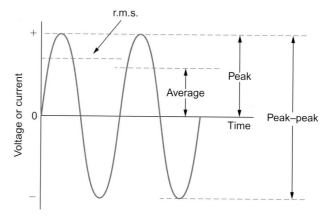

Figure 6.103 Two cycles of a sine wave voltage showing its r.m.s., peak and peak–peak values

Table 6.12 Multiplying factors for average, peak, peak–peak and r.m.s. values

Given quantity	Wanted quantity			
	Average	Peak	Peak–peak	r.m.s.
Average	1	1.57	3.14	1.11
Peak	0.636	1	2	0.707
Peak–peak	0.318	0.5	1	0.353
r.m.s.	0.9	1.414	2.828	1

UNIT 6

Example 6.56

A sinusoidal voltage has an r.m.s. value of 240 V. What is the peak value of the voltage?

The corresponding multiplying factor (found from Table 6.12) is 1.414. Hence,

$$V_{pk} = 1.414 \times V_{r.m.s.} = 1.414 \times 240 = 339.4 \, V$$

Example 6.57

An alternating current has a peak–peak value of 50 mA. What is its r.m.s. value?

The corresponding multiplying factor (found from Table 6.12) is 0.353. Hence,

$$I_{r.m.s} = 0.353 \times V_{pk-pk} = 0.353 \times 0.05 = 0.0177\,A$$

(or 17.7 mA).

Example 6.58

A sinusoidal voltage 10 V pk–pk is applied to a resistor of 1 kΩ. What value of r.m.s. current will flow in the resistor?

This problem must be solved in two stages. First we will determine the peak–peak current in the resistor and then we shall convert this value into a corresponding r.m.s. quantity.

$$\text{Since } I = \frac{V}{R} \text{ we can infer that } I_{pk-pk} = \frac{V_{pk-pk}}{R}$$

From which

$$I_{pk-pk} = \frac{10}{1000} = 0.01\,A \text{ (or 10 mA)}$$

The required multiplying factor (peak–peak to r.m.s.) is 0.353

Thus,

$$I_{r.m.s.} = 0.353 \times I_{pk-pk} = 0.353 \times 10 = 3.53\,mA$$

TYK 6.38

A sinusoidal voltage having an r.m.s. value of 220 V is applied to a resistor of 4 kΩ. Determine the peak value of the current flowing in the resistor.

TYK 6.39

A peak–peak voltage of 160 mV appears across a 50 Ω resistor. Determine the r.m.s. current flowing in the resistor.

Reactance

When alternating voltages are applied to capacitors or inductors the magnitude of the current flowing will depend upon the value of capacitance or inductance and on the frequency of the voltage. In effect, capacitors and inductors oppose the flow of current in much the same way as a resistor. The important difference is that the effective resistance (or reactance) of the component varies with frequency (unlike the case of a resistor where the magnitude of the current does not change with frequency).

Capacitive reactance

The reactance of a capacitor is defined as the ratio of applied voltage to current and, like resistance, it is measured in Ω. The reactance of a capacitor is inversely proportional to both the value of capacitance and the frequency of the applied voltage. Capacitive reactance can be found by applying the following formula:

$$X_c = \frac{1}{2\pi fC}$$

where X_c is the reactance (Ω), f is the frequency (Hz) and C is the capacitance (F).

Capacitive reactance falls as frequency increases, as shown in *Figure 6.104*. The applied voltage, V_c, and current, I_c, flowing in a pure capacitive reactance will differ in phase by an angle of 90° or $\pi/2$ radians (the **current leads the voltage**). This relationship is illustrated in the current and voltage waveforms (drawn to a common time scale) shown in *Figure 6.105* and as a **phasor diagram** shown in *Figure 6.106*.

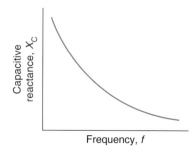

Figure 6.104 Variation of reactance with frequency for a capacitor

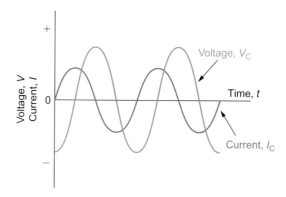

Figure 6.105 Voltage and current waveforms for a pure capacitor (the current leads the voltage by 90°)

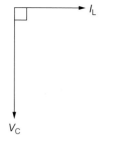

Figure 6.106 Phasor diagram for a pure capacitor

UNIT 6

Example 6.59

Determine the reactance of a 1 µF capacitor at (a) 100 Hz and (b) 10 kHz.

This problem is solved using the expression $X_c = \frac{1}{2\pi fC}$

(a) At 100 Hz

$$X_c = \frac{1}{2\pi \times 100 \times 1 \times 10^{-6}} = \frac{0.159}{10^{-4}} = 1.59 \times 10^3$$

Thus $X_c = 1.59 \text{ k}\Omega$

(b) At 10 kHz

$$X_c = \frac{1}{2\pi \times 1 \times 10^4 \times 10^{-6}} = \frac{0.159}{10^{-2}} = 0.159 \times 10^2$$

Thus, $X_c = 15.9 \,\Omega$

Example 6.60

A 100 nF capacitor is to form part of a filter connected across a 240 V 50 Hz mains supply. What current will flow in the capacitor?

First we must find the reactance of the capacitor:

$$X_c = \frac{1}{2\pi \times 50 \times 100 \times 10^{-9}} = 31.8 \times 10^3 = 31.8\,\text{k}\Omega$$

The r.m.s. current flowing in the capacitor will thus be:

$$I_c = \frac{V_c}{X_c} = \frac{240}{31.8 \times 10^3} = 7.5 \times 10^{-3} = 7.5\,\text{mA}$$

TYK 6.40

Determine the reactance of a 4.7 μF capacitor at (a) 100 Hz and (b) 10 kHz.

KEY POINT

The reactance of a capacitor is inversely proportional to frequency (in other words, it falls as the frequency increases). The reactance of an inductor is directly proportional to frequency (in other words, it increases as the frequency increases)

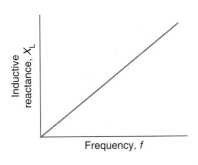

Figure 6.107 Variation of reactance with frequency for an inductor

Inductive reactance

The reactance of an inductor is defined as the ratio of applied voltage to current and, like resistance, it is measured in Ω. The reactance of an inductor is directly proportional to both the value of inductance and the frequency of the applied voltage. Inductive reactance can be found by applying the formula:

$$X_L = 2\pi f L$$

where X_L is the reactance in Ω, f is the frequency in Hz and L is the inductance in H.

Inductive reactance increases linearly with frequency as shown in *Figure 6.107*. The applied voltage, V_L, and current, I_L, developed across a pure inductive reactance will differ in phase by an angle of 90° or $\pi/2$ radians (the **current lags the voltage**). This relationship is illustrated in the current and voltage waveforms (drawn to a common time scale) shown in *Figure 6.108* and as a **phasor diagram** shown in *Figure 6.109*.

Example 6.61

Determine the reactance of a 10 mH inductor at (a) 100 Hz and (b) at 10 kHz.

(a) At 100 Hz

$$X_L = 2\pi \times 100 \times 10 \times 10^{-3} = 6.28\,\Omega$$

(b) At 10 kHz

$$X_L = 2\pi \times 10 \times 10^3 \times 10 \times 10^{-3} = 628\,\Omega$$

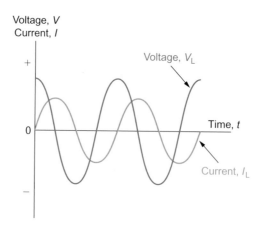

Figure 6.108 Voltage and current waveforms for a pure inductor (the voltage leads the current by 90°)

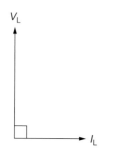

Figure 6.109 Phasor diagram for a pure inductor

Example 6.62

A 100 mH inductor of negligible resistance is to form part of a filter that carries a current of 20 mA at 400 Hz. What voltage drop will be developed across the inductor?

The reactance of the inductor will be given by:

$$X_L = 2\pi \times 400 \times 100 \times 10^{-3} = 251\,\Omega$$

The r.m.s. voltage developed across the inductor will be given by:

$$V_L = I_L \times X_L = 20\,\text{mA} \times 251\,\Omega = 5.02\,\text{V}$$

In this example, it is important to note that we have assumed that the DC resistance of the inductor is negligible by comparison with its reactance. Where this is not the case, it will be necessary to determine the impedance of the component (see the next section) and use this to determine the voltage drop.

TYK 6.41

Determine the reactance of a 60 mH inductor at (a) 100 Hz and (b) 10 kHz.

Impedance

Figure 6.110 shows two circuits which contain both resistance and reactance. These circuits are said to exhibit **impedance** (a combination of resistance and reactance) which, like resistance and reactance, is measured in Ω.

The impedance of the circuits shown in *Figure 6.110* is simply the ratio of supply voltage, V_S, to supply current, I_S. The impedance of the simple C–R and L–R circuits shown in *Figure 6.110* can be found by using the impedance triangle shown in *Figure 6.111*. In either case, the impedance of the circuit is given by:

$$Z = \sqrt{R^2 + X^2}$$

UNIT 6

KEY POINT

Impedance is the combination of resistance and reactance in an AC circuit. Impedance is measured in Ohms and is equal to the ratio of voltage supplied to the circuit to the current flowing in it

and the phase angle (between V_S and I_S) is given by:

$$\phi = \tan^{-1} \frac{X}{R}$$

where Z is the impedance (in Ω), X is the reactance, either capacitive or inductive (expressed in Ω), R is the resistance (in Ω) and ϕ is the phase angle in radians.

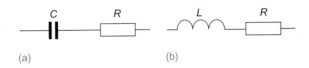

(a)   (b)

Figure 6.110 (a) C and R in series (b) L and R in series (note that both circuits exhibit a combination of resistance reactance – in other words, an **impedance**)

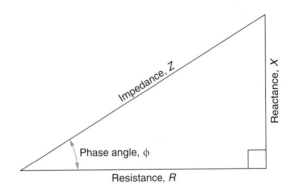

Figure 6.111 The impedance triangle

Example 6.63

A 2 μF capacitor is connected in series with a 100 Ω resistor across a 115 V 400 Hz AC supply. Determine the impedance of the circuit and the current taken from the supply.

First we must find the reactance of the capacitor, X_c:

$$X_c = \frac{1}{2\pi f C} = \frac{1}{6.28 \times 400 \times 2 \times 10^{-6}} = \frac{10^6}{5024} = 199 \,\Omega$$

Now we can find the impedance of the C–R series circuit:

$$Z = \sqrt{R^2 + X^2} = \sqrt{100^2 + 199^2} = \sqrt{49601} = 223 \,\Omega$$

The current taken from the supply can now be found:

$$I = \frac{V}{Z} = \frac{115}{223} = 0.52 \,\text{A}$$

TYK

TYK *6.42*

A 470 Ω resistor is connected in series with a capacitor of 8 μF across a 220 V, 50 Hz AC supply. Determine the impedance of the circuit and the current supplied to it.

TYK

TYK *6.43*

A 2 H inductor is connected in series with a resistor of 560 Ω across a 220 V, 50 Hz AC supply. Determine the impedance of the circuit and the current supplied to it.

KEY POINT

Power factor is the ratio of true power to apparent power in an AC circuit. When the power factor is unity (i.e. when it is 1) the true power is the same as the apparent power and all of the power supplied to the circuit is dissipated as heat. In this case, the circuit could be said to be 'purely resistive'. When the power factor is zero there is no true power and none of the power supplied to the circuit is dissipated as heat. In this case the circuit could be said to be 'purely reactive'

Power factor

The power factor in an AC circuit containing resistance and reactance is simply the ratio of true power to apparent power. Hence,

$$\text{Power factor} = \frac{\text{True power}}{\text{Apparent power}}$$

The **true power** in an AC circuit is the power which is actually dissipated in the resistive component. Thus,

$$\text{True power} = I_S^2 \times R (\text{W})$$

The **apparent power** in an AC circuit is the power which is apparently consumed by the circuit and is the product of the supply current and supply voltage (note that this is not the same as the power that is actually dissipated as heat). Hence,

$$\text{Apparent power} = I_S \times V_S (\text{volt} - \text{amperes})$$

Hence,

$$\text{Power factor} = \frac{I_S^2 \times R}{I_S \times V_S} = \frac{I_S^2 \times R}{I_S \times (I_S \times Z)} = \frac{I_S^2 \times R}{I_S^2 \times Z} = \frac{R}{Z}$$

From *Figure 6.111*, $\dfrac{R}{Z} = \cos\phi$

Hence, the power factor of a series AC circuit can be found from the cosine of the phase angle.

Example 6.64

A choke (a form of inductor) having an inductance of 150 mH and resistance of 250 Ω is connected to a 115 V 400 Hz AC supply. Determine the power factor of the choke and the current taken from the supply.

First we must find the reactance of the inductor,

$$X_L = 2\pi \times 400 \times 0.15 = 376.8 \, \Omega$$

We can now find the impedance from:

$$Z = \sqrt{250^2 + 377^2} = 452\,\Omega$$

Next we can determine the power factor from:

$$\text{Power factor} = \frac{R}{Z} = \frac{250}{452} = 0.553$$

Finally, the current taken from the supply will be:

$$I_S = \frac{V_S}{Z} = \frac{115}{452} = 0.254\,\text{A}$$

TYK

TYK 6.44

An inductor has a resistance of $200\,\Omega$ and an inductance of $150\,\text{mH}$. If the inductor is connected to a $110\,\text{V}$ $400\,\text{Hz}$ supply, determine (a) the current supplied, (b) the power factor of the inductor and (c) the phase angle.

Transformers

The transformer principle

The principle of the transformer is illustrated in *Figure 6.112*. The primary and secondary windings are wound on a common low-reluctance magnetic core. The alternating flux generated by the primary winding is therefore coupled into the secondary winding (very little flux escapes due to leakage). A sinusoidal current flowing in the primary winding produces a sinusoidal flux. At any instant the flux in the transformer is given by the equation:

$$\phi_{\text{max}} = \phi_{\text{max}}\,\sin(2\pi ft)$$

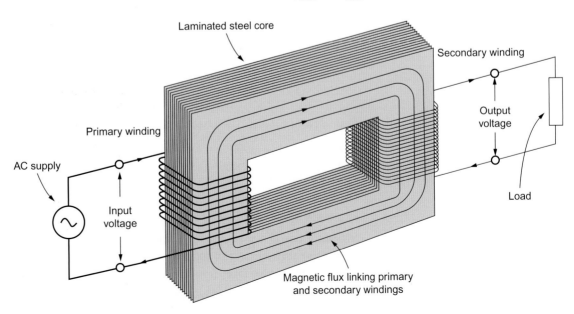

Figure 6.112 The transformer principle

where ϕ_{max} is the maximum value of flux (Wb), f is the frequency of the applied current (Hz) and t is the time (s).

The core of a transformer is effectively magnetized and demagnetized on every cycle of the AC supply. Any residual flux must be destroyed before the field can be built up in the opposite direction. This can result in a loss of energy which is referred to as **hysteresis** (the *B–H* curve that we met earlier effectively becomes a closed loop and the energy loss is proportional to the area of this loop).

The r.m.s. value of the primary voltage (see *Figure 6.113*), V_P, is given by:

$$V_P = 4.44 f N_P \phi_{max}$$

Similarly, the r.m.s. value of the secondary voltage, V_S, is given by:

$$V_S = 4.44 f N_S \phi_{max}$$

Now

$$\frac{V_P}{V_S} = \frac{N_P}{N_S}$$

where N_P/N_S is the **turns ratio** of the transformer.

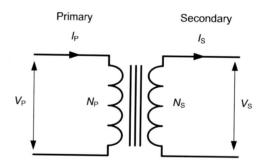

Figure 6.113 Primary and secondary voltages and currents

Assuming that the transformer is loss free, primary and secondary powers (P_P and P_S, respectively) will be identical. Hence,

$$P_P = P_S \text{ thus } V_P \times I_P = V_S \times I_S$$

Hence,

$$\frac{V_P}{V_S} = \frac{I_S}{I_P}$$

and

$$\frac{I_S}{I_P} = \frac{N_P}{N_S}$$

UNIT 6

Finally, it is sometimes convenient to refer to a **turns-per-volt** rating for a transformer. This rating is given by:

$$\text{Turns per volt} = \frac{N_P}{V_P} = \frac{N_S}{V_S}$$

Example 6.65

A transformer has 2000 primary turns and 120 secondary turns. If the primary is connected to a 220 V r.m.s. AC mains supply, determine the secondary voltage.

Rearranging $\dfrac{V_P}{V_S} = \dfrac{N_P}{N_S}$ gives:

$$V_S = \frac{N_S \times V_P}{N_P} = \frac{120 \times 220}{2000} = 13.2\,\text{V}$$

Example 6.66

A transformer has 1200 primary turns and is designed to operate with a 200 V AC supply. If the transformer is required to produce an output of 10 V, determine the number of secondary turns required. Assuming that the transformer is loss free, determine the input (primary) current for a load current of 2.5 A.

Rearranging $\dfrac{V_P}{V_S} = \dfrac{N_P}{N_S}$ gives:

$$N_S = \frac{N_P \times V_S}{V_P} = \frac{1200 \times 10}{200} = 60\,\text{turns}$$

Rearranging $\dfrac{I_S}{I_P} = \dfrac{N_P}{N_S}$ gives:

$$I_P = \frac{N_S \times I_S}{N_P} = \frac{60 \times 2.5}{1200} = 0.125\,\text{A}$$

Test your knowledge

TYK

TYK 6.45

A transformer has 600 primary turns and 80 secondary turns. If the primary is connected to a 110 V supply, determine (a) the turns ratio and (b) the secondary voltage.

KEY POINT

Transformers provide us with a convenient way of stepping up or stepping down an AC voltage. If a transformer is 'loss free' (in other words if no power is wasted in it) the power delivered to the primary will be the same as the power delivered from the secondary

Transformer applications

Transformers provide us with a means of coupling AC power or signals from one circuit to another. Voltage may be **stepped-up** (secondary voltage greater than primary voltage) or **stepped-down** (secondary voltage less than primary voltage). Since no increase in power is possible (transformers are passive components like resistors, capacitors and inductors) an increase in secondary voltage can only be achieved at the expense of a corresponding reduction in secondary current, and vice versa (in fact, the secondary power

will be very slightly less than the primary power due to losses within the transformer). Typical applications for transformers include stepping-up or stepping-down mains voltages in power supplies, coupling signals in AF amplifiers to achieve impedance matching and to isolate DC potentials associated with active components.

The electrical properties of a transformer are determined by a number of factors including the core material and physical dimensions. The specifications for a transformer usually include the rated primary and secondary voltages and current, the required power rating (i.e. the maximum power, usually expressed in volt–amperes, VA) which can be continuously delivered by the transformer under a given set of conditions, the frequency range for the component (usually stated as upper and lower working frequency limits), and the **regulation** of a transformer (usually expressed as a percentage of full load). This last specification is a measure of the ability of a transformer to maintain its rated output voltage under load.

Table 6.13 summarizes the properties of three common types of transformer. *Figure 6.114* shows a typical selection of small transformers with power ratings ranging from 0.1 to 100 VA whilst *Figure 6.115* shows the parts of a typical iron-cored power transformer prior to assembly.

Table 6.13 Properties of common types of transformer

Property	Transformer core type			
	Air cored	Ferrite cored	Iron cored (audio)	Iron cored (power)
Core material/construction	Air	Ferrite ring or pot	Laminated steel	Laminated steel
Typical frequency range (Hz)	30M to 1G	10k to 10M	20 to 20k	50 to 400
Typical power rating (VA)		1 to 200	0.1 to 50	3 to 500
Typical regulation				5% to 15%
Typical applications	RF tuned circuits and filters	Filters and HF transformers, switched mode power supplies	Smoothing chokes and filters, audio matching	Power supplies

Figure 6.114 A selection of transformers with power ratings from 0.1 to 100 VA

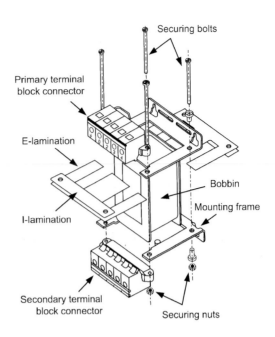

Figure 6.115 Parts of a typical iron-cored power transformer prior to assembly

Magnetic screening is sometimes used to reduce the effects of any leakage flux in the space surrounding a transformer. Materials used for screening must have a very high permeability in order to 'capture' the stray flux and channel it safely away. The most commonly used material is **mumetal** which is supplied in thin sheet or as a foil that can be wrapped around sensitive components.

TYK 6.46

Identify the components marked A–H in *Figure 6.116*.

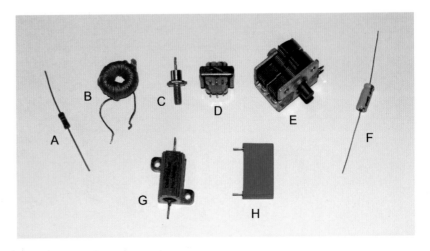

Figure 6.116 See TYK 6.46

UNIT 6

Rectifiers

Semiconductor diodes are commonly used to convert AC to DC, in which case they are referred to as **rectifiers**. The simplest form of rectifier circuit makes use of a single diode and, since it operates on only either positive or negative half-cycles of the supply, it is known as a **half-wave** rectifier.

Figure 6.117 shows a simple half-wave rectifier circuit. Mains voltage (220–240V) is applied to the primary of a step-down transformer (T1). The secondary of T1 steps down the 240V r.m.s. to 12V r.m.s. (the turns ratio of T1 will thus be 240/12 or 20:1). The transformer also provides **isolation** by avoiding any direct connection between the low-voltage DC output and the incoming AC supply.

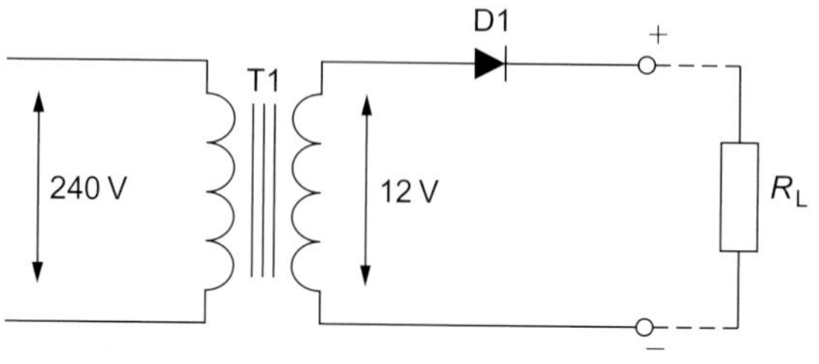

Figure 6.117 A simple half-wave rectifier circuit

Diode D1 will only allow the current to flow in the direction shown (i.e. from cathode to anode). D1 will be forward biased during each positive half-cycle (relative to common) and will effectively behave like a closed switch. When the circuit current tries to flow in the opposite direction, the voltage bias across the diode will be reversed, causing the diode to act like an open switch (see *Figures 6.118a and 6.118b*, respectively).

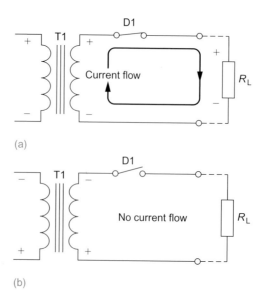

(a)

(b)

Figure 6.118 Half-wave rectifier circuit with (a) D1 conducting (positive-going half-cycles of secondary voltage) and (b) D1 not conducting (negative-going half-cycles of secondary voltage)

UNIT 6

The switching action of D1 results in a pulsating output voltage which is developed across the load resistor (R_L). Since the mains supply is at 50 Hz, the pulses of voltage developed across R_L will also be at 50 Hz even if only half the AC cycle is present. During the positive half-cycle, the diode will drop the 0.6–0.7 V forward threshold voltage normally associated with silicon diodes. However, during the negative half-cycle the peak AC voltage will be dropped across D1 when it is reverse biased. This is an important consideration when selecting a diode for a particular application. Assuming that the secondary of T1 provides 12 V r.m.s., the peak voltage output from the transformer's secondary winding will be given by:

$$V_{pk} = 1.414 \times V_{r.m.s.} = 1.414 \times 12\ V = 16.97\ V$$

The peak voltage applied to D1 will thus be approximately 17 V. The negative half-cycles are blocked by D1 and thus only the positive half-cycles appear across R_L. Note, however, that the actual peak voltage across R_L will be the 17 V positive peak being supplied from the secondary on T1, *minus* the 0.7 V forward threshold voltage dropped by D1. In other words, positive half-cycle pulses having a peak amplitude of 16.3 V will appear across R_L.

Example 6.67

A mains transformer having a turns ratio of 44:1 is connected to a 220 V r.m.s. mains supply. If the secondary output is applied to a half-wave rectifier, determine the peak voltage that will appear across a load.

The r.m.s. secondary voltage will be given by:

$$V_S = V_P/11 = 220/44 = 5\ V$$

The peak voltage developed after rectification will be given by:

$$V_{PK} = 1.414 \times 5\ V = 7.07\ V$$

Assuming that the diode is a silicon device with a forward voltage drop of 0.6 V, the actual peak voltage dropped across the load will be:

$$V_L = 7.07\ V - 0.6\ V = 6.47\ V$$

Smoothing circuits

Figure 6.119 shows a considerable improvement to the circuit of *Figure 6.117*. The capacitor, C1, has been added to ensure that the output voltage remains at, or near, the peak voltage even when the diode is not conducting. When the primary voltage is first applied to T1, the first positive half-cycle output from the secondary will charge C1 to the peak value seen across R_L. Hence, C1 charges to 16.3 V at the peak of the positive half-cycle. Because C1 and R_L are in parallel, the voltage across R_L will be the same as that across C1.

The time required for C1 to charge to the maximum (peak) level is determined by the charging circuit time constant (the series resistance

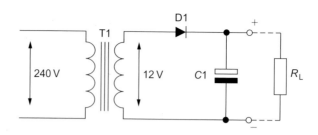

Figure 6.119 A simple half-wave rectifier circuit with reservoir capacitor

multiplied by the capacitance value). In this circuit, the series resistance comprises the secondary winding resistance together with the forward resistance of the diode and the (minimal) resistance of the wiring and connections. Hence $C1$ charges very rapidly as soon as $D1$ starts to conduct.

The time required for $C1$ to discharge is, in contrast, very much greater. The discharge time constant is determined by the capacitance value and the load resistance, R_L. In practice, R_L is very much larger than the resistance of the secondary circuit and hence $C1$ takes an appreciable time to discharge. During this time, $D1$ will be reverse biased and will thus be held in its non-conducting state. As a consequence, the only discharge path for $C1$ is through R_L.

$C1$ is referred to as a **reservoir** capacitor. It stores charge during the positive half-cycles of secondary voltage and releases it during the negative half-cycles. The circuit shown in *Figure 6.119* is thus able to maintain a reasonably constant output voltage across R_L. Even so, $C1$ will discharge by a small amount during the negative half-cycle periods from the transformer secondary.

Figure 6.120 shows the secondary voltage waveform together with the voltage developed across R_L with and without $C1$ present. This gives rise to a small variation in the DC output voltage (known as **ripple**). Since ripple is undesirable we must take additional precautions to reduce it. One obvious method of reducing the amplitude of the ripple is that of simply increasing the discharge time constant. This can be achieved either by increasing the value of $C1$ or by increasing the resistance value of R_L. In practice, however, the latter is not really an option because R_L is the effective resistance of

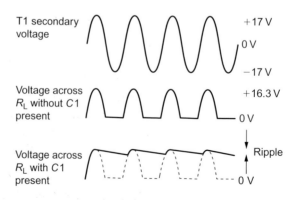

Figure 6.120 Waveforms for a half-wave rectifier circuit

the circuit being supplied and we do not usually have the ability to change it! Increasing the value of $C1$ is a more practical alternative for very large capacitor values.

Figure 6.121 shows a further refinement of the simple power supply circuit. This circuit employs two additional components, $R1$ and $C1$, which act as a filter to remove the ripple. The value of $C1$ is chosen so that the component exhibits a negligible reactance at the ripple frequency (usually 50 Hz for a half-wave rectifier or 100 Hz for a full-wave reJctifier operating from a standard UK AC mains supply). In effect, $R1$ and $C1$ act like a potential divider.

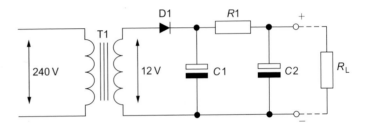

Figure 6.121 Half-wave rectifier circuit with R–C smoothing filter

Improved ripple filters

A further improvement can be achieved by using an inductor, $L1$, instead of a resistor in the smoothing circuit. This circuit also offers the advantage that the minimum DC voltage is dropped across the inductor (in the circuit of *Figure 6.121*, the DC output voltage is *reduced* by an amount equal to the voltage drop across $R1$).

Figure 6.122 shows the circuit of a half-wave power supply with an L–C smoothing circuit. At the ripple frequency, $L1$ exhibits a high value of inductive reactance while $C1$ exhibits a low value of capacitive reactance. The combined effect is that of an attenuator which greatly reduces the amplitude of the ripple while having a negligible effect on the direct voltage.

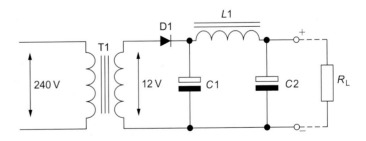

Figure 6.122 Half-wave rectifier circuit with L–C smoothing filter

Full-wave rectifiers

Unfortunately, the half-wave rectifier circuit is relatively inefficient as conduction takes place only on alternate half-cycles. A better rectifier arrangement would make use of both positive *and* negative half-cycles. These **full-wave rectifier** circuits offer a considerable improvement over their half-wave counterparts. They are not only more efficient but

are significantly less demanding in terms of the reservoir and smoothing components. There are two basic forms of full-wave rectifier; the bi-phase type and the bridge rectifier type.

Bi-phase rectifier circuits

Figure 6.123 shows a simple bi-phase rectifier circuit. Mains voltage (240 V) is applied to the primary of step-down transformer (T1) which has two identical secondary windings, each providing 12 V r.m.s. (the turns ratio of T1 will thus be 240/12 or 20:1 for *each* secondary winding).

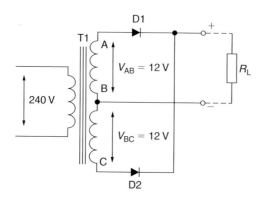

Figure 6.123 Bi-phase rectifier circuit

On positive half-cycles, point A will be positive with respect to point B. Similarly, point B will be positive with respect to point C. In this condition D1 will allow conduction (its anode will be positive with respect to its cathode) while D2 will not allow conduction (its anode will be negative with respect to its cathode). Thus D1 alone conducts on positive half-cycles.

On negative half-cycles, point C will be positive with respect to point B. Similarly, point B will be positive with respect to point A. In this condition D2 will allow conduction (its anode will be positive with respect to its cathode) while D1 will not allow conduction (its anode will be negative with respect to its cathode). Thus, D2 alone conducts on negative half-cycles.

Figure 6.124 shows the bi-phase rectifier circuit with the diodes replaced by switches. In *Figure 6.124a*, D1 is shown conducting on a positive half-cycle while in *Figure 6.124b* D2 is shown conducting. The result is that current is routed through the load *in the same direction* on successive half-cycles. Furthermore, this current is derived alternately from the two secondary windings.

As with the half-wave rectifier, the switching action of the two diodes results in a pulsating output voltage being developed across the load resistor (R_L). However, unlike the half-wave circuit the pulses of voltage developed across R_L will occur at a frequency of 100 Hz (*not* 50 Hz). This doubling of the ripple frequency allows us to use smaller values of reservoir and smoothing capacitor to obtain the same degree of ripple reduction (recall that the reactance of a capacitor is reduced as frequency increases).

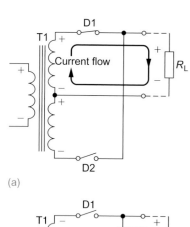

(a)

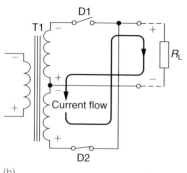

(b)

Figure 6.124 Bi-phase rectifier with (a) D1 conducting and D2 non-conducting. (b) Bi-phase rectifier with D2 conducting and D1 non-conducting

UNIT 6

As before, the peak voltage produced by each of the secondary windings will be approximately 17 V and the peak voltage across R_L will be 16.3 V (i.e. 17 V less the 0.7 V forward threshold voltage dropped by the diodes).

Figure 6.125 shows how a reservoir capacitor ($C1$) can be added to ensure that the output voltage remains at, or near, the peak voltage even when the diodes are not conducting. This component operates in exactly the same way as for the half-wave circuit, that is it charges to approximately 16.3 V at the peak of the positive half-cycle and holds the voltage at this level when the diodes are in their non-conducting states. The time required for $C1$ to charge to the maximum (peak) level is determined by the charging circuit time constant (the series resistance multiplied by the capacitance value). In this circuit, the series resistance comprises the secondary winding resistance together with the forward resistance of the diode and the (minimal) resistance of the wiring and connections. Hence, $C1$ charges very rapidly as soon as either D1 or D2 starts to conduct. The time required for $C1$ to discharge is, in contrast, very much greater. The discharge time contrast is determined by the capacitance value and the load resistance, R_L. In practice, R_L is very much larger than the resistance of the secondary circuit and hence $C1$ takes an appreciable time to discharge. During this time, D1 and D2 will be reverse biased and held in a non-conducting state. As a consequence, the only discharge path for $C1$ is through R_L. *Figure 6.126* shows voltage

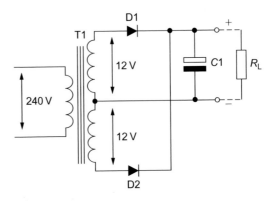

Figure 6.125 Bi-phase rectifier with reservoir capacitor

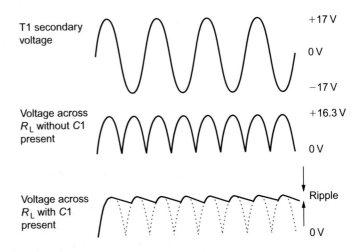

Figure 6.126 Waveforms for the bi-phase rectifier

waveforms for the bi-phase rectifier, with and without *C*1 present. Note that the ripple frequency (100 Hz) is twice that of the half-wave circuit shown previously in *Figure 6.120*.

Bridge rectifier circuits

An alternative to the use of the bi-phase circuit is that of using a four-diode bridge rectifier (see *Figure 6.127*) in which opposite pairs of diodes conduct on alternate half-cycles. This arrangement avoids the need to have two separate secondary windings.

A full-wave bridge rectifier arrangement is shown in *Figure 6.128*. Mains voltage (240 V) is applied to the primary of a step-down transformer (Tl). The secondary winding provides 12 V r.m.s. (approximately 17 V peak) and has a turns ratio of 20:1, as before. On positive half-cycles, point A will be positive with respect to point B. In this condition Dl and D2 will allow conduction while D3 and D4 will not allow conduction. Conversely, on negative half-cycles, point B will be positive with respect to point A. In this condition D3 and D4 will allow conduction while Dl and D2 will not allow conduction.

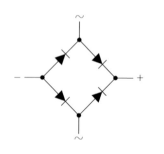

Figure 6.127 Four diodes connected as a bridge rectifier

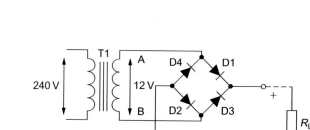

Figure 6.128 Full-wave bridge rectifier circuit

Figure 6.129 shows the bridge rectifier circuit with the diodes replaced by four switches. In *Figure 6.129a*, Dl and D2 are conducting on a positive half-cycle while in *Figure 6.129b* D3 and D4 are conducting. Once again, the result is that current is routed through the load *in the same direction* on successive half-cycles. As with the bi-phase rectifier, the switching action of the two diodes results in a pulsating output voltage being developed across the load resistor (R_L). Once again, the peak output voltage is approximately 16.3 V (i.e. 17 V less the 0.7 V forward threshold voltage).

Figure 6.130 shows how a reservoir capacitor (*C*1) can be added to maintain the output voltage when the diodes are not conducting. This component operates in exactly the same way as for the bi-phase circuit, that is it charges to approximately 16.3 V at the peak of the positive half-cycle and holds the voltage at this level when the diodes are in their non-conducting states. This component operates in exactly the same way as for the bi-phase circuit and the secondary and rectified output waveforms are the same as those shown in *Figure 6.126*. Once again note that the ripple frequency is twice that of the incoming AC supply.

UNIT 6

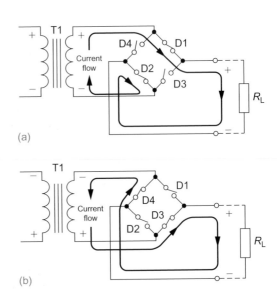

Figure 6.129 (a) Bridge rectifier with D1 and D2 conducting, D3 and D4 non-conducting. (b) Bridge rectifier with D1 and D2 non-conducting, D3 and D4 conducting

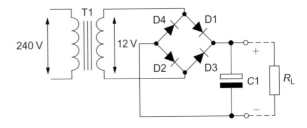

Figure 6.130 Bridge rectifier with reservoir capacitor

Finally, *R–C* and *L–C* ripple filters can be added to bi-phase and bridge rectifier circuits in exactly the same way as those shown for the half-wave rectifier arrangement (see *Figures 6.121 and 6.122*).

TYK 6.47

Sketch the circuit of a simple DC power source using a half-wave rectifier. Briefly explain the function of each component.

TYK 6.48

Explain the advantage of a full-wave rectifier circuit when compared with a half-wave rectifier circuit.

Electrical measurements

There are several test instruments in the workshop and laboratory that will allow you to make measurements on electrical and electronic circuits. This

final section is a brief introduction to these instruments and how to use them to make measurements.

Multimeters

For practical measurements on electrical and electronic circuits it is often convenient to combine the functions of a voltmeter, ammeter and ohmmeter into a single instrument (known as a multi-range meter or simply a *multimeter*). In a conventional multimeter as many as eight or nine measuring functions may be provided with up to six or eight ranges for each measuring function. Besides the normal voltage, current and resistance functions, some meters also include facilities for checking transistors and measuring capacitance. Most multi-range meters normally operate from internal batteries and thus they are independent of the mains supply. This leads to a high degree of portability which can be all-important when measurements are to be made away from a laboratory or workshop.

Analogue multimeters employ conventional moving coil movements (see *Figure 6.131*) and the display takes the form of a pointer moving across a calibrated scale. This arrangement is not so convenient to use as that employed in digital instruments because the position of the pointer is rarely exact and may require interpolation. Analogue instruments do, however, offer some advantages not the least of which lies in the fact that it is very easy to make adjustments to a circuit whilst observing the relative direction of the pointer; a movement in one direction representing an increase and in the other a decrease. Despite this, the principal disadvantage of many analogue meters is the rather cramped, and sometimes confusing, scale calibration. To determine the exact reading requires first an estimation of the pointer's position and then the application of some mental arithmetic based on the range switch setting.

Unlike their analogue counterparts, **digital multimeters** are usually extremely easy to read and have displays that are clear, unambiguous and capable of

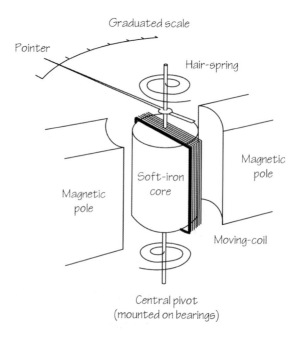

Figure 6.131 A moving coil meter movement

providing a very high resolution. It is thus possible to distinguish between readings that are very close. This is just not possible with an analogue instrument. Typical analogue and digital meters are shown in *Figure 6.132*.

Figure 6.132 Typical analogue and digital multimeters

Digital multi-range meters offer a number of significant advantages when compared with their more humble analogue counterparts. The display fitted to a digital multi-range meter usually consists of a 3½-digit seven-segment display – the ½ simply indicates that the first digit is either blank (zero) or 1. Consequently, the maximum indication on the 2V range will be 1.999V and this shows that the instrument is capable of offering a resolution of 1mV on the 2V range (i.e. the smallest increment in voltage increment that can be measured is 1mV).

The resolution obtained from a comparable analogue meter would be of the order of 50mV, or so, and thus the digital instrument provides a resolution that is many times greater than its analogue counterpart. *Figures 6.133* and 6.134 respectively show the controls and adjustments provided on typical analogue and digital meters. *Figure 6.135* shows how a DC voltage measurement is made using a typical analogue multimeter whilst *Figure 6.136* shows how a current measurement is made using a typical multimeter.

TYK *6.49*

Briefly explain the difference between analogue and digital multimeters. Which type of instrument offers the greatest resolution? Why is this?

TYK *6.50*

What indications are displayed on the analogue and digital multimeters shown in *Figure 6.137*?

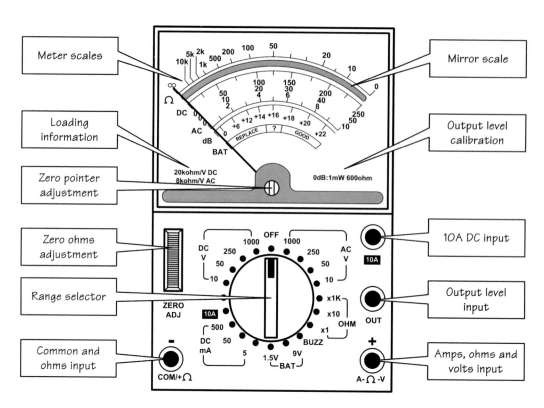

Figure 6.133 Controls and adjustments for a typical analogue multimeter

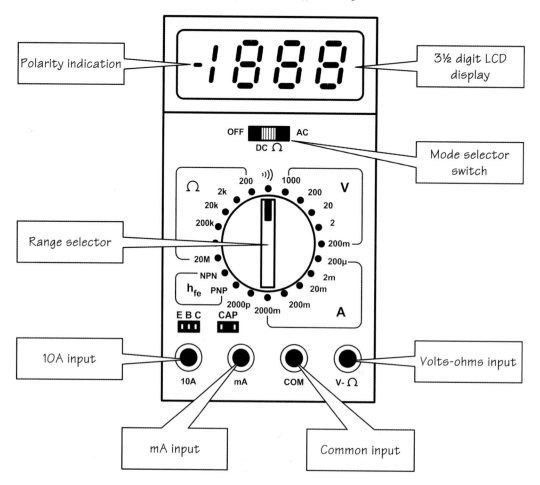

Figure 6.134 Controls and adjustments for a typical digital multimeter

UNIT 6

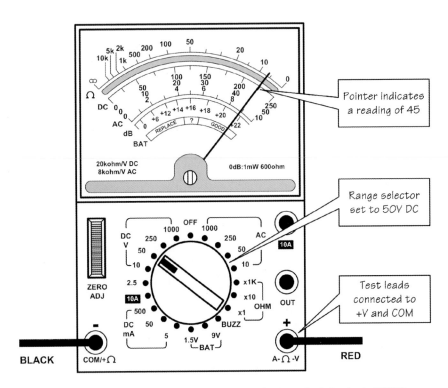

Figure 6.135 Using an analogue multimeter to measure DC voltage (the indication is 45 V DC)

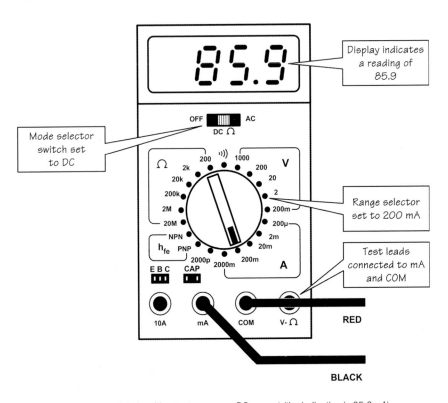

Figure 6.136 Using a digital multimeter to measure DC current (the indication is 85.9 mA)

Multimeter measurements

A typical multimeter measurement is shown in *Figure 6.138*. In this arrangement one meter is used to measure the supply current (note that the circuit must be broken and the meter inserted into it) whilst the second

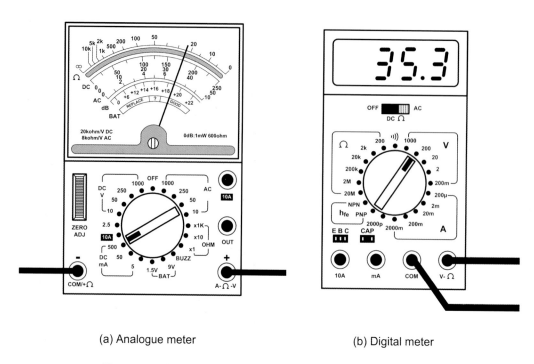

(a) Analogue meter

(b) Digital meter

Figure 6.137 See TYK 6.50

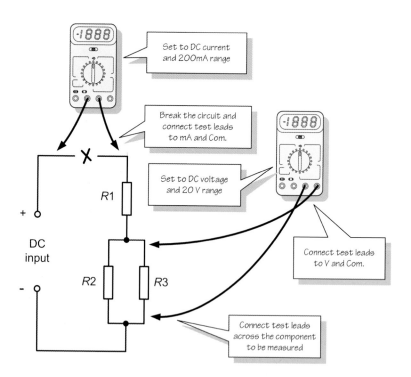

Set to DC current and 200mA range

Break the circuit and connect test leads to mA and Com.

Set to DC voltage and 20 V range

Connect test leads to V and Com.

Connect test leads across the component to be measured

DC input

R1

R2   R3

Figure 6.138 A typical multimeter measurement in which one meter is used to measure the supply current and a second meter is used to indicate the voltage drop across a resistor

meter is being used to measure the potential difference (voltage drop) across R3. In either case it is essential to make initial adjustments to the meter (range and test lead connections) BEFORE connecting the meter to the circuit and switching on the supply. It is also essential to observe Health and Safety precautions whenever working on 'live' electrical circuits. When in doubt you should always refer to your tutor!

Activity 6.5

The *internal resistance* of a battery or power supply limits the current that it can deliver to a load and has the effect of reducing the terminal voltage of the battery or power supply when supplying current (i.e. when 'on-load'). Devise an experimental procedure that will allow you to measure the internal resistance of a 9 V battery using a multimeter and suitably rated load resistor. Present your work in the form of a laboratory instruction sheet suitable for Level 3 students. Use practical measurements in order to verify that the experiment can be successfully carried out as you have planned. What was the measured value of internal resistance?

Oscilloscopes

An oscilloscope (see *Figure 6.139*) is an extremely comprehensive and versatile item of test equipment which can be used in a variety of measuring applications, the most important of which is the display of time related voltage waveforms.

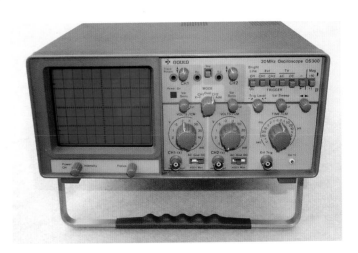

Figure 6.139 A typical two-channel bench oscilloscope

The oscilloscope display is provided by a ***cathode ray tube*** (CRT) that has a typical screen area of 8 cm × 10 cm. The CRT is fitted with a ***graticule*** that may either be integral with the tube face or a separate translucent sheet. The graticule is usually ruled with a 1 cm grid to which further bold lines may be added to mark the major axes on the central viewing area. Accurate voltage and time measurements may be made with reference to the graticule, applying a scale factor derived from the appropriate range switch.

A word of caution is appropriate at this stage, however. Before taking meaningful measurements from the CRT screen it is absolutely essential to ensure that the front panel variable controls are set in the ***calibrate*** (CAL) position. Results will almost certainly be inaccurate if this is not the case!

The use of the graticule is illustrated by the following example. An oscilloscope screen is depicted in *Figure 6.140*. This diagram is reproduced actual size and the fine graticule markings are shown every 2 mm along the central vertical and horizontal axes.

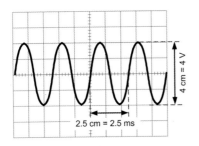

Timebase: 1 ms/cm
Vertical attenuator: 1 V/cm

Figure 6.140 Using an oscilloscope graticule

The oscilloscope is operated with all relevant controls in the 'CAL' position. The timebase (horizontal deflection) is switched to the 1 ms/cm range and the vertical attenuator (vertical deflection) is switched to the 1 V/cm range. The overall height of the trace is 4 cm and thus the peak–peak voltage is $4 \times 1V = 4V$. Similarly, the time for one complete cycle (period) is $2.5 \times 1ms = 2.5ms$. One further important piece of information is the shape of the waveform that, in this case, is sinusoidal. The typical front panel controls and adjustments for a general purpose bench oscilloscope is shown in *Figure 6.141*. The individual controls are summarised in Table 6.14.

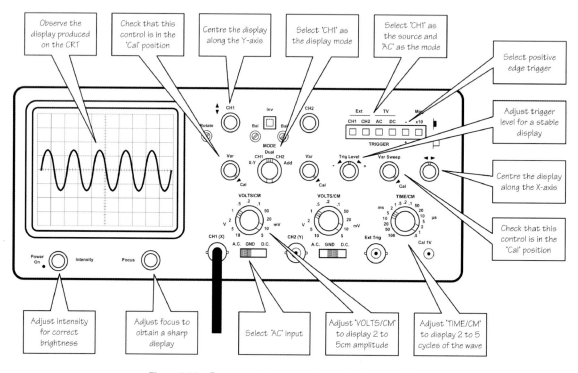

Figure 6.141 Front panel adjustments for a typical bench oscilloscope

TYK 6.50

What indications are displayed on the three oscilloscope displays shown in *Figure 6.142*? What is the phase relationship between the waveforms shown in *Figure 6.142(c)*?

TYK 6.51

Explain the function of each of the following oscilloscope controls:
(a) Brightness
(b) Focus
(c) Stability
(d) Trigger source
(e) Vertical attenuator.

Table 6.14 Oscilloscope controls and adjustments

Control	Adjustment
Focus	Provides a correctly focused display on the CRT screen
Intensity	Adjusts the brightness of the display
Astigmatism	Provides a uniformly defined display over the entire screen area and in both x and y directions. The control is normally used in conjunction with the focus and intensity controls
Trace rotation	Permits accurate alignment of the display with respect to the graticule
Scale illumination	Controls the brightness of the graticule lines
Horizontal deflection system	
Timebase (time/cm)	Adjusts the timebase range and sets the horizontal time scale. Usually this control takes the form of a multi-position rotary switch and an additional continuously variable control is often provided. The 'CAL' position is usually at one, or other, extreme setting of this control
Stability	Adjusts the timebase so that a stable waveform display is obtained
Trigger level	Selects the particular level on the triggering signal at which the timebase sweep commences
Trigger slope	This usually takes the form of a switch that determines whether triggering occurs on the positive or negative going edge of the triggering signal
Trigger source	This switch allows selection of one of several waveforms for use as the timebase trigger. The options usually include an internal signal derived from the vertical amplifier, a 50 Hz signal derived from the supply mains, and a signal which may be applied to an External Trigger input
Horizontal position	Positions the display along the horizontal axis of the CRT
Vertical deflection system	
Vertical attenuator (V/cm)	Adjusts the magnitude of the signal attenuator (V/cm) displayed (V/cm) and sets the vertical voltage scale. This control is invariably a multi-position rotary switch; however, an additional variable gain control is sometimes also provided. Often this control is concentric with the main control and the 'CAL' position is usually at one, or other, extreme setting of the control
Vertical position	Positions the display along the vertical axis of the CRT
AC-DC-ground	Normally an oscilloscope employs DC coupling throughout the vertical amplifier; hence a shift along the vertical axis will occur whenever a direct voltage is present at the input. When investigating waveforms in a circuit one often encounters AC superimposed on DC levels; the latter may be removed by inserting a capacitor in series with the signal. With the AC-DC-ground switch in the DC position a capacitor is inserted in the input lead, whereas in the DC position the capacitor is shorted. If ground is selected, the vertical input is taken to common (0 V) and the oscilloscope input is left floating. This last facility is useful in allowing the accurate positioning of the vertical position control along the central axis. The switch may then be set to DC and the magnitude of any DC level present at the input may be easily measured by examining the shift along the vertical axis.
Chopped-alternate	This control, which is only used in dual beam oscilloscopes, provides selection of the beam splitting mode. In the chopped position, the trace displays a small portion of one vertical channel waveform followed by an equally small portion of the other. The traces are, in effect, sampled at a relatively fast rate, the result being two apparently continuous displays. In the alternate position, a complete horizontal sweep is devoted to each channel alternately.

TYK

TYK 6.52

What adjustment should be made to an oscilloscope when it is to be used to display a small AC voltage superimposed on a much large DC voltage? Explain why this adjustment is necessary.

TYK 6.53

Explain why it is important to ensure that the variable controls of an oscilloscope are placed in the 'CAL' position before attempting to make an accurate measurement.

Oscilloscope measurements

Typical oscilloscope measurements are shown in *Figures 6.143* and *6.144*. In *Figure 6.143* the oscilloscope is being used to display the waveforms in a simple half-wave rectifier power supply. In *Figure 6.144* the oscilloscope is being used to measure the phase relationship between the voltages in a series C-R circuit. As with the multimeter measurements that we met earlier it is essential to make initial adjustments to the oscilloscope BEFORE connecting the oscilloscope to the circuit and switching on the supply. Once again, when in doubt you should refer to your tutor!

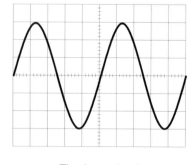

(a) Timebase: 1ms/cm
Y-attenuator: 1V/cm

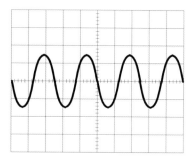

(b) Timebase: 20ms/cm
Y-attenuator: 50mV/cm

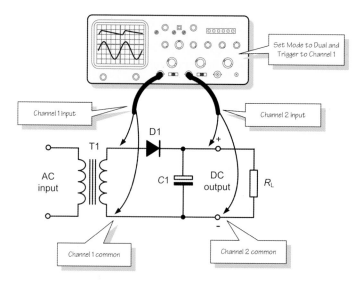

Figure 6.143 Oscilloscope measurements on a simple half-wave rectifier power supply

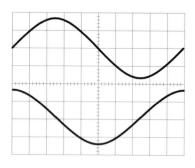

(c) Timebase: 100ms/cm
Y-attenuator: 50V/cm

Figure 6.142 See TYK 6.50

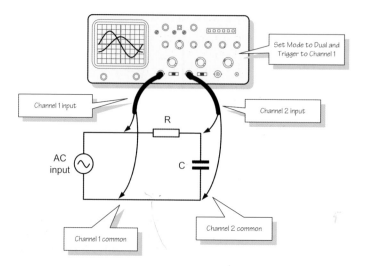

Figure 6.144 Oscilloscope measurements on a C-R series circuit

UNIT 6

Activity 6.6

Resonance occurs in an AC circuit when resistance, inductance and capacitance are present. In a series L-C-R circuit resonance occurs at a frequency where the reactance of the inductor is equal and opposite to that of the capacitor. Devise an experimental procedure that will allow you to investigate the voltages present in an L-C-R circuit at, or near, resonance. You will need to select suitable component values for L, C and R (keep this small) and use an AC signal generator with a variable frequency output. You will need to calculate suitable values for L and C based on the particular resonant frequency that you decide to use (around 1 kHz would be sensible). Present your work in the form of a laboratory instruction sheet suitable for Level 3 students. Use practical measurements in order to verify that the experiment can be successfully carried out as you have planned. What was the phase relationship between the circuit voltages at resonance? How large were the voltages that appear across the capacitor and the inductor when compared with the supply voltage? What can you learn from this?

Virtual instruments

In recent years a new type of electronic measuring instrument has become available. Rather than use conventional analogue, digital or CRT displays, these *virtual instruments* use plug-in adapters or USB-connected interfaces together with a PC (either desktop or laptop). The interface circuit captures a digital sample of the analogue input which can then be stored in memory and recalled for later display. Virtual instruments offer a number of advantages when compared with conventional test instruments including the ability to display waveform parameters (such as time, voltage, frequency and phase) as well as being able to store, recall and print waveform data. A typical virtual instrument display is shown in *Figure 6.145*.

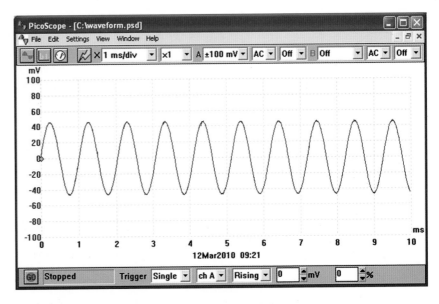

Figure 6.145   A typical display produced by a virtual oscilloscope

End of Unit Review

1. A power supply rated at 15 V, 0.25 A is to be tested at full rated output. What value of load resistance is required and what power rating should it have? What type of resistor is most suitable for this application and why?

2. Determine the resistance of the resistor network shown in *Figure 6.146* looking into terminal A and B with (a) terminals C and D left open circuit and (b) terminals C and D short circuit.

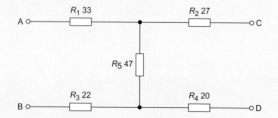

Figure 6.146 See End of unit review question 2

3. A resistor has a resistance of 218 Ω at 0°C and 225 Ω at 100°C. Determine the resistor's temperature coefficient and also determine its resistance at 25°C.

4. A resistor is marked with the following coloured bands: red, violet, orange, gold. What is the marked value and tolerance of the component? What current would flow in the component if it was connected across a 28 V power source?

5. Determine all currents and voltages in *Figure 6.147*.

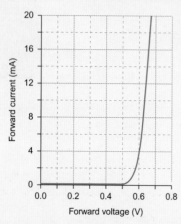

Figure 6.147 See End of unit review question 5

6. *Figure 6.148* shows the characteristics of a diode.

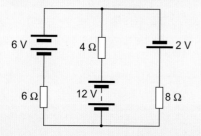

Figure 6.148 See End of unit review question 6

(a) What type of material is used in this diode? Give a reason for your answer.

(b) Use the characteristic shown in *Figure 6.148* to determine the resistance of the diode when (i) $V_F = 0.65$ V and (ii) $I_F = 4$ mA.

7. A parallel plate capacitor has plates of 0.02 m². Determine the capacitance of the component if the plates are separated by a dielectric of thickness 0.5 mm and relative permittivity 5.6.

8. A capacitor is required to store 0.5 J of energy when charged from a 120 V DC supply. Determine the value of capacitance required.

9. Capacitors of 20 μF and 80 μF are connected in series across a 50 V DC supply. Determine (a) the total charge and (b) the voltage drop across each capacitor.

10. An initially uncharged capacitor of 1 μF is charged from a 15 V DC supply via a 100 kΩ resistor. How long will it take for the capacitor voltage to reach 5 V?

11. A capacitor is initially to a potential of 200 V and then allowed to discharge through a 2 MΩ resistor. Determine the value of capacitance if it takes 50 s for the capacitor voltage to fall to reach 100 V.

12. A sine wave has a frequency of 250 Hz and an amplitude of 5 V. Determine its periodic time and r.m.s. value.

13. A sinusoidal voltage has an r.m.s. value of 22 V and a period of 16.7 ms. What is the frequency and peak–peak value of the voltage?

14. A sine wave has a frequency of 100 Hz and an amplitude of 20 V. Determine the instantaneous value of voltage (a) 2 ms and (b) 9 ms from the start of a cycle.

15. Determine the reactance of (a) a 33 mH inductor at 50 Hz and (b) a 220 nF capacitor at 5 kHz.

16. A 47 nF capacitor is connected across the 220 V 50 Hz mains supply. Determine the r.m.s. current flowing in the capacitor.

17. A 10 mH inductor of negligible resistance is used to form part of a filter connected in series with a 50 Hz mains supply. What voltage drop will appear across the inductor when a current of 1.5 A is flowing?

18. A 10 μF capacitor is connected in series with a 500 Ω resistor across a 110 V 50 Hz AC supply. Determine the impedance of the circuit and the current taken from the supply.

19. A choke having an inductance of 1 H and resistance of 250 Ω is connected to a 220 V 60 Hz AC supply. Determine the power factor of the choke and the current taken from the supply.

20. A transformer has 800 primary turns and 60 secondary turns. If the secondary is connected to a load resistance of 15 Ω, determine the value of primary voltage required to produce a power of 22.5 W in the load (assume that the transformer is loss free).

UNIT 6

Photo courtesy of iStockphoto, jacus, Image# 2396903

Many external factors, such as the economy, competitors, technological advances and regulatory agencies, affect the operation of an engineering company. Crash test dummies are used in the automotive sector to test new technology and materials, to provide optimum performance and meet safety requirements. It may take engineers hours to prepare a dummy for a crash, attaching instrumentation wires and calibrating the dummy with punishing blows to duplicate crash conditions.

Business Operations in Engineering

This unit is designed to provide you with an introduction to the business and commercial aspects of engineering. It aims to broaden and deepen your understanding of business, industry and the effects of engineering on the environment. It also aims to provide you with a firm foundation for employment in the engineering industry together with an understanding of the financial, legal, social and environmental constraints within which an engineering company operates.

When you have completed this unit you will understand how an engineering company is organized and you will be aware of the external factors and the economic environment in which it operates. You will also have an understanding of the impact of relevant legislation and the effect of environmental and social constraints on its operation.

To help you understand more about the financial aspects of running an engineering business, you will be introduced to the techniques used in the costing of an engineering operation including those that will tell you whether a business is operating at a profit or a loss.

This unit is assessed through a series of assignments and case studies, and it has strong links with the mandatory units Communications for Engineering Technicians (Unit 2) and Engineering Project (Unit 3). Wherever possible, you should apply the techniques that you have developed in the communications unit to work that you undertake in this unit. There are also links to several of the optional units including Commercial Aspects of Organizations Employing Engineers and Engineering Design.

Case studies (based on real or invented engineering companies) are an important part of this unit. When you carry out a case study you will be presented with sample data to analyse. You might find it useful to relate your experience of employment or work experience periods in industry to the case study as well as to work covered elsewhere in the unit.

BTEC National Engineering. DOI: 10.1016/B978-0-12-382202-4.00007-6

Engineering Companies

All engineering companies must operate as commercial enterprises in order to survive. In this section you will look at how an engineering company operates. You will learn about the various sectors in which engineering companies operate and the functions that are performed within a typical engineering company, such as research and development (R&D), design and manufacture. You will also learn about the various types of organization and how they differ. You will also gain an insight into how information flows within an engineering company. This section is important not only because it sets the scene for the sections that follow but also because it will help you to understand your eventual role within an engineering company. We start by looking at the areas within which engineering companies operate: we call these 'engineering sectors'.

KEY POINT

Engineering sectors define the broad areas within which an engineering company operates

Engineering sectors

Some of the engineering sectors, engineered products and engineering companies with which you are probably familiar include:

Chemical engineering	Fertilizers, pharmaceuticals, plastics, petrol, etc. Companies in this field include Fisons, Glaxo, ICI and British Petroleum.
Mechanical engineering	Bearings, agricultural machinery, gas turbines, machine tools and the like from companies such as RHP, GKN and Rolls-Royce.
Electrical and electronic engineering	Electric generators and motors, consumer engineering electronic equipment (radio, TV, audio and video), power cables, computers, etc., produced by companies such as GEC, BICC and ICL.
Civil engineering	Concrete bridges and flyovers, docks, factories, power stations, dams, etc., from companies like Bovis, Wimpey and Balfour-Beatty.
Aerospace engineering	Passenger and military aircraft, satellites, space vehicles, missiles, etc., from companies such as British Aerospace, Westland and Rolls-Royce.
Telecommunications	Telephone and radio communication, data communications equipment, etc., from companies such as Nokia, GEC, Plessey and British Telecom.
Motor vehicle engineering	Cars, commercial vehicles (lorries and vans), motorcycles, tractors and specialized vehicles from companies such as Rover, Vauxhall UK and McLaren.

As you work through this unit, it will help you to put things into context by relating the topics to those engineering companies with which you are familiar. This will give you an appreciation of the factors that affect their operation as well as the constraints under which they operate.

Some companies operate within more than one sector. For example, a company may produce products and provide services in both the electronic engineering and telecommunications sectors. Other companies may be active in one sector only and their products and services may only relate to that sector.

TYK

TYK 7.1

An engineering company specializes in the design and manufacture of wind generators. In which two sectors of engineering does this company operate?

Activity 7.1

Identify the sector(s) in which each of the following engineering companies operate:

1. **Perkins** (www.perkins.com)
2. **Thales Group** (www.thalesgroup.com)
3. **Dean and Dyball** (www.deandyball.co.uk)
4. **RPS Group** (www.rpsgroup.com)
5. **Bayer-Wood Technologies** (www.bayer-wood.co.uk)
6. **Smiths Group** (www.smiths-group.com)

Activity 7.2

Identify three engineering companies in your area that are active in three different engineering sectors. For each company, identify the range of products or services that it supplies and the nature of its business (e.g., manufacturing, maintenance, design, etc.). Present your work in the form of a data sheet for each company and include full company name, address and website.

Engineering functions

You need to be able to identify the various functions within an engineering company. These include both commercial and engineering aspects of the company's operation. The commercial functions include sales, marketing, distribution, commissioning, finance and purchasing. The engineering functions can include R&D, design, manufacturing, product development, quality and planning. Product support is another, extremely important, area of engineering. You should also be aware that non-engineering companies often require input from engineering services, such as maintenance

UNIT 7

of services and equipment. In addition to understanding the functions performed within an engineering company you must also be able to recognize the main responsibilities attached to key job roles within both the commercial and engineering functions.

The most important function within an engineering company (or any company for that matter) is overall management and control so this is where we will begin our investigation of how engineering companies operate.

Management

The production *workflow* and some of the functions in an engineering firm are shown in *Figure 7.1*. This production workflow starts with suppliers that provide an input to the various engineering processes. The output of the engineering processes is delivered to the customers. You may find this easier to recall by remembering the acronym *SIPOC*.

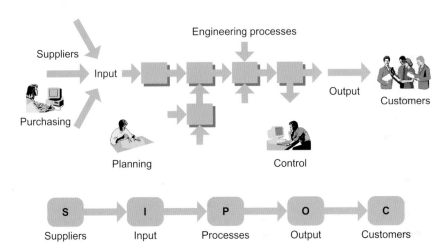

Figure 7.1 Simplified view of the flow of work in an engineering company from its suppliers to its customers

The three functions that we have included in the diagram (there are many more in a real engineering firm) operate as follows:

Purchasing	The purchasing function ensures that supplies are available as and when required by the engineering processes.
Planning	The planning function ensures that the correct engineering processes are in place and also that the workflow is logical and timely.
Control	The control function ensures the quality of the output and the cost-effectiveness of the processes.

Planning

The essential business activities performed in an engineering company can be grouped together under the general headings of *planning*, *controlling* and *organizing* (see *Figure 7.2*). The first of these activities, planning,

is absolutely fundamental to the correct functioning of an engineering company. If no planning is done then activities are almost certainly going to be very ineffective. What is planning? It is the sum of the following activities:

- setting the *goals* for an engineering company
- forecasting the *environment* in which the engineering company will operate
- determining the *means* to achieve goals.

Figure 7.2 Relationship between planning, controlling and organizing in an engineering company

Setting *goals* or *objectives* is the first step in planning. It determines the direction an engineering company is going. It encourages all engineering company members to work towards the same end otherwise members are likely to set their own objectives, which will conflict with each other. Good objectives make for rational engineering companies that are more coordinated and effective. The objectives must therefore be set by the most senior management group in an engineering company so that all of its staff can be given clear direction. If the goals are clearly stated, logical and appropriate for the business then they act both as motivators and yardsticks for measuring success.

Once the engineering company has clear direction the next step is to analyse the environment and forecast its effect on the business. For example, an engineering company that makes lawn mowers may set objectives so that, within 5 years, it will:

- achieve a 30% share of the market
- be acknowledged as the market leader
- be the accepted technological leader
- be highly competitive on price
- operate internationally.

When it forecasts its environment it may conclude that:

- new designs will be marketed by competitors
- new battery technology will become available to support cordless mowers
- there will be a sharp decline in demand for manual lawn mowers.

Its designs are technically very sound but are being threatened by new rotary 'hover' models that are proving attractive to customers. It has to decide how to deal with the threat, either to improve its existing design concept so that customers continue to find them attractive or to follow the new trend and produce products based on new design concepts.

Forecasting the environment allows the company to set new objectives and to prepare plans to meet its revised goals. Companies that fail to go through this process will go into decline in the long run because they are ignoring the changing world around them.

Once the goals are refined and changed in the light of environmental forecasting then plans can be made to achieve the goals. Some plans will not change that much while others will be dramatically affected by the changing environment. For this reason plans can be classified as follows:

- standing plans
- single use plans
- strategic plans
- tactical plans.

Standing plans are those that are used many times, and remain relatively unaffected by environmental change. Examples are employment, financial, operating and marketing policies and procedures. For example, hiring new employees involves standard procedures for recruitment and selection. Another example would be the annual routines for establishing budgets.

Single use plans are those that are used only once, such as those for the control of a unique project or specific budgets within an annual budget. (Budgets themselves are single use plans even though the procedures used for producing them are standing plans.)

Strategic plans are the broad plans related to the whole engineering company and include forecasting future trends and overall plans for the development of the engineering company. They are often in outline only and very often highly subjective, involving judgments made by top managers. For example, a plan may be made to build an entirely new factory based on forecasts of demand. This plan is strategic, and if the plan is wrong and the sales forecasts on which it is based do not materialize, then the results for a company could be devastating.

Tactical plans operate within the strategic plan. The new factory has to be brought into commission and production has to be scheduled and controlled. Plans for the latter are tactical, since they focus on how to implement the strategic plan.

TYK 7.2

Explain three different types of plan that may be used in an engineering company. Give an example of the use of each type of plan.

TYK *7.3*

Audio Power Systems (APS) manufactures a range of loudspeakers for audio hi-fi, music and sound-reinforcing applications. The board of directors at APS has set the following strategic goals for the company to be achieved over a 3-year period:
   (a) introduce two new loudspeaker designs each year
   (b) increase market share by 5% each year
   (c) double the number of European distributors by the end of the 3-year period.

Which three of the following objectives are most likely to be instrumental in delivering these strategic goals?
   (a) employing more staff in R&D
   (b) appointing an assistant to the marketing manager with a good command of European languages
   (c) introducing measures that will reduce pay costs by around 10%
   (d) increasing expenditure in targeted marketing campaigns
   (e) building a new anechoic chamber test facility.

Control

In an engineering business, control is required for a variety of reasons including ensuring that a project or product:

- remains within budget
- is delivered on time
- complies with relevant legislation
- meets appropriate quality standards
- operates according to the agreed specification
- fully satisfies the expressed and implied needs of the customer or client.

The pre-requisite of control is planning. Controlling involves comparing events with plans, correcting deviations and ensuring that the planned events happen. Sometimes deviations are so fundamental that they require a revision to the plan so that later events are controlled against a new plan. For example, the original sales forecast may turn out to be too optimistic, and production plans may have to be reduced to bring output into line with what sales are possible.

There are various ways in which control can be exercised. It can be *predictive* as in the case of a cash flow forecast. This forecast may indicate a shortfall of cash in August but a surplus in September. The finance manager may need to arrange additional finance with the bank in August and then in September he or she might deposit surplus funds onto the money market. The point here is that variances are predicted in advance, thereby promoting cash control.

In the case of monthly comparisons between budgeted expenditures and actual expenditures, an overspend might be revealed. This triggers action that holds back expenditure until spending comes back into line with budget. This is historical control since action is based on a report of events in the recent past.

Concurrent control is *real time* such as that which might occur in controlling a continuous process or a production line. In this case, the system has built in *feedback*, which enables it to remain in balance by regulating inputs and outputs. An example of concurrent control would be where a production process requires temperature regulation. The control system is designed to switch off heating when temperature reaches a threshold or switch on heating when it drops to a minimum level. The feedback is information on temperature. The same principle applies in some stock control systems where stocks are maintained at pre-determined minimum and maximum levels, with supplies being switched on and off to maintain equilibrium.

Test your knowledge

TYK

TYK 7.4

Explain why controls are required within an engineering company. Describe two examples of controls that should be in place in every engineering company.

> **KEY POINT**
>
> In an engineering firm, control is required to ensure that:
>
> - costs remain within budget
> - a project is delivered on time
> - stages in the project comply with relevant legislation
> - appropriate company and sector quality standards are met
> - customers' and clients' needs are satisfied and there is effective liaison at all stages of a project
> - modifications (and any other subsequent changes) meet the customer's specification and remain within budget

Leadership and direction

Planning and control activities are the tasks of engineering management. However, they are only achieved through people. People will work effectively if they are led and directed properly. This implies that top managers must be in touch with the engineering processes as well as business issues and they should be *visible*. They must have a clear vision for the future reinforced by specific objectives that are effectively communicated to their employees.

This approach to leadership is apparent in some of our best companies as exemplified by British Airways and Marks & Spencer. Such companies have a clear mission and objectives, and have a visible committed top management. This philosophy permeates the whole engineering company stimulating better performance from all employees.

Motivating good performance from all employees is the responsibility of all managers. What motivates individuals and groups within commercial engineering companies is a complex and important subject, the detail of which is well beyond the scope of this book. However, it is still worth saying that managers must discover what it is that will stimulate employees to work productively.

In general people respond best to *considerate styles* of management, whereby their personal contributions are fully recognized. It is also true that there has to be an atmosphere of discipline coupled with a work-oriented culture. The task has to be accomplished, and being considerate does not extend to the toleration of slack or sloppy practices and behaviour.

Allocation and supervision of work

This is the practical implementation of all that we have discussed in this section. An engineering company exists to fulfil the goals of its owners.

> **KEY POINT**
>
> Clear direction, sound and explicit guidelines, and proven work processes and techniques (all of which must be effectively communicated to employees) jointly contribute to the successful operation of an engineering business

UNIT 7

It has to function in a coordinated and rational way. The people who are its members have to work together and understand their specific roles and functions. They need to receive directions and work has to be allocated. There has to be supervision of these activities. An engineering company is analogous to a machine or a living organism. In order to function properly everything has to work together smoothly. The company's managers have the task of ensuring that this work takes place according to plan and within the engineering company's stated objectives.

Activity 7.3   Universal Diesels (case study)

Universal Diesels makes diesel engines for use in a variety of products and environments. It had one division that shipped 'knocked down' (KD) kits to overseas assemblers (these assemblers were foreign subsidiaries of the company). These firms assembled the kits into working engines and sold them on to other manufacturers for incorporation into their products. The kits were put together based on engineering specifications for each type of engine. A pre-production control department was responsible for translating the engineering specifications into packing lists used to prepare and pack the kits for export. Engineering specifications were constantly being updated to improve designs and to correct faults that came to light during service. Changes often had to be carefully planned so that the KD parts shipped were compatible and of the same quality level.

The KD operation received many complaints about the kits. Some assemblers said they could not assemble the engines because of specification errors or because the wrong parts were shipped. The pre-production control department was very sceptical about these complaints and was slow to respond. A proportion of the complaints were ill-founded so requests for urgent replacements were given a low priority and normal packing operations deadlines were considered more important than sending out urgent miscellaneous orders to enable the overseas assembly locations to complete their work.

The management were under heavy pressure to meet their own packing deadlines, which were constantly threatened by supply problems of all kinds and by frequent changes to orders from overseas assemblers. The kits normally took between 6 weeks to 3 months to reach their destination by sea freight, whilst miscellaneous orders had to be air freighted at high cost.

The problems continued to worsen as complaints from overseas companies grew. These companies tried to anticipate problems by comparing the packing documentation with the manufacturing engineering process sheets and faxing in queries on all discrepancies. This generated so many queries that it was impossible to respond quickly even if pre-production activities wished to. This actually generated even more inertia. In the end, head office appointed a senior manager as troubleshooter and sent him in to solve the problems.

1. What is wrong with the organization of this engineering company?
2. What role should the manufacturing engineering activity play here?
3. What should be the troubleshooter's first priority?
4. What do you think the troubleshooter should do in order to arrive at a long-term solution?

Discuss this in small groups and prepare a 10-minute presentation to the rest of your class or group giving details of your proposed solution.

Finance and accounting

The *finance director* will manage the company's cash flow, short- and long-term investments and supervise the company's accounting and budgetary system. His job is to ensure that the company has sufficient cash to support day-to-day operations. He should also ensure that cash that is not immediately required is made to work for the company. This is done by arranging short-term investments on the money markets or by switching funds into building society or bank deposit accounts. He may also be involved with international money flows and the 'hedging' of risk against exchange rate losses.

The *financial accountant* will be in charge of all recording in the financial accounting system. All business transactions are recorded in a system called *double entry accounts*. These records are called this because every business transaction has a twofold effect, and to make a complete record this twofold aspect has to be recorded. For example, if £100 was spent to buy raw materials, there has to be a record in the bank account of the money spent and then the existence of materials bought must be recorded in a purchases account. This method of recording enables the business to produce a profit and loss account and a balance sheet, which summarize all the transactions so that financial performance can be tracked and reported to shareholders and other interested parties, including the tax authorities.

The *management accountant* will administer the budgetary control and costing system. This system enables the business to forward plan its profit and loss and its overall financial position, as well as being able to control and report on costs of operation. Thus a budgeted profit and loss account and balance sheet is produced, called a *master budget*. This master budget summarizes the budgets for all cost centres or operating divisions. The management accountant monitors actual results against the budget and will use data from the double entry accounts system referred above. This monitoring enables departmental managers to correct deviations to budget, control and manage the cost of running their departments.

Part of management accounting is the process of *investment appraisal* to plan the purchase of fixed assets and to ensure that the best choices are made for new and replacement equipment. The management accountant will also prepare reports and analyses of various kinds for management.

The *cashier*, who is likely to report to the financial accountant, will deal with all bank and cash transactions. In a large company, the number of transactions is considerable, especially those that deal with receipts from customers and payments to suppliers. The work of this section is important because forecasting and monitoring cash flow is vital to the financial well-being of the business.

The *credit controller* will be concerned with authorizing new credit customers and controlling the amount of credit granted and reducing or preventing bad debts. A budget will be prepared giving planned levels for outstanding debtors. This figure really represents a short-term investment of capital in customers. This investment has to be managed within budgeted limits otherwise the company's finance costs would increase and there

would be further risks of non-payment. The credit controller would monitor the financial stability of existing customers or vet the standing of new customers. He might do this through a combination of bank references, credit agencies and studying the customers' own published accounts.

Other activities carried out in accounting include payment of wages and salaries, depreciation of fixed assets, maintaining shareholder records, paying shareholder dividends.

A typical engineering company structure for the accounting and finance function is given in *Figure 7.3*, which shows the departmental divisions indicated above.

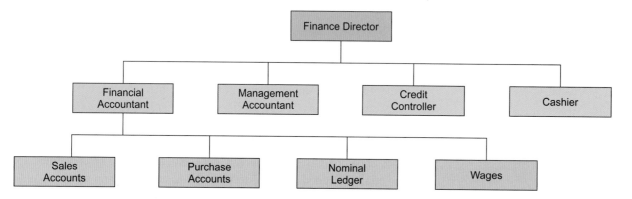

Figure 7.3 Organization of the finance function

Marketing

Marketing is said to be the most important function in the business, since if customers cannot be found for the company's products the company will go out of business, regardless of how financially well run or efficient it is.

Although sales is considered separately later, it is really a part of the marketing function. Marketing is all about matching company products with customer *needs*. If customer needs are correctly identified and understood, then products can be made, which will give the customer as much as possible of what he wants. Companies that view the customer as 'sovereign' are those companies that stay in business, because customers will continue to buy products that meet their requirements.

Hence marketing activities are centred around the process of filling customers' known needs, discovering needs the customer does not yet know he has and exploiting this by finding out how to improve products so that customers will buy this company's products in preference to other goods. Some of the most important activities are:

- market research
- monitoring trends in technology and customer tastes
- tracking competitors' activities
- promotion activities
- preparing sales forecasts.

UNIT 7

Remember that in some businesses the marketing activity is directed at end consumers, members of the public. This has to be done by national forms of advertising, such as TV commercials, direct mail, newspapers or through major retailers selling to the consumer. The methods used may be somewhat different if the customers are other companies. Although the principle of meeting customer needs is the same, the approach taken may be much more technical and may include the services of sales engineers to provide technical backup and advice. The publicity methods are more likely to be centred around trade fairs, exhibitions and advertisements in the trade press or technical journals for example. You should note these two distinct marketing approaches are, respectively, called consumer marketing and industrial marketing.

Sales

The sales department is concerned with advertising and selling goods. It will have procedures for controlling sales and the documentation required. The documents used are the same as for purchasing, described below, except from the supplier's viewpoint rather than from the customer's. It may employ *commercial travellers* or have a resident *sales force*. It is involved with many possible ways of publicizing the company's products such as trade fairs, wholesalers' displays, press and TV advertising, special campaigns, promotional videos, etc. It will also be concerned about the quality of goods and services as well as administering warranty and guarantee services, returns and repairs, etc.

Sales will maintain contacts with customers that will entail the following customer services:

- technical support
- after-sales service, service engineering
- product information, prices and delivery
- maintaining customer records.

A typical marketing and sales structure for a company engaged in industrial marketing is shown in *Figure 7.4*.

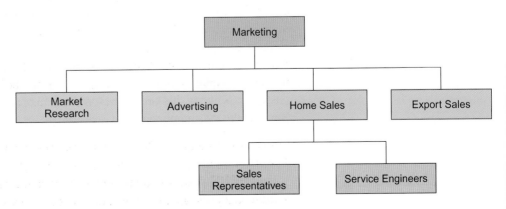

Figure 7.4 Organization of the marketing and sales functions

TYK 7.5

1. What is the most important function of marketing?
2. What is the purpose of market research?
3. Why do some companies use wholesalers rather than sell direct?
4. Name two ways of determining the price of a product.
5. What does a sales engineer do?

Distribution: consumer markets

For engineering companies operating in the consumer marketing field distribution can be accomplished through a variety of ways. This can include wholesalers, retailers, mail order or direct selling through the company's own retail outlets. Some companies may use all of these methods. We will examine the wholesale and retail systems, as well as the pricing aspects.

Retail outlets sell direct to the consumer. Most of these are shops or mail order businesses. Retailers fall into several types: hypermarkets, supermarkets, multiple shops, departmental stores, cooperative retail societies, independent retailers, voluntary retail chains, franchise outlets, discount stores, etc.

The purpose of retailing is to provide for the availability of goods close to where consumers live. Retailers also study consumer preferences and stock goods accordingly. They also keep manufacturers informed of what it is that consumers want so that supply matches demand.

If a retailer requires a large range of goods in relatively small quantities it is not very convenient to buy direct from manufacturers. Think of the number of different manufacturers that a small independent hardware store would have to deal with if it did deal direct with each manufacturer. Hence the continuing need for wholesalers, who stock goods from many manufacturers and can supply smaller quantities to retailers.

The *wholesaler* is a *middle man* and it is said that his presence puts up prices. This is not necessarily the case, since manufacturers can sell to him in bulk quantities and save on transport and administration costs. In effect wholesalers operate as intermediate storage depots for retailers and therefore provide a useful service. They can usually provide retailers with credit terms of trading, often enabling small businesses to sell before they have to pay for goods, or at least to reduce the impact of the cost of carrying a large range of stock items.

They can also act as a buffer to smooth out demand for manufacture. If demand is seasonal they can buy regularly through the year, thus making it easy for manufacturers to make goods in economic runs and then store stock to meet heavy demand, but which does not place excessive loads on the manufacturer's capacity.

UNIT 7

Wholesalers as such have been in decline in recent times, thus many manufacturers have started to deal direct, especially with large retailers, such as the supermarkets. However, they have had to take over the functions of storage, transport and dealing directly with retailers.

Companies may decide to deal directly with the public through mail order, thus bypassing the wholesaler and retailer. Mail order depends on a good postal service or the existence of transport operators who can provide a similar service. It has the advantage of being nation-wide or even international, thus extending the potential market enormously.

In some cases there are very large mail order retailers who buy from manufacturers and sell on to consumers. These companies sometimes operate normal retail outlets as well.

It is common for distributor prices to be expressed at a percentage discount from the price to be paid by the consumer. Thus manufacturers give discounts to wholesalers as an incentive to stock their goods and to provide a profit margin for them. A similar system will be used by the wholesaler when dealing with the retailer. However, if a price to the final consumer is not envisaged or fixed, then the situation is less clear and each party must charge a price, which his particular market will stand and depending on what quantities he needs to sell and what his actual costs are going to be.

In addition to wholesale and trade discounts, *quantity discounts* may be offered to encourage distributors to buy in large quantities, which may be more economical to supply and deliver, since there are economies of scale to be had, such as lower manufacturing, administrative and transport costs.

Further *incentives* may be offered in giving cash discounts. Cash discounts reward immediate or early payment. This may enable the manufacturer or wholesaler to reduce his need for working capital and reduce credit collection and control costs.

Discount structures are therefore used for many purposes. Firstly to increase sales, secondly to influence the pattern of sales and thirdly to reduce the costs of production and distribution. Sometimes discounts can produce more sales but have very little effect on profits since higher volumes and lower costs may not compensate for lower margins.

You should be aware that price is influenced by many factors:

- actual cost of manufacture
- what the market will stand
- what others are charging for similar products
- consumers' perceptions of quality and value.

The interaction of supply and demand is complex and outside the scope of this course. However, if supply exceeds demand, in general this exerts a downward pressure on prices, as manufacturers and distributors seek to sell goods they have made or bought. The costs of storage and distribution may be so high as to force sale at prices that might be below average cost. This is especially true of perishable goods and foodstuffs.

Alternatively, if demand exceeds supply that tends to bid up the price as consumers search for supplies. In some cases the increase in price then

serves to limit demand since some potential buyers drop out when the price goes too high, this then acts to dampen demand again and tends to bring equilibrium between supply and demand.

As you can appreciate this can get very complicated when manufacturers, wholesalers and retailers start to offer different discounts to try to influence events in their favour. Sometimes the competition is so cut-throat that the only winner is the consumer. In some cases the weaker players go under, leaving the more efficient firms to operate in a less hostile environment. Sometimes the bigger, stronger firms deliberately cut prices so low as to force others out of business and then exploit the consumer when they can dominate the market. However, this can go in reverse again if prices then go too high and attract new players into the market who then increase supply, which will produce a downward pressure again on prices, and so it goes on.

Distribution: industrial markets

For industrial market distribution the situation is more variable. Frequently the seller will have his own fleet of vehicles and may have warehousing facilities or geographically dispersed depots. An example might be a company that manufactures components for the motor industry. It may manufacture in one or more locations and have storage depots located near its customers. It may also deliver products direct to the motor manufacturer's plant either using its own transport or by using an *independent haulier*.

If the company makes products for international markets it may have to prepare and package products for sea or air freight. This could include using haulage contractors who will deliver direct into Europe using roll-on roll-off ferries.

Price structure tends to vary widely, but is usually based on negotiation between the seller and buyer. It may be done through a process of enquiry and quotation or may simply be based on price lists and discounts separately negotiated.

Research and development

New product design and development is often a crucial factor in the survival of a company. In an industry that is fast changing, firms must continually revise their design and range of products. This is necessary because of the relentless progress of technology as well as the actions of competitors and the changing preferences of customers.

A good example of this situation is the motor industry. The British motor industry has gone through turbulent times, caused by its relative inefficiency compared with Japan and Germany and also because the quality of its products was below that of its competitors. This difficult position was then made worse by having products that lacked some of the innovative features of the competition.

There are three basic ways of approaching design and development:

- driven by marketing
- driven by technology
- coordinated approach.

UNIT 7

A system driven by marketing is one that puts customer needs first, and only produces goods, which are known to sell. Market research is carried out, which establishes what is needed. If the development is technology driven then it is a matter of selling what it is possible to make. The product range is developed so that production processes are as efficient as possible and the products are technically superior, hence possessing a natural advantage in the market place. Marketing's job is therefore to create the market and sell the product.

Both approaches have their merits, but each of them omit important aspects, hence the idea that a coordinated approach would be better. With this approach the needs of the market are considered at the same time as the needs of the production operation and of design and development. In many businesses this inter-functional system works best, since the functions of R&D, production, marketing, purchasing, quality control and material control are all taken into account.

However, its success depends on how well the interface between these functions is managed and integrated. Sometimes committees are used as matrix structures or task forces (the latter being set up especially to see in new product developments). In some parts of the motor industry a function called *programme timing* coordinates the activities of the major functions by agreeing and setting target dates and events using network planning techniques.

The basic *product development process* is outlined as follows:

- idea generation
- selection of a suitable design
- preliminary design
- prototype construction
- testing
- design modification
- final design.

This is a complex process and involves cooperative work between the design and development engineers, marketing specialists, production engineers and skilled craft engineers to name some of the major players.

Ideas can come from the identification of new customer needs, the invention of new materials or the successful modification of existing products. Selection from new ideas will be based on factors like:

- market potential
- financial feasibility
- operations compatibility.

This means screening out ideas that have little marketability, are too expensive to make at a profit and which do not fit easily alongside current production processes.

After this, preliminary designs will be made within which trade-offs between cost, quality and functionality will be made. This can involve the processes of *value analysis* and *value engineering*. These processes look at

both the product and the methods of production with a view to maintaining good product performance and durability whilst achieving low cost.

Prototypes are then produced, possibly by hand and certainly not by using mass production methods. This is followed by rigorous testing to verify the marketing and technical performance characteristics required. Sometimes this process will involve test marketing to check customer acceptance of the new product.

Final design will include the modifications made to the design as a result of prototype testing. The full specification and drawings will be prepared so that production can be planned and started.

Test your knowledge

TYK

TYK 7.6

1. State the steps in the product development process.

2. Name and briefly describe three strategies for product development.

3. Describe the process of selection for new product ideas.

Manufacturing operations

The production, manufacturing or process of service delivery is at the heart of an engineering business. The company translates the designs for products or services that are based on market analysis into the products and services wanted by its customers. The production process can be seen within a framework of five main areas, which we will now discuss.

Process and facilities planning

Decisions have to be made in relation to location of the factory and the design and layout of production facilities. The design of production processes is interactive with product design, requiring close cooperation with research and product development functions.

Selecting the process of production is important and is strategic in nature. This means that it has a wide impact on the operation of the entire business. Decisions in this area bind the company to particular kinds of equipment and labour force because of the large capital investment that has to be made in order to make changes. For example, a motor manufacturer has to commit very large expenditures to lay down plant for production lines to mass-produce cars. Once in production the company is committed to the technology and the capacity created for a long time into the future. There are three basic methods for production processes:

- line flow
- intermittent flow
- project-based.

UNIT 7

Line flow is the type of system used in the motor industry for assembly lines for cars. It also includes continuous type production of the kind that exists in the chemicals and food industries. Both kinds of line flow are characterized by linear sequences of operations and continuous flows and tend to be highly automated and highly standardized.

Intermittent flow is the typical *batch production* or *job shop production*, which uses general-purpose equipment and highly skilled labour. This system is more flexible than line flow, but is much less efficient. It is most appropriate when a company is producing small numbers of non-standard products, perhaps to a customer's specification.

Finally, *project-based production* (or *one-off production*) is used for unique products that may be produced one at a time. Strictly speaking there is not a flow of products, but instead there is a sequence of operations on the product that has to be planned and controlled. This system of production is used for prototype production and in some engineering companies that produce major machine tool equipment for other companies to use in their factories.

TYK *7.7*

Explain the following types of engineering production:
(a) continuous flow
(b) intermittent or batch flow
(c) project-based.

TYK *7.8*

Which form of production, and why, is most appropriate to each of the following engineering processes:
(a) construction of a bridge
(b) manufacture of copper wire
(c) production of military aircraft
(d) manufacture of music CDs
(e) production of blank memory cards.

Capacity planning

Once facilities for production have been put in place the next step is to decide how to flex the capacity to meet predicted demand. Production managers will use a variety of ways to achieve this from maintaining excess capacity to making customers queue or wait for goods to having stocks to deal with excess demand. The process is complex and may require the use of forecasting techniques, together with careful planning.

Scheduling activities are different for each process method and require the use of a variety of techniques. The objectives of good scheduling are:

- meeting customer delivery dates
- correct loading of facilities
- planning starting times
- ensuring jobs are completed on time.

Later we will look at scheduling in relation to the management of engineering projects.

TYK 7.9

List four outcomes of good scheduling.

Inventory control

With any manufacturing facility good inventory control is an absolute essential. It is estimated that it costs up to 25% of the cost value of stock items per year to maintain an item in stock. Proper control systems have to be used to ensure that there is sufficient stock for production while at the same time ensuring that too much stock is not held. If stock levels are high there are costs associated with damage, breakage, pilferage and storage that can be avoided.

Workforce management

Workforce management is required in order to ensure that there is a suitably trained and experienced workforce that can apply the engineering processes, tools and facilities.

The important aspects of workforce management are:

- work and method study
- work measurement
- job design
- health and safety.

KEY POINT

Effective planning, allocation and supervision of work can considerably enhance the productivity of an engineering company

The production manager has to establish standards of performance for work so that the capacity of the factory can be determined and so that the labour costs of products can be calculated. Work study, method study and work measurement activities enable this to be done, as well as help to promote efficient and safe methods of working. The design of jobs is important in respect of worker health as well as effective work. Good job design can also make the work more interesting and it improves employee job satisfaction, which in turn can improve productivity.

UNIT 7

TYK 7.10

1. Explain the purpose of the following production functions:
   (a) manufacturing/production engineering
   (b) capacity planning
   (c) inventory control
   (d) quality control.

2. State and briefly describe the three methods of process design.

3. Name three techniques used in workforce management.

KEY POINT

Quality is generally defined as 'fitness for purpose'. In effect this means meeting the identified needs of customers. Thus it is really the customer that determines whether or not a company has produced a quality product, since it is the customer who makes an assessment of the perceived value of a product and is then either satisfied or dissatisfied

---

Quality control

Quality is a key objective for most engineering companies. It is especially important to the production function that is actually manufacturing the product for the customer.

Quality is generally defined as *fitness for purpose*. In effect this means meeting the identified needs of customers. Thus it is really the customer that determines whether or not a company has produced a quality product, since it is the customer who makes an assessment of the perceived value of a product and is then either satisfied or dissatisfied.

This does bring problems for manufacturers since customer perceptions of quality vary. As a consequence some customers will be more satisfied with a product than other customers (the ultimate test is probably whether a customer will make a repeat purchase or decide to purchase from another supplier). Because of the subjective nature of quality assessment, manufacturers often attempt to use more objective criteria for assessing fitness for purpose. This often includes:

- design quality
- conformance quality
- reliability
- service/maintenance.

Design quality is usually the joint responsibility of a company's marketing or customer liaison function and its R&D function. Design quality relates to the development of a specification for the product that meets a customer's identified needs.

Conformance quality means producing a product that conforms to the design specification. A product that conforms is a quality product, even if the design itself is for a cheap product. That may seem contradictory, but consider the following example. A design is drawn up for a budget camera, which is made from inexpensive materials and has limited capability. If the manufacture conforms to the specification then the product is of high quality, even though the design is of low quality compared with other more up-market cameras.

Reliability includes things like continuity of use measured by things like *mean time between failure (MTBF)*. Thus a product will operate for a specified time, on average, before it fails. It should also be maintainable when it does fail, either because it can easily and cheaply be replaced or because repair is fast and easy.

When an engineered product fails or becomes unreliable, it will require *service*. Service relates to after-sales service, the realization of guarantees and warranties, as well as the need for ongoing maintenance in order to ensure that performance remains within specification.

Quality control is concerned with administering *all* of these aspects. In the UK there are general standards for *quality systems*, the most relevant one here is BS 5750 and the international counterpart ISO 9000. The activities that make up a quality control system include the following:

- inspection, testing and checking of incoming materials and components
- inspection, testing and checking of the company's own products
- administering any supplier quality assurance systems
- dealing with complaints and warranty failures
- building quality into the manufacturing process.

Whilst many of these activities are performed in order to monitor quality *after* the event, others may be carried out to prevent problems *before* they occur and some may be carried out to determine causes of failure that relate to design rather than manufacturing faults.

KEY POINT

Engineering firms use quality control systems to ensure that the products and services that they provide conform to specification and fully meet the needs of their customers

TYK *7.11*

Explain what is meant by each of the following terms:
(a) design quality
(b) conformance quality
(c) reliability.

TYK *7.12*

List four activities that are performed within the context of quality control.

UNIT 7

Activity 7.4 Skylane Aircraft Manufacturing (case study)

Skylane Aircraft Manufacturing (SAM) builds a range of light aircraft. SAM is about to manufacture a new twin-engine light aircraft with customers being given the choice of seating configuration and type of engine to be used. The company is based at a UK regional airport in the East Midlands and it employs 40 production staff. A successful prototype has been produced and full production is expected to start with two aircraft a month increasing to one aircraft per week after 6 months.

SAM has established a market for the aircraft in the UK and in Europe where the aircraft is especially suited to leisure flying and is highly competitive with light aircraft imported from the USA.

SAM's Managing Director has asked you to advise the company on the following issues:

1. The most appropriate production process.
2. The advantages of production scheduling.
3. Quality control of incoming materials and components.
4. The reasons for introducing inventory control.

Present your answer in the form of a set of overhead projector transparencies accompanied by brief word-processed notes to be used at a presentation to SAM's board of directors.

Purchasing and supply

In large businesses purchasing is done by professional buyers, and is therefore a centralized activity. When a company has a large purchasing budget this makes economic sense, since the large purchasing power gives advantages in negotiating for keen prices, better delivery times or increased quality. In small businesses purchasing function is not centralized, usually because the operation is not large enough to support the employment of specialists. However, the basic principles of purchasing are the same, whatever the structure of the engineering company.

The main purchasing and supply functions are:

- researching sources of supply
- making enquiries and receiving quotations
- negotiating terms and delivery times
- placing contracts and orders
- expediting delivery
- monitoring quality and delivery performance.

The basic documents used for purchasing and supply are:

- requisitions from departments to buyer
- enquiry forms or letters to suppliers
- quotation document in reply to enquiry
- order or contract to buy
- advice note – sent in advance of goods
- invoice – a bill for goods sent to the buyer
- debit note – additional/further charge to invoice
- credit note – reduction in charge to invoice
- consignment note – accompanies goods for international haulage, containing full details of goods, consignee, consignor, carrier and other details
- delivery note – to accompany goods.

Purchasing procedures

Purchasing procedures involve raising a *purchase requisition* (i.e., a requisition to buy something). This may then require obtaining

quotations or estimates in order to choose the best supplier. Orders are sent to the chosen supplier. Goods are dispatched by the supplier, together with a delivery or advice note. When goods are received they are checked to see that the details on the delivery note agree with the actual goods received and that the goods have in fact been ordered. Goods not ordered may be refused. Accepted deliveries are signed for on the supplier's copy and given back to the driver. A goods received note is raised and sent to the purchasing department, so that the accounting function can be given confirmation of delivery before making payment against receipt of invoice.

These procedures are for the purpose of making sure that the goods are delivered on time, in the correct quantities, of the correct specifications and of the desired quality. Only if all are well is payment authorized.

A typical large, centralized purchasing engineering company would be as shown in the engineering company chart in *Figure 7.5*. It is common for there to be a purchasing manager and buyers who will specialize in particular types of commodity. Sometimes the products being purchased require that the buyer is technically qualified.

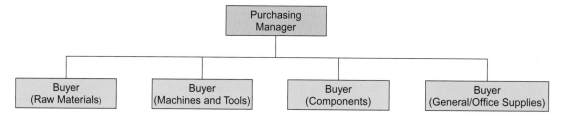

Figure 7.5 A centralized purchasing function

TYK *7.13*

1. Outline the basic documents for purchasing goods and the sequence in which they are used.

2. List the main functions of a purchasing department.

Organizational types and structures

There are many different types of engineering company, and many different organizational structure exist within them. Companies may be categorized in terms of size (e.g., micro, small, medium or large) as well as their status (e.g., sole trader, partnership, public (plc), private (Ltd), new, established, charitable and 'not for profit'). Most of these terms are self-explanatory and you should already be familiar with them.

It should be noted that there is some variation in the definition of the size of engineering companies. For example, the term *small and medium enterprise (SME)* could apply to a business enterprise of as few as 100 employees in Belgium or as many as 500 employees in Germany. However, within the

EU, and within the UK generally, the most commonly used definition of a *small enterprise* is an organization with fewer than 50 employees whilst a *medium enterprise* is taken to mean more than 50 employees and fewer than 250 employees. A *micro-business*, on the other hand, has less than 10 employees whilst a *large enterprise* has more than 500 employees. Different definitions of these terms exist in other countries, notably the USA. Finally, it is well worth noting that, in most economies, smaller enterprises predominate. In the EU, for example, SMEs comprise approximately 99% of all firms and employ between them about 65 million people.

Understanding how a particular engineering company operates is made easier by referring to its organizational structure. There are various ways of describing these structures and we will consider these one at a time. Note that the descriptions are not necessarily mutually exclusive, thus one description may also overlap with other descriptions.

Flat structures

Flat structures have few levels from top to the bottom. *Figure 7.6* shows a flat organizational structure in which six managers report directly to a *board of directors*. The six managers are described as *first line managers* because they directly supervise the activities of people who actually do the job. Such people are often described as *operatives*.

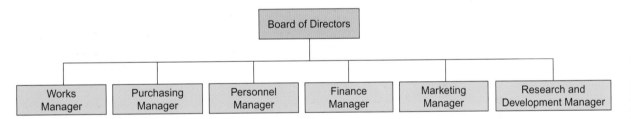

Figure 7.6 A flat company structure

In the 1990s the UK saw a significant move towards having flat organizations. The process has been described as *de-layering* and it is regarded as an organization structure that permits better communications both up and down the organizational levels. It is also seen as more cost-effective and responsive in dealing with demands placed on modern businesses.

Note that flat organizations are also hierarchical. A hierarchy is an organization with grades or classes ranked one above another. A flat organization also meets that criterion.

Tall structures

These are opposite to flat structures and contain many layers. *Figure 7.7* shows an example. Tall structures usually exist only in large organizations because of the necessity of dividing the tasks into chunks of work that can be handled by individual managers, departments and sections.

Tall structures are also hierarchies, only they contain many more levels than flat structures. These are structures with many or few layers showing

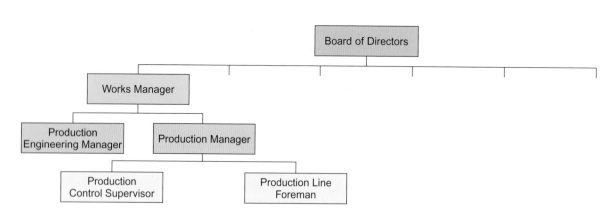

Figure 7.7 A hierarchical structure

rank order from top to bottom. They show a chain of command, with the most senior posts at the top and the most junior posts at the bottom. Plain hierarchies are the most common representations of organizations, as some aspect of hierarchy exists in all organizations.

An *organization chart* is a useful way of representing the overall structure, but it tells only part of the story. You should be aware that other documents are needed to fully understand how the organization works, as we observed earlier.

Hierarchical organizations can take many forms. We have already examined flat and tall structures. There are several other forms with which you should be familiar. One of the most common is the functional design. *Figure 7.8* shows a functional organization. The main functions of a commercial business are marketing, finance, purchasing and supply, manufacturing, R&D and personnel. Notice how each functional manager reports to the managing director who coordinates their activities. There are a number of advantages in functional structures:

- specialists can be grouped together
- it appears logical and easy to understand
- coordination is achieved via operating procedures
- suits stable environments and few product lines
- works well in small to medium size businesses.

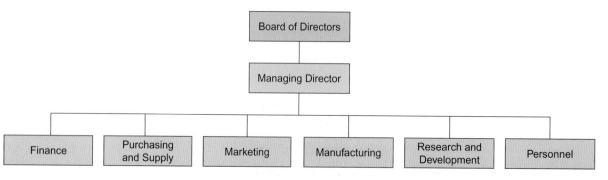

Figure 7.8 A functional structure

As a business grows the functional structure becomes less and less useful. This is because there are many more products and these may be manufactured in separate divisions of a company, especially if economies of scale are introduced into the manufacturing process.

Figure 7.9 shows an organization based on major product lines and is really a federal structure, which still has functional activities, but at a lower level in the organization, except for the R&D function, which is centralized. The managers of these operating divisions will control most of the functions required to run the business. In many conglomerate businesses this federal arrangement is achieved by having a holding company, which may be a *public limited company* (plc), which wholly owns a number of subsidiary companies, which are in effect divisions of the main business. *Figure 7.10* shows an example.

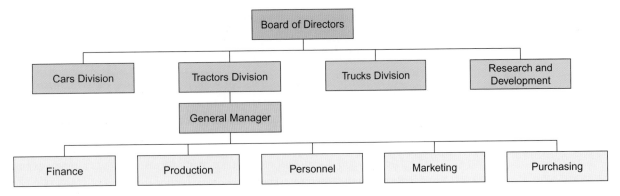

Figure 7.9 A federal or divisional structure

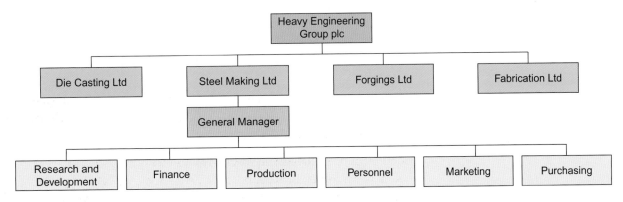

Figure 7.10 A conglomerate organization

An alternative to the product-based divisional design is one based on geographical divisions. *Figure 7.11* shows a geographical design, which still has many of the functions located at a head office, but which has branches dispersed around the country. These branches or divisions handle sales and manufacture, but are supported by head office for the other functions.

Matrix structures

A matrix structure, as its name implies, has a two-way flow of authority. Departmental or functional authority flows vertically and project

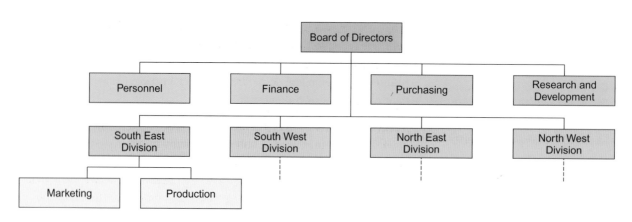

Figure 7.11 Geographical divisions

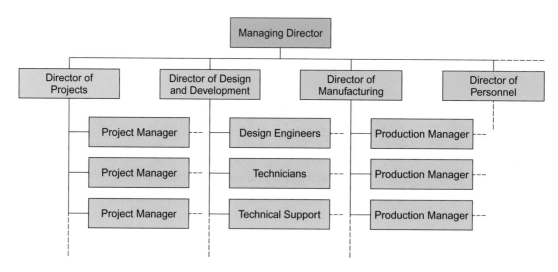

Figure 7.12 A matrix structure

management authority flows horizontally. Such a structure is shown in *Figure 7.12*. This depicts a company that designs and makes machines and tools to customer specifications.

Some engineering businesses operate as a series of projects. This is common, for example, in construction and in some types of engineering company, especially those that design one-off products or who design manufacturing equipment used by other manufacturers. As the chart shows, there are familiar functional divisions, and members of functional departments still report to their functional line manager. However, the project manager has the job of coordinating the work of functional specialists to ensure that the project is completed on time, to specification and within cost limits.

In the main, project managers do not have direct line authority, but have to influence and persuade others to achieve targets. They may have formal authority over project budgets and can set time schedules and decide what work is to be done, but little else. They also work as an interface with clients and subcontractors and their influence is often critical to the success of the

UNIT 7

project. Although they do not have formal authority over individual staff or their line managers, they nevertheless operate with the full support of senior managers. This means that functional specialists are obliged to provide the fullest cooperation and help otherwise they become answerable for failure to their own senior line managers.

The matrix system works very well in project-based industries, and that is why the design is used. It still retains many of the ingredients of other structures, and still has substantial hierarchical elements.

Hybrid structures

The structures discussed above are just examples of the main design principles for organizations. There are numerous variations and rarely do we find 'pure' forms of organization structure. We need to remember that organizations are created to serve the goals of their owners and that the precise structure will be designed to meet the needs of the business.

TYK 7.14

1. Explain the justification for (a) flat structures and (b) tall structures.

2. Explain why federal or division structures are used.

3. Explain the reason for matrix structures.

4. What is meant by a hierarchical structure?

TYK 7.15

Describe the role of a project manager in a large engineering company. Explain why this role is likely to be more effective in a matrix structure rather than in a tall hierarchical structure.

Activity 7.5

Draw a chart showing the organizational structure of your school or college. Use a CAD or drawing package to produce the chart and include departmental functions, job titles and names. Comment on the type of structure and suggest why it is appropriate in the case of your school or college.

Interfaces and information flow

The various functions within an engineering company, as well as individual teams within them, need to work together in order to function effectively and efficiently. This interaction might involve sharing information and making decisions about:

- processes and systems
- working procedures
- the people involved, including customers, suppliers and other employees.

You must be able to identify and describe the working relationships or interfaces that exist between departments within an engineering company.

You also need to be able to recognize types of information and the data that flows between the various departments within an engineering company. These can include:

(a) documents such as:
- design specifications
- purchase orders
- invoices
- production schedules
- quotations

(b) information about:
- stock levels
- work in progress
- resource utilization sales.

Note that all of the above may exist either as hard copy or may be in electronic form (such as a word-processed document, a hypertext document or a spreadsheet file).

Finance and accounting

The finance and accounting function interfaces with all other functions within an engineering company. Its recording and monitoring activities are central to, and have a major impact on, the whole business. In conjunction with the manufacturing function, *sales forecasts* will be used to prepare *production schedules*. Production may be sent to a warehouse and then put into the delivery and distribution system. From there the sales force will ensure that customers receive their orders when required. Alternatively delivery of specific customer orders may be made directly to customers.

Marketing

Marketing will supply information on prices. Prices may be determined primarily by the market rather than the cost of manufacture. Finance may provide cost information, but marketing may make the final pricing decisions.

Marketing will also identify customer needs and will liaise with product development activities on possible new products or modifications to existing products. R&D will initiate design studies and prototypes for new products and may supply some items for market testing. Engineers involved with design and development will be given information on customer needs and preferences and will be expected to produce designs, which meet those requirements.

There will be a need to communicate details of the costs of new products or redesigned products. The processes required to produce new components or whole new products will also require costing. R&D may specify the manufacturing process, but manufacturing engineering departments located at the production facility will implement them, and may also share in the costing process.

UNIT 7

Manufacturing and production

New products will have different characteristics, and perhaps be made from different materials when compared with previous products with similar functionality. This will require liaison between design engineers and manufacturing engineering on methods for production and in deciding what manufacturing equipment and machine tools are required. Detailed *process sheets* or *work instructions* may be required, which show how products are to be assembled or made.

Diagrams are extensively used as a means of communication in a manufacturing or production environment. These may include exploded diagrams used for assembly or maintenance, circuit diagrams used for electrical/electronic equipment, piping and hydraulic diagrams used for hydraulic equipment, and so on. Examples of these can be found in Unit 2.

Whilst the particular methods of production are the province of production management, the designer has to be aware of the implications for his design of different methods of manufacture, whether this be batch production, assembly lines or one-off projects. Detailed specifications of the new and changed product will be communicated and there may liaison on temporary and permanent deviations to original specifications in order to facilitate production.

When quality problems appear and are related to faulty design there will be liaison on ways in which design modifications can be phased into production as soon as possible.

There will be proposals for the replacement of machines and equipment used for manufacturing and production. This function may require quite sophisticated techniques for what is called 'investment appraisal' so that the company can choose the best methods of manufacture from several alternatives.

Also important is the control of raw materials and component stocks, especially the levels of *work-in-progress*. Finance will want to restrict stock levels to reduce the amount of capital tied up in stocks, whilst the production manager will be concerned with having sufficient stock to maintain production, but avoiding congestion of factory floor space.

Budgetary control of production cost centres will involve regular contact and advice from the finance function. Matters of interest will be costs of production, wastage rates, labour costs, obsolescent stock, pilferage, etc.

Purchasing

Specifications and drawings will be sent to the buyer for new products or machines for purchase. Problems of design and delivery will be discussed, and modifications to designs will be sent to suppliers through the buying department.

New product launches will be coordinated with R&D, the supplier, and of course, the manufacturing department. The buyer would be involved with supplies of new raw materials, new designs for components and will negotiate costs of tooling and long-term contracts.

Effective liaison is required with factory management on initial supplies for new materials or components. Buyers can usually assist production managers with quality and inspection problems and in dealing with return and replacement of defective materials. Sometimes buyers may be an interface with production and R&D in dealing with temporary or permanent deviations from the original engineering specifications. Chasing deliveries and ensuring supplies for factory use may be a major daily routine for some buying departments.

Systems for quality control may include some form of supplier quality assurance. The buying department will represent company interests to suppliers and may only use suppliers who have passed the company's quality assurance standards. The buyer will be involved with searches for new suppliers who can meet existing and new quality requirements.

Stock control systems used within the factory will affect the way purchasing is done. Economic order quantities may be established, which the buyer has to take into account when arranging supplies. Deliveries may have to be phased according to minimum, maximum and reorder stock levels. The buyer will need a clear understanding of the importance of deliveries, which enable the company to control its inventory costs, while at the same time ensuring a reliable supply of materials and components for production.

The company may operate a *just-in-time* system (*JIT*). JIT originated in Japan and it is a way of delivering supplies at the point in time they are required by production. JIT avoids the costs of holding buffer stocks of raw materials and components. It works well when suppliers are dependable and when transport systems are good. The buyer will liaise with the factory on the establishment and operation of the JIT for given products.

There will also be the routine matters of passing invoices for payment of goods or dealing with returns for credit so that the accounts department can pay for goods received. Materials purchasing will be subject to budgetary constraints like most other company activities. The purchasing department will be involved, either directly or indirectly in budgets for inventory levels, and in setting up minimum, maximum and reorder levels for stocks.

Monthly monitoring of inventory levels will be done by the accounting function and purchasing activities may be responsible for ensuring that stock of components and raw materials stay within agreed levels.

TYK 7.16

Explain the role of a buyer in an engineering company.

Activity 7.6   AccessTek (case study)

AccessTek is an engineering company based in Crawley, Sussex. The company manufactures entry control systems used in secure areas, such as airports, hospitals and defence establishments. The AccessTek system consists of a number

UNIT 7

of ruggedized keypads and card readers, which work together with a central control system based on a dedicated microcomputer system. The central station checks and validates entry data before opening remotely activated door locking mechanisms. The central station contains a database of authorized personnel and the system can be extended to permit iris and voice recognition in more secure applications.

AccessTek employs a team of seven field engineers who work together with the design and production departments to provide customers with individually tailored systems. The engineers also provide routine field service and maintenance. In addition to the team of field engineers, the engineering department employs a trainer who becomes involved with the final stages of installation and commissioning and who provides any necessary training for customers' personnel in the use of a system and in its administration. AccessTek provides 24/7 telephone support and guarantees a response time of less than 10 hours when field visits become necessary.

With the aid of a labelled diagram, identify the main communication links and information flows that should exist between the functions within AccessTek (marketing, design and development, production, and engineering) and with customers. For each link, give typical examples of the type of information that is transferred between the departments and also list ways in which the information can be transferred (e.g., e-mail, telephone, meeting, letter, etc.).

External Factors and the Economic Environment

External factors affect the operation of all engineering companies. Some factors may be foreseen in which case they are predictable (and can be planned for) but others are seemingly random occurrences. Some external factors can be regarded as *threats* whilst others can be regarded as *opportunities*. For example, the availability of a new manufacturing process that significantly reduces costs can be considered to be an opportunity. Falling consumer demand resulting from high interest rates can be considered a threat. Perhaps somewhat confusingly, some external factors can be considered to be either a threat or an opportunity depending upon whether or not a company is ready and able to respond!

External factors that you need to be aware of include:

- markets and the general state of the economy
- consumer demand
- demographic and social trends
- competitive products and services
- consumer confidence and customer/client relationships
- innovation and technological change.

It is worth noting that, to some extent, all of these factors are present to a varying degree all of the time.

Figure 7.13 shows factors that affect a typical engineering firm. Some factors are considered to be internal (and part of the task environment) whilst others are external (and considered to be part of the wider

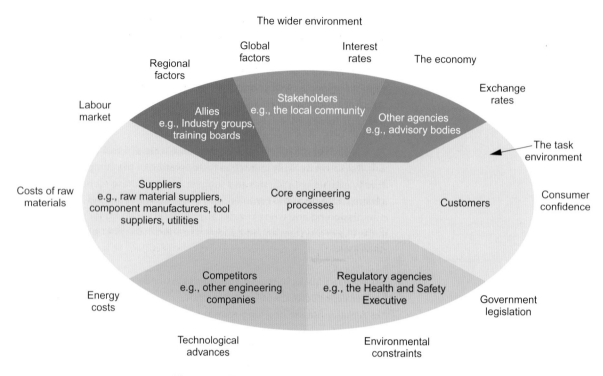

The wider environment

Global factors — Interest rates — The economy
Regional factors — Exchange rates
Labour market — Allies e.g., Industry groups, training boards — Stakeholders e.g., the local community — Other agencies e.g., advisory bodies — The task environment
Costs of raw materials — Suppliers e.g., raw material suppliers, component manufacturers, tool suppliers, utilities — Core engineering processes — Customers — Consumer confidence
Energy costs — Competitors e.g., other engineering companies — Regulatory agencies e.g., the Health and Safety Executive — Government legislation
Technological advances — Environmental constraints

Figure 7.13 Factors that affect a typical engineering firm. Note that the entire 'SIPOC' diagram (Figure 7.1) fits into the core of this diagram

KEY POINT

External factors affect the operation of all engineering companies. Some of these factors can be predicted and planned for whilst others cannot. Some factors can be considered to be opportunities and some can be considered to be threats

environment). Internal factors are relatively easy to influence. External factors, on the other hand, may be difficult if not impossible to influence. Note that *Figure 7.1* (shown on page 562) fits inside this diagram with the SIPOC stages fitting between 'Suppliers' and 'Customers' in the region that we have marked 'Core engineering processes'. Because it has far-reaching implications, it is important that you understand this diagram!

TYK *7.17*

Classify each of the following as a threat or an opportunity for a small engineering company:
● availability of low-cost business development loans
● interest rates climb to an all-time high
● high salaries offered by a major competitor
● a local college start a new course in advanced manufacturing technology.

TYK *7.18*

A change of Government may have consequences for the economy and in turn this may affect interest rates. Explain the likely consequences of (a) low and (b) high interest rates on the operation of a small engineering company.

UNIT 7

Activity 7.7   Digiplex (case study)

Digiplex, a small high-tech engineering company based in Kent, manufactures a range of colour photocopiers. The company has a customer base in the South East of England as well as northern France and Belgium. In common with other suppliers, Digiplex offers a standard warranty and an extended service contract of a 6-hour 'fix or replace' response time when one of its copiers breaks down.

Assess each of the following scenarios on a scale of 1 to 5 (1 = negligible and 5 = major) on its likely impact on the operation of the company. In each case briefly explain your answer and suggest which of these scenarios represent a potential threat to the company and which can be regarded as an opportunity.

Present your work in the form of a brief word-processed report to the company's chairman.

(a) An earthquake in the Far East results in the destruction of a major semiconductor manufacturing plant.
(b) A major fire causing long-term suspension of all services through the Channel Tunnel.
(c) Funded by a substantial European Community grant, a French manufacturer sets up a colour photocopier production plant in Calais.
(d) A substantial fall in price of desktop colour laser printers.
(e) A Japanese company patents a new low-cost, high-quality colour printing process.
(f) A series of strikes by workers in crosschannel ferry companies brings chaos to the channel ports.
(g) A progressive fall in interest rates results in an exceptionally large number of business start-ups in the South East of England.
(h) A major high-street chain of newsagents decides to provide low-cost in-store colour photocopying.
(i) Well known for its high salaries and excellent benefits packages, a North American electronic equipment manufacturer sets up its European headquarters on the same industrial estate.
(j) New European environmental legislation makes manufacturers of electronic equipment responsible for its disposal and/or recycling.

The economic environment

The economic environment is crucial in determining the financial context in which an engineering company operates. The profitability of an engineering company often goes hand-in-hand with the state of the economy as a whole but this is not always the case.

In order to understand the more important economic issues it is necessary to have a grasp of some basic concepts and terminology including *gross national product (GNP)*, *gross domestic product (GDP)* and *balance of payments*.

Figure 7.14 shows how the production of goods and services and their consumption can be regarded as forming a closed circle. The firms make the goods and provide the services consumed by the people living in the households. The people living in the households are the same ones that own and work in the production and service firms. The households create

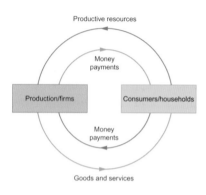

Figure 7.14 Circular flow of goods, services, resources and money

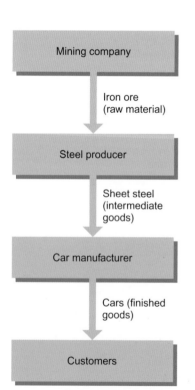

Figure 7.15 Intermediate and finished goods in a car production process

the demand for the goods and services. This, in turn, is satisfied by the production firms that pay the wages of the householders. Hence, the flow of goods, services and production resources are in one direction and this flow is balanced by the flow of money in the other direction.

It is worth noting that many goods and materials used in engineering processes are not consumed for their own sake but are actually used in the production of other goods. They are known as *intermediate goods*. For example, sheet steel is used in the production of cars, wooden planks in the construction of buildings, copper wire in the manufacture of transformers, and so on. You may now be wondering how this all works in the knowledge that GDP only accounts for goods that are in their finished state. What about intermediate goods?

To explain this, consider the use of sheet steel in the production of cars (see *Figure 7.15*). Here, the output of the steel producer is sheet steel and this is regarded as intermediate goods in the car production process. The steel producer purchases iron ore (a raw material) from a mining company and uses it to produce sheet steel. This is then sold to the car manufacturer who produces the finished product. Only the final sale of cars is included in the calculation of GDP.

An alternative but equivalent way of measuring GDP is to add up the value added at each stage of the production process. The value added by the steel producer is the difference between the value of output (sheet steel) and input (iron ore). The value added by the car manufacturer is the difference between the value of output (finished cars) and input (sheet steel and other materials).

It is important to note that there will often be several stages in the process of intermediate goods production but the sum of value added at each stage of production must always be equal to the value of the final output.

Gross national product

The total output of the UK can be measured by adding together the value of all goods and services produced by the UK. This figure is called the GNP. The word 'gross' implies that no deduction has been made for the loss in value of the country's capital equipment, which helps to make the national product, caused by normal wear and tear.

The word 'national' in this context does not mean that the GNP is the total output produced within the borders of the UK. You see, the gross product of the UK includes some output taking place and produced by resources within the UK but owned by people from other countries. Therefore, this particular element of output cannot be regarded as part of the national income. At the same time, some sources of output are located in other countries but owned by UK citizens. Therefore the GNP is defined as the value of the total output of all resources owned by citizens of the UK wherever the resources themselves may be situated.

Gross domestic product

If we measure the value of the total output actually produced within the borders of the UK, it is called the gross domestic product (GDP). So, the GDP is defined as the value of the output of all resources situated within the

UNIT 7

KEY POINT

GDP is the market value of everything produced within a country whereas GNP is the value of what's produced by the residents of a country even if they live abroad.

Gross Domestic Product (GDP)

less

Income payable to overseas divisions

plus

Income receivable from the rest of the world

equals

Gross National Income (GNI)

Figure 7.16 Relationship between GDP and GNI

UK wherever the owners of the resources happen to live. In many ways the GDP is a more important measure than the GNP.

The UK Government issues detailed statistics that indicate the performance of the different sectors of the economy. You can obtain more information from the National Statistics website (StatBase) at www.ons.gov.uk.

Gross National Income

Gross National Income (GNI) was traditionally called GNP, even though it was a measure of income rather than output (or production). The term GNP is no longer used within the official definition of economic accounts.

GNI is a measure of the income to the UK from production wherever in the world it occurs. For example, if a British-owned company operating in North America sends some of its profits back to the UK this *adds to* the UK's GNI. Similarly, a British subsidiary of a Japanese company sending profits to Japan will *reduce* the UK's GNI.

Strictly, GNI is GDP *plus* the net income from the rest of the world, see *Figure 7.16*.

Activity 7.8

The UK's GDP for the 11-year period from 1994 to 2004 is shown in *Table 7.1*. Plot this data as a line graph. Also determine:

(a) the percentage increase in GDP over the 10-year period from 1995 to 2004
(b) the average annual percentage increase in GDP over the 10-year period from 1995 to 2004
(c) the year in which the actual increase in GDP was least
(d) the year in which the actual increase in GDP was greatest.

Present your results in a brief written report. Hint: You may find a spreadsheet useful for determining annual changes and also for plotting graphs.

Table 7.1 UK GDP (at current basic prices) for the 11 years from 1994 to 2004

Year	GDP (£ million)
1994	680,978
1995	719,747
1996	765,152
1997	811,194
1998	860,796
1999	906,567
2000	953,227
2001	996,987
2002	1,048,767
2003	1,176,527
2004	1,225,339

Activity 7.9

Dragon Data Controls (DDC) is the UK subsidiary of a Japanese machine tool manufacturer. In 1 year, DDC receives a total income of £25 million from the sale of CNC equipment. Of this, £8 million worth of income comes from sales to other European countries. DDC returns £5 million of profit to its Japanese manufacturer. The company imports £7 million of components from Japan and its UK costs (employment and overheads) amounts to £13 million.

Explain the contribution that DDC makes to the UK's GNP and illustrate your answer using a pie chart. Present your work in the form of a brief word-processed report.

Gross domestic product per head

GDP per head is a measure of *productivity*. It relates output to the number of people employed producing that output. The formula used is:

$$\text{Output per head (or per capita)} = \text{output produced divided by the number of people producing it}$$

Put another way, we can write,

$$\text{Output per head} = \frac{\text{output}}{\text{employment}}$$

Often we are concerned not with absolute figures for output and employment but more with trends. For this reason, both the output and employment figures are usually quoted as *index* figures (i.e., they are stated relative to a base figure of 100 for a particular year). *Table 7.2* shows how the UK output per head and cost of labour changed over the 11-year period from 1990 to 2000.

Table 7.2 UK productivity and labour costs per unit for the period 1990 to 2000

Year	Output (manufacturing jobs)	Output (whole economy)	Unit labour costs (whole economy)
1990	83.7	86.7	90.5
1991	86.2	88.4	96.8
1992	91.7	91.3	99.1
1993	96.2	94.5	99.5
1994	100.5	98.2	98.4
1995	100.0	100.0	100.0
1996	99.4	101.6	101.8
1997	100.3	103.1	104.8
1998	101.2	104.6	108.2
1999	105.2	105.8	112.3
2000	111.3	108.2	114.8

KEY POINT

The circular flow of income and expenditure leads us to conclude that there are two ways of measuring this flow: by adding up all of the income or by adding up all of the expenditure. According to the rules of accounting these two must be equal!

Regional variations

There are significant variations in the economic performance of the various UK regions. *Table 7.3* shows how the GDP varies by UK region in 1996 and 1999. Given the different sizes and populations of the regions this table is not particularly meaningful in terms of the contribution of each person who lives and works in that region. To address this problem, *Table 7.4* shows how the GDP varies on a *per capita* (i.e., 'per head') basis.

Table 7.3 Regional GDP in the UK for 1996 and 1999

Region	GDP (£ million)	
	1996	1999
United Kingdom	674,029	787,386
North East	23,755	25,875
North West	68,937	77,562
Yorkshire and the Humber	50,043	57,554
East Midlands	44,184	50,906
West Midlands	54,851	63,495
East	66,484	81,793
London	99,490	122,816
South East	100,614	121,956
South West	50,128	58,151
England	558,483	660,108
Wales	27,017	30,689
Scotland	57,338	64,050
Northern Ireland	14,936	17,003

Table 7.4 Regional variation of GDP per capita in the UK for 1996 and 1999

Region	Per capita GDP (UK = 100)	
	1996	1999
North East	81.5	77.3
North West	89.2	86.9
Yorkshire and the Humber	88.8	87.9
East Midlands	95.4	93.6
West Midlands	92.2	91.7
East	112.5	116.4
London	126.1	130.0
South East	114.1	116.4
South West	92.5	90.8
Wales	82.6	80.5
Scotland	99.8	96.5
Northern Ireland	80.1	77.5

Activity 7.10

Present the data shown in Table 7.4 in the form of two bar charts in descending order of per capita GDP. Use the data to determine:

(a) the best and worst performing regions in each of the two years
(b) the region that has shown the greatest increase in per capita GDP over the 4-year period
(c) the region that has shown the greatest decline in per capita GDP over the 4-year period

Present your results in a brief written report. Hint: You may find a spreadsheet useful for determining the 4-year changes and also for plotting graphs.

Human Development Index

GDP (or *GDP per capita*) is often taken as an indicator of how developed a country is but its usefulness is somewhat limited as it only refers to economic welfare. In an attempt to put this right, the Human Development Index (HDI) combines indicators of adult literacy, average years of schooling and life expectancy together with income levels to give a better, though still not perfect, indicator of human development. For comparative purposes HDI is scaled to provide an index that ranges from 0 to 100; countries scoring more than 80 points are considered to have a high degree of human development whilst those scoring between 50 and 79 are said to be medium and those below 50 are categorized as low. *Table 7.5* provides some indicative HDI figures together with per capita GDP for each country. It is worth spending a little time looking at these figures.

Table 7.5 Comparative HDI and GDP for selected countries (2004 data)

Country	HDI	GDP per capita (US$)
Norway	96.3	54,360
Australia	95.5	32,030
Ireland	94.6	45,410
United States	94.4	39,430
Japan	94.3	36,170
United Kingdom	93.9	35,760
France	93.8	33,890
Germany	93.0	33,220
Poland	85.8	6,280
Croatia	84.1	7,800
Russia	79.5	4,080
Romania	79.2	3,280
Malawi	40.4	150
Mozambique	39.4	320
Niger	28.1	250

Direct employment

This is the term used to describe the mode of employment of people that are actually working to produce a product. For example, being directly

UNIT 7

employed on the manufacture of the furniture, the people working in a factory making furniture could all be regarded a direct labour.

Indirect employment

Following on from the previous example, the people concerned with the transportation of the production materials to the furniture factory and the finished furniture to the shops which sell the furniture, are regarded as indirect labour and are in indirect employment.

Note, however, that people may be directly employed by their own trade but indirectly by another. For example, the people felling the trees and producing the raw timber used to make the furniture are directly employed by the timber trade but are indirectly employed by the furniture making trade.

Exports

Exports are goods and services that we sell to other countries. Visible exports are the hard goods that are physically transported abroad and for which we receive payment. Invisible exports are those services that we provide for foreigners and for which they pay. Invisible exports include the payments we receive for insurance and financial services, technical or military training and the like. Tourism is also an important part of our invisible export trade. A German tourist in London will spend money that he has earned in Germany. This provides a useful income to the UK no different from, say, visibly exporting and selling the German a bottle of Scotch whisky in Berlin. Note that when we take our holidays abroad, the goods and services we buy and consume abroad are in fact imports to the UK.

Imports

Imports flow in the opposite direction to that of exports. Imports are the goods and services that we purchase from abroad.

Balance of payments

For a nation to remain financially sound, it must ensure that its expenditure does not exceed its income. The difference between the total income we earn from abroad for our exports is balanced against the total payment we must make abroad for our imports. We compare or *balance* the two figures and hopefully the export figure is at least as large as and preferably larger than the import figure.

Very often the UK balance of visible exports and imports shows an excess of import value over exports. However, we have a relatively healthy sale of our services abroad and our *invisibles* (or *invisible earnings*) usually gives us a favourable trading balance. The Government lets us know what our overseas trading position is by issuing monthly balance of payment figures.

Test your knowledge

TYK *7.19*

Define the terms (a) per capita GDP and (b) HDI.

TYK 7.20

Explain why GDP per capita is not a particularly good indicator of quality of life.

TYK 7.21

Which of the following groups of engineering products make a positive contribution to the UK's balance of payments?
(a) Imports of coal for UK power stations.
(b) Sales of American cars in the UK.
(c) UK sales of British-made, Japanese cars in the UK.
(d) European sales of British-made, Japanese cars.
(e) British cars sold in America.
(f) UK citizens buying British cars.

TYK 7.22

What are *invisibles*? List three examples of invisibles that contribute towards a more healthy balance of payments for the UK.

Local economy

For the first half of the twentieth century, engineering was generally located within cities. Since then there has been a tendency for any new engineering enterprise to be located in an industrial estate on the periphery of a town rather in the town itself. Usually this is because:

- the town centre is already too congested to allow for additional new industry;
- of the advantages of being located in a ready, purpose-built industrial accommodation on a site having good road links with the national motorway network;
- engineering activities that may involve noise and other pollutants are best kept away from the commercial and domestic centres of towns.

In general, the engineering industries that remained in the city centres have slowly become outdated and, in many cases, have closed down. The impact of this migration from the city centres to the suburbs has been to leave derelict buildings, unemployment and social deprivation for the city residents. For the outer suburbs receiving the new engineering industries, the impact has not always been positive. The decentralization of engineering from the city centres has contributed to urban sprawl, and this has led to conflict for land on the city's boundary between engineering, farming and recreation. Also, it has tended merely to move the problem of engineering pollution from the city centre to its suburbs.

UNIT 7

Regional economy

The regional economy comprises many local economies but the change in the engineering pattern is much the same. While there is still a great deal of engineering activity to be found in and around many large cities and built-up areas, there is a definite migratory move towards the small town and rural areas. This trend is to be found in most economically developed countries and has been a consistent feature of the last 25 years.

The eight *Regional Development Agencies* (RDAs) were established under the Regional Development Agencies Act 1998, and were formally launched in eight English regions on 1 April 1999. The ninth, in London, was established in July 2000 following the establishment of the Greater London Authority (GLA). Responsibility for sponsorship of the RDAs moved from the former Department for the Environment, Transport & the Regions to the Department of Trade and Industry (DTI) in 2001.

The primary role of the RDAs is as strategic drivers of regional economic development in their region. The RDAs aim to coordinate regional economic development and regeneration enables the regions to improve their relative competitiveness and reduce the imbalance that exists within and between regions.

Under the Regional Development Agencies Act 1998, each Agency has five statutory purposes, which are:

- to further economic development and regeneration
- to promote business efficiency, investment and competitiveness
- to promote employment
- to enhance development and application of skill relevant to employment
- to contribute to sustainable development.

The RDAs' agenda includes regeneration, taking forward regional competitiveness, taking the lead on inward investment and, working with regional partners, ensuring the development of a skills action plan to ensure that skills training matches the needs of the labour market.

National economy

At the national level the uneven spread of engineering is between the different regions. The processes that caused this variation are historic. Very often they are directly connected to the availability of natural resources. For example, in the nineteenth century, regions rich in coal were favoured with engineering expansion because of the local availability of coal to fire boilers to drive the steam engines that powered the factories. The technical skills acquired by the workers in the coal bearing regions were the same skills required for other industries and enterprises and cumulative expansion took place. This expansion, and the highly paid work it created, attracted labour from the less industrialized regions so exacerbating the regional disparities.

However, over the last 50 years there has been a shift of engineering away from the old industrial regions such as the Midlands and North East of England and parts of Scotland to more convenient locations such as the Thames Valley along the M4 motorway and along the M11 motorway north of London. The reasons for the regional shift are many and varied and include such factors as:

- because of its cost and record of causing pollution, coal is no longer a popular fuel;

> **KEY POINT**
>
> The RDAs aim to improve the quality of life in the regions that they serve. They aim to help businesses compete, revitalize urban and rural areas, promote sustainability, improve regional infrastructure and facilitate innovation. They also ensure that skills training in the region is closely matched to the labour market requirements

- with natural gas and electrical power being available almost anywhere in the country, new engineering activities can be located in regions having pleasant natural and social environments;
- because of the ubiquitous motor car, good roads and frequent air services, commuter and business communication to most regions is no longer a major problem;
- the availability of a pool of technologically skilled labour in places where high-technology companies are clustered together.

Within the UK there are significant differences in weekly earnings in different regions. The official statistics for 2003 are shown in *Table 7.6* whilst the increase in average weekly earnings (for all earners) from 1999 to 2003 is shown in *Table 7.7*.

Table 7.6 Gross weekly earnings by Government office region 2003

Region	Gross weekly earnings (£)		
	Men	Women	All earners
East	528.5	382.7	475.9
East Midlands	467.3	357.4	428.7
Great Britain	525.0	396.0	475.8
London	716.5	516.5	636.9
North East	437.8	347.3	402.1
North West	483.1	367.9	437.6
Scotland	483.7	372.4	436.8
South East	560.9	415.7	505.6
South West	485.1	364.7	440.6
Wales	448.3	357.3	414.5
West Midlands	477.1	363.7	435.8
Yorkshire and the Humber	463.8	360.4	425.5

Table 7.7 Increase in average weekly earnings (£) (all earners) from 1999 to 2003

Region	1999	2003
East	396.6	475.9
East Midlands	361.7	428.7
England	405.4	–
London	520.0	636.9
North East	349.6	402.1
North West	372.6	437.6
Northern Ireland	344.9	–
Scotland	364.9	436.8
South East	423.2	505.6
South West	364.9	440.6
Wales	353.6	414.5
West Midlands	375.6	435.8
Yorkshire and the Humber	361.0	425.5

Activity 7.11

1. Photocopy the map shown in *Figure 7.17* (your tutor may be able to supply you with this) and use it to identify and shade in the following economic regions:
   (a) Yorkshire and the Humber
   (b) East Midlands
   (c) West Midlands
   (d) London
   (e) South East
   (f) South West.

Figure 7.17 Map of the UK (see Activity 7.11)

2. Mark the following motorways on the photocopied map:
    (a) M1
    (b) M3
    (c) M4
    (d) M6
    (e) M5
    (f) M40
    (g) M45.
3. Which motorway(s) link the following pairs of economic regions:
    (a) East Midlands and West Midlands
    (b) South East and South West
    (c) London and Yorkshire and the Humber
    (d) London and East Midlands
    (e) London and West Midlands.
4. A steel producer in Sheffield supplies a transformer manufacturer in Coventry with sheet steel. The finished transformers are transported to Southampton for shipping to Bilbao in Spain.
    (a) Explain the effect this has on the GDP of various regions in the UK and name the economic regions concerned.
    (b) Explain the effect this has on the UK balance of payments and say whether it is positive or negative.
    (c) State which motorway routes are used for the transport of raw materials and which are used for the shipping of finished goods.

Activity 7.12

Find out the name of your local RDA by visiting `www.englandsrdas.com`. Follow the link to visit the website of your local RDA and use it to locate information on the Regional Economic Strategy for your area. Write a short essay (not more than 1000 words) that summarizes the key points of this strategy.

Activity 7.13

Plot the data from Table 7.7 in the form of a bar chart in ascending order of average weekly earnings. Use the data to determine:

1. The region in which average weekly earnings are the greatest.
2. The region in which average weekly earnings are the least.
3. The average weekly earnings in the North East expressed as a percentage of those in the South East.
4. The average weekly earnings in London expressed as a percentage of those in Yorkshire and the Humber.
5. A large production plant is to be set up in the North of England. In which one of the three regions is the labour cost likely to be the least?

Present your work in the form of a set of word-processed notes and an accompanying bar chart produced using a spreadsheet.

European Union

The performance of 16 European economies is shown in *Table 7.8*. Note from this the large variation in the number of people in employment and the sectors in which they are employed. Within the EU engineering activities

Table 7.8 Performance of 16 European economies in 1999

	Persons in employment in 1999 (thousands)	Percentage in 1999 employed in:			Unemployment rate in 1999 (percentages)	Long-term unemployed as a percentage of the unemployed 1999	GDP per head (PPS) Europe = 100 1998	Estimates of the percentage of GDP in 1996 derived from:		
		Agriculture	Industry	Services				Agriculture	Industry	Services
Europe	152,494	4.7	29.6	65.5	9.4	49.1	100	..	..	..
Austria	3,678	6.2	29.8	64.0	4.0	37.1	112	1.5	31.6	66.9
Belgium	3,987	2.4	25.8	71.8	8.8	59.3	111	1.4	27.3	71.3
Denmark	2,708	3.3	26.8	69.5	5.6	18.6	119	4.1	27.6	68.3
Finland	2,333	6.4	27.6	65.7	11.5	23.6	102	5.5	33.0	61.5
France	22,755	4.3	26.3	69.4	11.4	41.3	99	2.4	27.4	70.1
Germany	36,089	2.9	33.8	63.3	8.9	50.6	108	1.1	34.9	64.0
Greece	3,967	17.8	23.0	59.2	11.7	55.3	66	14.9	25.0	60.0
Ireland	1,593	8.5	28.3	62.5	5.9	56.0	108	7.2	40.4	52.4
Italy	20,618	5.4	32.4	62.2	11.7	60.8	101	3.5	28.9	67.6
Luxembourg	176	1.9	21.9	75.8	2.4	32.2	176	1.3	23.9	74.8
Netherlands	7,605	..	..	..	3.3	41.5	113	3.2	26.8	70.0
Portugal	4,830	12.6	35.3	52.1	4.7	39.9	75	4.1	33.9	62.0
Spain	13,773	7.4	30.6	62.0	16.1	45.0	81	4.5	29.0	66.5
Sweden	4,054	3.0	25.0	72.0	7.6	29.1	102	2.2	30.0	66.9
United Kingdom	27,107	1.6	26.0	72.3	6.1	30.3	102	1.8	29.8	68.4

have a similar varied pattern. The countries that tend to perform better in terms of industry and the economy tend to be those that were the first to become industrialized in the eighteenth and nineteenth centuries. Britain, Germany, France and Italy are predominant in Europe with the main concentration lying within a rough triangle formed by London, Hamburg and Milan. Ireland, Spain, Southern Italy and Greece lie outside this triangle and tend to be less industrialized.

Activity 7.14

Use the data in Table 7.8 to answer the following questions:

1. Which European country has the largest population?
2. Which European country has the smallest population?
3. In which European country was industry the most significant employment sector?
4. In which European country was industry the least significant employment sector?
5. Which European country had the highest unemployment rate?
6. Which European country had the lowest unemployment rate?
7. Which European country had the least per capita GDP?
8. Which European country had the greatest per capita GDP?

Global factors

The past 30 years has seen a shift in some of the major engineering activities that used to be concentrated in Europe, North America and Japan. In particular, much of the electronics and printing industries have migrated to the *Pacific rim* countries such as Hong Kong, Singapore, Taiwan, Thailand and more recently to Indonesia. The main reason for this shift is the low labour costs to be found in the Far East.

Another prime example of the shift of engineering activities out of Europe is that of shipbuilding. The UK's contribution in particular has fallen and is now virtually nonexistent except for the manufacture of oil platforms and ships for the Royal Navy.

However, the traffic in engineering activities has not been all negative. The Japanese, wanting to sell their cars in Europe, have established several engineering production plants in the UK. The firms of Toyota (Deeside and Burnaston), Nissan (Sunderland) and Honda (Swindon) are three good examples. All occupy rural sites and have access to skilled and well-educated workforces. Road communications are good and, in the case of Nissan, the site is in an assisted area where substantial Government grants are available.

Activity 7.15

A company based in the Far East has asked you to carry out some research in order to help them investigate some investment opportunities in Europe. They have asked you to produce a broad comparison of the performance of the various countries that comprise the EU.

UNIT 7

Photocopy the map of Europe shown in *Figure 7.18* (your tutor may be able to supply you with this).

(a) Shade in the countries that constitute the EU.
(b) Mark in the name of each EU country.
(c) List the countries of the EU in descending order of GDP. What is the position of the UK?

Produce your findings in the form of a set of overhead projector transparencies and handouts to be used at a board meeting.

Figure 7.18 Map of Europe (see Activity 7.15)

Legislation and Environmental Constraints

The Health and Safety at Work Act (1974) makes both the employer and the employee responsible for safety in the workplace. In addition, we mentioned the Control of Substances Hazardous to Health Act (1988) which is designed to safeguard all employees who work with substances hazardous to their health, such as chemicals and toxic dust. In this section we are going to introduce you to the health and safety issues that relate to specific engineering systems. You are expected to be aware of these and how they impact on the design, production and utilization of the systems to which they relate.

The Health and Safety at Work Act

The Health and Safety at Work Act (1974) is based on principles that are fundamentally different from any previous health and safety legislation. The underlying reasoning behind the Act was the need to foster a much greater awareness of the problems that surround health and safety matters and, in particular, a much greater involvement of those who are, or who should be, concerned with improvements in occupational health and safety. Consequently, the Act seeks to promote greater personal involvement coupled with an emphasis on individual responsibility and accountability.

You need to be aware that the Act applies to *people*, not to premises. The Act covers all employees in all employment situations. The precise nature of the work is not relevant, neither is its location. The Act also requires employers to take account of the fact that other persons, not just those that are directly employed, may be affected by work activities. The Act also places certain obligations on those who manufacture, design, import or supply articles or materials for use at work to ensure that these can be used in safety and do not constitute a risk to health.

Duties of employers

It is the *duty of the employer* to ensure, so far as is reasonably practicable, the health, safety and welfare at work of all the employees, also that all plant and systems are maintained in a manner so that they are safe and without risk to health. The employer is also responsible for:

- the absence of risks in the handling, storage and transport of articles and substances;
- instruction, training and supervision to ensure health and safety at work;
- the maintenance of the workplace and its environment to be safe and without risk to health;
- where appropriate, to provide a statement of general policy with respect to health and safety and to arrangements for safety representatives and safety committees.

UNIT 7

In addition to having responsibilities to the employee, the employer has responsibilities to other persons such as the general public. It is therefore also the *duty of the employer* to:

- conduct his undertakings in such a way so as to ensure that members of the public (i.e., those not in his employ) are not effected or exposed to risks to their health or safety;
- give information about such aspects of the way in which he conducts his undertakings to persons who are not his employees as might affect their health and safety.

Duties of employees

It is the *duty of every employee* whilst at work, to take all reasonable care for the health and safety of himself and other persons who may be affected by his act and omissions.

Employees are required to:

- cooperate with the employer to enable the duties placed on him (the employer) to be performed;
- have regard of any duty or requirement imposed upon his employer or any other person under any of the statutory provisions;
- not interfere with or misuse any thing provided in the interests of health, safety or welfare in the pursuance of any of the relevant statutory provisions.

It is the duty of *each person* who has control of premises, or access to or from any plant or substance in such premises (i.e., employer and employees), to take all reasonable measures to ensure they are safe and without risk.

Other relevant health and safety related legislation that you need to be aware of is as follows:

- Control of Substances Hazardous to Health (COSHH)
- Reporting of Injuries, Diseases and Dangerous Occurrences Regulations 1995 (RIDDOR)
- Electricity at Work Regulations 1989
- Asbestos at Work Regulations 2002
- Management of Health and Safety at Work Act 1999.

It is important to note that some industries and trade associations have introduced *Codes of Practice* which set minimum standards and govern the working practices of their members.

Figure 7.19 shows typical safety notices (together with relevant safety procedure documents) prominently displayed in an engineering workshop whilst *Figure 7.20* shows typical markings used for a fire exit.

Figure 7.19 Health and safety information prominently displayed in an engineering workshop

Figure 7.20 Typical fire exit markings

Activity 7.16

Consult the Health and Safety at Work Act and answer the following questions:

(a) What is an improvement notice?
(b) What is a prohibition notice?
(c) Who issues such notices?
(d) Who can be prosecuted under the Act?

Activity 7.17

Visit the COSHH section of the UK government's Health and Safety Executive website (you will find this at `www.hse.gov.uk/coshh`). View or download a copy of 'COSHH: a brief guide to the regulations' and use it to identify the eight steps that help to ensure that a company complies with the COSHH legislation.

Activity 7.18

Asbestos is a material that has only relatively recently been identified as a material that is dangerous to health. Use the Internet or your school or college library to answer the following questions:

1. What is asbestos?
2. Why is asbestos dangerous to health?
3. What diseases are caused by exposure to asbestos?
4. What occupations are particularly at risk from asbestos?
5. What legislation applies to the use and handling of asbestos?

Activity 7.19

Obtain a copy of 'RIDDOR Explained' (available online from the Health and Safety Executive website at `www.hse.gov.uk`) and use it to answer the following questions:

1. What is RIDDOR?
2. What does RIDDOR apply to?

UNIT 7

3. In what circumstances must a report be made?
4. Who should make the report?
5. How does RIDDOR apply to self-employed people?
6. Give four examples of major injuries that are reportable under RIDDOR.
7. Give four examples of dangerous occurrences that are reportable under RIDDOR.

Activity 7.20

Figure 7.21 shows the internal arrangement of a machine shop. List at least three safety and environmental hazards that are present in this environment. Draw up a 'Code of Practice' for people working in this environment and present your work in the form of an A3 poster that can be displayed on a wall.

Figure 7.21 An internal arrangement of a machine shop

Environmental constraints

Modern engineering processes and systems are increasingly designed and implemented to minimize environmental affects. Engineering companies must ensure that the negative effects of engineering activities on the natural and built environment are minimized. You need to be able to identify how individual engineering companies seek to do this, such as through:

- design of plant and products, which optimize energy use and minimize pollution
- good practices such as the efficient use of resources and recycling, and the use of techniques to improve air and water quality
- management review and corrective action
- relevant legislation and regulations.

Material processing

Many engineering activities involve the processing of materials. Such materials may appear in the product itself or may be used in the

manufacturing process. Some of these materials occur naturally and, after extraction from the ground, may require only minimal treatment before being used for some engineering purpose. Examples are timber, copper, iron, silicon, water and air. Other engineering materials need to be manufactured. Examples are steel, brass, plastic, glass, gallium arsenide and ceramic materials. The use of these materials produces effects: some beneficial and some not.

Economic and social effects stem from the regional wealth that is generated by the extraction of the raw material and its subsequent processing or manufacture into useful engineering materials. For example, the extraction of iron ore in Cleveland and its processing into pure iron and steel has brought great benefit to the Middlesbrough region. The work has attracted people to live in the area and the money they earn tends to be spent locally. This benefits trade at the local shops and entertainment centres and local builders must provide more homes, schools, and so on. The increased numbers of people produces a growth in local services, which includes a wider choice of different amenities, better roads and communications and arguably, in general, a better quality of life.

On the debit side, the extraction of raw materials can leave the landscape untidy. Heaps of slag around coal mines and steelworks together with holes left by disused quarries are not a pretty sight. In recent years much thought and effort has been expended on improving these eyesores.

Slag heaps have been remodelled to become part of golf courses and disused quarries filled with water to become centres for water sports or fishing. Disused mines and quarries can also be used for taking engineering waste in what is known as a landfill operation prior to the appropriate landscaping being undertaken.

Other potential problems can arise from having to transport the raw materials used in engineering processes from place to place. This can have an adverse affect on the environment resulting from noise and pollution.

Pollutants

Engineering activities are a major source of *pollutants* and many types of *pollution*. Air, soil, rivers, lakes and seas are all, somewhere or other, polluted by waste gases, liquids and solids discarded by the engineering industry. Because engineering enterprises tend to be concentrated in and around towns and other built-up areas, these tend to have high levels of pollution.

Electricity is a common source of energy and its generation very often involves the burning of *fossil fuels*: coal, oil and natural gas. In doing so, each year, billions of tonnes of carbon dioxide, sulphur dioxide, smoke and toxic metals are released into the air to be distributed by the wind. The release of hot gases and hot liquids also produces another pollutant: heat. Some electricity generating stations use nuclear fuel that produces a highly radioactive solid waste rather than the above gases.

The generation of electricity is by no means the only source of toxic or biologically damaging pollutants. The exhaust gases from motor vehicles,

oil refineries, chemical works and industrial furnaces are other problem areas. Also, not all pollutants are graded as *toxic*. For example, plastic and metal scrap dumped on waste tips, slag heaps around mining operations, old quarries, pits and derelict land are all *non-toxic*. Finally, pollutants can be further defined as *degradable* or *non-degradable*. These terms simply indicate whether the pollutant will decompose or disperse itself with time. For example, smoke is degradable but dumped plastic waste is not. *Figure 7.22* shows a typical example of an industrial process that generates airborne pollutants.

Figure 7.22 Generation of airborne pollutants by an engineering process

Carbon dioxide

Carbon dioxide in the air absorbs some of the long-wave radiation emitted by the earth's surface and in doing so is heated. The more carbon dioxide there is in the air, the greater the heating or greenhouse effect. This is suspected as being a major cause of global warming causing average seasonal temperatures to increase. In addition to causing undesirable heating effects, the increased quantity of carbon dioxide in the air, especially around large cities, may lead to people developing respiratory problems.

Oxides of nitrogen

Oxides of nitrogen are produced in most exhaust gases and nitric oxide is prevalent near industrial furnaces. Fortunately, most oxides of nitrogen are soon washed out of the air by rain. But if there is no rain, the air becomes increasingly polluted and unpleasant.

Sulphur dioxide

Sulphur dioxide is produced by the burning of fuels that contain sulphur. Coal is perhaps the major culprit in this respect. High concentrations of this gas cause the air tubes in people's lungs to constrict and breathing becomes increasingly difficult. Sulphur dioxide also combines with rain droplets eventually to form dilute sulphuric acid or *acid rain*. This is carried by the winds and can fall many hundreds of miles from the sulphur dioxide source. Acid rain deposits increase the normal weathering effect on buildings and soil, corrode metals and textiles and damage trees and other vegetation.

Smoke

Smoke is caused by the incomplete burning of fossil fuels, such as coal. It is a health hazard on its own but even more dangerous if combined with fog. This poisonous combination, called *smog*, was prevalent in the early 1950s. It formed in its highest concentrations around the large cities where many domestic coal fires were then in use. Many deaths were recorded, especially among the elderly and those with respiratory diseases. This led to the first Clean Air Act, which prohibited the burning of fuels that caused smoke in areas of high population. So-called smokeless zones were established.

Dust and grit

Dust and grit (or *ash*) are very fine particles of solid material that are formed by combustion and other industrial processes. These are released into the atmosphere where they are dispersed by the wind before falling to the ground. The lighter particles may remain in the air for many hours. They form a mist, which produces a weak, hazy sunshine and less light.

Toxic metals

Toxic metals, such as lead and mercury, are released into the air by some engineering processes and especially by motor vehicle exhaust gases. Once again the lead and mercury can be carried over hundreds of miles before falling in rainwater to contaminate the soil and the vegetation it grows. Motor vehicles are now encouraged to use lead-free petrol in an attempt to reduce the level of lead pollution.

Ozone

Ozone is a gas that exists naturally in the upper layers of the earth's atmosphere. At that altitude it is one of the earth's great protectors but when it occurs at ground level it is linked to pollution. *Stratospheric ozone* shields us from some of the potentially harmful excessive ultraviolet radiation from the sun. In the 1980s it was discovered that emissions of gases from engineering activities were causing a 'hole' in the ozone layer. There is concern that this will increase the risk of skin cancer, eye cataracts and damage to crops and marine life.

At ground level, sunlight reacts with motor vehicle exhaust gases to produce ozone. Human lungs cannot easily extract oxygen (O_2) from ozone (O_3) so causes breathing difficulties and irritation to the respiratory channels. It can also damage plants.

This ground level or *tropospheric ozone* is a key constituent of what is called photochemical smog or summer smog. In the UK it has increased by about 60% in the last 40 years.

Heat

Heat is a waste product of many engineering activities. A typical example being the dumping of hot coolant water from electricity generating stations into rivers or the sea. This is not so prevalent today as increasingly stringent energy saving measures are applied. However, where it does happen, river and sea temperatures can be raised sufficiently in the region of the heat outlet to destroy natural aquatic life.

Chemical waste

Chemical waste dumped directly into rivers and the sea, or on to land near water, can cause serious pollution that can wipe out aquatic life in affected areas. There is also the long-term danger that chemicals dumped on soil will soak through the soil into the ground water, which we use for drinking purposes and which will, therefore, require additional purification.

Radioactive waste

Radioactive waste from nuclear power stations or other engineering activities that use radioactive materials poses particular problems. Not only it is extremely dangerous to people – a powerful cause of cancer – but also its effects do not degrade rapidly with time, and remain dangerous for scores of years. Present methods of disposing of radioactive waste, often very contentious however, include their encasement in lead and burial underground or at sea.

Noise and light pollution

Where they have an adverse effect on the environment, noise and light produced by engineering activities can also be considered to be pollutants. Noise pollution may consist of low frequency vibration as well as the more obvious noise generated by engines and machinery. Light pollution may be produced when lights are left on all night or when flood lighting is used to illuminate engineering activities that are performed at night. Adequate lighting is, however, an essential requirement for health and safety and so it is sometimes necessary to balance the need to reduce light pollution with that of ensuring that engineering activities are carried out safely.

Derelict land

> **KEY POINT**
>
> Engineering activities produce a variety of pollutants. They may be classified as toxic or non-toxic, degradable or non-degradable

Derelict land is an unfortunate effect of some engineering activities. The term derelict land may be taken to mean land so badly damaged that it cannot be used for other purposes without further treatment. This includes disused or abandoned land requiring restoration works to bring it into use or to improve its appearance. Land may be made derelict by mining and quarrying operations, the dumping of waste or by disused factories from bygone engineering activities.

TYK 7.23

Name a pollutant that fits each of the following categories:
(a) toxic and degradable
(b) toxic and non-degradable
(c) non-toxic and degradable
(d) non-toxic and non-degradable.

Environmental legislation

Engineering activities can have harmful effects on the physical environment and therefore on people. In order to minimize these effects, there is a range

of legislation (rules and regulations) that all engineering companies must observe.

The appropriate United Kingdom Acts of Parliament include Deposit of Poisonous Wastes Act, Pollution of Rivers Act, Clean Air Act, Environmental Protection Act, and the Health and Safety at Work Act. Additionally, not only are there local by-laws to be observed there are also EU directives that are activated and implemented either through existing UK legislation in the form of Acts of Parliament or mandatory instructions called Statutory Instruments (SI).

New Acts and directives are introduced from time to time and industry needs to be alert to and keep abreast of these changes. Typical of these new initiatives is the European Electromagnetic Compatibility (EMC) legislation. This states that with effect from 1 January 1996, it is a requirement that all products marketed must conform with the new legislation. This new EMC legislation, at last, officially recognizes the well-known problem of unwanted electromagnetic wave radiation that emanates from most pieces of electrical equipment. The unwanted radiation can interfere with nearby electronic equipment causing it to malfunction.

In the case of UK Acts of Parliament, the above legislation is implemented by judgement in UK Courts of Justice in the normal manner but based on EU legislation, if more appropriate, or by judgement of the European Court of Justice.

The purpose of this legislation is to provide the following functions:

- *prevent* the environment being damaged in the first place;
- *regulate* the amount of damage by stating limits, for example, the maximum permitted amount of liquid pollutant that a factory may discharge into the sea;
- *compensate* people for damage caused, for example, from a chemical store catching fire and spreading wind borne poisonous fumes across the neighbourhood;
- *impose sanctions* on those countries or other lesser parties that choose to ignore the legislation;
- *define who is responsible* for compliance with legislation to persons who can be named and their precise area of responsibility documented.

For the purposes of showing understanding of the above, you are *not* expected to have a detailed understanding of the various Acts; however, you *should be aware of the general provisions* of the legislation and what it is trying to achieve. Your school, college or local library will be able to provide you with more details.

The effects of the above legislation on engineering activities has, in general, made them more difficult and more expensive to implement. A few simple examples of this follow.

- *Chemical factories* can no longer discharge their dangerous waste effluent straight into rivers or the sea without first passing it through some form of purification.

UNIT 7

- *Coal fuelled power stations* must ensure that their chimney stacks do not pollute the neighbourhood with smoke containing illegal limits of grit, dust, toxic gases and other pollutants. A system of smoke filtration and purification must be (expensively) incorporated.

- *Motor car exhaust gases* must be sufficiently free of oxides of nitrogen, carbon monoxide and other toxic gases. This can only be achieved by, among other things, replacing the crude petrol carburettor with a more sophisticated petrol injection system and fitting a catalytic converter in the exhaust system. All this has added to the price of the motor car and has made it more difficult for the DIY motorist to service their vehicle.

- All *electrical equipment*, including TVs, PCs, power hand tools, electro-medical machines, lighting and the like, must be tested and certified that they comply with the EMC legislation. So, in addition to the cost reducing any excessive radiation from the product itself, the purchase or hire of expensive EMC test equipment and the training of people to use must also be taken into account. Furthermore, because of delays in obtaining an official EMC examination and a supporting EC certificate, the introduction of new product designs can also be delayed and this may have adverse marketing effects.

Finally, it is important to be clear about the difference between *waste products* and *by-products*. The by-products from one process can be sold as the raw materials for other processes. For example, natural gas is a by-product of oil extraction and a useful fuel used in the generation of electricity. Waste products are those that cannot be sold and may attract costs in their disposal. Nuclear power station waste is a typical example.

TYK 7.24

Explain what is meant by EMC and why the legislation that relates to it is important.

TYK 7.25

Explain the effect of environmental legislation on the cost of motoring in the UK.

Activity 7.21

1. Find out what happens to the domestic waste produced in your locality.
2. What items of domestic waste are recycled?
3. Is any of the waste burnt to produce useful heat?
4. If the waste is transported to another site for processing, where does it go and what processes are used?
5. What arrangements are there for disposing of hazardous waste?
6. Present your findings in the form of a brief word-processed report. Illustrate your report with the aid of a simple flowchart.

Activity 7.22

Investigate the effects of each of the following engineering activities on the environment:

- manufacture of motor vehicles
- service and repair of motor vehicles
- road transport of bulk materials.

In your investigation consider the environmental effects of materials and consumables (such as fuel and lubricants) as well as the short- and long-term environmental effects of any waste products.

Present your work in the form of a brief written report.

Activity 7.23

Visit your local museum and obtain information concerning the engineering activities and companies that were active in your area (city, town or county) in:

- 1900
- 1950.

Compare these activities with those of the present day. Identify any new industries and explain why they have developed in your area. In the case of any industries or companies that have been active in your area for the full period, identify ways in which they have changed their operation in order to:

(a) remain competitive
(b) safeguard the environment.

Present your findings in the form of a brief class presentation with appropriate visual aids.

Costing Techniques

Any engineering company will incur a variety of costs. These will typically include:

- rent for factory and office premises
- rates
- energy costs (including heating and lighting)
- material costs
- costs associated with production equipment (purchase and maintenance)
- salaries and National Insurance
- transport costs
- postage and telephone charges
- insurance premiums.

Given the wide range of costs above, it is often useful to classify costs under various headings, including fixed and variable costs, overhead and direct costs, average and marginal costs, and so on.

In order to be able to control costs, it is, of course, vital to ensure that all of the costs incurred are known. Indeed, the consequences of not being fully aware of the costs of a business operation can be dire!

UNIT 7

This section examines a number of different methods used by businesses to determine the total cost of the product or service that they deliver. The prime objective of these techniques is that of informing commercial decisions such as:

- how many units have to be produced in order to make a profit?
- is it cheaper to make or buy an item?
- what happens to our profits if the cost of production changes?
- what happens to our profits if the cost of parts changes?

Costing methods

To ensure profitability, engineering companies need to have in place an effective costing system that takes into account the real cost of manufacturing the product or delivering the service that it provides. Without such a system in place it is impossible to control costs and determine the overall profitability of the business operation.

Cost accounting is necessary for a company to be able to exercise control over the actual costs incurred compared with planned expenditure. From the point of view of cost control, a costing system should not only be able to identify any costs that are running out of control but should also provide a tool that can assist in determining the action that is required to put things right.

Job costing

The most simple costing technique is known as 'job costing'. As its name implies, it relates to a unique operation, such as replacing a part or carrying out a modification to a product. Typical operations in which job costing is commonly used include:

- supplying a unique or 'one-off' item
- modifying a product in order to improve its performance
- adapting a product to meet a particular customer's requirements
- providing a service that is only required on an intermittent or irregular basis.

Job costing involves estimating the cost of materials used (sometimes referred to as a *bill of materials*) and then adding this to the cost of labour in order to determine the total amount that will be charged to the customer. Let us take an example:

A small company supplies and fits entertainment systems for use in coaches and buses. The company has been asked to supply a one-off entertainment system for use in a luxury coach. The quotation for parts and materials required for this job is as follows:

Item	Quantity	Price per unit	Cost
Display screens	2	£475	£950
DVD player	1	£275	£275
Rack unit	1	£125	£125
Cable	25m	£1 per metre	£25
Control box	1	£75	£75
Sundry items			£30
Total for parts and materials:			£1480

To this should be added the cost of labour. Let us assume that this amounts to 10 hours at £35 per hour (this figure includes the overheads associated with employment, such as National Insurance contributions). Hence the cost of labour is:

Item	Quantity	Price per unit	Cost
Labour	10 hours	£35/hour	£350
Total for labour:			£350

We can add the cost of labour to the total *bill of materials* to arrive at the final cost for the job, which amounts to £1830. Note that parts and materials *may* be supplied 'at cost' or 'marked up' by a percentage, which can often range from 10% to 50% (and sometimes more).

Total cost:	£1,830

Finally we would add value added tax (VAT) at the current rate to the total cost. This is the amount payable by the customer.

VAT: (at 17.5%)	£320.25
Amount payable:	£2,150.25

Contract costing

Large companies also use job costing when they produce a variety of different, and often unique, products at the same time. These products are often referred to as *custom built* and each is separately costed as a 'job' in its own right. This type of production is described as *intermittent* (and traditionally referred to as *job shop* production) to distinguish it from the *continuous* or *assembly-line* production associated with the manufacture of a large number of identical units.

In jobbing production, individual manufactured units are normally produced to meet an individual customer's requirements and production is not normally speculative. Costs are agreed before manufacturing starts and they form the basis of a contract between the manufacturer and the customer.

Parts costing

Parts costing is fairly straightforward and is simply a question of determining the cost of all of the physical parts and components used in a manufactured or engineered product. Parts costing works from the 'bottom up' – in other words, the cost of each individual component (i.e., the *per unit cost*) is determined on the basis of the given *standard supply multiple*. For example, if plastic case parts are purchased in quantities of 1000 then the per unit cost is simply the total cost of purchasing 1000 items from the supplier divided by the supply multiple (i.e., 1000). You should note that the cost of an individual item purchased from a supplier is usually very much greater than the per unit cost when the item is purchased in quantity. As an example of parts costing, consider the following example.

It is often useful to group together individual component parts under groupings of similar items. The reason for this is that such groupings tend

to be subject to the same fluctuation in cost. We can therefore quickly determine the effect of market fluctuations by examining the effect of changes on particular groups of parts.

TYK 7.26

Explain the following terms:
(a) job costing
(b) contract costing
(c) parts costing.

Activity 7.24

The bill of materials for the prototype of the new APS loudspeaker (see TYK 7.3 and *Figure 7.23*) is as follows:

Item	Quantity	Cost
18 mm plywood sheet	4 m^2	£2.50/m^2
20 × 20 mm timber batten	4 m	50 p/m
30 × 40 mm pine	1.8 m	£2/m
Grille and mounting hardware	1	£4.40
10 cm plastic port tubes	2	£1.00
Acoustic wadding	3 m	£1.50
Chrome corners	8	£0.60
Heavy-duty handles	2	£2.10
Screws and sundry items	1	£2.50
Jack socket and recess plate	1	£4.50
Low-frequency driver unit	1	£26.00
High-frequency driver unit	1	£16.00

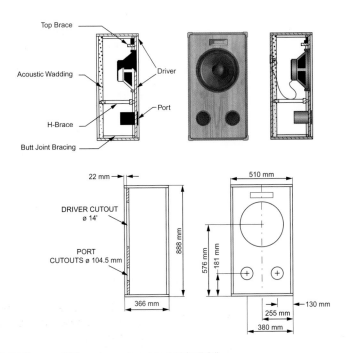

Figure 7.23 The new APS speaker system (see Activity 7.24)

Use a spreadsheet to determine the total cost of the parts used in the prototype loudspeaker. Also determine the total cost of the loudspeaker if the cost of timber increases by 20% and the supplier of the driver units offers a discount of 15%. Present your work in the form of two spreadsheet printouts.

Process costing

Process costing takes into account the cost of each manufacturing process and apportions part of the cost of each process to an individual product. Typical processes might be:

- forming, bending or machining of metal and plastic parts
- flow soldering of printed circuit boards
- heat treatment of metal parts
- paint spraying and finishing.

The following example illustrates the use of process costing. Note that, when determining the total cost of manufacturing a product, it is essential to take into account the notional cost of all of the processes involved. Alternatively, where there may be a number of products sharing the same process, it may be more convenient to amalgamate these costs into a production overhead cost. The danger with this, as we shall see later, is that we usually need to apportion costs more precisely so that we can determine the relative profitability of *each* product that we manufacture. If this is beginning to sound a little complex, an example will show you how process costing works.

Example 7.1

Let us assume that our loudspeaker manufacturer, APS, has invested in a flow soldering plant to automate the manufacture of their crossover units. The flow soldering plant operates at a rate of 60 units per hour and its operating cost (including capital cost recovery calculated over a nominal eight year asset life) amounts to £5000/week plus £10 material and energy costs per hour. Let us determine the cost of producing each crossover unit using the flow soldering process (this is known as the *unit cost* of the process) assuming that it operates for a total of 40 hours each week.

Based on 40 hours operation per week, the total cost of operating the flow soldering process for a week will be given by

$$Total\ cost = £5000 + (40 \times £10) = £5400$$

At 60 units per hour, the total weekly production will be given by

$$Total\ production = 60 \times 40 = 2400$$

The cost, per unit, will thus be given by

$$Cost\ per\ unit = \frac{£5400}{2400} = £2.25$$

TYK 7.27

Determine the cost per unit of the APS crossover if the flow soldering plant operates for 75 hours per week and the hourly operating costs increase to £15 per hour.

Absorption costing

One method of determining the total cost of a given product or service is that of adding the costs of overheads to the direct costs by a process of *allocation*, *apportionment* and *absorption*. Since *overheads* (or *indirect costs*) can be allocated as whole items to production departments, it is possible to arrive at a notional amount that must be added to the cost of each product in order to cover the production overheads.

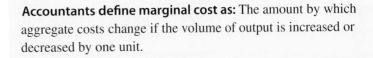

$$\text{Mark-up} = \frac{\text{Total of fixed and variable costs attributable to the product}}{\text{Total number of units produced}}$$

In absorption costing, each product manufactured is made (at least in theory) to cover all of its costs. This is achieved by adding a notional amount to the total unit cost of each product.

Marginal costing

Marginal costing provides us with an alternative way of looking at costs that provides an insight into the way costs behave by allowing us to observe the interaction between costs, volumes and profits. The marginal cost of a product is equal to the cost of producing one more unit of output.

> **Accountants define marginal cost as:** The amount by which aggregate costs change if the volume of output is increased or decreased by one unit.

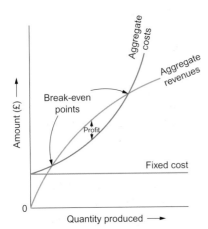

Figure 7.24 Marginal cost behaviour

KEY POINT

The marginal cost of a product is the cost of producing one more unit of output

Figure 7.24 shows the behaviour of costs based on the marginal cost concept. The vertical axis of *Figure 7.24* shows aggregate costs and revenues in £. The horizontal axis shows the units of output. The fixed costs are shown as a straight line across all levels of output. Fixed costs are costs which are unaffected by activity, at least in the short-term. *Figure 7.24* assumes that such costs do not vary for the whole range of output possible.

Fixed costs include things like:

- insurance premiums
- business rates
- subscriptions
- audit fees
- rental charges
- fixed elements of power and telephone charges.

There are a number of advantages of using marginal costing, notably:

- Marginal costing systems are simpler to operate than absorption costing systems because they do not involve the problems associated with overhead apportionment and recovery.
- It is easier to make decisions on the basis of marginal cost presentations. Where several products are being produced, marginal costing can show which products are making a contribution and which are failing to cover their variable costs.

The disadvantages of marginal costing include:

- The effect of time (and its effect on true cost) tends to be overlooked in marginal costing. For example, two jobs may have the same marginal cost but one may use significantly more plant time than another (thus increasing its true cost).
- There is a temptation to spread fixed costs and real dangers in neglecting these in favour of more easily quantified variable costs.

The choice of whether to use absorption costing or marginal costing is usually determined by factors such as:

- the system of financial control used within a company (e.g., *responsibility accounting* is consistent with absorption costing);
- the production methods used (e.g., marginal costing is easier to operate in simple processing applications whereas absorption costing is usually preferred when several different products require different plant and processing techniques);
- the significance of the prevailing level of overhead costs.

Activity-based costing

Activity-based costing is an attempt to assess the 'true' cost of providing a product or service. Knowledge of the 'true' cost is not only important in helping us to identify opportunities for cost improvement but it also helps us to make strategic decisions that are better informed.

Activity-based costing focuses on indirect costs (overheads). It does this by making costs that would traditionally be considered indirect into direct costs. In effect, it traces costs and overhead expenses to an individual *cost object*. The basic principles of activity-based costing are shown in *Figure 7.25*.

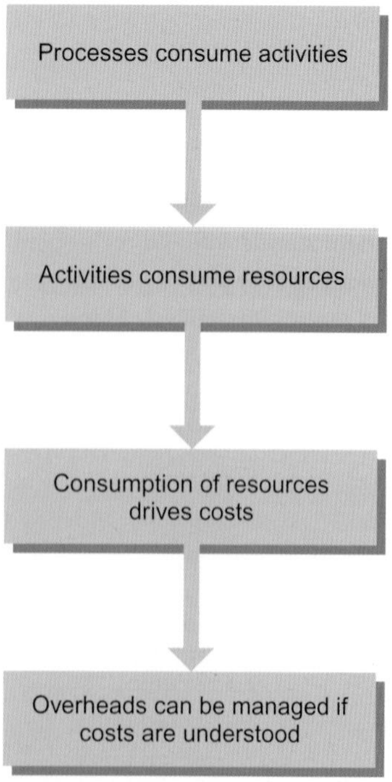

Activity-based costing is particularly useful when the overhead costs associated with a particular product are significant and where a number of products are manufactured in different volumes. Activity-based costing is particularly applicable where competition is severe and the margin of selling price over manufacturing cost has to be precisely determined.

The steps required to carry out activity-based costing are:

1. Identify the activities.
2. Determine the cost of each activity.
3. Determine the factors that drive costs.
4. Collect the activity data.
5. Calculate the product cost.

The use of activity-based costing is best illustrated by taking an example.

Processes consume activities

Activities consume resources

Consumption of resources drives costs

Overheads can be managed if costs are understood

Figure 7.25 Principles of activity-based costing

Example 7.2

Let us assume that APS has decided to go ahead with the production of its new loudspeaker but is also manufacturing a high-quality demountable loudspeaker stand. The company is interested in knowing how these two products compare using activity-based costing.

UNIT 7

Activity	Loudspeakers	Cost	Stands	Cost	Total
Setup	1 @ £45,000	£45,000	1 @ £25,000	£25,000	£70,000
Manufacture	3000 @ £20/unit	£60,000	1000 @ £12/unit	£12,000	£72,000
Assembly	3000 @ £10/unit	£30,000	1000 @ £6/unit	£6000	£36,000
Inspection	3000 @ £2/unit	£6000	1000 @ £2/unit	£2000	£8000
Packaging	3000 @ £2/unit	£6000	1000 @ £2/unit	£2000	£8000
Total		£147,000		£47,000	£194,000

The activity-based product cost for each loudspeaker amounts to £147,000/3000 = £49 whilst the activity-based product cost for each stand amounts to £47,000/1000 = £47. To these two costs we must add the direct (material) costs of each product. Assuming that this amounts to £60 for the loudspeaker and £20 for the demountable stand, we would arrive at a cost of £129 for the loudspeaker and £67 for the stand.

It is worth pointing out that traditional cost accounting would have arrived at two rather different figures for the costs of these two products. Let us assume that 7000 hours of direct labour are used in the manufacture of the loudspeakers and stands. Dividing the total overhead cost of £194,000 by this figure will give us the hourly direct labour cost of £27.71 per hour. If loudspeakers require 2 hours of direct labour and stands require 1.5 hours of direct labour the allocation of costs would be £55.42 per loudspeaker and £41.57 per stand. Adding the same direct (material) costs to this yields a cost of £115.42 for the loudspeaker and £61.57 for the stand.

Test your knowledge

TYK *7.28*

Explain, briefly, what is meant by:
(a) absorption costing
(b) marginal costing
(c) activity costing.

Activity 7.25

Voltmax specializes in the manufacture of chargers for rechargeable batteries. They produce two different chargers: one designed for standard Ni–Cd rechargeable cells and one for high-capacity Ni–MH cells. Both types of charger are packed in multiples of 10 before they are despatched to retail outlets. An analysis of the company's production reveals the following:

	Standard charger	High-capacity charger
Production volume	**10,000 units**	**5000 units**
Direct materials cost	**£5 per unit**	**£8 per unit**
Activity-based production analysis:		
Set-up costs	**£20,000**	**£20,000**
Manufacturing costs	**£3 per unit**	**£4 per unit**
Assembly costs	**£2 per unit**	**£3 per unit**
Packaging	**£1 per ten units**	**£1 per ten units**
Despatch/delivery	**£5 per ten units**	**£5 per ten units**

Use activity-based costing to determine the cost of each product.

Financial control methods

An engineering company will normally make use of a number of different control methods to ensure that its operation is profitable. These control methods include making forecasts of overall profitability, determining the contribution made by each individual activity towards overheads and fixed costs, and performing 'what-if' analysis to determine the effects of variations in cost and selling price. We shall start by describing the most simple method, break-even analysis.

Break-even charts

Break-even charts provide a simple (and relatively unsophisticated) method for determining the minimum level of sales that a company must achieve in order for the business to be profitable. Consider the simple relationship illustrated in *Figure 7.26*. Here total income has been plotted against total costs using the same scale for each axis. At point A, total costs exceed total income and the operation is not profitable, that is, it makes a *loss*. If we charge more for the product, whilst keeping the costs fixed, we would move from point A to B. At a certain point, total income exceeds total costs and we move into *profit*. Finally, let us assume that our total costs increase whilst the total income from sales remains unchanged. We would then move from profit (point B) to loss (point C).

The *break-even point* is the volume of sales at which the operation becomes profitable and it marks the transition from loss into profit. A *break-even chart* takes the form of a graph of costs plotted against volume of product sold. At this point, it is important to recall that the total costs of the business operation are the sum of the constant fixed and overhead costs with the variable costs of production. Thus,

$$\text{Total cost} = \text{Fixed cost} + \text{Overhead cost} + \text{Variable cost}$$

The income derived from the sale of the product (assuming a constant pricing structure) will simply be the product of the quantity sold (i.e., the volume of product sold) and the price at which it is sold (i.e., the per unit selling price). This relationship (a straight line) can be superimposed on the break-even chart and the point of intersection with the total cost line can be identified. This is the *break-even point* and the corresponding production volume can be determined from the horizontal axis, see *Figure 7.27*.

The *break-even quantity* can be determined from

$$\text{Break-even quantity} = \frac{\text{Fixed cost}}{\text{Selling price} - \text{Variable cost}}$$

Note that, in the above formula, *Selling price* and *Variable cost* are per unit.

It is also possible to use the break-even chart to determine the profit that would result from a particular production quantity.

Profit can be determined from

$$\begin{aligned}\text{Profit} = &(\text{Selling price} \times \text{Quantity sold}) \\ &- [\text{Fixed cost} + (\text{Variable cost} \times \text{Quantity sold})]\end{aligned}$$

Note that, in the above formula, *Selling price* and *Variable cost* are again per unit.

KEY POINT

The break-even point is the volume of output at which an engineering operation becomes profitable. In other words, it marks the transition from loss into profit

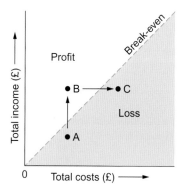

Figure 7.26 Total income plotted against total costs showing profit and loss regions

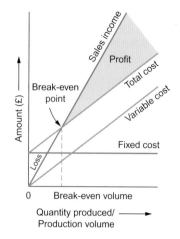

Figure 7.27 Fixed and variable costs plotted against production volume showing break-even

UNIT 7

Example 7.3

Let us assume that our loudspeaker manufacturer, APS, has established the following costs of manufacturing its demountable loudspeaker stand:

$$\text{Fixed costs} = £25,000$$
$$\text{Variable costs} = £20 \text{ per unit.}$$

If the stand is to be sold for £100, determine the break-even quantity and the profit that would be returned from sales of 10,000 units.

Now $\quad$ Break-even quantity $= \dfrac{\text{Fixed cost}}{\text{Selling price} - \text{Variable cost}}$

Thus $\quad$ Break-even quantity $= \dfrac{£25,000}{£100 - £20} = 312.5$ (i.e. 313 to achieve profit)

The profit based on sales of 10,000 units will be given by

Profit = (Selling price × Quantity sold) − [Fixed cost + (Variable cost × Quantity sold)]
Profit = (£100 × 10,000) − [£25,000 + (£20 × 10,000)] = £100,000 − £45,000

Thus the profit on 10,000 units will be **£55,000**.

TYK

TYK 7.29

Explain, briefly, what is meant by break-even and illustrate your answer using a graph showing costs plotted against production volume.

Activity 7.26

Voltmax (see Activity 7.25) has analysed its fixed and overhead costs relating to its standard charger which together amount to £100,000 whilst the variable costs of its standard charger amount to £12 per unit manufactured.

Construct a break-even chart and use this to determine:

1. the break-even production volume when the charger is sold at:
   (a) £20 per unit
   (b) £30 per unit
2. the profit for a production quantity of 10,000 units if the selling price is £30 per unit.

Present your work in the form of a brief word-processed report to Voltmax's board of directors.

TYK 7.30

The following figures relate to the production of a small component:

Fixed cost:	£30,000
Variable cost per unit:	£15
Selling price:	£25

Determine the break-even point in sales turnover.

It is important to realize that simple break-even analysis has a number of serious shortcomings. These may be summarized as follows:

- The sales income line (i.e., the product of the volume produced and its selling price) takes no account of the effect of price on the volume of sales. This is important as it is likely that the demand for the product will fall progressively as the selling price increases and the product becomes less competitive in the open market.
- The assumption that fixed costs remain fixed and variable costs increase linearly with production are somewhat dangerous. The reality is that both of these will change!

For the foregoing reasons it is important to regard break-even analysis as a 'rule-of-thumb' method for evaluating product pricing. Before making any business decisions relating to pricing and targets for production volume it is important to undertake further research into effect of pricing on potential sales as well as the pricing of competitive products.

Average cost

The fixed costs associated with production has to be shared between the entire volume produced. Hence, a proportion of the final cost of a product will be attributable to the fixed costs of manufacture. The larger the quantity produced, the smaller this proportion will be. In other words, the *average cost* of the product will fall as the volume increases. We can illustrate this in the form of a graph (see *Figure 7.28*). Note that,

$$\text{Average cost} = \frac{\text{Total cost}}{\text{Quantity produced}}$$

As an example of how this works, let us calculate the average costs.

> **KEY POINT**
>
> A proportion of the final cost of a product is attributable to the fixed costs of manufacture. Because of this, the average cost of a product falls as the volume produced increases

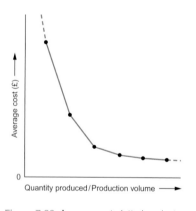

Figure 7.28 Average cost plotted against production volume

Example 7.4

Calculate the average costs of Voltmax's standard charger for production levels of 5000, 7500, 10,000 and 15,000 units. Based on the information given in Activity 7.25, Voltmax's fixed costs are £100,000 and the company's variable costs amount to £12 per unit.

Now

$$\text{Average cost} = \frac{\text{Total cost}}{\text{Quantity produced}}$$

and

$$\text{Total cost} = \text{Fixed cost} + \text{Variable cost}$$

Thus:

Quantity produced:	5000	7500	10,000	15,000
Fixed cost:	£100,000	£100,000	£100,000	£100,000
Variable cost:	£60,000	£90,000	£120,000	£180,000
Total cost:	£160,000	£190,000	£220,000	£280,000
Average cost:	£32	£25.33	£22	£18.67

This information is shown in graphical form in *Figure 7.29*.

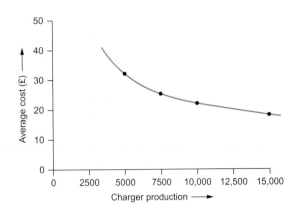

Figure 7.29 Average cost plotted against production volume for Voltmax's standard charger

Marginal cost

Once we establish a particular volume of production, the cost of producing one more unit is referred to as the marginal cost. The marginal cost of a product is that cost of the unit that results only from changes in those costs that do not vary with the amount produced. Marginal cost is *not* the same as average cost – the reason for knowing the marginal cost of a product is that it can help us decide whether or not to increase production from an existing level. This is best illustrated with another example.

Example 7.5

Voltmax has established a production level of 12,000 units for its standard charger. As before, the fixed cost of the company's standard charger manufacturing operation is £100,000 and the variable costs of producing the standard charger amount to £12 per unit. The charger is normally sold for £30 but a large high-street chain store has offered to take an additional 2000 units at a non-negotiable price of £20 per unit. We can use marginal costing to determine whether this proposition is financially sound.

The total cost associated with a production volume of 12,000 units is found from:

$$\text{Total cost} = \text{Fixed cost} + \text{Variable cost}$$

$$\text{Total cost} = £100,000 + (12,000 \times £12) = £100,000 + £144,000 = £244,000$$

Now the average cost (based on 12,000 units) will be given by

$$\text{Average cost} = \frac{\text{Total cost}}{\text{Quantity produced}} = £244,000/12,000 = £20.33$$

Based on an average cost of £20.33, a selling price of £20 per unit does not appear to be sound business sense. However, if we consider the marginal cost of the standard charger based on an *existing* production level of 12,000 units, we arrive at a rather different view. The rationale is as follows:

Let us assume a scenario in which we sell 12,000 standard chargers at £30 and 2000 standard chargers at £20. The total income produced will be given by

$$\text{Total income} = (12,000 \times £30) + (2000 \times £20) = £360,000 + £40,000 = £400,000$$

The total cost associated with producing 14,000 standard chargers will be:

$$\text{Total cost} = £100,000 + (14,000 \times £12) = £100,000 + £168,000 = £268,000$$

Thus the resulting profit will be given by

Profit = Total income − Total cost = £400,000 − £268,000 = £132,000

Had we decided not to accept the order for the extra 2000 units, we would have generated a profit given by

Profit = Total income − Total cost = £360,000 − £244,000 = £116,000

Thus, meeting the order for an additional 2000 units at £30 has helped to increase our profits by £16,000. The important thing to note here is that, although the selling price of £20 per unit is less than the average cost per unit of £20.33, it is actually greater than the marginal cost of £12!

Another way of looking at this: In marginal costing, we consider the cost of a product when all of the fixed costs are removed. We arrive at this figure by calculating the cost of producing just one more unit – the difference in the cost of this unit and the previously manufactured one is the variable cost attributable to just one unit (i.e., the variable cost per unit). This assumes that the variable cost per unit is the same for all volumes of production output. This will usually be true for significant production volumes (note how the average cost tends towards a fixed value as the quantity increases in *Figure 7.28*).

TYK 7.31

An electric screwdriver manufacturer has a total production volume of 6000 units. The fixed cost of the operation amounts to £150,000 whilst the variable cost amounts to £20 per unit. Determine the average cost per unit and the total profit if the electric screwdriver sells for £65.

Profitability

Profit, or *return on capital employed* (ROCE), is the aim of every business. Being able to make a realistic forecast of profits is an essential prerequisite to making a financial case for investment. It is also an essential ingredient in any business plan.

The need to maximize profits should be an important factor in decision making. Traditional theory assumes that a company will invest in the most profitable projects first, and then choose projects of descend profitability, until the return on the last project just covers the funding of that project (this occurs when the marginal revenue is equal to the marginal cost).

The process of choosing projects is, however, much more complex. It may, for example, involve strategic issues (such as the need to maintain a presence in a particular market or the need to develop expertise in a particular technology with the aim of improving profits at some later date). Furthermore, many companies do not have sufficient funds available to reach the marginal position. Instead, they will rely on one or two 'hurdle' rates of return for projects. Projects that do not reach these rates of return will be abandoned in favour of those that are considered 'profitable'.

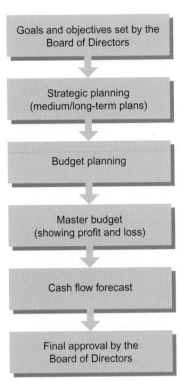

Figure 7.30 Process of financial planning and control

Financial planning

By now, you should have begun to understand how costing systems and techniques are applied in a typical engineering company. This next topic introduces you to some important aspects of financial planning and control.

Adequate financial planning is essential if a business is to achieve its objectives and profit targets. The basic procedure required to formulate a financial plan is as follows:

(a) Formulate company policy, profit targets and long-term plans.
(b) Prepare forecasts for sales, production, stocks, costs, capital expenditure and cash.
(c) Compile these separate forecasts into a master forecast.
(d) Consider all the alternatives available and select the plan that gives the best results, for example, in terms of profit and long-term financial stability.
(e) Review limiting factors and the principal budget factor. This process takes place concurrently with (d) and enables work to begin on the framing of the budgets in (f).
(f) Prepare individual budgets and finally the master budget, which includes a forecasted profit and loss account and balance sheet.

This process is illustrated in *Figure 7.30*.

Budget plans

The starting point of budget planning is with the board of directors (or equivalent body) who determines the scale and nature of the activities of the company. This policy and objective setting is done within the constraints that exist at the time. For example, plans may have to be made within the current capacities and capabilities of the company, since making changes to the location of operations, the size and composition of the workforce and the product range are usually long-term matters. The budget is essentially for the short-term, usually for 1 year, and created within the framework of long-term corporate planning. Successive budgets will be influenced by the preparation of long-term plans, but will always relate to the current period.

Some organizations prepare outline budgets over much longer periods perhaps for a 5–10-year horizon, but such budgets are really part of the long-term corporate planning activity and subject to major revision before being used as a basis for current period budgetary planning.

External factors will exercise considerable effects on the company in preparing its forecasts and budgets. Government policy, the proximity of a general election, taxation, inflation, world economic conditions and technological development will all combine to constrain or influence the budget planning process. Once the board of directors has settled on a policy within the prevailing situation, then the process of turning the policy into detailed quantitative statements can begin.

We normally assume a budget period of 1 year, which is usual for most industries. It is therefore recognized that the budget period is fixed in relation to the needs of the organization concerned and could be any

period ranging from 3 months to 5 years. The shorter the period the more accurate the forecasts will be and that is why most companies find that an annual budgeting procedure is a satisfactory compromise.

Business plans

On occasions, it is necessary to provide a detailed business plan in order to make a case for a particular business venture or project. Before a business plan is written, it is necessary to:

- clearly define the target audience for the business plan
- determine the plan's requirements in relation to the contents and levels of detail
- map out the plan's structure (contents)
- decide on the likely length of the plan
- identify all the main issues to be addressed (including the financial aspects).

Shortcomings in the concept and gaps in supporting evidence and proposals need to be identified. This will facilitate an assessment of research to be undertaken before any drafting commences. It is also important to bear in mind that a business plan should be the end result of a careful and extensive R&D project that must be completed before any serious writing should be started.

A typical business plan comprises the following main elements:

- An *introduction* which sets out the background and structure of the plan.
- A *summary* consisting of a few pages that highlight the main issues and proposals.
- A *main body* containing sections or chapters divided into numbered sections and subsections. The main body should include financial information (including a *profitability forecast*).
- *Market and sales projections* should be supported by valid market research. It is particularly important to ensure that there is a direct relationship between market analysis, sales forecasts and financial projections. It may also be important to make an assessment of competitors' positions and their possible response to the appearance of a rival product.
- *Appendices* should be used for additional information, tabulated data and other background material. These (and their sources) should be clearly referenced in the text.

The financial section of the plan is of crucial importance and, since it is likely to be read in some detail, it needs to be realistic about sales expectations, profit margins and funding requirements ensuring that financial ratios are in line with industry norms. It is also essential to make realistic estimates of the cost and time required for product development, market entry and the need to secure external sources of funding.

When preparing a plan it is often useful to include a number of 'what-if' scenarios. These can help you to plan for the effects of escalating costs, reduction in sales, or essential resources becoming scarce. During a *what-if analysis*, you may also wish to consider the *halve-double* scenario in which you

UNIT 7

examine the financial viability of the project in the event that sales projections are halved and costs and time are doubled. The results can be sobering!

When writing a business plan it is necessary to:

- avoid unnecessary jargon
- economise on words
- use short crisp sentences and bullet points
- check spelling, punctuation and grammar
- concentrate on relevant and significant issues
- break the text into numbered paragraphs, sections, etc.
- relegate detail to appendices
- provide a contents page and number pages
- write the summary last.

Finally, it can be useful to ask a consultant or other qualified outsider to review your plan in draft form and be prepared to adjust the plan in the light of comments secured and experiences gained.

TYK 7.32

Describe the main sections found in a business plan.

KEY POINT

A budget is a financial statement that relates to a particular period of time. A budget needs to be prepared and approved and is a means of meeting a company's objectives and financial goals

Budget plans

Budgets are used as a means of achieving planning and control objectives in most businesses and in many non-commercial organizations.

> **A budget has been defined as:** A financial or quantitative statement prepared and approved, prior to a defined period of time, of the policy to be pursued during that period for the purpose of attaining given objectives.

The benefits that derive from budgetary control arise from the ability to coordinate policy, plans and action and to be able to monitor the financial consequences of carrying out the plans.

An engineering company will prepare a number of budgets, each corresponding to a particular functional area. A named manager will normally control each budget although some managers may control several budgets according to the particular management organization employed within the company. In a typical engineering business you will find the following budgets:

- marketing budget
- manufacturing budget
- R&D budget
- administration budget
- capital expenditure budget
- cash budget.

Each of these budgets may be subdivided into further budgets. For example, the manufacturing budget may be subdivided into a budget for direct materials, a budget for direct labour and a budget for factory overheads (heating, lighting and other energy costs).

Each functional manager will forecast their own budget; however, there is a need for managers and departments to coordinate their budget activities. For example, the capital expenditure budget may reflect the purchase of major items of capital equipment (such as a forklift truck or an overhead crane) that will be shared by several departments.

Stock control

Stock control systems used within a manufacturing company will affect the way purchasing is done. Economic order quantities may be established, which the buyer has to take into account when arranging supplies. Deliveries may have to be phased according to minimum and maximum and re-order stock levels.

The firm's purchasing function will need a clear understanding of the importance of deliveries that enable the company to control its inventory costs, while at the same time ensuring a reliable supply of materials and components for the various production processes.

The company may operate a *JIT* system. It originated in Japan and it is a way of delivering supplies at the point in time they are required by production. JIT avoids the costs of holding buffer stocks of raw materials and components.

JIT works well when suppliers are dependable and when transport systems are good. The buyer will liaise with the factory on the establishment and operation of the JIT for given products.

There will also be the routine matters of passing invoices for payment of goods or dealing with returns for credit so that the accounts department can pay for goods received. Materials purchasing will be subject to budgetary constraints like most other company activities. The purchasing department will be involved, either directly or indirectly in budgets for inventory levels, and in setting up minimum, maximum and re-order levels for stocks.

Monthly monitoring of inventory levels will be done by the accounting function and purchasing activities may be responsible for ensuring that stock of components and raw materials stay within agreed levels.

Depreciation

Depreciation is the estimate of the cost of a fixed asset consumed during its useful life. If a company buys a car, to be used by a sales representative, for £15,000 it has to charge the cost in some reasonable way to the profit being earned. This process is essential, otherwise the whole cost of running the business cannot be obtained, and profit figures would be overstated. If it is estimated that the car will be worth £6000 in 3 years time when it is to be sold, then it could be charged to profit at (£15,000 − £6000)/3 = £3000 for each year of use.

UNIT 7

The way companies accumulate funds with which to replace fixed assets is to charge depreciation as an overhead cost. In the case of our car this recovers £9000 that together with the sale price of the used car generates a fund of £14,000 towards the purchase of a replacement. It is also possible that the sums so deducted from profit can be invested to offset inflation until the time comes to replace the asset.

Because depreciation is an estimate and is deducted from profit it has the effect of keeping the money available in the business. Many companies use the aggregate depreciation charged as the basis for the fund against which investment appraisal is done.

There are two main methods used to calculate depreciation. We will outline each in turn.

Straight line method

This charges an equal amount as depreciation for each year of the asset's expected life. It is called straight line method because if the annual amounts were plotted on a graph they would form a straight line. The formula is

$$d = \frac{p - v}{n}$$

where d = annual depreciation, p = purchase price, v = residual value and n = years of asset life. The method used for the car in the above example is straight line. It is a very popular and easy to use method.

Reducing balance method

In this method a fixed percentage is applied to the written down balance of the fixed asset. The formula for establishing this fixed percentage is

$$r = 1 - \sqrt[n]{\frac{v}{p}}$$

where r = the percentage rate, n = number of years, v = residual value and p = asset purchase price. Note that v must be a significant amount otherwise the rate will be very large and somewhat meaningless.

Using our car example above the reducing balance rate, r, is

$$r = 1 - \sqrt[3]{\frac{6000}{15,000}} = 1 - 0.7368 = 26.31\%$$

Applying this, the depreciation pattern is as follows:

Purchase price		£15,000
First year	26.31%	£3946
Reduced balance		£11,054
Second year	26.31%	£2908
Reduced balance		£8146
Third year	26.31%	£2145
Reduced value (approx)		£6001

The reducing balance system means that a greater sum is taken in earlier years, but reduces year on year since the percentage rate is applied to the reduced balance. This method is more complex but it is more logical because a new asset gives better service than an old asset and should suffer more depreciation in earlier years. Certainly, with regard to cars the early years' depreciation is very heavy in relation to resale prices.

TYK

TYK 7.33

Use the straight line model of depreciation to determine the asset value after 3 years of a CNC lathe purchased for £100,000 assuming that it has an asset life of 5 years.

KEY POINT

Depending on the return on investment, companies must decide whether to keep money in the bank or reinvest it in the business

Investment appraisal

The process of investment appraisal is necessary in order to select the best projects for investment, whether they be replacement machines for the factory, office equipment or new cars for managers and sales representatives.

Major factors to consider are:

- risk and return
- time scale
- time value of money
- evaluating alternative choices.

The first one, that of risk and return is important. A good general rule is to expect a high return if the investment is risky. That is why interest rates on mortgages are less than interest rates on loans for consumer durables. A bank will regard loans against property as low risk because a house is a better security than a refrigerator if something goes wrong.

For commercial businesses some capital investment projects carry low risk, such as those which merely involve a replacement of a machine that is being used to make a product for which there is a regular demand. However, the risk is much greater for investment in a new factory for a range of new products. These different risk profiles have to be taken into account when comparing different projects for investment.

Time scale is important too. If a project is going to take 5 years for completion, then many things could change before completion including inflation, Government action, market conditions, competitive pressures, etc. The investment decision must take account of all these risk factors as well and this is not an easy process, because of the need to forecast up to 5 years into the future.

Money has *time value*. This fact lies at the heart of the investment appraisal process. If you could choose to have £5 today or £5 in a year's time, which would you prefer and why?

Hopefully, you would take account of the risk of not getting your £5 in a year's time and would prefer to avoid risk and get it now. Also, you should have thought about the fact that if you had the £5 now, you could invest it for a year, so that it would be worth more than £5 in a year's time.

UNIT 7

The time value of money exists because of these two aspects: risk and return. Most people who decide to save money instead of spending it look for a secure and lucrative place to keep it until they wish to spend it.

If your risk preference is low you would probably put your money into a building society account. If you were more adventurous you might buy shares. The risks attached to buying shares are greater than a building society account, but the return may be greater.

The essence of the capital investment decision is no different to this. A company must decide whether or not to keep its money in the bank or invest in the business. If it chooses to invest in the business, the company must expect a better return than merely leaving it in the bank. It is for this reason that certain techniques for evaluating projects have arisen. There are numerous approaches, but the most theoretically sound method is called *Net Present Value* (NPV) technique.

Activity 7.27

Loudspeaker manufacturer, APS, has asked you to help it assess the financial implications of making two capital purchases: a new flow-soldering plant and a CNC table saw. The details of these two items are as follows:

Flow-soldering plant	Capital cost:	£90,000
	Asset life:	4 years
CNC table saw	Capital cost:	£75,000
	Asset life:	5 years

The company needs to know what the combined residual value of these two items will be after 3 years using the reducing balance method.

Write a brief word-processed report for the APS's board of directors and include all calculations.

Make or buy decisions

Make or buy decisions require the involvement of more than one functional area. They almost certainly require the involvement of the production function, production control and engineers from design and manufacturing.

Before an actual decision is made certain questions must be addressed:

(a) Could the item be made with existing facilities?
(b) If the answer to (a) is yes, is current production capacity adequate?

If the answer to (a) is no, then it indicates that new plant and equipment may be needed, or indeed that the company has no experience in this area of manufacture, which might therefore mean a move towards vertical integration, which involves a change to its basic business. The latter involves top-level policy decisions and will not yet even be considered at the functional level at all. Theoretically any item, whether already manufactured or purchased outside (i.e., *outsourced*) can be reviewed for alternative

sourcing, but in practice only those items will be considered for change that do not involve major changes to the business. Therefore, we will confine our analysis to decisions that can be made within the functional areas we described earlier.

Thus if a purchased item is being considered for manufacture then there must be suitable facilities available and sufficient capacity. If a manufactured item is being considered for purchase there must be alternative uses for the capacity or there should be an intention to reduce that capacity rather than leave it unused.

The other major factors to account for are:

- incremental costs of alternatives
- quality control problems
- multi-sourcing
- costs of tooling
- strategic importance.

Incremental cost is the additional or unavoidable cost incurred for an item purchased outside or being manufactured inside. The point to note is that many overhead costs will be unaffected by the make or buy decision. Thus overhead or indirect costs are almost certainly irrelevant to the decision and only the direct or marginal costs are important. Therefore comparison between the outside purchase price and the marginal or incremental manufactured price must be made to give a true picture.

If an item is being outsourced then quality control will need to be considered. Potential suppliers may not have in place suitable quality procedures or may not be part of any supplier quality assurance scheme. If multi-sourcing is necessary then this problem is even greater.

If an item for outside purchase is a special design to the company's own specification then capital investment costs will have to be accounted for. In these circumstances it is common for the customer and supplier to share the investment in tooling and this could be quite significant.

An item or component that is of strategic importance in the company's own manufacturing process presents risks if it is manufactured outside. This may be true if it is a crucial safety item or is something upon which other production processes depend. Thus a cessation of supply or quality problems could have serious consequences. Before arranging outside purchase these risks have to be evaluated and some form of risk reduction undertaken before committing to outside sourcing.

UNIT 7

TYK

TYK 7.34

Explain what is meant by a 'make or buy' decision. Also explain why such a decision can be important.

End of Unit Review

1. Identify and explain the effects of three external factors that affect engineering businesses.

2. Define and explain the terms:

(a) local economy

(b) regional economy

(c) national economy

(d) gross national product (GNP)

(e) gross domestic product (GDP)

(f) balance of payments.

3. Explain why Japanese firms sometimes find it advantageous to manufacture their products in the UK as opposed to Japan. Also explain the effect that this has on the UK's GNP.

4. Explain the difference between GNP and GDP.

5. The annual percentage change in GDP for the UK and Germany over an 8-year period is shown in the table below:

Year	1991	1992	1993	1994	1995	1996	1997	1998
Germany	5.0	2.2	−1.1	2.3	1.7	0.8	1.5	2.2
UK	−1.5	0.1	2.3	4.4	2.8	2.6	3.5	2.2

(a) Compare the performance of Germany and the UK between 1991 and 1993. Suggest reasons for these trends in relation to the national economies of the two countries.

(b) Determine the average percentage increase in GDP for each country over the full 8-year period. Which country has performed best over this period?

6. (a) Distinguish between visible and invisible exports.

(b) Give two examples of invisible exports.

7. Give two examples of the effect of relevant UK and EU environmental legislation on a typical engineering company.

8. (a) Explain the term 'by-product' in relation to an engineering process.

(b) Give two examples of by-products produced as a result of typical engineering processes and name the processes.

9. Describe, briefly, the effects of the following waste products produced as a result of engineering activities:

(a) carbon dioxide

(b) sulphur dioxide

(c) ozone

(d) smoke.

10. Engineering companies must organize their activities in such a way as to comply with relevant environmental legislation. Give two examples of environmental legislation that relates to engineering activities. In each case, describe typical measures taken to ensure that the legislation is complied with.

11. Explain why quality control is important in an engineering firm. Give examples of four typical activities that are performed as part of a quality control system.

12. Explain what is meant by 'JIT' production.

13. Define the following terms:

(a) fixed cost

(b) overhead cost

(c) variable cost.

14. The following figures relate to the production of a small component:

Fixed cost:	£25,000
Variable cost per unit:	£12
Selling price:	£26

Determine the break-even point in sales turnover.

15. Explain the concept of 'marginal cost'.

16. Sketch a graph showing how costs and revenues vary with units of output. Mark the break-even points on your graph.

17. Explain what is meant by incremental cost in relation to a 'make or buy' decision.

18. Determine the residual asset value after 3 years of a CNC milling machine purchased for £120,000 if the machine has an asset life of 5 years.

19. Explain three factors that must be considered in investment appraisal.

20. State two advantages and two disadvantages of outsourcing the manufacture of a component used in an engineered product.

Engineering Design

Design is the process of converting an idea or market need into the detailed information necessary to manufacture a product or deliver a service. Successful design involves combining a number of skills together with the use of available technologies, materials and processes and then being able to deliver the product or service at a cost that will make it competitive with other products and services.

Being able to design something is a fundamental engineering skill but simply being able to design something is not the end of the story. Equally important are the skills that you will need in order to be able to communicate your design to other people; even the most basic design will be hopelessly flawed if you cannot explain to people what it is about and how to make it! As an engineer, you might be involved with the design of an engineered product that varies from something as basic as a screwdriver to something as complex and sophisticated as a military aircraft. This unit will help you understand the design process and provide you with a variety of skills that will help you communicate your ideas to other people.

BTEC National Engineering. DOI: 10.1016/B978-0-12-382202-4.00008-8

The Design Process

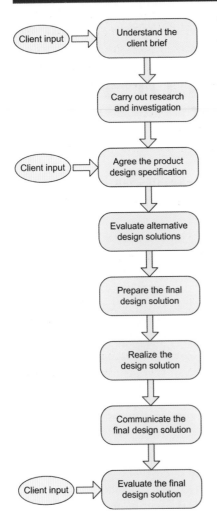

Figure 8.1 Stages in the design process

The design process is the name given to the various stages that we go through when we design something. Each stage in the process follows the one that goes before it and each stage is associated with a particular phase in a **design project**. You also need to be aware that design is not always about creating a brand-new product or service. Instead, it is often about improving or modifying an existing product. When designing a brand-new product or service, the design process will have more stages because we usually need to consider a wider range of options and alternative solutions than when we are simply modifying or redesigning an existing product.

A typical design project involves the following tasks:

- understanding and describing the problem
- developing a design brief with the client
- carrying out research and investigation
- generating ideas using techniques such as brainstorming and mind mapping
- investigating solutions and applying scientific principles
- developing an agreed set of design specifications
- communicating the design solution using appropriate engineering drawings
- realizing the design solution
- evaluating the design solution.

We shall look at what actually makes up some of these tasks later on. For now, you just need to be aware that design involves a series of inter-related tasks (some involving client input) and that each of these tasks forms a stage in reaching your eventual goal: an engineered product or the delivery of an engineered service (*Figure 8.1*).

Activity 8.1

According to the Royal Academy of Engineering, engineering design encompasses three key stages of realization:

1. need – all design begins with a clearly identifiable need
2. vision – all designs arise from a creative response to a need
3. delivery – all designs result in a system, product or project that meets the need.

Download a copy of the Royal Academy paper on Engineering Design Principles from the Academy's website at www.raeng.org.uk/education/vps/pdf/design_principles.pdf and use it to investigate two of the following case studies:

- the Tsing Ma bridge – designed by Mott MacDonald
- an intelligent prosthesis – Chas. A. Blatchford and Sons
- the Trent aero engine – Rolls-Royce
- a breath actuated metered dose – IVAX Pharmaceuticals.

Answer each of the following questions in relation to the two case studies that you have chosen:

1. What need did the design satisfy?
2. What features of the design were particularly innovative?
3. How was the design project delivered?

What makes a successful design?

Not all engineering designs are successful and some are spectacular failures such as the recall in February 2010 of several million Toyota vehicles directly attributable to inadequate design of the accelerator pedal. Most people agree that a commercially successful product design can be recognized by having several readily observable qualities. These include a product that:

- fully meets the needs and expectations of customers and clients
- is in demand and delights customers and clients
- is straightforward to manufacture using commonly available materials and well-understood processes
- can be brought to the market quickly and cost efficiently
- is both clearly distinguishable from, and superior to, similar products from competitors
- is likely to generate repeat business
- ensures profitability and generates the expected financial return
- enhances a company's reputation and place in the market.

Many of the above qualities go hand-in-hand. A product that delights clients and customers will most certainly also enhance the reputation of the company that has produced it. However, not all of the above qualities will necessarily apply to a particular product design. For example, a product sold below market price or one that uses very high-quality materials may satisfy clients and customers but in the end might not be profitable!

> **KEY POINT**
>
> Realizing an engineering design involves three key stages: need, vision and delivery

> **KEY POINT**
>
> A successful design can be recognized by a number of qualities and paramount amongst them is a high level of customer satisfaction

TYK 8.1

List each of the main stages in the design process.

TYK 8.2

List five ways in which we might recognize a successful design.

Activity 8.2

Brand Engineering manufactures a wide range of electrical switches, circuit breakers and fuses which are used in telecommunications and avionic equipment. The company has recently carried out a customer satisfaction survey (see page 644) and the Production Manager has asked you to analyse the data and report your findings to the company's Board of Directors. Prepare a briefing document for the Production Manager that makes at least three specific recommendations for further investigation. Use not more than a single word-processed A4 page.

UNIT 8

Product line	Total number of customers	Number satisfied	Number dissatisfied	Don't know
Switches	77	58	15	4
Fuses	125	119	3	3
Circuit breakers	59	39	11	9

Product line	Annual turnover 2007 (£ million)	Annual turnover 2008 (£ million)	Annual turnover 2009 (£ million)
Switches	2.8	2.8	2.7
Fuses	3.3	4.3	4.9
Circuit breakers	1.7	1.4	1.2

The customer/client relationship

It is widely recognized that the relationship between an engineering company and its client/customers is crucial in ensuring the long-term profitability and ultimately the survival of a company. This relationship is one that often takes time to develop and usually involves customer/client interaction at various levels and at various times. Many companies have established roles for *key account managers* where an individually named person within a company carries overall responsibility for maintaining an effective and profitable relationship with one or more of its customers or clients.

Internal and external customers

Customers and clients are not always external (i.e., outside a company). In many large organizations products and services (including the design function) can be sourced within a company itself and delivered from one department, profit or cost-centre, to another. Internal customers are just as important as external ones and it is important that everyone involved recognizes this fact!

Customer requirements

Customer requirements are often clearly stated before a design project begins. This statement of requirements usually takes the form of a specification but sometimes the client or customer's requirements are implicit. For example, the use or application of a product should be self-evident and it should also be safe to use and fit for purpose in all normal circumstances. So, for example, when a customer requires a car jack to be designed it should go without saying that the jack should be suitable for safely raising any standard vehicle without fear of collapse.

When dealing with a client it is important to ensure that you meet the client's expressed as well as their implied needs. If you have any doubt about this it is always best to seek clarification from your client. It is also worth remembering that one client may be different from another even though the purpose, function and application of a product may be the same. For example, a customer in the USA will want to operate an electrical

appliance from a 60 Hz 115 V AC supply whilst one in the UK will wish to use the same appliance with a 50 Hz 230 V AC supply.

Customers may require widely different volumes of a product, from a 'one-off' custom-built prototype, small batch or high volume production. They may also wish to be involved with the production and manufacturing process or have no part in it. In order to have *traceability*, they may wish to have parts, components and materials inspected and/or certified before use. They may also wish to impose their own quality assurance standards or simply receive confirmation of the standards that are currently used within an engineering company.

Design triggers

The design process usually begins when there is an identified need or when there is a problem to solve. The problem could be based around developing a new product or service or it could involve modifying or improving an existing product or service. You cannot begin the design process unless you have a clear understanding of what it is that the client needs or what the problem is that you are being asked to solve.

The need for a design may also be triggered by a number of other factors including the results of market research, the need for innovation or the availability of new manufacturing processes such as *computer integrated manufacture* (CIM). In addition, two main strategies are employed by engineering companies when they seek to develop new products. These techniques are usually referred to as 'technology push' and 'demand pull' and because they have a direct impact on the design process it is worth exploring them in a little more detail.

TYK 8.3

Explain why customer relationships are important.

TYK 8.4

(a) Explain what is meant by a 'design trigger'.
(b) List three different types of design trigger.

Technology push and demand pull

Technology push is product development that is driven by ideas or capabilities created by the development organization in the absence of any specific need that customers may have. In technology push situations, innovations are created and then appropriate applications or user populations are sought that fit the innovation.

Demand pull, on the other hand, is product development that is driven by user needs and requirements (i.e., demand) rather than by technologies, ideas or capabilities created by the manufacturing company itself. Note that demand pull is sometimes also known as market pull.

By responding quickly to clients' needs and providing a quick return on investment, market pull is often regarded as a 'cash cow'. Technology pull, on the other hand, can involve a significant 'up front' investment that may only yield a return at some time in the future. Many successful companies use a mixture of both market pull and technology pull strategies. The balance between the two strategies will then usually depend on just how much risk a particular company is prepared to take at any particular time!

Technology-driven companies look beyond the limitations imposed by current technologies and manufacturing capabilities. They assume that the needs of tomorrow's clients and customers will not be met with currently available technologies. These companies are often prepared to develop completely new technologies and manufacturing capabilities without necessarily knowing what the market interest will be for any products that are developed from these technologies manufacturing capabilities. Once the new technologies and manufacturing capabilities are developed, these companies believe that a search for market opportunities will yield commercially successful products.

Some companies adopt an approach that is predominantly **market-driven** while others are mainly technology-driven. Some claim that they are market-driven but are actually sales- or manufacturing-driven. Such companies tend to only develop products in response to the requirements of their existing client base and also within their existing manufacturing capabilities. Successful market-driven companies need to be fully aware of their marketplace. They need to ensure that new products are developed that will meet the needs of existing clients as well as new customers. Market-driven companies will need to grow their market share by concentrating on using the technologies they already have or those that can be easily and cost-effectively acquired.

After a while, a successful technology-driven product will often become pulled by the needs of the marketplace. Until this happens, the product is unlikely to be a huge commercial success. Ideally, the transition from technology push to market pull will happen as quickly as possible but that is not always the case.

Companies that are mainly market-driven can often become vulnerable in a market downturn. This is because there is a very real risk of losing a market to one or more aggressive competitors. With no new technology products to bring to the market, market-driven companies are often unable to diversify and so must work doubly hard in order to hold onto their existing markets. Conversely, companies that are mainly technology-driven can be exposed to the risk of not finding a market for the new technology products that they have developed. They may have invested significant time, energy and capital in the development of a technology that might yield no return. Companies prepared to accept this risk tend to be larger, well-established companies able to make a significant investment in research and development. Success is not guaranteed and a long-term strategy is essential (*Figure 8.2*).

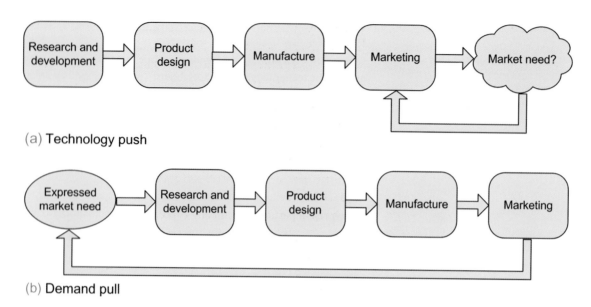

(a) Technology push

(b) Demand pull

Figure 8.2 Technology push and demand pull

Activity 8.3

Use library and Internet resources to research the effect of technology push and demand pull on the design and development of mobile phones in the UK over the last 20 years. Be sure to include mention of some or all of the following technologies/capabilities in your report and whether you would consider them to be examples of technology push or demand pull:

- Bluetooth
- Enhanced Data Rates for GSM (EDGE)
- General Packet Radio Service (GPRS)
- Global System for Mobile Applications (GSM)
- High-Speed Downlink Packet Access (HSDPA)
- Second Generation mobile (2G)
- Short Message Service (SMS)
- Third Generation mobile (3G)
- Wireless Application Protocol (WAP) (*Figure 8.3*).

Figure 8.3 First, second and third generation mobile phones

UNIT 8

Activity 8.4

Scotchgard™ is a product that renders a fabric material impervious to dirt and grease. The technology behind Scotchgard™ was developed by 3M in the 1950s. At that time no commercial market applications had been identified for the technology. Use library and/or Internet resources to investigate the development of Scotchgard™ and present your findings in the form of a technology magazine article of not more than 750 words.

TYK 8.5

Explain what is meant by:
(a) technology push
(b) demand pull.

TYK 8.6

State two advantages and two disadvantages of product development strategies based on:
(a) technology push
(b) demand pull.

Product design specifications (PDS)

Before you can begin a design, you need to have a good understanding of what it is that your client needs. Normally you can do this by developing a design brief with your client. However, on its own, the design brief does not include sufficient detail to check that each aspect of the design criteria has been met. For this, you need a much more detailed document known as a product design specification (PDS). Later we will look at what exactly should go into a PDS but for now we will concentrate on the design factors. These vary from product to product but they typically include:

- functionality (what the product should do)
- physical size and weight (how big it should be and how much it should weigh)
- aesthetics (what the product should look like)
- ergonomics (how easy it will be to use the product)
- reliability (how reliable the product will be)
- maintainability (how easy will it be to repair if the product goes wrong?)
- manufacturability (how easy will it be to build and assemble the product?)
- compatibility (what other products/systems must it work with?)
- efficiency (how much energy does the product use and how much is wasted?)
- cost (how much does the product cost to manufacture, operate and maintain?)
- compliance (with standards and other relevant legislation).

Having considered each of these factors in relation to a particular design requirement, a designer can arrive at a set of design criteria which gives details of the performance of the product against each of the relevant factors. It is worth noting that not all of the design factors will be relevant to a particular product design. For example, maintainability might be very important in the case of a car or motorcycle but it will be irrelevant to a disposable product such as a toner cartridge for a laser printer.

To help you appreciate the thought process that an engineering designer has to go through and the questions that they need to answer before finalizing a particular set of design criteria, here is a list of the design factors that you might need to consider if you were about to design a battery charger:

Application

It helps to know what the charger is to be used for, what type of battery it is to work with and any unique features or characteristics that it might be required to have. You will also need to know about the intended target markets for the product and what their needs and expectations are.

Physical dimensions and weight

What should the physical size and weight of the charger be? This could be important in determining the technology used in the charger's design (e.g., whether it should be based on conventional linear technology or whether it should use much lighter and more compact switched mode circuitry).

Enclosure material and finish

What type of enclosure will be appropriate for the finished charger? Should the enclosure be made from steel, aluminium extrusion, or moulded ABS plastic? Should the enclosure be anodized, given a paint finish or should it use a coloured plastic material?

Supply voltage

Will the battery charger be used in the UK or should it be suitable for wider European/international use with a range of input voltages? If the product is to be used in the USA then it will need to be able to operate with a 110V AC supply. Alternatively, if the charger is to be used in a vehicle or light aircraft it will need to be used with a nominal 12 or 24V DC supply.

Mounting

For smaller low power chargers (say up to 20W) it often makes sense to use a simple wall-outlet mounted unit or simply mould the unit into a mains lead assembly. Such units can be relatively inexpensive but higher powers (above 20W or so) will need a separate enclosure fitted with input and output connectors as well as any necessary controls and indicators.

Energy efficiency

In many cases energy efficiency will not usually be important but for some applications (such as battery chargers that operate from solar photovoltaic panels or from hand-operated generators) high efficiency is crucial in order to

avoid wasted power. In larger battery chargers there will also be a need to avoid excessive heat dissipation and so, once again efficiency might be important.

Battery type

Different types of battery have different charging requirements. For example, lead-acid batteries need a constant voltage charging source but nickel-cadmium (Ni-Cd) batteries need constant current charging. You will need to refer to the literature supplied by the individual battery manufacturer in order to determine the appropriate charging method. This will be essential if you want to ensure optimum performance and also to maximize battery life. Note also that the performance of most batteries tends to vary with ambient temperature. Compensation will therefore be required in applications where performance has to be maintained over a wide range of temperatures.

Charging rate

Charging time and current is usually dependent on the battery type as well as the charger power. Generally a faster charger is more expensive because it requires a larger power supply but once again it is essential to ensure that you comply with the individual battery manufacturer's recommendations. Failure to observe the charging recommendations can result in significantly reduced battery life and in extreme cases may result in an unacceptable risk of fire or explosion!

Battery discharge and cycling options

With Ni-Cd batteries it is desirable to ensure that a battery is fully discharged before commencing a full charging cycle. This helps to maximize the battery's service life (i.e., the total number of charge/discharge cycles).

Controls and indicators

In many cases you will need to provide the end-user with a means of knowing what the state of charge is and whether or not a battery has reached a fully charged state. There are various ways of doing this, including the use of LED indicators, LCD screens, meters and lamps. In order to set different charging rates, adjustable controls, using switches, buttons and variable resistors, may also be required.

Protection

There will invariably be a need to protect both the charger and the battery from component failure, inadvertent misconnection (e.g., reverse polarity) or a short-circuit at the output. Once again, various methods can be employed ranging from simple fusing arrangements to fast-acting electronic cut-outs. There may also be a requirement to protect a charger from excessive temperature rise. This can easily be achieved by the use of a thermal cut-out. Supply borne surges, that can damage electronic charging circuitry, can be eliminated by the use of voltage-dependent surge suppressors.

Environmental conditions

This may not be too much of a concern for equipment that is designed for use in a domestic or laboratory environment but could be vitally important when a charger is to be used in a harsh environment such as a military aircraft, ship or space vehicle. In these environments the charger might

need to operate in a range of ambient temperatures, humidity and also in the presence of shock and vibration.

Safety requirements

Most customers will have a requirement for safety agency approval. All products used in Europe will require Certification Europe (CE) conformance marking and similar standards will apply in most other countries. In all cases it will be necessary to ensure that safety standards of insulation and the use of appropriately rated components is observed.

Electromagnetic compatibility

Depending on the market area, compliance with relevant electromagnetic compatibility (EMC) and electromagnetic interference (EMI) standards may be required. For example, any product containing a microprocessor or a switching power supply will be subject to EMC regulation in Europe and also FCC regulations in the USA. Both may require third-party verification that the relevant EMC and EMI standards can actually be complied with.

Aesthetics and ergonomics

You will need to consider what the charger will look like and whether it might need to be given 'visual appeal'. This will be relatively unimportant for industrial and commercial applications but could be important in the case of equipment for domestic and entertainment applications. Ergonomic considerations apply to how easy and intuitive the equipment is to use. You will need to think about carrying and handling arrangements as well as accessibility and visibility of controls and indicators. You should also consider use by people with visual or other impairment. For example, audible warnings may not be heard by a person who is deaf but they might also not be heard when the equipment is used in a noisy environment.

Target price

Cost is affected by various factors including the power, size and technology used in the charging circuit as well as the enclosure and finish.

Maintenance, servicing and disposal

You may need to consider what features should be incorporated in your design in order to facilitate the maintenance, service and eventual disposal of your battery charger. Low cost chargers, such as those that are supplied in moulded enclosures may not be considered as being maintainable. The reason for this is simply because the cost of maintenance and repair could easily exceed the cost of replacing the entire unit. With larger, more sophisticated and more expensive chargers it is often worth considering how you will provide access to internal parts and components for replacement in the event of failure. It would also be worth ensuring that you only use commonly available parts that are easily replaceable. Parts and circuit boards should carry identification and test points should be provided in order to facilitate the connection of test equipment. In order to avoid inadvertent misconnection, connector pairs should be polarized and non-interchangeable with other connectors. Finally it is important to think about what happens to the equipment when it reaches the end of its

service life. This is particularly important in the case of batteries and other electronic waste which must no longer be disposed with normal domestic waste.

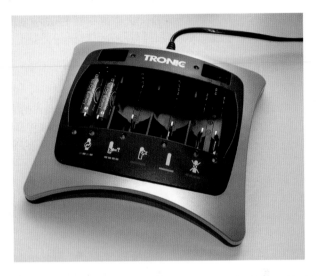

Figure 8.4 Example of a well-designed battery charger for Ni-Cd and nickel metal hydride batteries of various types. The charger automatically checks the state of each cell and incorporates controlled discharge facilities. Note also that the designer has addressed ergonomic and aesthetic requirements of the design brief

TYK 8.7

State the terms that are used to describe each of the following:
1. The various stages that are followed when preparing a design.
2. A list of the factors that need to be taken into account in the design of an engineered product.
3. The ability of a product or service to meet a given standard or specification.

TYK 8.8

List at least six factors that should be considered when arriving at a set of criteria for each of the following engineered products:
1. A wheelchair for a disabled person.
2. A portable MP3 player.
3. A garden wheelbarrow.
4. A supermarket trolley.
5. An adjustable spanner.

Activity 8.5

Think about an engineered product that you use in everyday life such as a car, computer or multimedia player. Write down a list of three or four features or improvements that you would like to incorporate in the product. For each feature or improvement suggest what modifications or changes would have to be made. Present your work in the form of a single A4 page and include hand-drawn sketches where appropriate.

Activity 8.6

In relation to the previous activity, select just one of the features or improvements that you have identified for your chosen engineered product. Now try to think of the problem that has led you to suggest that this feature needs some attention. Summarize the problem in a single paragraph using no more than three sentences.

Legislation and Standards and Other Design Constraints

In order to bring a product to the market a manufacturer must demonstrate that the product is safe to use, would not adversely affect the operation of any other product or service, and can be safely disposed of. In some instances, failure to meet the relevant standards and directives can result in costly and protracted legal action.

Standards and legislation

Standards are published specifications that establish a set of common definitions and criteria that should be complied with. Standards are designed to be used consistently, as a set or rules, guidelines and definitions. Standards are applied to many materials, products and services in order to improve the effectiveness and reliability of products and services. Laws and regulations often refer to underpinning standards. Compliance with the relevant standards then becomes a mandatory requirement.

European standards

Equipment designed for sale or use in any European Union country must comply with the relevant *directives*. These define a set of requirements but leave it to the standards (primarily European harmonized standards) to define the technical requirements. To put it simply, a manufacturer (or the manufacturer's European representative) must first ensure that a product complies with any applicable directives before *CE marking* the product.

Displaying the CE mark on a product (and/or its packaging) is mandatory for most types of product. The CE mark indicates that the product complies with the relevant European Commission (EC) Directives. You will probably not be surprised to learn that there is even an EU Directive that governs how the CE mark is used (*Figure 8.5*).

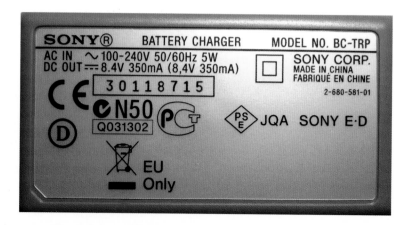

Figure 8.5 Example of CE marking. The product that carries this marking is a camcorder battery charger. Note that the CE mark appears to the left of the charger's serial number

UNIT 8

You need to be aware that the CE mark does not indicate conformance with a particular standard. Instead, the CE mark indicates that a product complies with all of the standards that might apply to it. This means that, for example, a child's toy and a microwave oven will both carry the CE mark even though they meet a totally different set of standards.

It is important to note that many products need to demonstrate compliance with more than one EU Directive and engineering companies need to be fully aware of all of the directives that could potentially apply to the products that they design and manufacture. Furthermore, since legislation is constantly changing, it is necessary to keep a watchful eye on any new or revised directives. Here are some of the EU Directives and associated UK regulations that you need to be aware of.

The EMC Directive

The EMC Directive of the EC has widespread implications for any engineered product that uses electricity or electronics. The Directive states that products must not emit unwanted electromagnetic pollution which might otherwise cause interference to other appliances and services. Equally important is that the Directive also states that products must themselves be immune to a reasonable amount of interference. As with all CE mark directives, the primary purpose of the EMC Directive is the creation of a single market for electrical goods throughout Europe. The protection requirements of the Directive are the means by which this is achieved, not the fundamental objective. In contrast to all the other CE mark directives, the EMC Directive's primary requirement is the protection of the electromagnetic spectrum, not the safety of equipment.

The Low Voltage Directive (LVD)

The Low Voltage Directive (LVD) relates to electrical equipment installation including cables, flexible leads and wiring. The Directive applies to equipment that operates from 50 to 1000 V AC or from 75 to 1500 V DC.

The Machinery Safety Directive (MSD)

The Machinery Safety Directive (MSD) relates to a wide range of products that comprise a number of linked parts (at least one of which moves) and a source of energy that is other than human. Note that there is a list of exceptions to this Directive and it includes tractors, military and police vehicles as well as freight and passenger carrying vehicles.

The Waste in Electrical and Electronic Equipment Directive (WEEE)

The Waste in Electrical and Electronic Equipment Directive (WEEE Directive) aims to minimize the impact of electrical and electronic goods on the environment, by increasing reuse and recycling and reducing the amount of WEEE going to landfill. It seeks to achieve this by making producers responsible for financing the collection, treatment and recovery of waste electrical equipment, and by obliging distributors to allow consumers to return their waste equipment free of charge.

The RoHS Regulations

RoHS stands for the 'restriction on the use of certain hazardous substances in electrical and electronic equipment'. RoHS is a European Directive which aims to control the use of certain hazardous substances in new electrical and electronic equipment (EEE). The RoHS Regulations require manufacturers to limit the amount of any hazardous materials contained in products and to be able to demonstrate that products are compliant. The RoHS Directive applies to products placed on the European market. In the UK the RoHS Directive is implemented through the RoHS Regulations.

British Standards

The British Standards Institution (BSI) was the first national Standards making body in the world. Independent of Government, BSI is a non-profit making organization. BSI is globally recognized as an independent and impartial body serving both the private and public sectors, working with manufacturing and service industries, businesses and Governments to facilitate the production of British, European and International Standards.

British Standards are created by appropriately qualified and experienced people who are brought together (for the specific purpose of creating a standard) by BSI. They discuss and agree on the details that will form a British Standard. Once a consensus has been reached, a draft of the new standard is released, and anyone with an interest is invited to comment. After all comments have been reviewed, the new standard is published as a British Standard. BSI's registered certification mark, the Kitemark and its Registered Firm symbol are instantly recognized by customers and suppliers, respectively. They show that a company is committed to improving the quality and reliability of its products (*Figure 8.6*).

> **KEY POINT**
>
> BSI is the UK's National Standards Body (NSB) and it represents the UK interests across all of the European and International Standards committees

> **KEY POINT**
>
> The CE mark shows that a particular product complies with the relevant European Directive and so can legally be sold in any EU state

> **KEY POINT**
>
> Compliance is about being able to demonstrate that a set of prescribed criteria are met (there may be several ways of demonstrating compliance with a particular EU Directive)

Figure 8.6 This mains extension reel carries both a CE mark (left) and the BSI Kitemark (right)

TYK 8.9

Explain how standards are used and why they are important.

UNIT 8

TYK 8.10

Sketch (a) the CE mark and (b) the BSI Kitemark. What do each of them indicate?

TYK 8.11

Briefly explain the role of the BSI in the preparation and dissemination of new standards.

TYK 8.12

Figure 8.7 shows the marking on an RF-switched mains outlet.
1. What British Standard is referred to?
2. What is the power rating of the socket?
3. What radio frequency is used to control the socket?
4. Can the product be legally used in the UK? How do you know?
5. Can the product be disposed of along with normal domestic waste? How do you know?

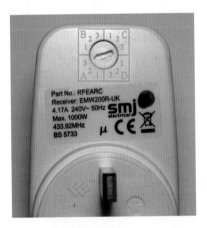

Figure 8.7 See TYK 8.12

Activity 8.7

Dolphin Electronics, a supplier of marine electronic equipment, has asked you to advise the company on the design of a VHF radio transceiver for yachting and small boat enthusiasts. The radio is to be used in inshore waters and on lakes and rivers and it will need to comply with relevant UK and international legislation. Investigate the standards and legislation which might apply to this type of equipment (e.g., output power, frequency range) and present your work in the form of a briefing paper for the Design Team at Dolphin Electronics. Make sure that you list your information sources and any relevant legislation.

Activity 8.8

A variety of hazardous materials are used in some engineered products and their associated manufacturing processes. Identify three such materials and explain what they are used for. For each material summarize any legislation that might apply to its use and disposal. Present your findings in the form of a brief PowerPoint presentation. Do not forget to mention the references sources that you have used!

Activity 8.9

Use library and/or Internet resources to investigate the disposal of electronic waste (such as computers, printers and monitors) in the UK. How should such waste be disposed of and what should happen to it? What markings should be included on electronic equipment? Present your work in the form of a brief article (not more than 750 words) for your local newspaper. Your article should be suitable for a non-technical audience.

Activity 8.10

Figure 8.8 shows the marking on the packaging of a domestic food mixer. Explain the meaning of the RoHS marking. What does this tell you about the appliance? Visit the RoHS website at www.rohs.gov.uk and use this to obtain a copy of the leaflet 'What is RoHS?'. Use the leaflet to answer the following questions:

1. What hazardous substances are restricted under RoHS?
2. What organization is responsible for enforcing RoHS?
3. What Statutory Instrument (SI) underpins RoHS?
4. Does RoHS apply to imported or badged equipment?
5. What does a manufacturer need to do in order to demonstrate compliance with RoHS?

Figure 8.8 See Activity 8.10

Environmental constraints and sustainability

The impact on the environment is increasingly becoming an important consideration in the design and manufacture of engineering products. In addition, issues relating to energy efficiency and sustainability need to be taken into account. A good example of how these issues have impacted on engineering designers is the introduction of flexible-fuel, dual-fuel and electrically powered vehicles. Another example of how environmental constraints have affected the design and manufacture of engineered products is the introduction of low-energy light bulbs. There has been considerable investment which has improved the performance and reduced the cost of low-energy lamps and they are now commonplace in homes and offices in the UK and in the rest of the world.

The Environmental Protection Act (EPA) 1990

The Environmental Protection Act (EPA) 1990 seeks to prevent the pollution from emissions to air, land or water as a result of manufacturing

and production processes. Authorization to operate the relevant processes must be obtained from the enforcing authority which, for the more heavily polluting industries, is HM Inspectorate of Pollution. Control of pollution to air from the less heavily polluting processes is through the local authority.

Regulations also place a ***duty of care*** on all those involved in the management of waste, be it collecting, disposing or treating ***controlled waste*** which is subject to licensing.

In addition to extending the existing ***Clean Air Acts*** by including new measures to control nuisances, the EPA introduces litter control, amends the Radioactive Substances Act 1960, regulates the introduction of genetically modified organisms, controls the import and export of waste and regulates the supply, storage and use of polluting substances. The EPA also requires local authorities to maintain a register of contaminated land.

TYK *8.13*

What is a 'duty of care' and why is it important?

Activity 8.11

Use library and/or Internet resources to investigate the introduction of flexible-fuel vehicles (FFV) and dual-fuel vehicles. Which manufacturers have developed FFV and dual-fuel models? In which countries have they been introduced and why? Present your answer in the form of a brief illustrated article for a car enthusiast's magazine.

Activity 8.12

Use library and/or Internet resources to investigate the introduction of low-energy light bulbs. Explain the system used for energy rating and give examples that contrast the performance of low-energy lamps with conventional incandescent light bulbs. Explain how the improved performance is achieved and describe the technology that is used. Present your answer in the form of a brief illustrated article for a DIY enthusiast's magazine (*Figure 8.9*).

Manufacturing constraints

Manufacturing constraints relate to a number of aspects that can impact on the design of a product. For example, a design might fully satisfy the original design criteria and a prototype might work well but it might not be possible to manufacture the product in any quantity due to a shortage of material, labour or processing plant. In fact, there is little point in pursuing a design that you know cannot be manufactured in the required quantities. It would be far better to modify the design by selecting alternative materials and/or processes so that quantity manufacture is possible.

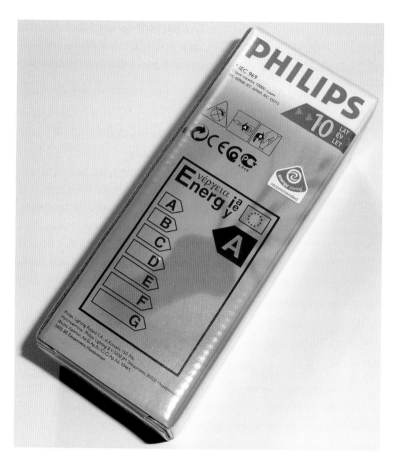

Figure 8.9 Packaging for a low-energy light bulb conforming to IEC 969

Another important manufacturing constraint is cost. This must take into account not just the cost of materials and labour but the initial cost associated with set-up and tooling as well as the cost of supporting and maintaining a product. The designer needs to be aware of the cost-effectiveness of their design as well as operating within the allocated budget. A detailed product costing (including quantities) must always be an essential part of the design process. There is little point in producing a design that will be rejected on grounds of cost.

When selecting materials it is important to ensure that they are safe to work with and will not be hazardous to the health of those that are involved with the manufacturing process as well as those that might use (or even misuse!) the product after it has been manufactured.

The Care of Substances Hazardous to Health Act (COSHH)

Some manufacturing processes involve substances that might potentially be hazardous to the health of those involved with the manufacturing process and possibly also to the health of the end-user of the product if the product is misused in some way. Other manufacturing processes may create hazardous by-products and/or waste materials and can also result in health problems unless effective controls and precautions are taken. Many products are immediately recognizable as hazardous. Such substances include paint, bleach, asbestos, dust, hydraulic fluids.

UNIT 8

COSHH covers chemicals, products containing chemicals, fumes, dusts, vapours, mists and gases, and biological agents (germs). If the packaging has any of the hazard symbols then it is classed as a hazardous substance.

Exposure to hazardous substances can be controlled by various measures including:

- finding alternative material and processes that do not use hazardous substances
- understanding the health hazards and their effects
- carrying out a detailed risk assessment
- reducing exposure by providing control measures and making sure they are used
- providing information, instruction and training for employees and others
- providing monitoring and health surveillance in appropriate cases
- effective planning for emergencies.

KEY POINT

A variety of different constraints will impinge on the design of any engineered product. They include the need to comply with relevant standards and legislation as well as the ability to manufacture, maintain and support the product on a cost-effective basis

TYK 8.14

What substances are covered by COSHH?

TYK 8.15

What are control measures and what control measures can be used to reduce the risk of exposure to hazardous substances?

Activity 8.13

Visit the Health and Safety Executive website (www.hse.gov.uk) and investigate the signs that are used to identify hazardous substances. Prepare an A4 handout showing the signs (both existing signs used in the UK and the new international signs, where appropriate) for each of the following hazardous substances:

(a) toxic substance
(b) compressed gas
(c) harmful substance
(d) flammable substance
(e) explosive material
(f) corrosive material
(g) oxidizing material
(h) irritant.

Activity 8.14

Bob North worked as a wireman for a manufacturer of electrical switchgear until he was forced to take early retirement in 2004 due to ill health. His illness was later attributed to chronic asthma contracted as a result of prolonged exposure to solder fumes. A Health and Safety Executive (HSE) investigation revealed that rosin-based solder was being used in the company and that fume extraction equipment had been faulty since 2001 and had not been repaired or replaced after that time. What control measures should have been in place that could have avoided Bob North's condition? Present your answer in the form of a brief report (of not more than 750 words) to the company's Health and Safety Officer.

Preparing Design Proposals

As part of the assessment of this unit you will need to show that you can produce an effective design proposal that fully meets the requirements of a PDS. This is a skill that can be acquired with practice and it involves three basic stages:

- analysing (and fully understanding) the PDS
- carrying out research and examining alternative design solutions
- producing and evaluating the design proposal (this will involve access to design reference material, catalogues, drawings, etc.).

We will start this section by returning to the all-important PDS and looking at how this informs the process of preparing and delivering a design proposal.

The product design specification (PDS)

Earlier we mentioned that the PDS is the key document that informs the design process and ensures that all of the design criteria and client/customer's needs are met. In this section we shall expand on this theme by providing you with information that will help you to prepare your own PDS. We will start by looking again at the design brief and how this helps a designer understand the client or customer's requirements.

The design brief

A design brief is a statement that identifies what is required in order to satisfy a given need or solve a design problem. The design brief must not be vague or too long. It must also avoid any preconceived ideas of what the design solution might be. The design brief should not actually suggest the solution to the problem. Instead, it is for you, the designer, to suggest ways of solving the problem which fulfils the design brief and solves the client's problem in the most effective and appropriate way.

The wording of a design brief is important. You should only use simple words and you will need to agree these with your client before you start to look for ways of solving the design problem. If the client provides you with the design brief (rather than you having to develop it with your client) you need to make sure that you understand it fully before going any further.

Here is an example of a very specific design brief:

Design an electrically-driven pump that will raise $0.25\,\text{m}^3$ of water through a vertical height of $2.5\,\text{m}$ in a time of $5\,\text{min}$.

I hope that you have spotted that this design brief is *very* precise! In fact, it is so precise that it quotes *figures* for the volume of water, the height that it must be raised and the time that it must take to do it. These figures are part of the specification for what the engineering system must do.

TYK *8.16*

What is a 'design brief' and why is it important?

TYK *8.17*

What would you expect to see in a 'design brief' for a pair of headphones for use with a CD or MP3 player? List at least five key points.

The design specification

A specification is a set of detailed requirements that must be satisfied by the design solution. The specification must be written down. It must be accurate and precise. A PDS will help a design engineer by giving them a very clear indication of the required performance of the product or service that is being designed. It is therefore important to establish the design specification with your client at a very early stage in any design project.

The design specification is something that can be *measured*. The performance of your chosen design solution will need to be measured and compared with the design specification. A successful design solution is one that fully meets or conforms to the design specification. Note that it may well be necessary to go back to your client if, for some reason, you find that your design solution only partially satisfies the design specification. Your client will have a view as to whether this is a problem that can be tolerated or whether it must be resolved.

British Standard PD 6112 identifies the items that should be included in a design specification. These include the following.

Title

The name or brief description of the product or service that you are designing.

History and background information

A general description of the design problem including, where appropriate, details of the target market for the product.

Design criteria/scope

A description of each of the required features of the product or service. These may include characteristics such as size (e.g., 'must be compact', 'must be at least 2 m wide'), surface finish (e.g., 'must be smooth', 'must be water resistant') and performance (e.g., 'must be fast', 'must be powerful', 'must be efficient'). The scope of the specification may also include any special needs of the people using the product, for instance if the users are elderly or have specific disabilities.

Definitions

Definitions should be included for any terms that your client may be unfamiliar with or that may have a special meaning. For example, a 'fuel strainer' may be defined as 'a fine gauze mesh that prevents dirt particles entering a carburettor'.

Conditions of use

You will need to give an indication of how, when and in what situation the product or service will be used. For example, 'continuous use', 'light domestic applications', 'commercial and industrial use', 'intermittent use', and so on.

Characteristics

A simple break down of the main components of the product, what their function is and any other features that the client needs to be aware of.

Reliability

An estimate of the working life of the product (e.g., 4000 hour, 10,000 operations and so on).

Servicing features

A description of the maintenance requirements and the provision for repair or replacement of materials and parts that might wear out (e.g., coolant, lubricating oil and so on) or that may need repair if the product fails or breaks down.

Other requirements that usually need to be included in a specification might take into account some or all of the design factors that we met earlier, including the following.

Ergonomics

Details about how users will interact with the product or service.

Aesthetics

Information relating to the appearance (e.g., shape, colour, surface finish) of the product or service.

Safety

Any specific safety features that relate to the product or its use.

Economics

The cost of producing, marketing and supplying the product or service when compared with the income that is generated from its sale or use.

Manufacture

Information relating to how the product will be made or assembled (e.g., 'using pre-fabricated parts assembled on-site', 'using fully automated assembly techniques', 'supplied in knock-down form for self-assembly' and so on).

Technical specifications

In addition to design specifications you will often see technical specifications that give key performance data for a product, such as supply voltage, power, efficiency, etc. For example, the technical specification for a variable speed cordless drill might be as follows:

Voltage	18 V
Battery capacity	2.6 Ah
Charge time	1 h
Chuck capacity	1.5–13 mm
Power output	400 W
No load speed	0–500 and 0–1700 rpm
Maximum torque	44 Nm
Drilling capacity	38 mm (wood), 13 mm (steel)
Weight	2.3 kg
Dimensions	Length 210 mm, height 239 mm, width 95 mm

KEY POINT

To check that an engineered product works as planned, it is necessary to measure its performance and then compare it with the performance that you expect. This should be written down in the form of a detailed technical specification. A specification is vitally important; without it you have no means of knowing whether a product satisfies a particular set of design criteria

TYK 8.18

Explain what is meant by a technical specification and how this differs from a PDS.

TYK 8.19

Explain how a technical specification is used in the design and testing of a typical engineered product.

TYK 8.20

List four items that should be included in the technical specification for:
(a) a soldering iron
(b) a car battery
(c) a fuel pump.

Activity 8.15

Obtain a full set of technical specifications for two comparable multi-speed bench drills designed for use by DIY enthusiasts. Explain, using your own words, what each specification means and compare the features, performance and price of the two drills. Which drill would you recommend, and why would you recommend it? Present your work in the form of a review for a DIY magazine.

Activity 8.16

Caravex, a major supplier of caravan equipment and accessories, have asked you to advise them on the design of an inverter that will provide AC mains power for use in a tent or caravan. The inverter is to provide sufficient power to operate a small TV receiver or laptop computer and should be robust and compact. Develop an outline technical specification and design proposal for a suitable inverter. Present your work in the form of a brief PowerPoint presentation and include relevant charts and diagrams.

Activity 8.17

Design a lightweight demountable mast that will support an antenna system weighing 1.5 kg at a height of 4.5 m above the ground. The mast system is to be self-supporting and have a footprint not exceeding 4 m^2. The mast must be capable of rapid assembly and disassembly, incorporate a means of checking that the mast is vertically aligned, and its disassembled length should be less than 1.6 m. Sketch three candidate design solutions and identify the optimum design solution, giving reasons for your choice. Specify suitable materials (and finishes, where appropriate) for each of the individual parts and components.

Activity 8.18

The quoted specification for a high-quality domestic loudspeaker is as follows:

Impedance	$8\,\Omega$ at 1 kHz
Frequency response	60 Hz to 12 kHz at $\pm\,3$ dB
Dimensions (W $\times$ D $\times$ H)	35 cm $\times$ 50 cm $\times$ 75 cm
Drivers	1 $\times$ mid-range and 1 $\times$ tweeter
Finish	teak veneer

Identify at least three important items that are missing from this specification and explain why they should be included.

Activity 8.19

Investigate typical technical specifications for any two of the following engineered products:

(a) a pillar drill
(b) a fuel pump
(c) an electrical tester
(d) a battery charger.

Present your findings in the form of a brief 'fact sheet' for each product. Make sure that you explain any technical terms mentioned in the specification.

UNIT 8

Design resources and reference material

At various stages in the design process a designer will need to have access to a variety of design resources and reference material. These include:

- manufacturers' catalogues (e.g., fixings and fittings, seals, drive belts, gears, electronic components and so on)
- application notes (e.g., transistors, pumps, bearings, motors and so on)
- databases (e.g., physical properties of metals, alloys and composite materials)
- datasheets and design tables (e.g., structural beam sections, surface coatings, batteries, wires and cables and so on).

Several examples of typical design reference material were given earlier in Unit 2 so please refer to this chapter for further information.

Activity 8.20

The Airbus A321 is the largest aircraft of the A320 family. The major design change incorporated in the A321 is a stretched fuselage and a strengthened undercarriage which handles the increased weight. The aircraft also has an improved fuel system and larger tyres for enhanced braking. The A321-200 operates with a maximum take-off weight of 93,000 kg and it can accommodate a maximum of 220 passengers.

Visit the Goodyear Aviation website (www.goodyearaviation.com) and download a copy of the latest Goodyear Aviation Tire Data Book, then answer the following questions:

1. Use the Data Book to locate a specification for an aircraft tyre to be used on the main landing gear of an A321 aircraft (see *Figure 8.10*). In relation to the recommended Goodyear tyres, state:
   (a) the dimensions of the tyre when inflated
   (b) the maximum rated speed
   (c) the rated inflation pressure required for the tyre to support its rated load
   (d) the approximate load required to bottom the tyre on the rim at rated inflation pressure
   (e) the Goodyear part number for the tyre.
2. With the aid of a sketch, explain the meaning of each of the following tyre specifications:
   (a) cross-section width
   (b) shoulder width
   (c) section height
   (d) outside diameter
   (e) rim ledge diameter.
3. On what does the static load radius of a tyre depend? How is static load radius calculated?
4. Calculate the static load radius of an aircraft tyre having a mean overall diameter of 1.8 m and a rim flange diameter of 0.8 m when the tyre is experiencing 50% deflection.

Figure 8.10 Main landing gear of an Airbus A321

Producing and Presenting a Final Design Solution

In this section we shall be looking at how the design for an engineered product or service is planned and developed through all of the stages from the initial design brief right through to the final design proposal presented to the customer or client. We shall start by introducing the need for a plan and the design strategy that underpins it.

Planning

Planning is an important task and it needs to be carried out in detail and also before any of the research, design work or prototype manufacture is carried out. An effective plan will save both time and money and it will ensure that resources are not wasted. A plan defines a strategy for a project in terms of the individual tasks and activities that take place and the resources that will be required. It is also important to share this plan with others that are involved and obtain their agreement to supply the required input and resources within the timescale defined in the plan.

Design strategies

What is a design strategy? It is basically having a design method and it consists of two things:

- a framework of intended actions within which to work
- some form of management control and review function that will enable you to adapt and modify your actions as the problem unfolds and the design progresses.

Using a design strategy may seem to divert effort and time from the main task of designing, but this may not be a bad thing, as the purpose of a design strategy will make you think of the way design problems will be dealt with. It also provides you with an awareness as to where the design team is going and how it intends to get there. The purpose then of having a strategy is to ensure that design activities are realistic with respect to the constraints of both time and resources within which the design team has to work. In a manufacturing company the most used strategy is a sequence of actions that have been previously applied to an already existing product. For instance, to design a variation of an already designed electromechanical device or engineering service, the strategy most likely to be used for the new variation will be the same tactics and design methods used for the previous design. This would therefore be making use of a 'pre-established' strategy. The relevant tactics would be drawn from conventional techniques and rational methods already familiar to the design team. This type of strategy applies to innovative designs.

It is not always possible to have a design strategy, as would be the case in research design situations but having no particular plan of action would be a type of strategy in itself. This could be referred to as an 'inventive' strategy. The type of final design may be purely inventive, where no previous market exists.

Often the designers may not know when or what the final outcome may be, although hopefully they may achieve some degree of success in designing a material, product or engineering service that can be commercially exploited. The relevant tactics would be mainly creative.

The two strategies mentioned are extreme forms. In all probability, most designs require a compromise between the two; certain parts of the project design may need the inventive strategy if it calls for unknown areas of engineering design. The *pre-established strategy* is predominantly a convergent design approach, whereas the *inventive strategy* is predominantly a divergent approach. Usually the aim of a design strategy is to converge onto a final, evaluated and detailed design.

Sometimes in reaching that final design it may be necessary in some areas of the design to diverge, so as to widen the search for new ideas and solutions. Therefore the overall design process is mainly convergent, but has elements of divergent thinking.

Convergent thinkers are usually good at detail design, and evaluating and choosing the most suitable solution from a range of options. On the other hand, divergent thinkers are best at conceptual design problems and are able to produce a wide range of alternative solutions.

Planning a design

Having established a design strategy we can begin to construct an outline plan for the design project. Most design projects involve a number of consecutive stages and they typically involve the following tasks:

- clarifying objectives
- establishing functions
- setting requirements
- generating alternatives
- evaluating alternatives
- improving details.

It is worth looking at each of these stages in a little more detail, expanding on the scope of each stage as well as its aim and the typical methods employed.

Clarifying objectives

This stage would in all probability have been dealt with during the design brief and design proposals stages, when the objectives of the design would be stated. But you would still need to clarify them before design can commence.

Aim

To understand the client's needs and to clarify the design objectives and sub-objectives, and the relationships between them.

Method

This is best achieved by carrying out a needs analysis in conjunction with the client and potential users.

Establishing functions

Although the client may have specified the functions expected from the design, the designer may find a more radical or innovative solution by reconsidering the level of the problem definition. They may be able to offer the customer a better solution to the functional problems of the design in excess to the expected at no extra cost.

Aim

To establish the functions required.

Method

Break down the overall function into sub-functions. The sub-functions will comprise of all the functions expected within the product.

Setting requirements

Design problems are usually all set within certain limits, these limits may be cost, weight, size, safety or performance, etc., or any combination of them. Quantify the requirements in the form of a design specification.

Aim

To produce an accurate specification of the performance of the designed product.

Method

Identify the required performance attributes, these may well have been considered at the design feasibility study stage or may have been specified by the client.

Generating alternatives

Even if you think you have a good design solution (or can simply modify or adapt an existing product) always look further, if time and costs permit, for alternative solutions. A radical approach can often generate new and better solutions than the one that may result from making small changes to an existing product.

Aim

To have a choice of solutions to allow comparisons of ideas in solving the design requirements.

Method

It would help to draw several design layout drawings or models available to enable discussions to take place with the design team and the client.

Evaluating alternatives

When some alternative design proposals have been thought about and maybe some design layouts have been produced, the evaluation of the alternatives can be discussed.

Aim

To evaluate the alternatives, choose the ones that fully satisfy the client's needs and meet all of the identified requirements. Note, also, that you will

normally be looking for the most cost-effective solution and your client may be willing to compromise on performance if the cost can be reduced. The final decision will normally rest with the client.

Method

Compare the value of the alternative design proposals against the original proposal agreed with the customer on the basis of performance and cost.

Improving details

There are two main reasons for improving details, they are either aimed at increasing the product value to the customer or reducing the cost to the producer.

Aim

To increase or maintain the value of the product to the customer at the same time reducing its cost to the producer. This may follow directly from the adoption of an alternative solution but may also result from making minor changes to components, or finish.

Method

This is normally approached using two methods, one is called *value engineering* and the other is known as *value analysis*.

Developing ideas

In everyday life, ideas often seem to flow naturally. When designing an engineered product or service this is not always the case. Furthermore, if you can only come up with a limited number of ideas (say one or two) you might need to generate more ideas to provide you with a wider range of alternatives or options.

In order to generate ideas we can make use of one or two tried and tested ideas. The first of these is called *brainstorming*. In brainstorming a group of people sit around and fire ideas at one another. There are several basic rules for brainstorming:

- Everyone in the group should contribute and has an equal right to be heard.
- All ideas (however unlikely or preposterous) must be treated with equal respect.
- Everything should be written down so that no ideas are lost (usually one member of the group is made responsible for this and ideas are recorded on a flip chart so that all can see what has been written down).
- Adequate time should be set aside for the brainstorming exercise and there should be no interruptions.
- It is important to avoid probing ideas too deeply. This can be left until a later stage.
- Agree, at the end of the session, a selection of ideas (typically three or four) that should be considered as candidates for carrying forward to the next stage of the process. These are the ideas that the group considers (by poll, if necessary) to be the most feasible in terms of satisfying the design

brief. Do not, at this stage, reject the other ideas because you might need to come back to these later!

- At first sight, some ideas may be considered less credible or less serious than others by some of the members of the group (we often describe such ideas as being off-the-wall). Nobody in the group should be made to feel bad or inferior if other members of the group consider their ideas strange or unworkable. Some of the most innovative engineering designs have resulted from brainstorming sessions that have unearthed ideas that, at first sight, have been considered unworkable by the majority of those involved.

Another technique that is used to generate ideas is **mind mapping**. A mind map is a sketch or drawing that allows you to identify all the factors that need to be taken into account when developing a solution to a design brief. The name of the product or service appears at the centre of the mind map and each of the solutions and other factors are placed around it. The map can then be progressively expanded as more detail is added.

There are a number of advantages of using a mind map to generate ideas and to understand the relationship between them. These include:

- the design brief (or design problem) appears in the centre of the map so it is very easy to see how all of the potential solutions and any other factors relate to it
- the links that exist between solutions and other factors can be immediately recognized
- the map can be easily grasped without having to read a lot of words
- it is easy to extend a map or add more information to it
- a mind map can help to stimulate thought and aid understanding
- it is often easier and faster to create a mind map than spend a lot of time putting your ideas in writing!

KEY POINT

Brainstorming and mind mapping are techniques that are frequently used to generate ideas

A further technique is that of carrying out research into similar or competing products or engineering services in order to determine how the product or service works and how it can be improved. This exercise is sometimes also carried out as part of the initial market research.

The next stage in the design process requires a number of potential solutions to the design problem to be identified and investigated. You would normally produce a range of different solutions including those that may rely on different technology, different materials or different operating principles.

At this point you will be able to identify production constraints for each of your design solutions including availability (and cost) of materials, components and resources and the means of processing or manufacturing the product. You should also consider issues relating to health and safety and any legislation that may have an impact on the manufacture, use or disposal of the product.

Having developed a range of solutions you will be in a position to make an informed decision about which one should be your final design solution. When carrying out this formative evaluation you will sometimes find that the choice will be obvious (e.g., the alternative solutions may be

prohibitively expensive or use dangerous materials or processes in their manufacture) however, you may sometimes find that it is quite difficult to arrive at a final choice. In such cases you might wish to assemble a ***design review panel*** of colleagues and/or potential users in order to help you arrive at a final solution.

Drawings

Engineering designers need to communicate their ideas by using various types of drawing and diagram as appropriate to the type of information that they need to convey (*Figure 8.11*). The main types of drawing are:

- block diagrams
- general arrangement drawings
- detail drawings
- schematic diagrams
- layout drawings

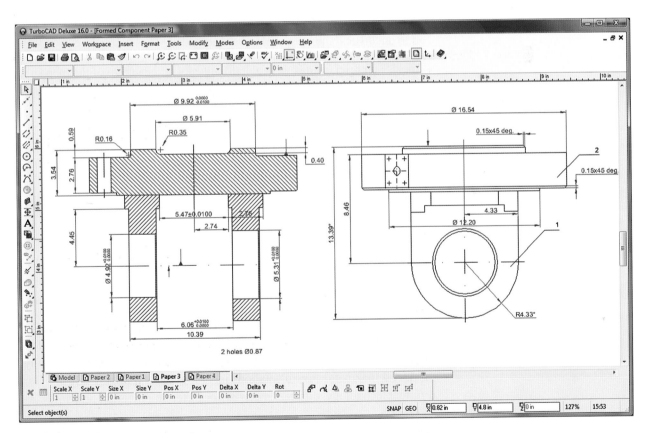

Figure 8.11 An example of a CAD-produced engineering drawing used to communicate an engineering design

In addition, exploded views (*Figure 8.12*) are frequently used to show the relationship between different parts used in an assembly. This type of diagram is extremely useful when assembling or disassembling an engineered product.

All engineering drawings should be produced using appropriate drawing standards and conventions for the following reasons.

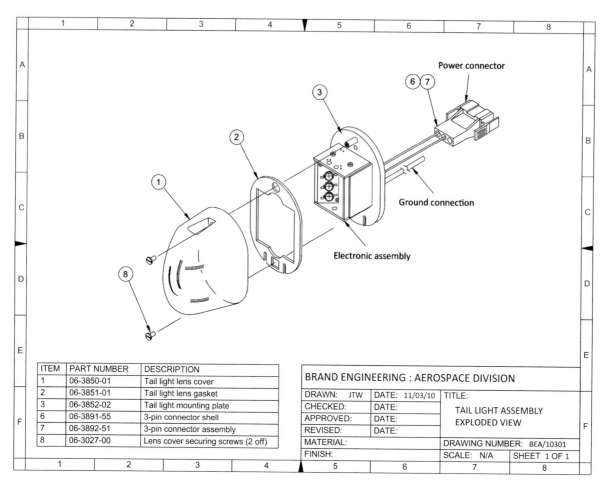

Figure 8.12 A typical exploded view showing the arrangement of components and parts in an engineered product

Time

It speeds up the drawing process, making life easier and reducing the lead-time required to get a new product into production.

Appearance

The consistent use of standards and conventions not only serves to make your drawings look more professional but it also improves the 'image' of yourself and your company. Badly presented drawings can send out the wrong messages and can call your competence into question.

Portability

Drawings produced to international standards and conventions can be read and correctly interpreted by any trained engineer, anywhere in the UK or abroad. This avoids misunderstandings that could lead to expensive and complex components and assemblies being scrapped and dangerous situations arising. (Note that problems can still sometimes occur due to text and written notes that may be language dependent.)

Drawing conventions

Drawing conventions used by engineers in the UK are specified in BS 8888:2008, *Technical Product Specification* (TPS). This is produced in

KEY POINT

BS8888:2008 Technical Product Specification describes and explains the drawing conventions used by engineering designers in the UK

several parts and is harmonized with the appropriate International Standards Organization (ISO) standards. A student guide to the TPS is also available from BSI. This is known as PP 8888-2:2007 and is designed specifically for engineering students in further and higher education.

The TPS enables product designers and manufacturers using British and international standards to have a common technical language. It provides engineers and others involved in the design and manufacturing process with a precise and unambiguous means of describing a technical product, which can avoid misunderstanding in the communication of manufacturing requirements.

TYK 8.21

What type of drawing would be needed for each of the following design tasks:
1. the brake hydraulic system used in a sports car
2. the physical location of components on a printed circuit board
3. a turned and threaded brass collar
4. the electrical connections to a transformer
5. the assembly of parts used in an electric motor.

3D modelling and visualization

Modern CAD software allows much more than just simple 2D and 3D drawing. Engineers now have access to several sophisticated software packages that allow permit solid modelling and visualization of a design prior to manufacture. Using such techniques it is also possible to examine how components and parts fit together and how they interact with one another. It is also possible to run software applications that can simulate variations in physical, electrical and mechanical parameters (such as mechanical or thermal stress or how the operation of an electronic circuit is affected over a range of temperatures and supply voltages). Materials and surface textures can also be applied so that it is possible to gain an impression of what a fully finished component will look like (*Figure 8.13*).

TYK 8.22

What is 3D visualization and why is it useful?

Summarizing the design brief

When presenting the final design solution it is important to summarize the design brief. This will have shaped the design process and it will have been key to the development of your design proposal. When summarizing your

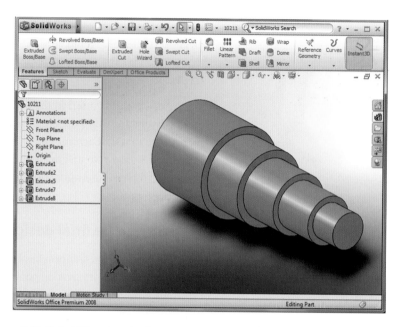

Figure 8.13 Typical 3D product visualization (the component can be rotated and viewed from any direction)

design brief you should review it against each of the bullet points listed below:

- the function of the product
- who will be using the product
- what the product should look like
- the materials it should be made from
- the technology needed to produce it
- the production cost (or service delivery cost)
- the number required (and thus the scale of production)
- the timescale for the product
- health and safety issues
- issues relating to quality and 'fitness for purpose'
- any constraints that impact on design, manufacture, use and eventual disposal of the product.

The design brief will have been developed into a formal design specification. The specification should be quantified (in terms of parameters such as size, weight, energy consumption, efficiency, etc.) and it should be measurable. The design specification needs to address the function of the product, its form, size and shape, ergonomic features, and so on. Note that the design specification may specify particular parameters in different ways such as:

Input	13.8 V DC
Minimum output	40 W
Maximum weight	25 kg
Minimum turning circle	3.75 m
Speed range	270–350 rev/s
Height	1.5 m ± 0.05 m
Frequency	1.75 kHz ± 2.5%
Supply voltage	200–240 V AC

You will need to include a full design specification which includes meaningful parameters against which the performance of the product can be measured.

The design brief should also provide an indication of what materials and processes should be used in the manufacture of the product. There may be good reasons for using different materials in the production of a prototype from those that will actually be used when the product goes into production. If this is the case you will need to provide further details together with reasons and possible implications of using different materials in the eventual production process.

Presenting the final design solution

Once the final design solution has been agreed, you will require a detailed set of drawings, diagrams, component lists and assembly instructions. Your drawings will incorporate, or be supplemented by, design notes that explain how the design operates, and how it is to be manufactured. Where numerical values are included as part of the proposed solution, you will need to demonstrate how you applied your scientific and mathematical understanding to drive at the design values. For example, if you are designing an electric water pump, you will need to relate the energy input and motor efficiency to the head of water that the pump will raise. Your design calculations need to be checked carefully for accuracy and validity.

Finally, you will need to communicate your final design solution using the drawings and diagrams that you have prepared together with the component and parts lists and other written information. Depending on your design solution you may need to produce any or all of the following types of document:

- freehand sketches
- general arrangement drawings
- detail drawings
- schematic/circuit diagrams
- flowcharts
- 3D models and visualization of surfaces and coatings
- exploded views
- software simulation
- component/parts lists
- materials specifications
- costings and cost estimates
- design calculations
- descriptions of manufacturing processes
- information relating to compliance with standards, approvals and quality assurance.

Design documentation

When you work on a design project it is a good idea to keep a design diary or *design portfolio* containing all of the notes, sketches and drawings that you use at each phase of the design process. A design portfolio should contain notes, sketches and drawings showing what was done, why it was

done, when it was done and who did it. The design portfolio will become invaluable when it comes to presenting your design solution to other people and it should normally contain the following.

A description of the problem

The problem is the task set (or the need identified) by the client (or that you have identified for yourself. This section should be quite brief and will normally just be a few sentences that describe the problem. However, in some cases you may wish to add some sketches or drawings to enhance or clarify your description.

A statement of the design brief

The design brief (or client brief) states what needs to be done to solve the problem. It should be written from the perspective of the client. In other words, it should say what the client is looking for. You will normally wish to agree the design brief with your client so it is important to get the wording right. Once again, this section is usually just text but there may be occasions when you might wish to use a drawing or a sketch to clarify and enhance your wording.

Results of the research and investigation

You will need to summarize the results of your research and investigation using charts, diagrams, tables and any other method that helps convey the results to your readers. Since charts and diagrams can usually be easily understood this is often the best way to represent the outcome of your research and investigation. One important point is that you should always quote the source of any data that you use. You should also provide a copy of any market research that you carried out.

The investigation may also consider other aspects such as identifying primary and secondary markets for the product or service, existing products and services that may be similar or competing in the same market, materials and manufacturing constraints, ergonomics and aesthetics, issues relating to standards and health and safety.

A summary of the candidate solutions

List each of the candidate solutions that you arrived at and explain how you arrived at them. Give brief details of any brainstorming sessions that you held or include a mind map if you used one.

The design criteria and the design specification

Include the design criteria that you established as well as the detailed design specification.

The candidate solutions

You should describe the design ideas for each of our candidate solutions and the process by which you arrived at the final design solution. If you used an evaluation matrix this should be included here. You should include details and sketches of the ideas that you rejected. Although these ideas may not be taken forward, they give an idea of the process that you went through and, at some time in the future, you or your client may want to return to them.

The final design solution

You will need to describe your final design solution in detail. To do this you will need to use detailed sketches and drawings. These will help to convey your ideas to the client and they may also assist with marketing the product or service that you have designed.

Your working and presentation drawings

Your working drawings must be precise and drawn to scale. They must give specific information about dimensions and the materials that should be used. You may wish to include drawings of individual parts as well as details showing how your product should be assembled. Dimensions should be clearly indicated and all other relevant information should be included. One or more presentation drawings should be included to show what the finished product will look like. You should choose an appropriate drawing technique for this (perspective, isometric, oblique, 3D rendered CAD, etc.).

Information relating to manufacturing the product or supplying the service

You will need to provide details showing how you envisage that your product will be manufactured or how your service will be delivered. For example, you may need to supply a parts list, a component list or a cutting diagram. Much will depend on the individual product or service. Finally you should provide a simple step-by-step explanation of what needs to be done to assemble or construct your designed product. A series of sketches might be useful here.

A final evaluation of your work

Your design folder will not be complete without a full evaluation of the work that you have done. You should also include comments and feedback that you received from your client as well as information on any modifications that were incorporated as the work went on. In many design projects your client will insist on holding a regular review meeting in order to discuss progress with the project. If your client does not ask for this, you might find it useful to organize a presentation of your work or provide a brief written report at various stages of the project.

You need to justify your final design solution by demonstrating that you have been able to comply with each of the essential design criteria as well as meeting the design specification that you agreed with your client. You also need to say how you complied with any relevant standards or legislation.

You need to evaluate your work with a critical eye and indicate where there are any particular strengths or weaknesses. You should also provide full details of any testing or measurement that you carried out. If your measurements did not confirm that you have met the design specification you need to suggest why this is and what should be done to ensure that the design specification is met.

Finally, you need to ask yourself whether your final design solution meets your client's needs and expectations. You should also comment on any improvements or modifications that could be made to improve the product or service or make it more cost-effective.

TYK *8.23*

What is a design portfolio and what should it contain?

Delivering your presentation

Your formal presentation should be a verbal presentation supported by appropriate visual aids (e.g., a PowerPoint presentation or overhead projector transparencies) or a brief technical report (included sketches and other visual aids). In either case, your presentation should be delivered in a way that is appropriate to your audience. This will invariably mean that you should keep your presentation brief and to the point. At the same time, you should ensure that you have covered all the main points that make up your design solution. In any event, your presentation must be interesting and appropriately paced so that the attention of the audience does not wander. If you are delivering a verbal presentation it is also important to be a good listener and be able to respond to any questions or queries that are raised by your audience.

Fortunately, much of the material that you need to include in your verbal presentation will already exist. All you need to do is to summarize it and present it in a way that your audience can quickly and easily grasp. In most cases, you will wish to supply your audience with notes or printed handouts. These can be based on PowerPoint screens of overhead projector transparencies and can be augmented with sketches and presentation drawings, as appropriate.

If you choose to include a brief technical report, this can be assembled from the material that you have collected as you worked through the various stages of the project. Typical section headings might include the following.

Summary
A brief overview for busy readers who need to quickly find out what the report is about.

Introduction
This sets the context and background and provides a brief description of the problem that you have solved and may also include a statement of the design brief.

Main body
A comprehensive description of the design solution including how and why it was chosen together with details of the research and investigation that you carried out. You should also include (and comment on) the design criteria and the final design specification. Your design solution should be presented together with sketches, drawings, 3D models and visualization (as appropriate).

Evaluation

A detailed evaluation of the work that you carried out including any problems that you encountered and how you solved them.

Recommendations

This section should provide information on how the design solution should be implemented (including, for example, information on manufacture or assembly). You may also wish to include any modifications or changes that you would recommend.

Conclusions

You should end your report with a few concluding remarks about your design solution and how effective you think it is likely to be at solving your client's problem.

References

This section should provide readers with a list of sources for further information relating to the scientific principles or technology used, including (where appropriate) relevant standards and legislation.

It is important to take care when you present your work in the form of a written report. To avoid confusion, the normal conventions of grammar and punctuation must be used. Words must be correctly spelt so do make use of a dictionary or spellchecker if you are uncertain about spelling. Never use jargon terms and acronyms unless you are sure that your audience will understand them. Layout is important so use numbered sections and paragraphs and try to keep your sentences short and to the point.

The assessment of this unit is based on the design, development, prototype manufacture and testing, and evaluation of your engineered product. Your work should be submitted in the form of a portfolio which contains evidence of the work that you have carried out. In addition, your teacher or tutor will provide you with a 'witness statement' detailing the quality of your oral presentation. This final activity takes the form of a 'checklist' for the contents of your portfolio. You should check that each item is present before you submit your design portfolio:

- details of the original client brief
- the needs analysis and list of requirements
- the final design specification
- alternative solutions considered
- the final design solution (and why it was chosen)
- detailed engineering drawings, diagrams and schematics
- evidence of project planning (e.g., a Gantt chart)
- details of prototype manufacture and assembly
- photographs of the completed prototype
- details of testing, evaluation and recommendations for any modifications necessary prior to production
- a copy of your presentation materials (including any notes or technical report)
- a contents sheet.

KEY POINT

A design portfolio brings together all of the documentation that is used in the design process

UNIT 8

Activity 8.21

Produce a final design solution for the lightweight demountable mast in Activity 8.17. Present your work in the form of a PowerPoint presentation and include full details of materials and processes as well as any design constraints. Also include an outline costing based on an initial production of 100 units.

Activity 8.22

Produce a final design solution for the high-quality domestic loudspeaker in Activity 8.18. Present your work in the form of a PowerPoint presentation and include full details of materials and processes as well as any design constraints. Also include an outline costing based on an initial production of 100 units.

Production planning

Having produced a detailed specification together with a comprehensive set of drawings we are ready to move on to the next stage: planning the production process. What we mean by this is the sequence of operations and processes that allow us to manufacture the product or, in the case of an engineered service, supply the service.

The sequence of operations is just as important as the processes that are involved. Also important is the choice of materials, components and parts. The production planning process involves considering a number of factors including:

- the available resources
- the materials, parts and components to be used
- the processes that are to be used
- the available equipment and machinery
- the sequence of production
- the arrangements for inspection and quality control
- the arrangements for servicing, maintenance, recycling and disposal
- health and safety factors.

To help you understand what production planning is about consider the manufacture of an automatic greenhouse watering system that allows water to flow into a container whenever the soil in the container dries out. The system is to comprise a moisture detecting probe, connecting cable, electronic control unit and an electrically operated solenoid valve fitted with tap and hosepipe connectors. The control unit is to derive its supply from a standard 220 V AC mains supply via a sealed power unit which converts the incoming mains supply to 24 V DC for powering the solenoid, control circuit and LED indicators.

The main components for the automatic watering system are listed below:

- moisture probe*
- probe cable fitted with hermetically sealed connectors*
- electronic controller (printed circuit board)*

- solenoid valve
- solenoid valve housing fitted with tap and hosepipe connectors*
- ABS enclosure for the electronic controller, indicators and electrical connectors
- mains power unit (220 V AC input and 24 V DC output).

Note that components marked * are to be manufactured. The remaining items are to be bought-in.

The production sequence for the automatic watering system is as follows:

- manufacture the printed circuit board
- obtain electronic parts and solder into printed circuit board
- drill the ABS enclosure to accommodate the printed circuit board mounting pillars, LED indicators and connectors
- assemble the printed circuit boards and other parts in the ABS enclosure and solder the interconnecting wires
- manufacture the solenoid valve housing and fit the solenoid valve and hose connectors
- manufacture the moisture probe and probe cable
- assemble the automatic watering system by connecting the moisture probe and solenoid valve
- connect the mains power unit and test the watering system.

Note that the moisture probe, solenoid valve housing and electronic controller can all be manufactured independently and stored in batches prior to final assembly. Note also that individual parts and sub-systems should be separately tested and inspected prior to final assembly. Finally it is important to realize that although a designer should give some indication of how a product could be manufactured, the task of developing and finalizing a production plan is usually left to those that are actually responsible for the production process (i.e., the Production Manager).

KEY POINT

Production planning is used to determine the sequence of operations required to manufacture an engineered product. Designers should be fully aware of how a product will be manufactured even though they may not be accountable for the manufacturing process

Test your knowledge

TYK

TYK 8.24

List six issues that should be considered when producing an outline production plan for an engineered product.

Activity 8.23

Produce an outline production plan for the lightweight demountable mast in Activity 8.17. Hint: Start by identifying the parts, components and processes used.

Activity 8.24

Produce an outline production plan for the high-quality domestic loudspeaker system in Activity 8.18. Hint: Start by identifying the parts, components and processes used.

End of Unit Review

1. Summarize each of the main stages in the design and development of an engineering product.

2. Explain the three key stages of design realization.

3. Explain the purpose of a PDS. Who should be involved with producing a PDS?

4. What is the difference between a client and the end-user of an engineered product? Give an example of an engineered product where the client and the end-user are the same.

5. List four characteristics normally associated with a 'successful design'.

6. What is the difference between an internal and an external customer? Give an example of an engineered product that might be designed for an internal customer.

7. Explain what is meant by demand pull and technology push. Give an example of each.

8. Describe, with the aid of an example, the contents of a typical design brief.

9. Describe two methods of generating ideas. In what circumstances might these methods be useful as part of the design process?

10. Why is it important to have a detailed design specification? List six items that should normally form part of a design specification.

11. Explain the significance of (a) the CE Mark and (b) the BSI Kitemark.

12. Explain the role of a National Standards Body (NSB).

13. Explain why energy efficiency is important in the design of many engineered products.

14. Describe three constraints that impact on the design of a typical engineered product.

15. Describe four different types of design resource and explain how each is used to inform the design process.

16. Explain the purpose of a design portfolio. What items should be included in it?

Units

The International System of Units (SI) is now the accepted international standard for all units of scientific measure. It has been legally accepted in America even though it has not been fully adopted, the American preference still being for the old English or imperial system of measurement. All scientific concepts and science teaching in this country, as well as throughout Europe and many other areas of the world, have legally adopted the SI system and so these units will be used throughout this book for scientific and technological applications.

The SI consists of three main groups: seven base units, two supplementary units and a number of derived units. *Table A1.1* shows the seven base units by *SI name* and *symbol*. The full scientific definition of these units, as defined by the International Committee for Weights and Measures (CIPM), is given below. You might find these definitions rather strange. They are, however, true and accurate definitions of the quantities we know so well! Do not worry too much if you cannot understand these definitions at this time. Treat this appendix on units *as a source of reference* that is always here for you when needed.

Table A1.1 SI base units

Basic quantity	Dimension	SI unit name	SI unit symbol
Mass	M	Kilogram	kg
Length	L	Metre	m
Time	T	Second	s
Electric Current	I	Ampere	A
Temperature	Θ	Kelvin	K
Amount of Substance	N	Molelier	mol
Luminous Intensity	J	Candela	Cd

Kilogram

The kilogramme or kilogram is the unit of mass; it is equal to the mass of the international prototype of the kilogram, as defined by CIPM.

Metre

The metre is the length of the path travelled by light in a vacuum during the time interval of 1/299792458 s.

Second

The second is the duration of 9192631770 periods of radiation corresponding to the transition between the two hyperfine levels of the ground state of the caesium 133 atom.

Ampere

The ampere is that constant current which if maintained in two straight parallel conductors of infinite length, of negligible circular cross section, and placed 1 m apart in a vacuum, would produce between these conductors a force equal to $2 \times 10^{-7} \, Nm^{-1}$ length.

Kelvin

The Kelvin, unit of thermodynamic temperature, is the fraction 1/273.16 of the thermodynamic temperature of the triple point of water.

Molelier

The mole is the amount of substance of a system which contains as many elementary particles as there are atoms in 0.012 kg of carbon 12. When the mole is used, the elementary entities must be specified and maybe atoms, molecules, ions, electrons, or other particles, or specified groups of such particles.

Candela

The candela is the luminous intensity, in a given direction, of a source that emits monochromatic radiation of frequency 540×10^{12} Hz and that has a radiant intensity in that direction of 1/683 watt per steradian.

In addition to the seven base units there are the two supplementary units: the *radian* for plane angles and the *steradian* for solid three-dimensional angles. Remember that these units are ratios, so have no units.

Table A1.2 SI supplementary units

Supplementary unit	SI unit name	SI unit symbol
Plane angle	Radian	^{c} or rad
Solid angle	Steradian	sr

The SI derived units are defined by simple equations relating two or more base units. The names and symbols of some of the derived units may be substituted by special names and symbols. Some of the derived units, which you may be familiar with, are listed in *Table A1.3* with their special names as appropriate.

Table A1.3 SI derived units

Name	Symbol	Physical quantity
Coulomb	C	Quantity of electricity, electric charge
Farad	F	Electric capacitance
Henry	H	Electrical inductance
Hertz	Hz	Frequency
Joule	J	Energy, work, heat
Lux	Lx	Luminance
Newton	N	Force, weight
Ohm	Ω	Electrical resistance
Pascal	Pa	Pressure, stress
Siemans	S	Electrical conductance
Tesla	T	Induction field, magnetic flux density
Volt	V	Electric potential, electromotive force
Watt	W	Power, radiant flux
Weber	Wb	Induction magnetic flux

You will be introduced to many of these units as you progress through the chapters on mechanical principles and electrical principles, where you will be asked to use them, when solving problems. The SI is a decimal system of units where fractions have been eliminated, so multiples and sub-multiples are formed from a series of prefixes. That is, we multiply the original unit by powers of ten for units greater than one and by decimal fractions of ten for numbers less than one. Some of these multiples (with which I am sure you are familiar) are detailed in *Table A1.4*.

Table A1.4 SI prefixes

Prefix	Symbol	Multiply by
peta	P	10^{15}
tera	T	10^{12}
giga	G	10^{9}
mega	M	10^{6}
kilo	K	10^{3}
hecto	H	10^{2}
deca	Da	10^{1}
deci	d	10^{-1}
centi	c	10^{-2}
milli	m	10^{-3}
micro	μ	10^{-6}
nano	n	10^{-9}
pico	p	10^{-12}
femto	f	10^{-15}

So, for example: 1 millimetre = 1 mm = 10^{-3} m, 1 cm^3 = $(10^{-2}$ m$)^3$ = 10^{-6} m^3 and 1 mm = 10^{-6} m. Note the way in which powers of ten

are used. The above examples show us the correct way for representing multiples and sub-multiples of units.

Finally, you should be aware of some commonly used, legally accepted, *non-SI units*. These are detailed in *Table A1.5*.

Table A1.5 Non-SI units

Name	Symbol	Physical quantity
Ampere-hour	Ah	Electric charge
Day	D	Time, period
Degree	°	Plane angle
Electron volt	eV	Electric potential
Kilometre per hour	kph	Velocity
Hour	h	Time, period
Litre	L, l	Capacity, volume
Minute	min	Time, period
Metric tonne	t	Mass

Note that velocity in kilometres per hour is a derived unit that is made up from the units of length (distance) and time.

Answers to Test Your Knowledge Questions

Unit 2

2.1

1. First edition 2002, second edition 2007, third edition 2010
2. See page iv
3. Mike Tooley and Lloyd Dingle
4. ISBN 9780123822024
5. Newnes (an imprint of Elsevier).

2.2

(a) 1994
(b) Texas Instruments
(c) Four
(d) No internal connection (or 'not connected')
(e) 20
(f) SN54ALS08 and SN54AS08
(g) 7 V
(h) −65°C to +150°C
(i) 5 V, 25°C.

2.3

Material, length and diameter, thread, head type, finish, stock code, supply multiple, price.

2.4

The order should specify a quantity of 15 diecast boxes, catalogue reference DB65-02P. It is important to remember that the dimensions specified in the short-form catalogue will be external dimensions and the DB65-01P will have insufficient internal clearance.

2.5

Assuming that a CD-ROM has a capacity of 700 Mb then 248 data sheets can be stored.

2.6

Because several students may have the same last name (hence the key may not be unique) the database requires unique keys for each record.

2.7

(a) A *field* is an individual item of data stored within the record for a particular item. Fields can be used for storing text or numerical data. For example, the database record for a machined bolt might have several fields including one to indicate the type of thread and another to indicate the length of the bolt.
(b) A *record* comprises a complete set of data for a particular item. For example, the database record for a diecast box might comprise of separate fields including the length, width and height of the box, its material and its surface finish.
(c) A *key* is a unique item of data used to identify a particular record. No two records must have the same key. For example, a car tyre would have a record that included several fields (width, tread pattern, diameter, etc.) together with a unique stock code which would be used as the key.

(d) In a *relational* database you are able to display reports that bring together data from various sources. This allows you to view the data in various ways (not just as single records). For example, you could display a list of all of the components that come from a particular supplier or you could display a list of the component parts manufactured in brass.

2.8

Records of components and materials.

2.9

Date of purchase (<10 years from current date); contact name, contact title, company name, address line 1, address line 2, town/city, county/state, post/zip code, country.

2.10

Contact name, contact salutation, contact title, telephone number, fax number, e-mail address.

2.11

Advantages: occupies less space than printed publication, can be searched electronically; *Disadvantages*: cannot be kept up to date, amount of data limited to 700 Mb.

2.12

URL stands for 'uniform resource locator'. A URL is used to identify a particular piece of information (usually a web page) that can be accessed via the World Wide Web. Every location on the World Wide Web must have a unique URL.

An example URL for an engineering company is: http://www.britten-norman.com (Britten-Norman is an aircraft manufacturer based on the Isle of Wight).

The URL of the web page shown in *Figure 2.9* is http://www.howard.co.uk/index.htm.

2.13

1. An academic institution located in the UK.
2. A commercial business based in the UK.
3. An organization based in the UK.

2.14

1. 2002/03
2. 2006/07
3. 27

4. 50%
5. 105
6. 17.5

2.15

1. 160 times
2. 12
3. 2
4. 240%

2.16

1. Processor
2. floppy disk
3. 10%
4. 50%
5. £25

2.17

1. 78 MPa achieved after about 30 h.
2. 15 MPa achieved after 4 h.
3. It is a roughly linear relationship and this suggests that the bond strength is directly proportional to the curing time.
4. Bond strength reaches a maximum value which is not exceeded for any further increase in curing time.
5. 21 h.

2.18

1. 0.2 mA
2. 0.33 V
3. 200 Ω
4. 103 Ω

2.19

1. (a) 400,000, (b) 550,000
2. 2004
3. 2003, 2004 and 2006
4. 2002 and 2005
5. 2.4 million

2.20

(a) Figure 2.48a–d and f is shown in first angle projection. Figure 2.48e is shown in third angle projection.
(b) Figure 2.48b denotes first angle and Figure 2.48e denotes third angle.

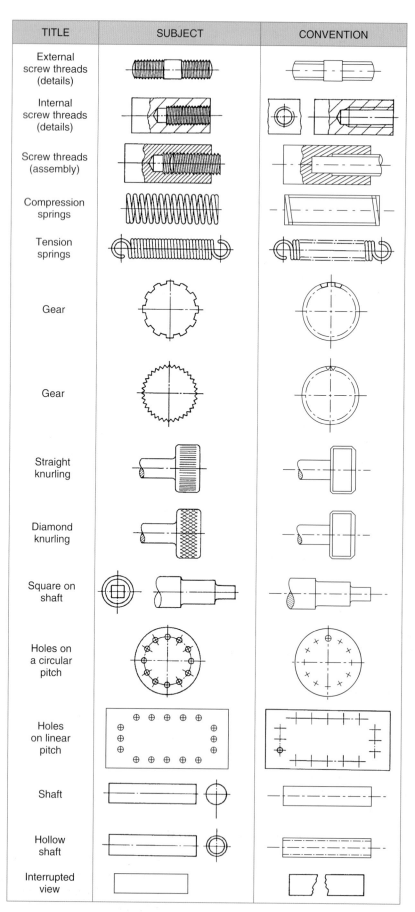

Figure A2.2.1 Answer to TYK 2.24

2.21

1. A2, correct
2. G1, correct
3. Incorrect, should be K2
4. E2, correct
5. Incorrect, should be D1
6. B5, correct
7. Incorrect, should be D1
8. D1, correct.

2.22

1. Correct
2. Correct
3. Incorrect, arrow should be a dot
4. Incorrect, arrow should be a dot
5. Correct.

2.23

Above floor, assembly, chamfer, CSK, cylinder, DIA, FIG, HEX, material, MAX, MIN, RAD, required, right hand, specification, diameter.

2.24

See *Figure A2.2.1*.

2.25

See *Figure A2.2.2*.

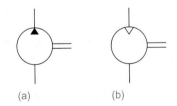

(a)   (b)

Figure A2.2.2 Answer to TYK 2.25

2.26

(a) 3/2 DCV
(b) Flow is from 1 to 2 in the position shown
(c) Flow is from 1 to 3 when actuated.

2.27

Motorized 3/2 DCV.

2.28

(a) Cooler
(b) Pilot controlled DCV
(c) Flow restrictor (affected by viscosity)
(d) Hydraulic pump (variable)
(e) Filter or strainer
(f) Air dryer
(g) Pneumatic actuator
(h) Single ended linear actuator with spring return
(i) 4/2 push-button directional control valve
(j) Motorized pneumatic pump.

2.29

(a) Battery (single cell)
(b) Variable capacitor
(c) Earth
(d) Lamp
(e) Bell
(f) Fuse
(g) Fixed resistor
(h) Variable potentiometer
(i) Iron cored inductor
(j) Meter or motor
(k) Galvanometer
(l) Milliammeter
(m) Single-pole single-throw switch
(n) Diode
(o) Light emitting diode
(p) PNP transistor
(q) Amplifier.

Unit 3

3.1

Entries may include:

Illustrations: sketches, drawings, charts, tables, photographs, circuit diagrams, mind maps, brainstorming ideas, presentation slides, etc.

Lists and records of correspondence: e-mails, web visits, telephone calls, faxes, letters, interviews, visits, meetings, etc.

Lists, information sources and data: initial ideas, project title and selection criteria, references, bibliographic reading, books, learned journals, web visits, equipment, resources, page dumps, survey questions and results, experimental results, field data, project brief, specification, etc.

Planning: project planning, resource and time planning, technical planning, bar charts, time lines, Gantt charts, program evaluation and review technique (PERT) planners, computer diary print-offs, project planning software printouts, updates and amendments to plans.

Implementation and evaluation: evaluation matrices, selection criteria, experimental methods, statistical analysis, draft calculations, build and test data and analysis, etc.

General: Index, sectioning, colour markers, contact details, etc.

3.2

Omissions and points for clarification:

- Exact BS/ISO standards (e.g., BS EN 61000-4-9:1993) minimum manufacturing standards for portable electrical equipment BS EN 60065, 60952, etc.
- Exact nature of chassis and cover materials (e.g., tough polymer or light alloy?).
- Price range to be set will impact on quality and ruggedness of the components used for construction.
- Conventional colour coding of connectors or otherwise?
- Proven and reliable technology to be used, confirm for what aspects of design, the essential or all?
- Accessibility of fuse to be determined to conform to health and safety standards, confirm testing and maintenance standards for equipment, so that user manual can be written.
- Confirm that integrated electric power connector keeps build and production within limits.

This list is neither exclusive nor exhaustive but suggests typical student answers.

3.3

(a) Possible solutions:
  - extend length of handbrake level
  - reposition handbrake assembly
  - extend length of foot pedals and have two position steering wheel (allowing seat to be further back)
  - lower handbrake assembly
  - alter range of seat adjustment
  - convert to dashboard push/pull lever handbrake.
(b) Selection criteria:
  - cost
  - time
  - scale of problem (e.g., does it affect many or few)
  - particular model run time (old new, to be replaced, not to be replaced, etc.)

- simplicity/difficulty of modification
- effect on range of seat adjustment
- effect on ease of operation of handbrake
- effect on all sizes of driver
- ergonomics/aesthetics.

(c) A matrix of selection criteria against design alternatives is then set up and the alternatives should be ranked. There is no 'ideal' solution to this problem, compromises would need to be made based on the importance of the selection criteria in any one particular case.

3.4

Project planning is different from other forms of planning and scheduling simply because the set of activities that constitute a project are unique and only occur once. Whereas, for example, production planning and scheduling relate to a set of activities that may be performed a large number of times.

3.5

PERT requires that project activities should be discrete and have definite start and end points. Each activity is placed into a diagram that shows the interrelationship between them and the effect that altering one or more of these activities will have on the whole plan. The technique is very effective when planning project times and resources that involve a very large number of interrelated activities.

3.6

PERT allows us to identify the path through the network for which the total activity times are the greatest – this is the critical path. The Critical Path Method (CPM) is widely used in the engineering industry and is really just a subset of the PERT method.

The critical path is important for the following reasons:

- Since it represents the most time-critical set of activities, the total time to reach the project goal can *only* be reduced by reducing time spent in one or more of the activities along the critical path. In other words, the critical path highlights those activities that should be critically reviewed to see whether they can be shortened in any way. Putting extra resources into one or more of the critical path activities can usually be instrumental in reducing the overall project time.
- The critical path is unique for a particular set of activities and timings. If any of these are changed (e.g., by directing extra resources into them) a new

critical path will be revealed. We can then apply PERT evaluation to this new critical path, critically reviewing the activities that it points us to and by repeating the process we can continue to reduce project time.

3.7

(a) (i) 1 day, (ii) 1 day, (iii) 1 day.
(b) 8 days.
(c) The critical path then changes and follows the route of activities 0, 1, 2, 4, 5, 6 and 7.

3.8

Critical path is C–D–E–F and expected time to completion is 8 units.

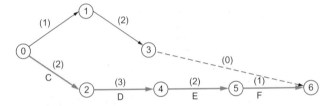

Figure A2.3.1 Network diagram answer to TYK 3.8

3.9

Critical path is B–C–F and expected time to completion is 7 units.

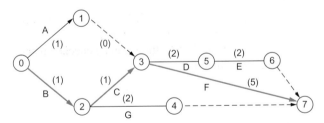

Figure A2.3.2 Network diagram answer to TYK 3.9

3.10

(a) Draft spec and budget, specify materials/parts, design cross-over, design enclosure
(b) Build enclosure
(c) Construct cross-over
(d) Order/await drivers

3.11

(a) Approaching 40 MPa
(b) Around 55 MPa.

3.12

(a) 528571
(b) 442857

3.13

Other important selection criteria, include:

- reliability
- cost
- performance
- minimum greenhouse gas emissions
- fuel economy
- ease of maintenance
- passenger capability
- airline knowledge of operating type
- ground logistics and turnround time
- airfield infrastructure for type.

3.14

Key points to consider:

- Know own project subject matter.
- Decide on types of audio-visual aids to use.
- Prepare presentation slides/PowerPoint pages, etc., well in advance, having determined the form and length of your presentation from your tutor.
- Frame content of presentation around the headings: introduction, proposed outcome(s), achievements, results/evaluation, conclusion.
- Limit slides/transparencies to a manageable number.
- Practice presentation well before the day, ensure timing and amend accordingly.
- Ensure familiarity with audio-visual aids that will be used.
- Determine the presentation skills being assessed.
- Ensure all props/equipment is working and on display if appropriate.
- Speak up, speak clearly, do not rush, do not mumble, look at audience.
- Be adequately prepared to answer questions.

Unit 4

4.1

1. (a) $2 + x - x^2$, (b) $3 - 7a + 2a^2$,
   (c) $ab - 4a - 3b + 12$, (d) $3x^2 + 7x - 6$

2. (a) $x^2 - 3xy + 2y^2$, (b) $3p^2 - 3pq + 2p^3 - 2p^2q$,
   (c) $2xy^2 - 2xyz - y^2z + yz^2$
3. $(ab^2c)(c + a^2c + 1)$
4. (a) $(2,8)$, $(4,4)$, $(2,2,2,2)$; (b) (n, n); (c) (wx, yz), (wxy, z)
5. (a) $6a^2 + 4a - 2$, (b) $4 - x^4$, (c) $3a^3b + a^2b^2 - 2ab^3$
6. (a) $(x + 3)(x - 1)$, (b) $(a + 3)(a - 6)$,
   (c) $(2p + 3)(2p + 4)$.

4.2

1. (a) $a^5b^{-1}c^3d$, (b) $4(6x^3y^2 - xy^2)$
2. (a) 32, (b) $\dfrac{8}{27}$, (c) b^2
3. (a) 70, (b) 10
4. (a) $\dfrac{3}{4}a^{-6}b^2$, (b) d
5. $a^2b^2c^2d^2$

4.3

1. $u = v - at$
2. $v = \sqrt{u^2 + 2as}$
3. $m = \dfrac{F}{a + g}$
4. $A_1 = \dfrac{\rho_2 A_2 v_2}{\rho_1 v_1}$
5. 20,000
6. $r = \sqrt{\dfrac{v}{\pi h}}$
7. $R_2 = \dfrac{R_1 R_4}{R_3} = 2.25$
8. 0.017
9. $a = \dfrac{s}{n} - \dfrac{1}{2}(n - 1)d$
10. $x = \dfrac{ac + bc + b^2}{c + b}$
11. $C = 4.8343 \times 10^{-6}$
12. $\dfrac{1}{2(x^2 - 1)}$

4.4

1. (a) $x = 1$, (b) $p = 4$, (c) $x = -2$, (d) $x = 4$,
   (e) $a = -15$, (f) $p = -\dfrac{1}{3}$
2. $v = 45$ when $i = 3$
3. (a) $x = 1$, (b) $p = \pm 4$, (c) $x = -2$, or 3,
   (d) $p = -8$ or 4, (e) $x = -\dfrac{1}{2}$, or $\dfrac{3}{2}$, (f) $b = -1$ or $\dfrac{4}{3}$

4. (a) $x = -\dfrac{2}{3}$ or $x = \dfrac{1}{2}$, (b) $x = -1.35$, or $x = 0.21$,
   (c) $x = -8$, or $x = -2$, (d) $x = 2.62$, or $x = 0.38$,
   (e) $x = 3.303$ or $x = -0.303$, (f) ± 1.58,
5. (a) $x = 1$, $y = 2$, (b) $x = 2.5$, $y = 1$, (c) $x = 2$, $y = 3$,
   (d) $p = 2$, $q = 3$
6. 13
7. (a) $x = 4$, $y = 1$, (b) $x = 7$, $y = 3$

4.5

1. 14
2. $\mu = 0.4$
3. $I = 0.017R^2$

4.6

1. From *Figure 4.14*, we have $\sin A = \dfrac{a}{c}$, $\cos A = \dfrac{b}{c}$,
   $\tan A = \dfrac{a}{b}$ and so $\dfrac{\sin A}{\cos A} = \tan A$.
2. (a) 0.7071, (b) 0.7071, (c) 1, (d) 0.7071, (e) 0.2147
   (f) 572.96
3. (a) 45°, 60°, 30°, (b) 116.58°, 30°, 45°, (c) 45°,
   89.98°, −26.565° or 153.435°
4. Angles: 90°, 55°, 35°; sides: 6, 8.57, 10.46
5. Angles: 75°, 75°, 30°; sides: 10, 10, 5.176
6. Length of hypotenuse $= 10.82$ cm, area of
   triangle $= 27$ cm^2
7. Vertical height $= 6.928$ cm, area $= 27.713$ cm^2
8. 46.74 m
9. (a) 27.05 m, (b) 58 m
10. Force F_1 horizontal $= 21.65$ kN, vertical $= 12.5$ kN
    and F_2 horizontal 10 kN and vertical
    component $= 17.32$ kN; these can be checked using
    Pythagoras' theorem

4.7

1. (a) *Sine*: first/second, first/second, third/fourth;
   (b) *Cosine*: first/fourth, second/third, first/fourth;
   (c) *Tangent*: first/third, first/third, second/fourth;
   all respectively.
2. (a) The *sine rule* is used when one side and any two
   angles are known or if two sides and an angle (not the
   angle between the sides) are known. (b) The *cosine
   rule* is used when three sides are known or two sides
   and the included angle are known.
3. $\angle A = 45.1°$, $\angle B = 37°$, $\angle C = 97.9°$, $a = 37.2$ cm,
   $b = 31.6$ cm, $c = 52.0$ cm
4. $\angle A = 93.82°$, $\angle B = 56.25°$, $\angle C = 29.93°$,
   area $= 29.9$ cm^2
5. (a) 16.9 cm^2, (b) 31.5 cm^2, (c) 15.7 cm^2

4.8

1. (a) 72.19°, (b) 286.48°, (c) 76.20°
2. (a) 1.45, (b) 3.30, (c) 5.15
3. (a) 2.73, (b) 3.2
4. 6.597 m²
5. 0.362 cm²

4.9

1. The tangent function is not continuous, being undefined at intervals of $\theta = -\frac{\pi}{2}, \frac{\pi}{2}, \frac{3\pi}{2}$, etc., while the sine function is continuous. The range of possible values of tan θ is unlimited, while for the sine function it is limited between ± 1. The tangent function is periodic but the period in this case is π not 2π as is the case with the sine function.

2. Your graph of $y = 2\sin \theta$ will have twice the amplitude of the graph of $y = \sin \theta$, that is ± 2, while the graph of $y = \sin 2\theta$ will have twice the frequency of $y = \sin \theta$, that is it will complete two cycles in 2π radian and its time period will be halved.

3. The difference in displacement between the time when these functions first reach their first maximum. The cosine function is phase advanced, by $\frac{\pi}{2}$, when compared with the sine function, that is it reaches its first maximum 90° or $\frac{\pi}{2}$ radian before the sine function.

4. (a) 1, $\sqrt{2}$, 2, (b) 0.5, 0, -1, (c) $\frac{\sqrt{3}}{4}$, 0.5, 0

5. The graph obtained is $y = \tan \theta$ and it has the characteristics described in the answer to TYK 4.12 Question 1.

4.10

1. 6 cm
2. 2545 cm²
3. 0.194 m²
4. 157 cm³
5. 47.1 cm²
6. 10 mm
7. 12.73 cm
8. 4.34 × 10⁻³ m³

   $4.34 \times 10^{-3}\,\text{m}^3$
9. 1.309 × 10⁻⁴ m³

   $1.309 \times 10^{-4}\,\text{m}^3$
10. $V = 154.985\,\text{cm}^3$ and $S = 150.796\,\text{cm}^2$

4.11

1. Business and administration = 29.24% (105.2°), humanities and social science = 42.28% (152.2°), physical and life sciences = 15.74% (56.7°), technology = 12.74% (45.9°).

2.

x	35	36	37	38	39	40	41	42	43	44	45
f	1	2	4	5	7	5	4	7	2	2	1

3. (a) The histogram and frequency distribution produced should be based on the data tabulated here, the percentage height of column relates to average for class interval.

Class interval	62	67	72	77	82	87
Percentage	6.67	18.33	30	28.67	11.67	6.67
Cumulative frequency	4	15	33	49	56	60

(b) The 50th percentile estimate from the cumulative frequency plot is 71.

4.12

1. 127
2. Mean = 20, median = 8.5 and mode = 9
3. Mean = 11.4, median = 11.5 and bimodal = 9, 12
4. Mean = 38.6 cm
5. Mean = 169.075 mm, median value (50th percentile) = 168.45 mm
6. Mean = 10.008 mm
7. (a) Mean = 3.42%, (b) modal value = 3.38%, (c) median (50th percentile) = 3.4625%

4.13

1. (a) The differential calculus may be defined as that branch of mathematics concerned with finding out how things change with respect to variables such as time, distance or speed, especially when these changes are continually varying. Its functions in engineering include: the study of motion and the way this motion in machines, mechanisms and vehicles, varies with time, the way in which pressure, density and temperature change with height or time and how electrical quantities vary with time, such as electrical charge, alternating current, electrical power, etc. (b) The integral calculus has two primary functions. It can be used to find the length of arcs, surface areas or volumes enclosed by a surface. Its second function is that of anti-differentiation.

2. By finding the gradient of functions we are able to find out how the variables in the function change with respect to each other. To find the rate of change of the

dependent variable y with respect to the independent variable x at a particular point (value) we need to find the gradient of the curve at that point (i.e., graphically differentiate the function).

3. Gradient $= -4$ at $x = -1$ and 8 at $x = 2$.
4. Gradient $= 3$ at $x = 3$.
5. $8x - 3$ is the derived function of $y = 4x^3 - 3x + 6$ which is also the gradient of the function at the point (x, y).
6. $\dfrac{dy}{dx}$ and $\dfrac{d^3y}{dx^3}$, respectively.

4.14

1. (a) $\dfrac{dy}{dx} = 3x^2$, (b) $\dfrac{dy}{dx} = 30x^4$,

   (c) $\dfrac{ds}{dt} = 2t + 4$, (d) $\dfrac{dA}{dr} = 2\pi r$,

   (e) $\dfrac{dy}{dx} = x^{\frac{-1}{2}}$ or $\dfrac{1}{\sqrt{x}}$, (f) $\dfrac{dp}{dq} = 2q + q^{-2}$

2. (a) $\dfrac{dy}{dx} = 2x - 2$, (b) $\dfrac{ds}{dt} = 15t^4 - 9t^2 - 4$,

   (c) $\dfrac{dp}{dq} = -3q^{-4} + \dfrac{1}{2}q^{\frac{-1}{2}} + 0.56q^{-1.56}$,

   (d) $\dfrac{dy}{dx} = 30x^{0.5} + 1.75x^{-1.7}$

3. (a) 24.2 m/s, (b) 62 m/s

4.15

1. (a) $\dfrac{dy}{dx} = \cos x$, (b) $\dfrac{dy}{d\theta} = -\sin\theta$,

   (c) $\dfrac{dp}{d\beta} = 3\cos\beta$, (d) $\dfrac{dy}{d\theta} = 8\cos 2\theta + 9\sin 3\theta$

2. (a) $\dfrac{dy}{d\theta} = 2e^{2\theta}$, (b) $\dfrac{dy}{dx} = -9e^{-3x}$, (c) $\dfrac{dt}{d\theta} = -e^{-3\theta}$

3. (a) $\dfrac{dy}{dx} = \dfrac{3}{x}$, (b) $\dfrac{dp}{d\theta} = 0.48e^{2\theta} - \dfrac{6}{\theta}$

4.16

1. (a) $\dfrac{x^4}{4} + c$, (b) $x^3 + c$, (c) $2x^{\frac{1}{2}} + c$, (d) $4\sqrt{x} + c$

2. (a) $x^3 + 4x^2 + 5x + c$, (b) $\dfrac{x^4}{16} - 6x^{\frac{1}{2}} + \dfrac{2}{3}x^{\frac{3}{2}} + c$,

   (c) $\dfrac{x^3}{3} + 2x^2 + 4x + c$

3. (a) $2.5\cos\theta + c$, (b) $\dfrac{1}{12}\sin 4x + c$,

   (c) $-\dfrac{3}{2}\cos 2\theta - \dfrac{4}{3}\sin 3\theta + c$

4. (a) $-e^{-0.5x} + c$, (b) $6(x\ln x - x) + c$,
   (c) $3\ln x + c$

5. (a) $10\dfrac{5}{12}$, (b) $\dfrac{1}{15}$

6. (a) $v = 1.5t^2 + 4t + 8$, $s = 0.5t^3 + 2t^2 + 8t$, and $s = 780$ m

7. $12\frac{2}{3}$ square units

Unit 5

5.1

1. Size, direction, point of application.
2. (a) A scalar has size (magnitude) only (e.g., time, mass). (b) A vector quantity has both magnitude and direction (e.g., force, weight, velocity).
3. In its simplest sense, a force is a push or a pull exerted by one object on another. It is fully defined by the relationship $F = ma$ and differs from weight force in that for weight $W = mg$ the acceleration is that due to gravity, which has an average value of $9.81\,\mathrm{Nm}^{-2}$ on earth.
4. Compression, tension.
5. (a) 1.1772 MN, (b) 624 kN.
6. (a) 360 N, (b) -360 N (where the minus means in the opposite sense).

5.2

1. In the triangle rule, the vector forces are added head to tail in the parallelogram rule, they are added tail to tail and then positioned parallel.
2. The minus sign reverses (turns through $180°$) the direction of the vector.
3. (a) The equilibrant is that single force that when added to a system produces equilibrium. (b) The resultant is that force which when acting alone produces the same effect as the other forces acting together.
4. Resultant $= 5.93$ kN acting at $61°$ from the horizontal and the equilibrant $= 5.93$ kN acting at $241°$ from the horizontal, both in an anticlockwise direction.
5. Resultant $= 10.56$ kN acting at $41°$ clockwise from the horizontal and the equilibrant $= 10.56$ kN acting at $139°$ anticlockwise from the horizontal.

5.3

1. In a concurrent force system all the lines of action of the forces within the system pass through the same point. In a non-concurrent system this is not the case, so in this system there may be a turning moment as well as a resultant and equilibrant.
2. No, this is not true. Although equal and opposite their lines of action do not necessarily pass through the same point. A couple is formed causing the body to rotate clockwise or anticlockwise.
3. The moment of a force is defined as the product of a force and the perpendicular distance that the force is applied from the pivot or axis of the line of action of the force.
4. Simply because the distance = 0 from the pivot and so the product (turning moment) must be zero.
5. By using the trigonometric ratios to find the vertical component of the force.
6. The sum of the clockwise moments equals the sum of the anticlockwise moment and upward forces equal downward forces.
7. (a) A couple occurs when two equal and opposite forces acting in opposite directions have their lines of action parallel but not coincident. (b) Torque is the turning moment of a couple and is equal to the force multiplied by the radius.

5.4

1. At right angles or at 90° to each other.
2. Positive forces are to the right and upwards, clockwise moments are positive and vice versa.
3. Refer to *Figure 5.20*.
4. Resultant = 5.8 kN acting at 199° anticlockwise from the horizontal and the equilibrant = 5.8 kN acting at 19° clockwise from the horizontal.
5. Resultant = 11.04 KN acting horizontally to the right, turning moment about A = 16.07 kNm clockwise.
6. Resultant = 16.92 kN.
7. Resultant = 15.8 kN, 0.85 m from right side wheel towards centre.

5.5

1. F = 30 kN
2. Resulting moment = 26.03 kNm (clockwise)
3. R_A = 18.8 kN and R_B = 13.2 kN

5.6

1. (a) Tensile stress set up by forces tending to pull the material apart, (b) shear stress produced by forces tending to cut through the material (i.e., tending to make one part of the material slide over the other), (c) compressive stress produced by forces tending to crush the material, all stresses defined by force over area.
2. Hooke's law states that within the elastic limit of a material the change in shape is directly proportional to the applied force producing it.
3. (a) The elastic modulus is a measure of the stiffness of a material and is defined as *stress/strain* = E with units of GPa. (b) The shear modulus is a measure of rigidity (resistance to bending) of a material and is defined as *Modulus of rigidity (G)* = *shear stress* (τ)/ *shear strain* (γ). It has the same unit as the modulus of elasticity (i.e., GPa).
4. (a) 240,000 N/m², (b) 228,000,000 N/m², (c) 600,000,000 N/m² or 600 MN/m², (d) 3300 N/m², (e) 10,000,000,000 N/m².
5. The factor of safety takes account of a certain factor of ignorance when designing engineering artefacts for a certain purpose. It is a factor over and above the normal working stresses that takes account of material weaknesses, excessive loading, unknown operating loads and manufacturing material defects.
6. 137.5 MPa.
7. (a) This is the point at the end of the elastic range (the elastic limit), where any extra stressing of the material will cause plastic deformation. (b) This point signifies the greatest tensile strength of the material. (c) This is the range where permanent deformation takes place after release of the load. (d) This deformation starts off as general deforming all the material and then becomes local when 'necking' takes place.

5.7

1. Inertia is that force resisting change in momentum (i.e., resisting acceleration has units of force, the Newton).
2. Since momentum is the mass multiplied by velocity, then change in moment is given by $mv - mu$ and rate of change of momentum is $(mv - mu)/t$ or $[m(v - u)]/t$, which is the same as ma, and $F = ma$. Hence the relationship.
3. A body moving with constant velocity that is not acted upon by a force is in equilibrium.
4. Momentum = mass × velocity and has units kg m/s = N s.

5.8

1. (a) 166.67 m/s, (b) 504 kph.
2. Velocity.

3. Velocity and acceleration.
4. Time has unit of the second, acceleration due to gravity m/s² and displacement is measured in metres (m).

5.9

1. Acceleration.
2. Distance.
3. Area under graph, the time interval.
4. Zero, vt
5. $\frac{1}{2}vt$.
6. Uniformly retarded motion.
7. $s = ut + \frac{1}{2}at^2$.
8. Variable.
9. $a = 5\,\text{m/s}^2$, $s = 40\,\text{m}$.

5.10

1. $3.75\,\text{m/s}^2$
2. $3000\,\text{kg}$, $160{,}000\,\text{kg}$
3. $100\,\text{m}$
4. $6.4\,\text{s}$
5. (a) $160\,\text{kN}$ (b) $128\,\text{kN}$

5.11

1. Work may be defined as the product of the force required to overcome resistance and the distance moved in the direction of the force (i.e., $W = F \times d$). It has units of Nm or Joules.
2. The work done against gravity is given by WD = mgh it has units of Nm or Joules.
3. Energy may neither be created nor destroyed only changed from one form to another.
4. (a) Mechanical to electrical, (b) chemical to mechanical, (c) chemical to electrical, (d) electrical to sound (vibration) energy.
5. KE = $1/2\,mv^2$ where m = mass in kg, v = velocity in ms⁻¹. Rotational KE = $1/2I\omega^2$ where I= mass moment of inertia. The moment of inertia of a rotating mass I can be defined in general terms by the expression $I = Mk^2$ where M = the total mass of the rotating body and k = the radius of gyration, that is the radius from the centre of rotation where all of the mass is deemed to Act.
6. Instantaneous power = force used × velocity and average power = work done divided by the time taken, both have units of Nm/s or J/s = Watt.

7. Machine B is more powerful. It delivers more work per unit time, that is, 1548 W compared with machine A which delivers 1500 W.
8. 59.38 MJ.
9. Average power = 2050 W.
10. 22.15 m/s.

5.12

1. Impulse = applied force × time = change in linear momentum or impulse = $Ft = m(v - u)$. The units are the same as those for momentum (i.e. kg m s⁻¹ or N s).
2. In an inelastic collision, when impulsive forces occur, some energy is lost through change into heat and sound and momentum is not entirely conserved; this is accounted for by the coefficient of restitution (e).
3. 400 m/s.
4. 434 kN.

5.13

1. Frictional resistance depends on: the materials involved, the condition of the two surfaces and the forces holding the surfaces in contact.
2. This statement is false, because it does not hold true for very low and very high speeds.
3. They are related by the fact that the coefficient of friction is equal to the tangent of the angle of friction (i.e., $\mu = \tan \phi$ where ϕ is the angle of friction).
4. Refer to *Figure 5.55*.
5. 235.4 N.
6. $\mu = 0.19$.

5.14

1. Inertia resistance, weight and frictional resistance.
2. Refer to *Figure 5.58*.
3. (a) 3814.4 Nm, (b) 5086 W.
4. (a) 9.884 kN, (b) 1.9768 MJ, (c) 59.304 kW.

5.15

1. Kg/m³.
2. The relative density of solids and liquids uses the density of water as a base 1000 kg/m³, while for gases air is used as the base 1.2256 kg/m³.
3. It will reduce. The heat energy gives more energy to the water molecules and they may become more separated at a molecular level thus less concentrated, reducing density.
4. It is a ratio.
5. Density = 720 kg/m³, relative density = 0.72.

5.16

1. Hydrostatic pressure is the pressure created by stationary or slow moving bulk liquid, atmospheric pressure is that due to the weight of the air in the atmosphere acting on the earth's surface, and absolute pressure = gauge pressure + atmospheric pressure.
2. (a) 2.034 MPa, (b) 90 MPa, (c) 2.6 MPa.
3. 106732.8 Pa.
4. (a) 949.002 kN m^{-2}, (b) 447.325 kN m^{-2}, (c) 7101.325 kN m^{-2}, (d) 202.325 kN m^{-2}.
5. 18 m.

5.17

1. The simplest type of mercury barometer consists of a mercury filled tube, which is inverted and immersed in a reservoir of mercury. The atmospheric pressure acting on the mercury reservoir is balanced by the pressure $\rho g h$ created by the mercury column. Thus the atmospheric pressure can be calculated from the height of the column of mercury it can support.
2. (a) 40024.8 Pa, (b) 141349.8 Pa.
3. Archimedes' principle states that when a body is immersed in a fluid it experiences an upthrust, or apparent loss of weight, equal to the weight of the fluid displaced by the body. This equality relationship enables us to see that when the body is immersed in the fluid, it floats when the upthrust force (equal to the weight of the fluid displaced) equals the weight of the body. If the weight of the body is less that the weight of the fluid it displaces then it will rise (upthrust). Thus this principle enables us to determine why and when airships, balloons, ships and submarines will float, that is why and when they are buoyant.
4. (a) 86.083 kN, (b) 75.047 kN.
5. (a) 58.86 N, (b) 6 kg.
6. (a) Centroid is simply the centre of mass, (b) the total thrust force that Acts on an immersed surface is the product of the pressure at the centroid (centre of mass) and the total area of surface in contact.
7. Net thrust = 13080 Pa, line of action of thrust force from base = 4 m, anticlockwise.

5.18

1. (a) Laminar flow is flow in which the fluid particles move in an orderly manner and retain the same relative positions in successive cross-sections, (b) turbulent flow is flow in which the fluid particles may move perpendicular as well as parallel to the surface of the body and undergo eddying or unsteady motions, (c) a stream tube or tube of flow is considered to be an imaginary boundary defined by streamlines drawn so as to enclose a tubular region of fluid.

2. The mass flow rate is given by the equation $\dot{m} = \rho_1 A_1 v_1 = \rho_2 A_2 v_2$ for incompressible flow. The density does not change from one section to the next, so that $\rho_1 = \rho_2$ and therefore we have $\dot{Q} = A_1 v_1 = A_2 v_2$ which is the volume flow rate. Thus both the mass flow rate and volume flow rate relationships obey the continuity equation for incompressible flow.
3. (a) 0.047 m^3/s, (b) 40.055 kg/s, (c) 4.16 m/s.

5.19

1. Temperature is a measure of the quantity of energy possessed by a body or substance. It measures the vibration of the molecules that form the substance.
2. (a) 203 K, (b) 15°C.
3. Alcohol, resistance or thermocouple thermometers could be chosen but for the range, the best degree of accuracy and large scale differences make the alcohol thermometer the best choice.
4. 34 × 10^{-6}.
5. 11.9 × 10^{-6}.

5.20

1. Heat flow $\dot{Q}$ = energy/time or $\dot{Q} = J/s$ = Power in Watts. Thus when heat flows into or out of a thermal system, power is either being absorbed or given out. Heat energy flowing into a system is considered to be positive and heat flow from a system is negative.
2. A quantity of heat energy has the symbol Q; note that there is no dot above the Q; the dot signifies heat flow (the answer given above to Question 1).
3. Refer to the start of heat energy transfer mechanisms for a full explanation.
4. Heat transfer by convection needs a medium through which to transfer heat energy using both convection currents and conduction, while heat transfer by radiation may travel through free space (a vacuum) as is the case with the heat transferred to earth by the sun.
5. Engineering thermodynamics is concerned with the relationship between heat, work and the properties of systems. As engineers we are concerned with the machines (engines) that convert heat energy from fuels into useful mechanical work, hence the importance of the subject.

5.21

1. A modern idea of heat is that it is the energy in transition and cannot be stored by matter. Heat Q,

may be defined as the transient energy brought about by the interaction of bodies by virtue of their temperature difference when they communicate. Matter possesses stored energy but not transient (moving) energy such as heat or work. Heat energy can only travel or transfer from a hot body to a cold body. When heat flows from a hot body to a cold body thermal power is being transferred, thus heat flow is thermal power.

2. Thermodynamic systems contain particular amounts of a thermodynamic substance, such as fuel, air and other vapours or gases that are surrounded by some form of identifiable boundary, for example, fuel/air mixture within a piston, etc.

3. A thermodynamic system must contain: a working fluid or substance, a heat source and system boundaries, which may or may not be fixed.

4. Intensive properties are independent of the mass of the working substance (e.g., pressure, temperature, volume). Extensive or specific properties are dependent on the mass of the working substance (e.g., specific internal energy, specific volume (i.e., per unit mass, as can be seen by their units)).

5. Overall this statement is false. The first part is true, there is no mass transfer of the working fluid on a closed system, however the system boundaries are not necessarily rigid, for example, a piston is a closed system but it moves to provide power.

6. The essential difference is that the working fluid in an open system may cross the system boundaries (e.g., an aircraft gas turbine engine) whereas in a closed system no working fluid crosses the system boundaries.

7. $Q - W = \Delta U$ where Q is the heat energy, W is the work done on or by the system and ΔU is the change in the internal energy of the working fluid in the system that results from the heat and work transfers, to and from the system, all terms are in Joules or Nm.

8. $W = -5\,kJ$, that is, 5 kJ of work is *done on* the system.

9. Simply that the efficiency can never be more than or equal to 100%, in real systems there will always be losses due to unavoidable heat transfers in the form of friction, etc.

10. Thermal efficiency = 43.7%.

5.22

1. The specific heat capacity of a substance is that required to raise the temperature of 1 kg of the substance by 1 K. In all materials the specific heat capacity varies with temperature but in solids this variation is very small, in gases this variation can

be large and is dependent on the process, that is, whether the variation takes place at constant pressure or constant volume, where for gases two different specific heat capacities must be used.

2. Because, during the process the heat energy supplied also does work, as well as increasing the temperature of the working fluid, hence the specific heat capacity is greater for a constant pressure process, when compared with that for a constant volume process.

3. Thermal energy $Q = mc\Delta T$, where Q = heat energy in J, m = mass of the substance or working fluid in kg, c = specific heat capacity in J/kg K and ΔT is the temperature difference that occurs during the process.

4. Thermal energy = 46.8 kJ and thermal power = 390 W.

5. 900 J/kg K.

6. 1.004 MJ.

5.23

1. A perfect gas is simply one that obeys the gas laws; air is an example of an ideal gas, since it very closely approximates to the gas laws.

2. The volume.

3. The volume is reduced threefold.

4. The unit for R is J kg^{-1} K^{-1} (i.e., the Joule per kilogramme Kelvin). So once this constant has been multiplied by the mass of the particular gas and by the temperature change it has been subjected to, the quantity of heat energy in Joules can be established.

5. $V = 80\,cm^3$.

6. (a) 13.30 bar, (b) 136.5°C.

7. 022 m^3.

8. (a) $R = 297$ J/kg K, (b) $c_p = 1040$ J/kg K.

Unit 6

6.1

1. Any two from copper, aluminium and silver (or other metals).

2. Any two from plastics, rubber, ceramics, glass.

3. Silicon, germanium.

6.2

Silver. This material has the lowest resistivity and the highest relative conductivity of those listed in Table 6.1.

6.3

0.56 Ω

6.4

1. 33.3 mA
2. 6.72 V
3. 3.3 kΩ

6.5

1. 15 Ω
2. 29.4 mW
3. 675 mW
4. 58 mA

6.6

288 W, 1.73 MJ

6.7

Red, red, red, red

6.8

(a) 270 kΩ ± 5%
(b) 10 Ω ± 10%
(c) 6.8 MΩ ± 5%
(d) 0.39 Ω ± 5%
(e) 2.2 kΩ ± 2%

6.9

44.65–49.35 Ω

6.10

407.2 Ω

6.11

98.7 kΩ

6.12

$I_1 = 0.5$ A, $I_2 = 1$ A, $I_3 = 0.5$ A, $V_1 = 6$ V, $V_2 = 6$ V, $V_3 = 9$ V

6.13

66.7 Ω

6.14

10 Ω

6.15

Figure A2.6.1 Answer to TYK 6.15

See *Figure A2.6.1*.

6.16

1. Doping
2. Depletion region
3. Forward bias
4. Reverse bias
5. Forward threshold voltage

6.17

Table A2.6.1

Capacitor type	Ceramic	Electrolytic	Polyester	Mica	Polystyrene
Capacitance range (F)	2.2 p to 100 n	100 n to 10 m	10 n to 2.2 μ	0.47 to 22 k	10 p to 22 n
Typical tolerance (%)	±10 and 20	−10 to 50	±10	±1	±5
Typical voltage rating (V)	50–200 V	6.3–400 V	100–400 V	350 V	100 V

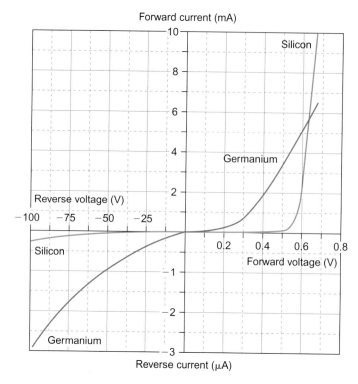

Figure A2.6.2 Answer to TYK 6.17

6.18

625 kV/m

6.19

10.34 mC

6.20

2.67 V

6.21

1.36 J

6.22

611 pF, 8 kV

6.23

Any three chosen from those shown in Table A2.6.1

6.24

(a) 5 μF, 150 μC
(b) 45 μF, 450 μC

6.25

(a) 3 μF, 100 V; 6 μF, 50 V
(b) 3 μF, 15 μJ; 6 μF, 7.5 μJ

6.26

(a) 132 s
(b) 18.3 s

6.27

39.2 μF

6.28

12 A

6.29

$6\,\mu\text{Wb}$

6.30

$1 \times 10^6\,\text{A/Wb}$

6.31

$0.5\,\text{A}$

6.32

$313\,\text{mV}$

6.33

$0.48\,\text{J}$

6.34

$136\,\text{mH}$

6.35

$0.067\,\text{T}$

6.36

(a) $45.45\,\text{Hz}$
(b) $5\,\mu\text{s}$

6.37

(a) $110\,\text{V}$
(b) $-94\,\text{V}$

6.38

$78\,\text{mA}$

6.39

$1.13\,\text{mA}$

6.40

(a) $339\,\Omega$
(b) $3.39\,\Omega$

6.41

(a) $37.7\,\Omega$
(b) $3.77\,\text{k}\,\Omega$

6.42

$616\,\Omega, 0.36\,\text{A}$

6.43

$842\,\Omega, 0.26\,\text{A}$

6.44

(a) $0.26\,\text{A}$
(b) 0.47
(c) $62°$

6.45

(a) $7.5{:}1$
(b) $14.7\,\text{V}$

6.46

(a) Resistor (low-power carbon film)
(b) Inductor (ferrite cored toroidal)
(c) Diode
(d) Miniature transformer
(e) Air-spaced variable capacitor
(f) Capacitor (axial lead electrolytic)
(g) Resistor (high-power metal clad)
(h) Capacitor (non-polarized polyester)

6.47

T1 is a transformer that steps down the incoming AC mains voltage (220–240 V) in order to produce a much lower voltage (typically 12–20 V). T1 also provides isolation so that no part of the output is connected directly to the AC mains supply (this could be dangerous).

D1 is a semiconductor (silicon) rectifier diode which only allows the current to flow one way. D1 is forward

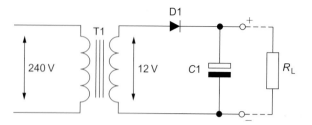

Figure A2.6.3 Answer to TYK 6.47

biased during each positive half cycle allowing current to flow which then charges the reservoir capacitor, C1, and supplies the load R_L.

C1 stores up a charge on the positive half cycles (when D1 is conducting) and releases its charge on the negative half cycles (when D1 is not conducting). This helps to maintain the output voltage developed across the load.

6.48

Because conduction takes place only on alternate half cycles, half-wave rectifiers are relatively inefficient (they only work half of the time). A better rectifier arrangement would make use of both positive and negative half cycles. These full-wave rectifier circuits offer a considerable improvement over their half-wave counterparts. They are not only more efficient but less demanding in terms of the reservoir and smoothing components. There are two basic forms of full-wave rectifier: the bi-phase rectifier and the bridge rectifier.

Unit 7

7.1

Electrical engineering and mechanical engineering.

7.2

Any three selected from:

Standing plans: These are plans that are used many times, and remain relatively unaffected by environmental change. Examples are employment, financial, operating and marketing policies and procedures. For example, hiring new employees involve standard procedures for recruitment and selection. Another example would be the annual routines for establishing budgets.

Single use plans: These are plans that are used once only, such as those for the control of a unique project or specific budgets within an annual budget. (Budgets themselves are single use plans, even though the procedures used for producing them are standing plans.)

Strategic plans: These are broad plans related to the whole engineering company and include forecasting future trends and overall plans for the development of the engineering company. They are often in outline only and very often highly subjective, involving judgments made by top managers. For example, a plan may be made to build an entirely new factory based on forecasts of demand. This plan is strategic, and if it is wrong and the sales forecasts on which it is based do not materialize, the results for a company could be devastating.

Tactical plans: These are plans that operate within the overall strategic plan. The new factory has to be brought into commission and production has to be scheduled and controlled. Plans for the latter are tactical, since they focus on how to implement the strategic plan.

7.3

(a), (b) and (d).

7.4

In an engineering business, control is required for a variety of reasons including ensuring that a project or product:

- remains within budget
- is delivered on time
- complies with relevant legislation
- meets appropriate quality standards
- operates according to the agreed specification
- fully satisfies the expressed and implied needs of the customer or client.

Two examples of the controls that might be in place are financial controls (to ensure that a project remains within budget) and quality controls (to ensure that a product meets with appropriate standards).

7.5

1. The most important function of marketing is matching company products and services to the needs of its customers and clients.
2. The purpose of marketing research is determining what the customer or client needs as well as discovering needs the customer or client does not yet know they have. It can also involve finding out how to improve products so that customers and clients will

buy a company's products and services in preference to those offered by other suppliers.

3. Some companies choose to use wholesalers rather than sell direct for the following reasons:
   - The company can sell to a wholesaler in bulk quantities and thus reduce costs of transport and administration
   - Wholesalers can operate intermediate storage depots between manufacturers and retailers thereby reducing the need for complex distribution systems and avoiding the need to store large quantities of finished goods on company premises
   - Wholesalers can often provide retailers with credit terms of trading, thus enabling small businesses to sell before they have to pay for goods (or at least to reduce the impact of the cost of carrying a large range of stock items)
   - Wholesalers can Act as a buffer to smooth out demand for manufacture. If demand is seasonal they can buy regularly through the year, thus making it easy for manufacturers to make goods in economic runs and then store stock to meet heavy demand, but which does not place excessive loads on the manufacturer's capacity.

4. Any two selected from trade fairs: exhibitions and advertising.

5. A sales engineer provides technical backup and advice as well as identifying and specifying new/ replacement products.

7.6

1. The steps in the product development process are:
   - idea generation
   - selection of a suitable design
   - preliminary design
   - prototype construction
   - testing
   - design modification
   - final design.

2. Three strategies for product development are:

   Marketing driven: These are strategies that put the needs of customers and clients first, and only produce goods which are known to sell. Market research is carried out which establishes what is needed.

   Technology driven: These are strategies that are based on selling what it is possible to make. The product range is developed so that production processes are as efficient as possible and the products are technically superior, hence possessing a natural advantage in the market place.

Coordinated approaches: These combine the merits of both marketing-driven and technology-driven strategies. With this approach the needs of the market are considered at the same time as the needs of the production operation and of design and development. In many businesses this inter-functional system works best, since the functions of R&D, production, marketing, purchasing, quality control and material control are all taken into account.

3. Ideas can be selected from the identification of new customer needs, the invention of new materials or the successful modification of existing products. Selection from new ideas will be based on factors like:
   - market potential
   - financial feasibility
   - operations compatibility.

   This means that ideas are rejected if they:
   - have little market appeal
   - are too expensive to make at a profit
   - do not fit easily alongside current production processes.

7.7

(a) *Continuous flow* (or *line flow*) is the type of system used in the motor industry for assembly lines for cars. It also includes continuous type production of the kind that exists in the chemicals and food industries. Both kinds of line flow are characterized by linear sequences of operations and continuous flows and tend to be highly automated and highly standardized.

(b) *Intermittent flow* (or *batch flow*) is the typical batch production or job shop production which uses general-purpose equipment and highly skilled labour. This system is more flexible than line flow, but is much less efficient than line flow. It is most appropriate when a company is producing small numbers of non-standard products, perhaps to a customer's specification.

(c) *Project-based production* (or *one-off production*) is used for unique products that may be produced one at a time. Strictly speaking there is not a flow of products, but instead there is sequence of operations on the product that has to be planned and controlled. This system of production is used for prototype production and in some engineering companies that produce major machine tool equipment for other companies to use in their factories.

7.8

(a) Project-based (b) continuous flow (c) batch (d) batch (e) continuous flow

7.9

Four outcomes of good scheduling are:

- meeting customer delivery dates
- correct loading of facilities
- planning starting times
- ensuring jobs are completed on time.

7.10

1. (a) *Manufacturing/production engineering*: Translates designs for products into finished products. Usually involves physical manufacture and/or assembly and testing of a finished product.

   (b) *Capacity planning*: This involves matching the production capacity to meet predicted demand. Production managers will use a variety of ways to achieve this from maintaining excess capacity to making customers queue or wait for goods to having stocks to deal with excess demand. The process is complex and may require the use of forecasting techniques, together with careful planning.

   (c) *Inventory control*: This involves ensuring that proper control systems are in place to ensure that there is sufficient stock for production while at the same time ensuring that too much stock is not held. If stock levels are high there are costs associated with damage, breakage, pilferage and storage that can be avoided.

   (d) *Quality control*: This involves ensuring that products and services are fit for purpose and meet the identified needs of customers. Quality control involves various aspects including:

   - inspection, testing and checking of incoming materials and components
   - inspection, testing and checking of the company's own products
   - administering any supplier quality assurance systems
   - dealing with complaints and warranty failures
   - building quality into the manufacturing process.

2. Process design involves making decisions in relation to location of a production plant as well as the design and layout of production facilities. The design of production processes is interactive with product design, requiring close cooperation with research and product development functions. Selecting the process of production has a wide impact on the operation of the entire business. Decisions in this area bind the company to particular kinds of equipment and labour force because of the large capital investment that has to be made in order to make changes. Once in production the company is committed to the technology and the capacity created for a long time into the future. There are three basic methods for production processes:

   - continuous or line production processes
   - intermittent or batch production processes
   - project-based production processes.

3. Three techniques used in workforce management are:

   - work and method study
   - work management
   - job design.

7.11

(a) *Design quality* is usually the joint responsibility of a company's marketing or customer liaison function and its research and development function. Design quality relates to the development of a specification for the product that meets a customer's identified needs.

(b) *Conformance quality* means producing a product that conforms to the design specification. A product that conforms is a quality product, even if the design itself is for a cheap product. That may seem contradictory, but consider the following example. A design is drawn up for a budget camera, which is made from inexpensive materials and has limited capability. If the manufacture conforms to the specification then the product is of high quality, even though the design is of low quality compared with other more up-market cameras.

(c) *Reliability* includes things like continuity of use measured by things like *mean time between failure (MTBF)*. Thus a product will operate for a specified time, on average, before it fails. It should also be maintainable when it does fail, either because it can easily and cheaply be replaced or because repair is fast and easy.

7.12

Any four selected from:

- inspection, testing and checking of incoming materials and components
- inspection, testing and checking of the company's own products
- administering any supplier quality assurance systems
- dealing with complaints and warranty failures
- building quality into the manufacturing process.

7.13

1. The basic documents used for purchasing and supply (in order of use) are:
   - requisitions from departments to buyer
   - enquiry forms or letters to suppliers
   - quotation document in reply to enquiry
   - order or contract to buy
   - advice note – sent in advance of goods
   - invoice – a bill for goods sent to the buyer
   - debit note – additional/further charge to invoice
   - credit note – reduction in charge to invoice
   - consignment note – accompanies goods for international haulage, containing full details of goods, consignee, consignor, carrier and other details
   - delivery note – to accompany goods.
2. The main functions of a purchasing department are:
   - researching sources of supply
   - making enquiries and receiving quotations
   - negotiating terms and delivery times
   - placing contracts and orders
   - expediting delivery
   - monitoring quality and delivery performance.

7.14

1. (a) *Flat structures* permit better communications both up and down the organizational levels. They are also seen as more cost effective and responsive in dealing with demands placed on modern businesses.
   (b) *Tall structures* divide an organization into a number of areas that can be handled by different departments. The advantages of this type of structure are as follows:
      - specialists can be grouped together
      - it appears logical and easy to understand
      - coordination is achieved via operating procedures
      - suits stable environments and few product lines
      - works well in small to medium size businesses.
2. As a business grows the functional structure becomes less and less useful. This is because there are many more products and these may be manufactured in separate divisions of a company, especially if economies of scale are introduced into the manufacturing process. Federal divisional structures allow the different divisions within a large company to operate separately. Managers of each operating division will control most of the functions of that particular part of the business. Centralized functions (such as R&D, purchasing and supply, etc.) may still be present within a federal structure providing a resource that can be accessed by each division when required.
3. Matrix structures are well suited to project-based industries where project managers have the job of coordinating the work of functional specialists in order to ensure that a project is completed on time, to specification and within cost limits.
4. A hierarchical structure is one in which the various positions within a company are ascribed different levels within the organization. This establishes a set of clearly defined reporting lines in which it is possible to trace management accountability and responsibility. Hierarchical structures are often described with the aid of an organizational chart which can be either tall or flat depending upon the size and scope of the organization concerned.

7.15

The project manager has the job of coordinating the work of functional specialists to ensure that projects are completed on time, to specification and within cost limits. Project managers do not normally have direct line authority, but they do need to exert influence and persuade others to achieve targets. They may have formal authority over project budgets and can set time schedules and decide what work is to be done, but little else. They also work as an interface with clients and subcontractors and their influence is often critical to the success of the project. Although they do not have formal authority over individual staff or their line managers, they nevertheless operate with the full support of senior managers. This means that functional specialists are obliged to provide the fullest cooperation and help, otherwise they become answerable for failure to their own senior line managers. The matrix system more readily permits project managers to access the various functions within a company without having to move up and down the lines of responsibility that exist within hierarchical structures.

7.16

The buyer would be involved with supplies of new raw materials, new designs for components and will negotiate costs of tooling and long-term contracts. Buyers can usually assist production managers with quality and inspection problems and in dealing with return and replacement of defective materials. Sometimes buyers may Act as an interface with production and R&D in dealing with temporary or permanent deviations from the original engineering specifications.

Chasing deliveries and ensuring supplies for factory use may be a major daily routine for some buyers. The buyer may also be responsible for ensuring that purchased goods and materials meet the required quality assurance standards. Buyers may also be involved with searches and the tendering process for new suppliers who can meet existing and new quality requirements.

The buyer may have to determine the economic order quantities when arranging supplies. The buyer also needs to have a clear understanding of the importance of deliveries which enable the company to control its inventory costs, while at the same time ensuring a reliable supply of materials and components for production. With a just-in-time (JIT) operation, the buyer must liaise with the production facility in order to ensure that JIT requirements are met. The buyer is also concerned with inventory levels and in setting up minimum, maximum and reorder levels for stocks.

7.17

In order of bullet points: opportunity, threat, threat, opportunity.

7.18

(a) Reduced costs and increased sales (increased profitability).
(b) Increased costs and reduced sales (reduced profitability).

7.19

(a) Gross domestic product per capita is a measure of a nation's productivity. It relates output to the number of people employed producing that output. The formula used is:

Output per capita = total output produced divided by the number of people producing it

Often we are concerned not with absolute figures for output and employment but more with trends and comparisons. For this reason, both the output and employment figures are often quoted as index figures relative to a base figure of 100 for a particular year.

(b) HDI combines indicators of adult literacy, average years of schooling and life expectancy together with income levels to give an indicator of human development. HDI is scaled to provide an index on a scale of 0–100; countries scoring more than 80 points are considered to have a high degree of human development whilst those scoring between 50 and 79 are said to be medium and those below 50 are categorized as low.

7.20

GDP per capita is not a particularly good indicator of quality of life because is only refers to a nation's economic welfare. It does not take into account factors such as adult literacy, average years of schooling and life expectancy.

7.21

(d) and (e).

7.22

Invisibles (or invisible earnings) are income derived from the sale of services (rather than physical goods) to other countries. Invisibles help UK to show a health balance of payments (i.e., a favourable trading balance when the balance of visible exports and imports shows an excess of import value over export value). Examples of invisibles are insurance services, consultancy and legal services.

7.23

(a) Smoke
(b) Mercury
(c) Heat
(d) Plastics

7.24

EMC stands for 'electromagnetic compatibility' and it relates to the interference caused by and experienced by electrical and electronic equipment such as computers, television receivers, power tools, electromedical equipment, welding equipment and lighting.

All electrical and electronic equipment tends to radiate unwanted electromagnetic energy and this can have an effect on other electrical/electronic equipment. Conversely, sensitive electronic equipment can be susceptible to the electromagnetic noise and interference radiated by other equipment. Thus EMC is important in terms of both radiation and susceptibility, and strict controls are enforced to ensure that equipment is

designed, used and maintained so that it neither radiates interference nor is susceptible to it.

In order to confirm that an item of electrical/electronic equipment meets the strict requirements of current EMC legislation, all equipment must be supplied with an EC certificate of conformity.

7.25

Vehicle exhaust gases must be sufficiently free of oxides of nitrogen, carbon monoxide and other toxic gases. This can be achieved by replacing the crude petrol carburettor with a more sophisticated petrol injection system and fitting a catalytic converter in the exhaust system. All this has added to the price of the motor car and has made it more difficult for the DIY motorist to service their own vehicle.

7.26

(a) *Job costing* relates to a unique 'job' or operation, such as replacing a part or carrying out a modification to a product. Typical operations in which job costing is commonly used include:

- supplying a unique or 'one-off' item
- modifying a product in order to improve its performance
- adapting a product to meet a particular customer's requirements
- providing a service that is only required on an intermittent or irregular basis.

Job costing involves estimating the cost of materials used (sometimes referred to as a 'bill of materials') and then adding this to the cost of labour in order to determine the total amount that will be charged to the customer. Let us take an example:

(b) *Contract costing* is similar to job costing but used by larger companies when they are involved with a project for an individual customer or client. Production is designed to meet the individual customer's requirements and is not speculative (in other words, the work does not start until it is agreed with the customer or client). Specifications, costs and charges are agreed before manufacturing starts and they form the basis of a contract between the supplier and the customer.

(c) *Parts costing* is usually straightforward and simply involves determining the cost of all of the physical parts and components used in a manufactured or engineered product. Parts costing works from the 'bottom up' – in other words, the cost of each individual component (i.e., the 'per unit cost')

is determined on the basis of the given 'standard supply multiple'. For example, if plastic case parts are purchased in quantities of 100 then the per unit cost is simply the total cost of purchasing 100 items from the supplier divided by the supply multiple (i.e., 100).

7.27

£1.36

7.28

(a) *Absorption costing* involves determining the total cost of a given product or service is that of adding the costs of overheads to the direct costs by a process of allocation, apportionment and absorption. Since overheads (or indirect costs) can be allocated as whole items to production departments, it is possible to arrive at a notional amount that must be added to the cost of each product in order to cover the production overheads.

$$\text{Mark-up} = \frac{\text{Total of fixed and variable costs attributable to the product}}{\text{Total number of units produced}}$$

In absorption costing, each product manufactured is made (at least in theory) to cover all of its costs. This is achieved by adding a notional amount to the total unit cost of each product.

(b) *Marginal costing* is an alternative costing method that provides us with an alternative way of looking at costs that provides an insight into the way costs behave by allowing us to observe the interaction between costs, volumes and profits. The marginal cost of a product is equal to the cost of producing one more unit of output.

Accountants define marginal cost as:

The amount by which aggregate costs change if the volume of output is increased or decreased by one unit.

Fixed costs take into account:
- insurance premiums
- business rates
- subscriptions
- audit fees
- rental charges
- fixed elements of power and telephone charges.

(c) *Activity costing* tries to assess the 'true' cost of providing a product or service. Knowledge of the

'true' cost is not only important in helping us to identify opportunities for cost improvement but it also helps us to make strategic decisions that are better informed.

Activity-based costing focuses on indirect costs (overheads). It does this by making costs that would traditionally be considered indirect into direct costs. In effect, it traces costs and overhead expenses to an individual 'cost object'.

Activity-based costing is particularly useful when the overhead costs associated with a particular product are significant and where a number of products are manufactured in different volumes.

Activity-based costing is particularly applicable where competition is severe and the margin of selling price over manufacturing cost has to be precisely determined.

The steps required to carry out activity-based costing are as follows:

- identify the activities
- determine the cost of each activity
- determine the factors that drive costs
- collect the activity data
- calculate the product cost.

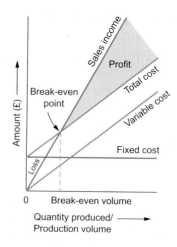

Figure A2.1.1 Answer to TYK 7.29

7.29

Break-even is the volume of sales for a product or service at which the operation becomes profitable and it marks the transition from loss into profit. A *break-even chart* takes the form of a graph of costs plotted against volume of product sold. At this point, it is important to recall that the total costs of the business operation are the sum of the constant fixed and overhead costs with the variable costs of production. Thus:

$$\text{Total cost} = \text{Fixed cost} + \text{Overhead cost} + \text{Variable cost}$$

The income derived from the sale of the product (assuming a constant pricing structure) will simply be the product of the quantity sold (i.e., the volume of product sold) and the price at which it is sold (i.e., the per unit selling price). This relationship (a straight line) can be superimposed on the break-even chart and the point of intersection with the total cost line can be identified. This is the *break-even point* and the corresponding production volume can be determined from the horizontal axis (*Figure A2.1.1*).

7.30

3000

7.31

£45; £120,000

7.32

The main sections of a business plan are usually as follows:

- An *introduction* which sets out the background and structure of the plan.
- A *summary* consisting of a few pages that highlight the main issues and proposals.
- A *main body* containing sections or chapters divided into numbered sections and subsections. The main body should include financial information (including a *profitability forecast*).
- *Market and sales projections* should be supported by valid market research. It is particularly important to ensure that there is a direct relationship between market analysis, sales forecasts and financial projections. It may also be important to make an assessment of competitors' positions and their possible response to the appearance of a rival product.
- *Financial details*: The financial section of the plan (following on from market and sales projections) is of crucial importance. Because it is likely to be scrutinized in detail, it needs to be realistic about sales expectations, profit margins and funding requirements ensuring that financial ratios are in line with industry norms. It also

needs to include realistic estimates of the cost and time required for product development, market entry and the availability of external sources of funding.

- *Appendices* should be used for additional information, tabulated data and other background material. These (and their sources) should be clearly referenced in the plan.

7.33

£40,000

7.34

A make or buy decision involves deciding on whether to manufacture an item or component or to purchase it from another manufacturer or supplier. Make or buy decisions require the involvement of more than one functional area within a company and typically involve the production and production control functions as well as engineers from design and manufacturing and also advice from those involved with specifying and purchasing component parts and materials.

Before a make or buy decision is made various questions must be addressed:

(a) Could the item be made with existing facilities?
(b) If the answer to (a) is yes, is current production capacity adequate?

If the answer to (a) is no, then it indicates that new plant and equipment may be needed, or indeed that the company has no experience in this area of manufacture, which might therefore mean a move toward vertical integration, which involves a change to its basic business. The latter involves top-level policy decisions and will not yet even be considered at the functional level at all. Theoretically any component part, whether already manufactured is outsourced (i.e., purchased from elsewhere) can be reviewed for alternative sourcing, but in practice only those items will be considered for change that do not involve major changes to the business.

If a component part is being considered for manufacture then there must be suitable facilities available and sufficient capacity. Conversely, if a manufactured item is being considered for purchase there must be alternative uses for the capacity or there should be an intention to reduce that capacity rather than leave it unused.

Other factors that might need to be taken into account include:

- incremental costs (e.g., will any additional costs be incurred with the decision?)
- quality assurance (e.g., will a bought-in part meet existing quality requirements?)
- multiple sourcing (e.g., will the decision make the company dependent on only one source for the component?)
- costs of tooling (e.g., if the component is to be made in-house will there be any additional costs for tooling?)
- strategic issues (e.g., is the decision in accordance with company strategy and could it be advantageous to a competitor?).

Index